NONCLASSICAL ION CHANNELS IN THE NERVOUS SYSTEM

Methods in Signal Transduction

Series Editors: Joseph Eichberg, Jr. and Michael X. Zhu

The overall theme of this series continues to be the presentation of the wealth of up to date research methods applied to the many facets of signal transduction. Each volume is assembled by one or more editors who are pre-eminent in their specialty. In turn, the guiding principle for editors is to recruit chapter authors who will describe procedures and protocols with which they are intimately familiar in a reader-friendly format. The intent is to assure that each volume will be of maximum practical value to a broad audience, including students and researchers just entering an area, as well as seasoned investigators.

Lipid-Mediated Signaling Transduction, Second Edition
Eric Murphy, Thad Rosenberger, and Mikhail Golovko

Calcium Entry Channels in Non-Excitable Cells
Juliusz Ashot Kozak and James W. Putney, Jr.

Autophagy and Signaling
Esther Wong

Signal Transduction and Smooth Muscle
Mohamed Trebak and Scott Earley

Polycystic Kidney Disease
Jinghua Hu and Yong Yu

New Techniques for Studying Biomembranes
Qiu-Xing Jiang

Ion and Molecule Transport in Lysosomes
Bruno Gasnier and Michael X. Zhu

Neuron Signaling in Metabolic Regulation
Qingchu Tong

Non-Classical Ion Channels in the Nervous System
Tian-Le Xu and Long-Jun Wu

For more information about this series, please visit: https://www.crcpress.com/Methods-in-Signal-Transduction-Series/book-series/CRCMETSIGTRA?page=&order=pubdate&size=12&view=list&status=published,forthcoming

NONCLASSICAL ION CHANNELS IN THE NERVOUS SYSTEM

Edited by
Tian-Le Xu and Long-Jun Wu

CRC Press
Taylor & Francis Group
Boca Raton London New York

CRC Press is an imprint of the
Taylor & Francis Group, an **informa** business

First edition published 2021
by CRC Press
6000 Broken Sound Parkway NW, Suite 300, Boca Raton, FL 33487-2742

and by CRC Press
2 Park Square, Milton Park, Abingdon, Oxon, OX14 4RN

© 2021 Taylor & Francis Group, LLC

CRC Press is an imprint of Taylor & Francis Group, LLC

Library of Congress Cataloging-in-Publication Data

ISBN: 978-0-367-62395-1 (hbk)
ISBN: 978-0-367-62397-5 (pbk)
ISBN: 978-1-003-10926-6 (ebk)

Typeset in Times
by MPS Limited, Dehradun

Contents

v

Preface

Ion channels generate bioelectricity and impact nearly every aspect of human life—transmission of nerve impulses, muscle contractions, learning and memory, vesicle trafficking, as well as our bodies' perception of various internal and external stimuli. In last few decades, scientists have been decoding the structures and functions of classical ion channels, with a tremendous upsurge in basic research on two archetypal classes—voltage-gated ion channels and ligand-gated ion channels. In addition to classical ion channels, living organisms express many atypical ion channels with unique biophysical and functional properties in the evolutionary process. In recent years, many new ion channel families have been identified and characterized, bringing about the blossom of exciting research on nonclassical ion channels. Nonclassical ion channels refer to those newly discovered and less well-understood channels, for example, acid-sensing ion channels (ASICs) that respond to proton elevation, transient receptor potential (TRP) channels that are sensitive to a wide range of temperatures and chemical stimuli, mechanically activated ion channels, and two-pore domain K^+ (K2P) channels that are subject to a broad array of chemical and physical regulations. While investigating deeper into the structure, assembly, and regulations of these nonclassical ion channels, the significance of their roles has been discovered in modulating normal functions, such as memory, social behaviors, and food intake, as well as dysfunctions of the central and peripheral nervous systems, including autism, stroke, chronic pain, anxiety, depression, and addiction. Importantly, the functions of many nonclassical ion channels are not limited to the nervous systems; they are found in all body systems.

Studying nonclassical ion channels is not just about finding the ligand for an orphan receptor or revealing the structure of these membrane proteins. Instead, it is fascinating to uncover how they play important roles in both physiological and pathological conditions, thus providing a better understanding of the human body function and new strategies for treating human diseases. Take ASIC as an example: Krishtal and colleagues first recorded a pH-dependent inward current in rats' primary sensory neurons in 1980s, but it was not until 1997 that Lazdunski group . successfully cloned the first ASIC gene. After the ASIC protein structure was solved by Gouaux group in 2007, our knowledge of ASICs has extended to their function in synaptic plasticity and acidosis-induced cell death. Targeting ASICs in ischemic stroke has now become promising for neuroprotection.

The field of nonclassical ion channels is vast and it is expanding tremendously. This book introduces the frontiers of this exciting field ranging from basic biology, structures, and regulations, to state-of-art techniques and experimental models. The main purpose of this book is to summarize recent advances in our understanding of the biology and working principles of a large collection of nonclassical ion channels in neurological function and disorders. The book is designed to aim at researchers and students who are interested in ion channel research and neurobiology of diseases. It is divided into 19 chapters, logically presenting main lines of recent research on nonclassical ion channels from basic biology, to functional significance, to their roles in neurological disorders. For example, chapters on ASICs address the proton sensing and signaling mediated by the channel, physiological function of ASIC subunits in synaptic plasticity and learning/memory, and pathological implications of ASICs in ischemic cell death. Moreover, roles of other nonclassical ion channels (P2X, TRP, Piezo, Hv1, K2P channels, lysosome ion channels) in health and disease are substantially discussed in this book. There are also chapters addressing the circuit mechanisms underlying the vital brain functions related to social communications, anxiety, depression, and food intake. A particular chapter reviews the previous studies using the non-human primate animal model to tackle the mysteries of brain function.

Finally, we wish to express our sincere thanks to all authors for their timely contribution to this volume and Ms. Ann B. Peterson at Mayo Clinic for proofreading some chapters. We would also like to acknowledge the professionalism of CRC Press/ Taylor & Francis Group for their efforts in bringing this volume into completion. We hope that this book will provide an initial detailed account of the most recent developments in the field of nonclassical ion channels, awaiting many more exciting discoveries in the years to come.

Editors

Tian-Le Xu, PhD, Principal Investigator and Professor, Center for Brain Science, Shanghai Children's Medical Center, and, Department of Anatomy and Physiology, Shanghai Jiao Tong University School of Medicine. Dr. Tian-Le Xu obtained his PhD of Neurobiology from Fourth Military Medical University. He did his postdoc training at Kyushu University and then joined University of Science and Technology of China as Professor. He was a principal investigator at Institute of Neuroscience, Chinese Academy of Sciences. Dr. Xu's research focuses on neuronal signaling and related neural disorders including chronic pain and ischemic stroke, with an emphasis on the role of acid-sensing ion channels (ASICs) in these processes. The research from Dr. Xu's lab has unveiled rich and new knowledge on the regulation and diverse functions of ASICs in health and disease.

Long-Jun Wu, PhD, Professor and Consultant, Department of Neurology, Mayo Clinic. Dr. Long-Jun Wu received his PhD in Neurobiology and Biophysics from the University of Science and Technology of China. After postdoc trainings at University of Toronto and Harvard Medical School, Dr. Wu was Instructor at Harvard Medical School and then Assistant Professor at Rutgers University. Dr. Wu's recent research primarily focuses on the neuroimmune interaction, particularly the function of microglial ion channels and receptors, in normal and diseased brain. By understanding and manipulating microglial functions, Dr. Wu's research aims to develop potential therapeutics targeting microglia in the treatment of various neurological disorders, such as epilepsy, chronic pain, stroke, neurodegeneration, and autoimmune neurology.

Contributors

Xian-Ping Dong
Departments of Physiology and
Biophysics, Dalhousie University
Halifax, Canada

Bo Duan
Department of Molecular, Cellular, and
Developmental Biology
University of Michigan
Ann Arbor, MI, USA

Zahra Farzinpour
Hefei National Laboratory for Physical
Sciences at the Microscale
CAS Key Laboratory of Brain Function
and Disease
School of Life Sciences
Division of Life Sciences and Medicine
University of Science and Technology
of China
Hefei, China

Mahar Fatima
Department of Molecular, Cellular, and
Developmental Biology
University of Michigan
Ann Arbor, MI, USA

Jun Gao
Department of Neurobiology
School of Basic Medical Science
Nanjing Medical University
Nanjing, China

Neng Gong
Key Laboratory of Primate
Neurobiology
CAS Center for Excellence in Brain
Science and Intelligence Technology
Institute of Neuroscience
Chinese Academy of Sciences
Shanghai, China

Xue Gu
Center for Brain Science of
Shanghai Children's Medical Center
Department of Anatomy and Physiology
Shanghai Jiao Tong University School of
Medicine
Shanghai, China

Peng Huang
Collaborative Innovation Center for
Biomedicine
School of Clinical Medicine
Shanghai University of Medicine and
Health Sciences
Shanghai, China

Peng Jiang
Department of Cell Biology and
Neuroscience
Rutgers University
Piscataway, NJ, USA

Robert Juniewicz
Department of Cell Biology and
Neuroscience
Department of Biological Sciences
Rutgers University
Piscataway, NJ, USA

Wei-Guang Li
Center for Brain Science of Shanghai
Children's Medical Center
Department of Anatomy and Physiology
Shanghai Jiao Tong University School of
Medicine
Shanghai, China

Michael Lin
Department of Cell Biology and
Neuroscience
Rutgers University
Piscataway, NJ, USA

Ze-Jie Lin
Center for Brain Science of Shanghai
Children's Medical Center
Department of Anatomy and Physiology
Shanghai Jiao Tong University School
of Medicine
Shanghai, China

Jingyi Liu
Department of Molecular, Cellular, and
Developmental Biology
University of Michigan
Ann Arbor, MI, USA

Ming-Gang Liu
Center for Brain Science of Shanghai
Children's Medical Center
Department of Anatomy and Physiology
Shanghai Jiao Tong University School
of Medicine
Shanghai, China

Hui Lu
GW Institute for Neuroscience
Department of Pharmacology and
Physiology
School of Medicine and Health Sciences
The George Washington University
Washington, DC, USA

Vincent R. Mirabella
Department of Neuroscience and Cell
Biology
Child Health Institute of New Jersey
Rutgers Robert Wood Johnson Medical
School
New Brunswick, NJ, USA

Madhuvika Murugan
Department of Neurology
Mayo Clinic
Rochester, NY, USA
Department of Neurosurgery
Rutgers University
Piscataway, NJ, USA

Zhiping P. Pang
Department of Neuroscience and Cell

Biology, Child Health
Institute of New Jersey
Rutgers Robert Wood Johnson Medical
School
New Brunswick, NJ, USA

Kenneth G. Paradiso
Department of Neuroscience and Cell
Biology
Child Health Institute of New Jersey
Rutgers Robert Wood Johnson Medical
School
New Brunswick, NJ, USA

Ritika Raghavan
Department of Cell Biology and
Neuroscience
Rutgers University
Piscataway, NJ, USA

Xiao Su
Department of Neuroscience and Cell
Biology
Child Health Institute of New Jersey
Rutgers Robert Wood Johnson Medical
School
New Brunswick, NJ, USA

Maharaib Syed
Department of Cell Biology and
Neuroscience
Rutgers University
Piscataway, NJ, USA

Jin Wang
Department of Basic Medicine and
Clinical Pharmacy
China Pharmaceutical University
Nanjing, China

Jun Wang
Department of Neurobiology and
Department of Neurology
Second Affiliated Hospital
Zhejiang University School of
Medicine
NHC and CAMS Key Laboratory of
Medical Neurobiology

MOE Frontier Science Center for Brain
Research and Brain-Machine Integration
School of Brain Science and Brain
Medicine
Zhejiang University
Hangzhou, China

Si-Yu Wang
Department of Basic Medicine and
Clinical Pharmacy
China Pharmaceutical University
Nanjing, China

Yi-Zhi Wang
Center for Brain Science
Shanghai Children's Medical Center
Department of Anatomy and Physiology
Shanghai Jiao Tong University School of
Medicine
Department of Neurology
Northwestern University
Feinberg School of Medicine
Chicago, IL, USA

Long-Jun Wu
Departments of Neurology, Immunology,
and Neurosciences
Mayo Clinic
Jacksonville, FL, USA

Yi Wu
Collaborative Innovation Center
for Biomedicine
School of Clinical Medicine
Shanghai University of Medicine and
Health Sciences
Shanghai, China

Bailong Xiao
State Key Laboratory of Membrane
Biology
Tsinghua-Peking Center for Life
Sciences
Beijing Advanced Innovation Center
for Structural Biology
IDG/McGovern Institute for Brain
Research School of Pharmaceutical

Sciences
Tsinghua University
Beijing, China

Han Xu
Department of Neurobiology and
Department of Neurology of the Second
Affiliated Hospital University School
of Medicine
NHC and CAMS Key Laboratory of
Medical Neurobiology
MOE Frontier Science Center for Brain
Research and Brain-Machine Integration
School of Brain Science and Brain
Medicine
Zhejiang University
Hangzhou, China

Mengnan Xu
Departments of Physiology and
Biophysics Dalhousie University
Halifax, Canada

Pan Xu
GW Institute for Neuroscience
Department of Pharmacology and
Physiology
School of Medicine and Health Sciences
The George Washington University
Washington, DC, USA

Tian-Le Xu
Center for Brain Science
Shanghai Children's Medical Center
Department of Anatomy and Physiology
Shanghai Jiao Tong University School of
Medicine
Shanghai, China

Huaiyu Yang
Shanghai Key Laboratory of Regulatory
Biology
School of Life Sciences
Institute of Biomedical Sciences
East China Normal University
Shanghai, China

Qian Yang
Department of Neurobiology and
Department of Neurology of the Second
Affiliated Hospital
Zhejiang University School of Medicine
NHC and CAMS Key Laboratory of
Medical Neurobiology
MOE Frontier Science Center for Brain
Research and Brain-Machine Integration
School of Brain Science and Brain
Medicine
Zhejiang University
Hangzhou, China

Xiao-Na Yang
Department of Basic Medicine and
Clinical Pharmacy
China Pharmaceutical University
Nanjing, China

Ye Yu
Department of Basic Medicine
and Clinical Pharmacy
Pharmaceutical University
Nanjing, China

Yuanlei Yue
GW Institute for Neuroscience
Department of Pharmacology and
Physiology
School of Medicine and Health
Sciences
The George Washington University
Washington, DC, USA

Wei-Zheng Zeng
Center for Brain Science
Shanghai Children's Medical Center
Department of Anatomy and Physiology
Shanghai Jiao Tong University School
of Medicine
China School of Life Sciences
Westlake University
Westlake Laboratory of Life Sciences
and Biomedicine
Westlake Institute for Advanced Study
Hangzhou, China

Hui-Xin Zhang
Department of Neurobiology
School of Basic Medical Science
Nanjing Medical University
Nanjing, China

Shaoying Zhang
Shanghai Key Laboratory of Regulatory
Biology
School of Life Sciences
Institute of Biomedical Sciences
East China Normal University
Shanghai, China

Wen Zhang
National Institute on Drug Dependence
Peking University
Beijing, China

Xiaobing Zhang
Department of Psychology
Florida State University
Tallahassee, FL, USA
Hangzhou, China

Zhi Zhang
Hefei National Laboratory for Physical
Sciences at the Microscale
CAS Key Laboratory of Brain
Function and Disease
School of Life Sciences
Division of Life Sciences and Medicine
University of Science and Technology
of China
Hefei, China

Michael X. Zhu
Department of Integrative Biology and
Pharmacology
McGovern Medical School
The University of Texas Health Science
Center at Houston
Northwestern University
Feinberg School of Medicine
Houston, TX, USA

1 Endogenous Activation and Neurophysiological Functions of Acid-Sensing Ion Channels

Wei-Zheng Zeng, Yi-Zhi Wang, and Tian-Le Xu

CONTENTS

1.1 INTRODUCTION

Cells constantly generate chemical components for signal transduction and energy production. Cellular pH change is among the most fundamental chemical changes, though the pH sensing and its physiological mechanisms are not well understood. It has been known that extracellular pH changes modulate the activities of various ion channels and transporters, thus passively affecting cell signal transductions.

Acid-sensing ion channels (ASICs) are proton-gated cationic channels mainly expressed in neurons. To date, six ASIC isoforms rooted from four genes have been identified, *ACCN1*, *ACCN2*, *ACCN3*, and *ACCN4* (ASIC1a, 1b, 2a, and 2b are splice isoforms) (1). As primary pH sensors, ASICs are specifically gated by protons, enabling them to convey unique properties of pH signaling. In the peripheral nervous system, activation of ASICs induces pain (2), while in the central nervous system it modulates synaptic plasticity, influencing the generation of memories (3), fear conditioning (4,5), and extinction (6). Pathological activation of ASICs in tissue acidosis contributes to neuronal damage induced by ischemia (7,8) and neuro-inflammatory processes (9).

A hallmark of mammalian ASICs is that protons act not only as agonists but also as strong inhibitors by inducing two types of desensitization: low-pH (<pH 6.0) desensitization that shuts channels from the open state, and steady-state desensitization at mild acidic pH (~pH 7.0) that prevents the channels from opening by further acidification (10). Unlike in pathological environments (e.g. ischemic stroke), the pH alterations under physiological conditions are confined in a small range and from limited sources. Given that tissue acidosis is slow (from days to weeks) and mild (from pH 7.4 to 7.0) (11), endogenous ASICs might need to be activated in an unexpected fashion to overcome these two obstacles. In the first part, we will review the potential endogenous proton sources for activating ASICs. Accumulating evidence shows that the function of ASICs varies from central nervous system (CNS) and peripheral nervous system (PNS) regions by triggering different signaling cascades. In the second part, we will summarize the recent advances regarding the neurophysiological functions of ASICs. We hope to provide an overview of the concepts, outstanding questions, and latest progress in the study of ASICs.

1.2   ENDOGENOUS CONDITIONS THAT MAY ACTIVATE ASICs

ASICs are primarily gated by extracellular protons *in vitro*. Although some non-proton ligands have also been uncovered to directly gate ASICs (2,12), the endogenous conditions that activate ASICs remain elusive. How sensitive are ASICs to endogenous pH changes? *In vitro* data from cultured neurons and heterologous expression systems show that pH values ≤7.1 can activate ASICs (13,14), suggesting that these channels likely respond to very subtle pH reductions. Is proton the sole endogenous ligand of ASICs? Does the *in vivo* pH fall to levels sufficient to activate ASICs? In mammalian cells, the production of protons (or pH reduction) is mainly accomplished by two mechanisms: *de novo* production of acidic species by energy metabolism and net transmembrane efflux of protons through proton-permeable ion channels or pumps (11). Notably, accumulating evidence has shown that ASIC-like currents can also be amplified near neutral pH or directly induced at neutral pH by small molecules, for example, ATP (15) and GMQ (2), respectively. Given that certain endogenous mediators are co-released with H^+ under pathological conditions involving tissue acidosis, such as ischemic stroke and inflammatory pain, we speculate that these mediators may work synergistically with protons upon ASICs (15–17). This section focuses on endogenous proton production pathways that likely cause ASIC activation *in vivo*.

1.2.1 METABOLIC PRODUCTION OF PROTONS

Various metabolic reactions such as glycolysis in the cytoplasm and ATP production in mitochondria generate acidic species (11). Metabolic acidosis occurs when either a net increase in the production of nonvolatile acidic species or a loss of bicarbonate from the body overwhelms the mechanisms that regulate acid-base homeostasis, which is referred to as acute or chronic metabolic acidosis, respectively (18). Acute metabolic acidosis, which mainly consists of diabetic ketoacidosis and lactic acidosis (18), most likely results in local acidification that activates ASICs. Extensive studies have identified several endogenous proton sources that act on ASICs. However, more information is needed to clarify the regulation of proton sources during acidosis in the future.

1.2.1.1 Carbon Dioxide

Carbon dioxide (CO_2) is an important environmental cue associated with many biological activities such as respiration, photosynthesis, and decomposition of organic matter. CO_2 may act as a signaling molecule by itself or it can change pH homeostasis by generating metabolites, such as carbonic acid and bicarbonate ions, when dissolved in aqueous fluid. The CO_2 signal is critical in many invertebrates as it regulates insect innate behaviors, such as seeking food and hosts (19,20). However, with the exception of studies on the sensory system (21,22), the role of the CO_2 signal in mammalian nervous systems remains poorly understood. It is important to determine the sensor(s) and the downstream signaling pathway(s) of CO_2 in mammalian brain. Inhalation of CO_2 can reduce brain pH by 0.1–0.2 unit due to the catalytic function of carbonic anhydrase that converts CO_2 and water to carbonic acid, which subsequently produces bicarbonates, carbonates, and protons. With the ability to sense the physiological pH reduction and the wide expression in nervous systems, ASICs are well suited to be the sensors and downstream targets of the CO_2 signal.

The usage of CO_2 in humans was first documented nearly a century ago. It was found that CO_2 inhalation could trigger panic attacks in humans (23–25), whereas it terminated seizures in patients with epilepsy. However, to our knowledge, the links between CO_2 inhalation and physiological phenomena remained largely unknown until recently. Exquisite work by Wemmie and colleagues demonstrated that acid-sensing ion channel 1a (ASIC1a) plays a critical role in sensing CO_2 in CNS (26). The observation that CO_2 triggered fear in mice resembles that in humans, indicating that CO_2 may act on the amygdala, which plays a critical role in both innate and acquired fear behaviors. Coincidently, ASIC1a is particularly abundant in mouse amygdala, raising the possibility that the inhaled CO_2 reduced brain pH, which evoked fear behavior via activation of ASIC1a in this brain area. To support this, eliminating or inhibiting ASIC1a abolished proton-activated current in neurons and in turn diminished the CO_2-evoked fear. CO_2 inhalation was also found to inhibit seizures in humans.

Studies have found that CO_2 reduces cortical pH within seconds of inhalation and that breathing CO_2 increases brain acidosis during a chemoconvulsant-evoked seizure. Subsequent findings demonstrate that acidosis activates inhibitory interneurons through ASIC1a to increase inhibitory tone (27) and CO_2 inhalation requires ASIC1a to interrupt seizures in a mouse model. These findings suggest that ASIC1a is an emerging CO_2 sensor in the CNS, while the downstream consequences following ASICs activation depend on where they localize and in which cell types they are expressed.

1.2.1.2 Lactate

Lactate is produced by glycolysis during periods of intense physical activity or pathophysiological conditions related to insufficient oxygen supply, like in the cases of tissue ischemia (14,15), hypoxia, and incision injury (28). The accumulation of lactate during pathophysiological conditions leads to tissue acidosis, which may activate or regulate ASICs. Indeed, several lines of evidence have shown that ASICs are the sensors for tissue acidosis caused by lactate accumulation from muscle to brain (13,14,29,30). As the most sensitive subunit of the ASIC family (open at pH 7.0, half-activation at pH 6.5), ASIC3 is highly expressed in sensory neurons innervating skin, skeletal muscle, and the heart. This expression pattern suggests ASIC3 to be a promising sensor for lactate acidosis. For example, in myocardial ischemia, insufficiency of oxygen supply leads to accumulation of lactate and pH reductions both intracellularly (31) and extracellularly (32). Cardiac sensory neurons (metaboreceptors) are reported to respond to ischemic pH reduction (≤ 7.0), which was mostly carried out through ASIC3. However, although ASICs are proposed to detect lactic acidosis and to transduce cardiac ischemic pain, the pathological pH reduction alone (7.3 to 7.0) seems to have little effect on activating ASICs. It appears that some mediators are involved to amplify the response of ASICs to pathological pH reduction. An important finding was that lactate can enhance the proton-activated currents on sensory neurons, probably through a mechanism involving extracellular calcium chelation (17,33). Moreover, the effect of lactate enhancement on ASICs is conserved, revealing a gating mechanism of ASICs by catalyzing the relief of Ca^{2+} blockade (33). These studies indicate that lactate may amplify the gating effect of protons and be relevant in disorders related to acidosis.

Although it has long been considered a metabolic dead end, growing evidence has indicated that lactate is an emerging critical energy source in the brain (34–36) and it plays an indispensable role in brain functioning. Since neurons but not glia account for most of the brain's energy expenditure, it is hypothesized that there is an energy transfer from astrocytes to neurons especially through lactate (37). The hypothetical energy transfer forms the core of the so-called astrocyte-neuron lactate shuttle model (38,39). Interestingly, lactate is transported exclusively by monocarboxylate transporters (MCTs), which cotransport protons across the cell membrane (Figure 1.1). Very recently, Suzuki and colleagues demonstrated that astrocytic glycogen-derived lactate is involved in long-term memory formation and the maintenance of long-term potentiation (LTP) of synaptic strength in rat brains (40). In their study, a learning task triggers astrocytic glycogen breakdown

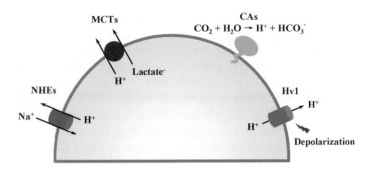

FIGURE 1.1   Proton carriers that regulate intercellular pH fluctuations. The plasma membrane localized Na^+/H^+ exchangers (NHEs) are the predominant pH-regulatory transporters responsible for intracellular alkalinization. Monocarboxylate transporters (MCTs) that cotransport lactate and protons represent new fundamental proton resources between neurons and glia. Mammalian cells are continuously exposed to CO_2, which can be catalyzed to carbonic acids and in turn controls local pH changes. Some isoforms of CAs are located in the plasma membrane, where they can be anchored to the extracellular surface by a glycosylphosphatidylinositol anchor (CA4) or a transmembrane segment (CA9, CA1,2 and CA14). Hv1 is a voltage-gated proton channel widely expressed in mammalian cells, which is responsible for oxidative burst during phagocytosis, pH-dependent activation of spermatozoa, as well as acid secretion by tracheal epithelium.

and the release of lactate via MCT1 and MCT4 in the rat hippocampus. Lactate is then taken up by neurons via MCT2 and triggers an increase in the expression of the activity-dependent gene Arc as well as the phosphorylation of CREB (a nuclear transcription factor) and cofilin (an actin-binding protein), which are all known to contribute to memory consolidation processes. However, the mechanisms by which the transport of lactate from astrocytes to neurons causes these effects are yet to be determined. One possibility is that the transport of protons and lactate from astrocytes to the extracellular space activates ASICs on neurons, especially during the burst of MCT function associated with brain activation such as visual stimulus or a learning task. This concept is worth testing since it will provide a link between brain metabolism and neuronal activity. MCT-derived protons that may activate ASICs will also be discussed later.

1.2.2   ACIDIFICATION NICHE AND PROTON GENERATORS

While global changes in intracellular or extracellular pH are most likely harmful and unlikely to serve as signals, transient and localized pH changes may play such a role. Several lines of evidence show that localized proton gradients exist within (41,42) and between (43,44) cells, which are likely to have consequences on receptors, channels, and enzymes involved in proton signaling in the region. Whether ASICs can be activated in any cellular microenvironment remains an open question. Synaptic cleft seems a promising niche. First, synaptic transmission involves the corelease of neurotransmitters and protons from synaptic vesicles, which generate local acidification. Second, ASICs have been found in the post-synaptic compartment (45). Intriguingly, recent breakthrough in imaging probes via pHluorin modification enables a precise measurement of pH drops of synaptic cleft (ΔpH ~ −0.2–0.6) during synaptic transmission. Interestingly, strong evidence showed that presynaptic stimulation induces ASIC excitatory postsynaptic currents at the mouse calyx of held synapse, amygdala, striatum, and anterior cingulate cortex (ACC) (3,45–47). These encouraging findings indicate that protons and ASICs are a neurotransmitter/receptor pair critical for CNS signal transmission.

Unlike the synaptic cleft that is acidified during synaptic transmission, proton channels/pumps are widely expressed among tissues and may fulfill the criteria to activate non-synaptic ASICs (48). Supporting this hypothesis, recent works utilizing the light-gated proton pump archaerhodopsin-3 (Arch) demonstrated that localized proton transient activates ASICs cell-autonomously or from the neighboring cells (49,50). These findings provide the proof-of-concept that proton generators may activate ASICs. The next question is whether any endogenous proton channels/pumps activate ASICs, which is still under vigorous exploration. We review the potential candidates below (Figure 1.1).

1.2.2.1   Na$^+$/H$^+$ Exchangers (NHEs)

Na$^+$/H$^+$ exchangers (NHEs) are secondary active transporters that mediate the equal exchange of extracellular Na$^+$ for intracellular H$^+$ across the plasma membrane, thereby regulating the Na$^+$/H$^+$ homeostasis of the cytoplasm or organellar lumen (51). To date, eight NHE family members have been identified in mammals. NHE1–5 are expressed on the plasma membrane in various cell types, while NHE6–8 reside on intracellular organellar membranes of the endosomal/trans-Golgi network (51). The entire NHE isoforms share a characteristic secondary structure composed of an NH$_2$-terminal 10–12 transmembrane ion exchange domain and a large hydrophilic COOH-terminal regulatory domain (52). NHE1 is the most ubiquitously expressed isoform on the plasma membrane and it regulates a number of cell behaviors, including adhesion, shape determination, migration, and proliferation (53). Thus, it will be primarily emphasized in the following text. The activation of NHE1 is often regulated by extracellular signals such as growth factors, hormones, Ca^{2+} and osmotic cell shrinkage, as well as post-translational modifications including phosphorylation and the binding of regulatory proteins that converge to the C-terminal regulatory domain (53). In addition to an established role in intracellular pH and cell volume homeostasis, NHE1 can serve as a structural anchor for actin filaments (54) and as a plasma membrane scaffold in the assembly of signaling complexes, which are independent of its ion exchanger function (55).

Several lines of evidence show that the activity of NHE1 is tightly related to pH microenvironment gradient and has multiple roles in cell behaviors. One example is found in metastasis (56). Magnetic resonance imaging (MRI) studies have revealed that tumor cells have alkaline intracellular pH (pH$_i$) values (7.12–7.65 compared with 6.99–7.20 in normal tissues) and acidic interstitial extracellular pH (pH$_e$) values (6.2–6.9 compared with 7.3–7.4) (57), creating a reversed pH gradient across the cell membrane, which increases as the tumor progresses. Recent investigations have highlighted the fundamental role of NHE1 and MCT (also known as H$^+$/lactate cotransporter) in acidification of the extracellular environment during malignant invasion of tumors. Moreover, the pH change was shown to be driven primarily by the stimulation of NHE1 activity, as the increase in pH_i was completely and reversibly inhibited by incubation with the NHE1 inhibitor, 5-N,N-dimethylamiloride (DMA) (58).

Second, the activity of NHE1 in the proton microenvironment gradient has also been implicated in signal transduction. Recently, two studies have demonstrated that protons, released via *C. elegans* intestinal NHE PBO-4 (also known as NHX-7, the ortholog of mammalian NHE1), act on a proton-gated cation channel (comprised of PBO-5 and PBO-6 subunits) located in the adjacent muscle cells to induce contraction during the defecation cycle (43,44). These studies provide direct evidence supporting the idea that localized H^+ can operate as signal molecules. Notably, NHE1 is the most abundant NHE isoform in the CNS, which overlaps with the expression pattern of ASIC1a. Like the neuronal protective role of ASIC1 deletion, genetic ablation of NHE1 also led to decreased neuronal death after *in vitro* and *in vivo* ischemia (59). Meanwhile, the activity of NHE1 is significantly increased during ischemic conditions (60). It is reasonable to hypothesize that the increased activity of NHE1 in the brain may activate ASICs to induce neuronal death during pathological conditions such as ischemia.

1.2.2.2   Hydrogen Voltage-Gated Channel 1 (Hv1)

Hv1 is a voltage-gated proton channel with highly selective permeation to H^+. Although proton currents were identified in certain kinds of cells and found to be associated with the NADPH oxidase (NOX) decades ago (61,62), the genes for the proton channels were not discovered until 2006 when two groups cloned human and murine Hv1 genes (63,64). The activation of Hv1 is caused by membrane depolarization and strongly dependent on the pH at both sides of the membrane. Within the physiologically achievable pH range, the channel opens only when the proton gradient is outward, making it extrude H^+ under most circumstances (65). Hv1 is widely expressed in immune cells including microglia, macrophages, monocytes, granulocytes (65), spermatozoa (66), and respiratory epithelial cells (67). The highest level of native expression of Hv1 was found in human eosinophils, where the current density of H^+ reaches 200 pA/pF (65). Hv1 has multiple functions in different cell types. With the use of an Hv1-deficient mouse line, Hv1 had been found to modulate B-cell activation (68), basophil histamine release (69), and acid secretion from airway epithelium (67) and human spermatozoa (66). The well characterized function of Hv1 includes reactive oxygen species (ROS) production in phagocytic cells as a part of innate immunity, where NOX transfers electrons across the plasma membrane, rapidly depolarizing and acidifying phagocytic cells. Given that excessive depolarization and intracellular acidification inhibit further expulsion of electrons, a charge-compensating mechanism is needed to maintain NOX activity (61). By sensing voltage and pH gradients, Hv1 is ideally suited to the task of charge compensation for NOX activation.

Microglia are the resident macrophages of the brain, which share many phenotypic markers and are capable of performing bactericidal functions like peripheral macrophages. In the normal brain, microglia appear to be in a functionally resting state, in which they are highly ramified, extending processes far into the surrounding brain tissue to act as antennae. A variety of stimuli, including neuronal injury, trauma, ischemia, inflammation, infection, and several neurological diseases, lead to microglial activation, which proceeds through a series of steps that include changes in morphology and expression of surface antigens, release of cytokines and cytotoxins, antigen presentation, and phagocytosis (70). However, the presence of Hv1 in mammalian brain neurons and glia has been debated (71,72). Recently, a systematic work by Wu and colleagues demonstrated that mouse and human brain microglia, but not neurons or astrocytes, express large Hv1-mediated currents. Moreover, Hv1 is required for NOX-dependent ROS generation in brain microglia and its activity increased in both *in situ* and *in vivo* ischemia models. Mice lacking Hv1 were protected from NOX-mediated neuronal death in ischemic conditions (73). These findings implicate that Hv1-dependent ROS production is responsible for a substantial fraction of brain damage at early time points after stimulation. Interestingly, given that mice lacking ASIC1 also show less neuronal death in ischemic models, it remains an open question whether activated microglia act as a mobile proton source that targets ASICs in neurons to induce further damage. To fill this gap, we found that opening of the Hv1 channel by depolarization can release protons to activate neighboring ASICs in a heterologous system (50). These results provide *in vitro* evidence to support protons as a signaling messenger between microglia and neurons. Future efforts will be focused on gathering *in vivo* activation evidence.

1.2.2.3 Carbonic Anhydrases (CAs)

The carbonic anhydrases (CAs) are a family of enzymes that reversibly catalyze CO_2 hydration to H^+ and HCO_3^- with catalytic rates of up to 10^6 s^{-1} (74). To date, at least 14 different isoforms are found in mammals, while their function is likely to differ depending on the site of expression. Some of these isozymes are cytosolic (CA1, CA2, CA3, CA7 and CA8); others are membrane-bound, where they can be targeted to the plasma membrane by a glycosylphosphatidylinositol (GPI) anchor (CA4) or a transmembrane segment (CA9, CA12 and CA14); CA5 is mitochondrial localized and CA6 is secreted in the saliva and milk (48,74). CAs have been shown to reversibly generate and buffer localized pH gradients in the CNS (75,76), CO_2 sensing systems (77,78) and tumor cells (79). Small and rapidly diffusible buffers facilitate the displacement of proton equivalents in cells. As CO_2 is an uncharged acid equivalent, it can diffuse without transiently binding to immobilized charged macromolecules. CAs also speed up the dissipation of localized pH gradients by accelerating the conversion of CO_2 to H_2CO_3 (41).

A notable role of plasma membrane CAs is producing highly localized proton signals in the CO_2 sensing systems (77). Carbonation sensing is conserved from insects to mammals. Because CO_2 works in both gas and dissolved forms (converted to H^+ and HCO_3^-), it evokes diverse responses in different tissues and cell types. In fruit flies, *Drosophila*, gaseous CO_2 is detected by taste receptor cells (TRCs) on the antenna, whereas dissolved CO_2 is detected by those on the proboscis (80–82). In mice, ingested CO_2 is sensed by TRCs on the tongue and soft palate, the same group of cells that sense sour (83). A membrane GPI-anchored CA4 is found to be specifically expressed on the surface of sour-sensing neurons and provides localized protons to activate sour TRCs (77). Consistently, atmospheric CO_2 is sensed by mouse olfactory subsystem necklace glomeruli, which also express cytosolic CA2 (84). These findings demonstrate an important role of CAs in the mammalian CO_2 sensing. How does the activity of CAs couple to sensory neuron activation? The postulate that CAs present in gustatory and olfactory cells cooperate with pH-sensitive receptors seems to provide a possible explanation.

1.3   ASICS AND NEUROPHYSIOLOGICAL FUNCTIONS

ASICs are widely expressed in both the CNS and PNS depression-related behaviors, while only three subunits, ASIC1a, ASIC2a, and ASIC2b, are identified in CNS (1,85,86). The ASIC1a subunit is localized throughout the brain and spinal cord, especially in those regions with high synaptic density, such as hippocampus, amygdala, striatum, and spinal dorsal horn (SDH) (1,3,85,87). ASIC1b is specifically expressed in nociceptors (85). The expression patterns of ASIC2a and ASIC2b are less investigated. However, recent studies show that ASIC2a (or ASIC2b) is always co-expressed with ASIC1a to form heteromeric channels in CNS (85,88–90), suggesting a similar expression pattern like ASIC1a. On the other hand, ASIC1a is not always co-expressed with ASIC2a (or ASIC2b); for example, only ASIC1a subunit was detected in fast-spiking basket cells (91). Unlike these subunits, ASIC3 is only expressed in PNS, mainly in sensory neurons, such as dorsal root ganglia (DRG) and trigeminal nerves (10,85,92), which contribute to acidosis-induced nociception.

Accumulating evidence shows that ASIC1a channels are mainly expressed on neural soma and dendrites (1,3,85,93). Moreover, they were found to be present in synaptosome-enriched subcellular fractions and colocalize with postsynaptic proteins, such as postsynaptic density-95 (PSD-95), protein interacting with C-kinase-1 (PICK1), β-actin, and A kinase anchoring protein 79/150 (AKAP79/150), suggesting that ASIC1a channels are enriched in dendritic spines (85,94,95). It is interesting that ASIC1a is not directly bound to PSD-95, but rather associated with PSD-95 in dendritic spines via binding to the ASIC2a subunit (89). Although ASIC3 was also found to interact with PSD-95 (96), there is no information about its precise subcellular localization. Moreover, intracellular ASIC1a was recently identified in the mitochondrial membrane (mtASIC1a) of the mouse cortical neurons (97,98). Considering the key role of mitochondria in neuronal functions (99,100), mtASIC1a may contribute to neural physiological functions in a whole new way. In the following section, we present previous results on the neuronal physiological functions of ASICs in terms of synaptic roles and region-specific differences.

1.3.1   SYNAPTIC DEVELOPMENT

Because homomeric ASIC1a channel is Ca^{2+}-permeable (although with a very low efficiency) (1,101), and it is universally and highly expressed across nervous systems (10,94), the role of ASIC1a in synaptic function has attracted much more attention than the other five subunits in past decades (45,85,94,95). Some studies showed that deletion of the *Asic1a* gene reduced dendritic spine density of cultured hippocampal slices, whereas overexpressing ASIC1a led to the opposite effect (102). One of the possible explanations is that Ca^{2+} influx via ASIC1a increased Ca^{2+}/calmodulin-dependent protein kinase II (CaMKII) phosphorylation levels and caused spine remodeling (102). Recently, we found that ASIC1a plays a key role in regulating spine maturation in medium spiny neurons (MSNs) (3). Although deletion of the *Asic1a* gene significantly increases spine density of MSNs, they are more immature spines. In ASIC1a-null striatum, the PSD structure becomes smaller and its protein composition, including some key postsynaptic receptors such as NMDAR, is altered. The CaMKII-ERK pathway, which can be activated by either NMDAR- or ASIC1a-mediated Ca^{2+} signaling, is disrupted in $Asic1a^{-/-}$ MSNs. This pathway is essential for dendritic spine maturation. Therefore, all these changes of ASIC1a-null dendritic spines may result from reduced Ca^{2+} signaling. As a result, ASIC1a KO mice show impaired motor learning behavior, which can be rescued by overexpression of ASIC1a in the striatum.

It was also observed that dendritic spine density was decreased in cultured hippocampal slices from $Asic2^{-/-}$ mice (89). One reasonable mechanism is that deletion of the *Asic2* gene reduced the expression of ASIC1a at the dendritic spine, which consequently caused a decreased Ca^{2+} signaling and spine density (89,102). However, a recent study showed that disrupting ASIC1a in nucleus accumbens (NAc) increased dendritic spine density and miniature excitatory postsynaptic currents (mEPSCs) (95). The inconsistency may be due to the use of different experimental systems (cultured brain slice vs specific brain tissue) and/or different brain regions (hippocampus vs NAc). It is still too early to decide on the exact role of ASIC1a in dendritic spine development. Most research results concerning this topic are descriptive and the reported findings are rather conflicting. Thus, more studies with deeper insights are needed to clarify the underlying molecular mechanisms. Additionally, although ASIC3 was observed to interact with PSD-95 in PNS (96), it remains unclear whether ASIC3 is also distributed at spines and contributes to spine development. Given that ASIC1a can form heteromeric channels with ASIC3 (1), it is highly possible that ASIC3 may also regulate spine density in an ASIC1a-dependent manner.

However, it remains unclear how postsynaptic ASICs are activated during spine development, before mature synapse structures are formed and acidic contents can be released from the presynaptic side of the synapse (103,104). Thus, ASICs do not seem to regulate *de novo* dendritic spine formation. By contrast, they may be more likely involved in synapse pruning. The ASICs localized to redundant and misconnected spines most likely do not receive enough presynaptic acidic stimulation, and therefore remain silent. Only in the correctly connected synapses, ASIC-related signaling pathway(s) can be activated and contribute to spine development, leading to more efficient synaptic configurations (95,102). Therefore, although ASIC gene disruption only led to mild spine loss (or increase) in the mouse brain, it may impair the efficiency of the synaptic network and therefore cause neurological defects.

As energy production centers in neurons, mitochondria play critical roles in synapse development (99,100,105). Both axons and dendrites contain lots of mitochondria, but dendritic spines do not (with one exception of olfactory bulb dendritic spines, which contain mitochondria) (100,105). Despite this, mitochondria are always distributed closely to dendritic protrusions (105). Increases in mitochondrial number (e.g. by overexpression of dynamin-related protein-1) or up-regulation of mitochondrial activities (e.g. by creatine treatment) can significantly increase spine density (99). Conversely, reduced mitochondrial number and activities can cause spine destruction (e.g. by amyloid beta treatment; 106–108). Although the underlying mechanism remains to be determined, one promising hypothesis is that spine development requires more ATP and Ca^{2+}, which are provided by mitochondria. Recently, mtASIC1a was identified in mouse brain mitochondria and found to regulate mitochondrial pathophysiological function(s) (97,98). Accumulating evidence shows that mtASIC1a can be activated by an intracellular pH drop and regulates mitochondrial ionic balance (98,109). In $Asic1a^{-/-}$ brains, mitochondria were larger and the Ca^{2+} retention ability (CRC) was higher than wide type ones (97). Moreover, by two-dimension gel electrophoresis and mass spectrometry analysis, it was noticed that the expression levels of seven mitochondrial proteins were changed due to $Asic1a$ gene deletion (97). Thus, the alteration of mitochondrial function may provide a possible explanation for the mild spine loss in $Asic1a^{-/-}$ brains. Additionally, we found that several important postsynaptic proteins were down-regulated after $Asic1a$ gene deletion (3). Although it remains to be determined whether the altered expression of these proteins was the reason for or the result of the spine loss, the findings definitely indicate an important role of ASIC1a in dendritic spine development.

1.3.2 SYNAPTIC PLASTICITY

The fluid content of synaptic vesicles is acidic (~pH 5.5), which is released into synaptic cleft together with neurotransmitters (104). The released protons are either neutralized by the extracellular matrix or undergo re-uptake by some kinds of proton exchangers (104). Neuronal activities are always accompanied by pH fluctuations in the synaptic cleft, affecting synaptic activities (110,111). Thus, a plausible hypothesis has been raised that synaptically localized ASICs are activated during neurotransmission and contribute to synaptic functions. In other words, due to the existence of postsynaptic ASICs, protons may serve as a unique neurotransmitter to alter postsynaptic function. A recent study has provided interesting evidence to support this hypothesis, by showing that presynaptic stimulation transiently decreased extracellular pH value in the synaptic cleft and that ASIC1a contributed to mEPSCs in the lateral amygdala (45). ASIC1a is also reported to contribute to spontaneous inhibitory postsynaptic current (IPSC) of principal cells in rat basolateral amygdala (112). Consistently, in another study, virus-based overexpression of ASIC1a in $Asic1a^{-/-}$ NAc decreased mEPSC frequency (95). Either changing the pH buffering capacity or deletion of the $Asic1a$ gene altered the high frequency stimulation (HFS)-induced LTP (45), suggesting that ASIC1a contributes to synaptic plasticity. These findings are consistent with a previous study that showed a defect of HFS-LTP in hippocampal slices from global $Asic1a$ gene-deleted mice (94). Recently, we showed that ASIC1a in excitatory neurons of ACC is a major player in cortical LTP and chronic pain (47). Activation of ASIC1a in ACC stimulates protein kinase C-gamma, which promotes GluA1 insertion into the postsynaptic membrane. Moreover, ASIC1a also contributes to long-term depression (LTD) at mouse insular synapses (113). Interestingly, compared to ACC, genetic deletion of ASIC1a or pharmacological inhibition of ASIC1a only significantly impairs LTD induction without affecting LTP in the insular cortex. This strongly suggests that ASIC1a regulates distinct signaling pathways in different brain regions.

As discussed below, ASIC-null mice exhibit many neuronal dysfunctions in multiple areas, indicating extensive contributions of ASICs in regulating synaptic plasticity (85,94). Among them, it is interesting that deletion of the *Asic1a* gene significantly increased presynaptic release probability in hippocampal neurons (114), consistent with a previous finding that presynaptically released protons suppressed neurotransmitter release (111). These results raised a possibility that ASIC1a may also localize at axon terminals, which could be activated by presynaptically released protons and consequently inhibit neurotransmitter release. Presynaptic mitochondria play important roles in neurotransmission (e.g. by providing the energy needed for synaptic vesicle release and recycling) (100,105). Thus, the recently identified mtASIC1a could also contribute to presynaptic functions indirectly, via regulation of mitochondrial activities. On the other hand, a recent report showed that the HFS-induced LTP was intact when ASIC1a expression in mouse CNS was specifically disrupted by Cre-mediated excision via the nestin promoter (nestin-Cre *Asic1a$^{-/-}$*) (115). This inconsistency may come about because of the different gene deletion strategies used (cell-restricted deletion vs global disruption of the *Asic1a* gene). Interestingly, LTP in the amygdala was actually blocked in the nestin-Cre *Asic1a$^{-/-}$* mice (116), further indicating that ASIC1a may function differently in different brain regions.

1.3.3 REGIONAL SPECIFIC FUNCTIONS OF ASICS

ASICs are widely expressed across the CNS and PNS (10,85). Making use of genetic deletion mutants in rodents, an increasing number of neurophysiological functions of ASICs have been revealed, such as fear conditioning, anxiety, sensory mechanotransduction, and retinal functions (1,85,87). However, as described earlier, some of the previous findings have been challenged with the implementation of alternative new strategies. Specifically, the role of ASIC1a in the hippocampus requires further exploration. Here, we focus only on the well-confirmed physiological roles of ASICs.

1.3.3.1 Amygdala

The amygdala plays a central role in forming memories induced by strong emotions, such as conditioned (acquired) and unconditioned (innate) fear (117). ASIC1a is abundantly expressed in the amygdala (5,10), suggesting a possible contribution in fear-related responses. Overexpression of human ASIC1a in mouse brains using the pan-neuronal synapsin 1 promotor significantly enhanced both acid-evoked currents in neurons and fear conditioning (118). On the contrary, deletion of the *Asic1a* gene impaired cue and context fear conditioning, but baseline fear on the elevated plus maze remained unchanged (5,26,119). Moreover,

ASIC1a-null mice also showed reduced unconditioned fear in the open field test, during acoustic startle, and in response to predator odor (5,26,119). Further study showed that restoration of ASIC1a in mouse basolateral amygdala (BLA) successfully rescued fear conditioning, but not unconditioned fear response, suggesting that BLA may be the site where ASIC1a takes part in fear memory (26,118,119). However, we recently found that extinction of conditioned fear only requires ASIC1a in the ventral hippocampus (vHPC), but not ASIC1a in the dorsal hippocampus, medial prefrontal cortex, or BLA (6). Additionally, ASIC1a also contributes to anxiety- and depression-related behaviors (85,112,120).

As discussed above, brain pH homeostasis is crucial for neuronal physiological functions (86,110,121), which is maintained by pH buffering systems (121). Critical among these systems is the CO_2/HCO_3^- system (121). In a reversible way CO_2 is dissolved in water to form carbonic acid, which is rapidly dissociated into H^+ and HCO_3^-, leading to pH reduction (121). Long time ago, people have observed that breathing CO_2 caused panic attacks in patients with panic disorder and those patients with increasing hypercarbia tended to be extremely anxious (26). Recently, via MRI technology, it was observed that inhaled CO_2 significantly reduced the pH in mouse and human brains (122). Amygdala-expressed ASIC1a could detect inhaled CO_2-induced pH reduction and play an important role in CO_2-induced fear responses (26). Both genetic deletion of *Asic1a* gene and pharmacological inhibition of amygdala-expressed ASIC1a suppressed these activities (26). Localized restoration of ASIC1a in amygdala rescued the deficit of CO_2-induced fear/panic behaviors in ASIC1a-null mice (26). It is possible that besides contribution to fear memory, amygdala also functions as a brain chemosensor, directly sensing fear-evoked stimuli such as inhaled CO_2 and initiating behavior responses, and amygdala-expressed ASIC1a is required for these processes (26). However, a more recent study showed that CO_2 inhalation could still evoke fear/panic behaviors in humans with BLA damage (123).

ASIC1a contributes to both mEPSC and mIPSC (45,95,112), which means that activation of ASIC1a will change both excitatory and inhibitory inputs at the same time. Considering that ASIC1a is expressed in both inhibitory and principal neurons (85,91), functional alterations of ASIC1a may change excitatory/inhibitory (E/I) balance of neuronal network, which may lead to varying complicated consequences. For example, CO_2 inhalation-evoked fear/panic behaviors are mainly due to the excitatory function of ASIC1a (E/I balance tends to E) (26,45,116). On the other hand, the observation that blockade of ASIC1a increases anxiety most likely resulted from an inhibitory role of ASIC1a (E/I balance tends to I) (112). The report showing that activation of ASIC1a terminated seizures offers another piece of evidence supporting the inhibitory role of ASIC1a (27).

1.3.3.2   Retina

The retina is one of CNS regions with a high level of metabolic activity, which leads to transient interstitial acidification around optic nerves during normal neurotransmission (124,125). This pH alteration was shown to contribute to the fine tuning of visual perception and regulate the adaptation to ambient light intensities (126,127). By *in situ* hybridization and immunohistochemistry, ASIC1a was detected in the outer and inner nuclear layers (cone photoreceptors, horizontal cells, some amacrine and bipolar cells) and in the ganglion cell layer (128–132). The electroretinography (ERG) examination showed that both *in vivo* pharmacological inhibition of ASIC1a by the specific inhibitor psalmotoxin 1 (PcTX1) and *in vivo* knockdown of ASIC1a by antisense oligonucleotides significantly decreased the photopic a- and b-waves and oscillatory potentials (OP), suggesting that ASIC1a plays an important role in normal retinal functions such as processing light-induced visual signals (128). In this particular study, as soon as 4 hours after intravitreal injection, PcTX1 significantly reduced the photopic a- and b-waves and OP (128). This fast effect suggested that ASIC1a directly participated in neural network activity in the process of cone phototransduction (128). Both ASIC2a and ASIC2b were found to be expressed in the outer nuclear layer (photoreceptors) and at lower levels in some cells of the distal and proximal inner nuclear layer (130,133). Interestingly, deletion of the *Asic2* gene led to increased ERG a- and b-waves, an opposite consequence of ASIC1a inhibition, suggesting an enhanced gain of phototransduction in ASIC2-null retina (133). The ASIC3 subunit is also present in the rod inner segment of photoreceptors, retinal ganglion cells, horizontal cells, and some amacrine cells, which all contribute to cone phototransduction (130). However, deletion of the *Asic3* gene did not significantly affect the pH sensitivity of retina (134), suggesting a different role of ASIC3 in the retina from ASIC1a, ASIC2a, and ASIC2b. Additionally, retinal morphological change has not been observed with *Asic1, Asic2*, and *Asic3* genes deleted separately in mice (132–134), while the ultrastructure has not been examined.

In the retina, photon signals are firstly transmitted from the light-sensitive photoreceptors (rods and cones) to bipolar cells, by which they are relayed to the output-collecting ganglion cells (124,125). Although ASIC subunits 1a, 2a, 2b, and 3 were found to distribute nearly across the whole retina (128,133,134), unfortunately, it remains unknown to which step(s) of photon signal transduction ASICs contribute. Thus, the functional alterations seen with ASIC deletion and inhibition may represent the summed consequences of changes in multiple interwoven circuits. Despite this, there is a possible explanation based on the channel properties of different ASICs. At least five forms of ASICs are expressed in the retina, including homomeric ASIC1a and ASIC2a channels, and heteromeric ASIC1a/2a, ASIC1a/2b, and ASIC2a/2b channels (ASIC3 only contributes to a limited extent to pH detection) (130,135). The pH value required for half-maximum activation (pH_{50}) of mouse homomeric ASIC2a channels (and heteromeric ASIC2a/2b channels) is quite low (~3.87) (1). Such a severe acidosis condition barely occurs under physiological conditions, except for pathological conditions like retinal ischemia (121,136). Thus, unlike homomeric ASIC1a channels, whose pH_{50} is around 6.37 (1), homomeric ASIC2a channels may not be activated during normal retina activities. In ASIC1a-null retinas, homomeric ASIC2a and heteromeric ASIC2a/2b channels are the main forms of ASICs, which fail to be activated by physiological acidification, leading to a reduced cone phototransduction (128). On the contrary, in ASIC2-null retinas, homomeric ASIC1a channels are the dominant forms of ASICs, with an increased sensitivity to pH fluctuations (pH_{50} of heteromeric ASIC1a/2a channels is around 6.06), resulting in an enhanced cone phototransduction (133). Moreover, because homomeric ASIC1a channels are the only Ca^{2+}-permeable ASICs, the acidification-induced Ca^{2+} signaling may be amplified in ASIC2-null retina, which may be another reason for the enhanced rod phototransduction. Therefore, ASIC2a and 2b may contribute to the pH sensitivity and Ca^{2+} signaling in retinas indirectly by regulating subunit compositions of ASICs. This hypothesis requires testing by electrophysiological experiments in future studies.

1.3.3.3   Dorsal Root Ganglia (DRG)

Besides the retina, all six ASIC subunits have been identified in other sensory neurons such as trigeminal ganglia (TG), nodose ganglia (NG), and DRG, where the ASICs contribute to mechanotransduction (1,85,135). Unlike in the CNS systems, most studies here focused on ASIC3, which was specifically enriched in PNS (1,85,135). Nearly 20 years ago, ASIC3 was first cloned from rat DRG and named as a dorsal root acid-sensing ion channel (DRASIC) (10). As the most sensitive ASICs to extracellular pH (can be activated by pH 7.2) (1), this channel distributes in nearly all types of DRG neurons (both large and small diameter ones) and plays an important role in sensory transduction (1,85,135,137,138). ASIC3-null mice showed impaired sensory detection from hairy skin (138). In rodents, pharmacological blockade of ASIC3 (e.g. APETx2) or suppression of ASIC3 expression at the periphery nerves reduces the pain behaviors (138,139). On the contrary, activation of ASIC3 by non-proton ligand GMQ significantly enhanced pain behaviors (2). Furthermore, recently, our lab found that serotonin can amplify the activity of ASIC3-containing channels (ASIC3 homomeric channels and heteromeric channels, like ASIC3/1a), leading to enhanced pain behaviors (140). Additionally, inflammatory factors, including lactic acid, ATP, arachidonic acid, agmatine, and hypertonicity, also sensitize ASIC3 (141). Interestingly, results from studies of *Asic3* gene deletion in animals are conflicting. Both reduction (138) and enhancement of sensory afferent/pain behaviors (142,143) were observed, suggesting a complicated role of ASIC3 in sensory circuits.

Although ASIC1 is also highly expressed in sensory neurons, its role in sensory afferent appears less important as compared with ASIC3 (85,135). Subcutaneous injection or local administration of PcTx1, the specific inhibitor of homomeric ASIC1a, failed to affect pain behaviors of rodents (87), suggesting peripheral homomeric ASIC1a may not contribute to nociception. However, it is also possible that peripheral ASIC1a is involved in nociception as a subunit of heteromeric channels, like ASIC3/1a. Some reports showed that $Asic1a^{-/-}$ mice (global gene deletion) exhibited an increased sensitivity to mechanical stimuli when compared to control ones (144). However, no change was observed in $Asic1a^{-/-}$ mice of the sensitivity of cutaneous mechanoreceptors and the behavioral responses to mechanical and thermal stimuli (145). Spinal cord ASIC1a was reported to play a role in nociception (87). Thus, the increased sensitivity to mechanical stimuli may be due to the deletion of ASIC1a in CNS (e.g. spinal cord) but not that in DRG neurons. Some other ASIC subunits have been reported to contribute to sensory detection in PNS (85,135). For example, ASIC2a forms heteromeric channels with ASIC3 in mouse cardiac DRG neurons, contributing to chest pain and other cardiac diseases (90). ASIC2b is also highly expressed in NG and aortic baroreceptor terminals and plays an important role in baroreceptor and autonomic control of circulation (146). As promising targets, the exploration of ASICs has been also expanded to many non-neuronal tissues and with different animal models, which reveal many essential functions of ASICs (147,148). For example, in *C. elegans*, ASICs in enteric serotonergic neurons mediate food responses and control foraging behaviors (149).

1.4 SUMMARY AND OUTLOOK

In summary, we presented previous literature regarding the putative sources of protons capable of activating endogenous ASICs, including metabolic production of protons, acidification niche, and proton generators (NHEs, Hv1, CAs etc.). We also reviewed the functions of ASICs in synapse development, synaptic plasticity, and their roles in the amygdala, retina, and DRG. Indeed, our knowledge of the structure, gating, and functions of ASICs has expanded in the past few years (150). However, more studies are certainly warranted to enrich our understanding of the cell biology and clinical translational potential of ASICs. Here, we would like to point out some unresolved issues or future research directions, in hopes of stimulating more insightful studies in this field.

1. Identifying novel physiological and pathophysiological functions of ASICs. Investigations should be undertaken to address the precise role of different ASIC isoforms, particularly the much less-characterized ASIC2a and ASIC2b, in execution of higher-level behavioral functions, such as social communication, decision-making, and reward/aversion evaluation. ASIC1 and ASIC2 are widely distributed in the CNS. However, the isoforms-specific functions of ASICs are still far less studied with many brain areas remaining unexplored, such as habenula, inferior olive, periaqueductal gray, superior colliculus etc. (151).

2. Monitoring the subcellular dynamics of pH fluctuations in freely moving animals. Protons are considered an important extracellular transmitter that plays key roles in inter-neuronal and neuronal-glial communications. However, the highly temporal and spatial dynamics of proton signaling pose a great difficulty in capturing local proton transmission in a specific brain region in freely behaving animals. The current pH measurement tools lack sufficient spatial and/or temporal resolutions to track the fast pH fluctuations, particularly in subcellular domains of the cells. Future development of more sensitive pH-detecting probes for *in vivo* imaging will help to elucidate the details of proton signaling and its pathophysiological roles in the brain.

3. Imaging the trafficking of ASICs *in vivo*. ASICs, like other classical ion channels, undergo constitutive and regulated tracking across the plasma membrane or intracellular compartments. Recently, evidence has been

accumulated to demonstrate the vital role of membrane insertion or endocytosis of ASICs in regulating its pathophysiological functions (152). However, there remains a challenge to elucidate the exact mechanisms of ASIC trafficking between the plasma membrane and intracellular compartments with excellent spatial and temporal resolutions. To fully address this issue, more elaborate molecular probes and microscopy approaches for ASICs should be developed to reveal the contribution of ASIC trafficking to synaptic transmission as well as the pathogenesis of various neurological diseases.

4. One crucial question in the ASIC field is to identify novel non-proton ligands of the channels. Considering the strong pH buffering system in the body and the desensitization of ASICs by prolonged protonation, it is highly possible that some unidentified non-proton ligands of ASICs exist in our body to help drive the activation of the channel under those conditions with less protons or severe acidosis. Our previous work reported GMQ as an important non-proton ligand of ASIC3 in the peripheral sensory neurons (2). Interestingly, Deval and colleagues identified lysophosphatidylcholine and arachidonic acid as endogenous activators of ASIC3 in the absence of any

TABLE 1.1
Involvement of ASICs in Pathological Conditions Related to Acidosis

Pathological Conditions	Mediators	Subunits	Localization	Ref
Primary hyperalgesia Secondary hyperalgesia	ATP, lactate, 5-HT, AA, etc.BDNF.	ASIC3ASIC1a	Sensory neuronSpinal dorsal corn	154,155,156, 157
Acid induced cutaneous pain	Protons	ASIC3	Skin innervated sensory neuron	138,158,159
Retinal function	Protons	ASIC1ASIC2ASIC3		134
Arthritis	Protons	ASIC3		160,161
Brain ischemia Muscle/Cardiac Ischemia	Lactic acid	ASIC1ASIC3	Brain Cardiac innervated sensory neuron	8,16215
Postoperative pain	Lactic acid, 5-HT, AA	ASIC3	Skin/muscle innervated sensory neuron	29
Seizures	Lactic acid, CO_2	ASIC1a	Brain	27
Anxiety and panic	CO_2	ASIC1a	Amygdala	26,120
Breath control	CO_2	ASIC1	Lateral Hypothalamus Nucleus of the solitary tract	163,164

Abbreviations: AA, arachidonic acid; BDNF, brain-derived neurotrophic factor.

extracellular acidification (153). For ASIC1a, MitTx, the prevalent components from the Texas coral snake, was shown to be a potent activator of ASIC1a homomers (12). Unfortunately, no endogenous activator for ASIC1a has been identified yet, which requires future extensive studies to tackle the problem.

5. Acidosis is one of the most common features of several neurological diseases such as stroke, chronic pain, Parkinson's disease, multiple sclerosis, epilepsy, etc. ASICs, as important sensors of acidosis, have shown promising translational potential as new therapeutic targets for a wide range of acidosis-associated diseases (Table 1.1). Thus, one exciting area for future research is to identify and validate new small molecules or peptides that target ASICs for the treatment of disorders with severe acidosis. This kind of translational study is of primary significance in developing new pharmacological agents for better treatment of acidosis-related diseases. More knowledge on the structure, function, and regulation of ASICs needs to be gained by a combination of traditional and cutting-edge approaches to translate this promise into reality.

ACKNOWLEDGEMENTS

This work is supported by grants from the National Natural Science Foundation of China (31930050, 81961128024, 81730095), the Science and Technology Commission of Shanghai Municipality (18JC1420302), and the Shanghai Municipal Science and Technology Major Project (2018SHZDZX05), and an innovative research team of high-level local universities in Shanghai.

REFERENCES

1. Wemmie, J. A., Price, M. P., and Welsh, M. J. (2006) Acid-sensing ion channels: advances, questions and therapeutic opportunities. *Trends Neurosci* **29**, 578–586.
2. Yu, Y., Chen, Z., Li, W. G., Cao, H., Feng, E. G., Yu, F., Liu, H., Jiang, H., and Xu, T. L. (2010) A nonproton ligand sensor in the acid-sensing ion channel. *Neuron* **68**, 61–72.
3. Yu, Z., Wu, Y. J., Wang, Y. Z., Liu, D. S., Song, X. L., Jiang, Q., Li, Y., Zhang, S., Xu, N. J., Zhu, M. X., Li, W. G., and Xu, T. L. (2018) The acid-sensing ion channel ASIC1a mediates striatal synapse remodeling and procedural motor learning. *Sci Signal* **11**, eaar4481.
4. Coryell, M. W., Ziemann, A. E., Westmoreland, P. J., Haenfler, J. M., Kurjakovic, Z., Zha, X. M., Price, M., Schnizler, M. K., and Wemmie, J. A. (2007) Targeting ASIC1a reduces innate fear and alters neuronal activity in the fear circuit. *Biol Psychiatry* **62**, 1140–1148.
5. Wemmie, J. A., Askwith, C. C., Lamani, E., Cassell, M. D., Freeman, J. H., Jr., and Welsh, M. J. (2003) Acid-sensing ion channel 1 is localized in brain regions with high synaptic density and contributes to fear conditioning. *J Neurosci* **23**, 5496–5502.
6. Wang, Q., Wang, Q., Song, X. L., Jiang, Q., Wu, Y. J., Li, Y., Yuan, T. F., Zhang, S., Xu, N. J., Zhu, M. X., Li, W. G., and Xu, T. L. (2018) Fear extinction requires ASIC1a-dependent regulation of hippocampal-prefrontal correlates. *Sci Adv* **4**, eaau3075.

7. Duan, B., Wang, Y. Z., Yang, T., Chu, X. P., Yu, Y., Huang, Y., Cao, H., Hansen, J., Simon, R. P., Zhu, M. X., Xiong, Z. G., and Xu, T. L. (2011) Extracellular spermine exacerbates ischemic neuronal injury through sensitization of ASIC1a channels to extracellular acidosis. *J Neurosci* **31**, 2101–2112.
8. Xiong, Z. G., Zhu, X. M., Chu, X. P., Minami, M., Hey, J., Wei, W. L., MacDonald, J. F., Wemmie, J. A., Price, M. P., Welsh, M. J., and Simon, R. P. (2004) Neuroprotection in ischemia: blocking calcium-permeable acid-sensing ion channels. *Cell* **118**, 687–698.
9. Friese, M. A., Craner, M. J., Etzensperger, R., Vergo, S., Wemmie, J. A., Welsh, M. J., Vincent, A., and Fugger, L. (2007) Acid-sensing ion channel-1 contributes to axonal degeneration in autoimmune inflammation of the central nervous system. *Nat Med* **13**, 1483–1489.
10. Waldmann, R., Champigny, G., Bassilana, F., Heurteaux, C., and Lazdunski, M. (1997) A proton-gated cation channel involved in acid-sensing. *Nature* **386**, 173–177.
11. Zeng, W. Z., and Xu, T. L. (2012) Proton production, regulation and pathophysiological roles in the mammalian brain. *Neurosci Bull* **28**, 1–13.
12. Bohlen, C. J., Chesler, A. T., Sharif-Naeini, R., Medzihradszky, K. F., Zhou, S., King, D., Sanchez, E. E., Burlingame, A. L., Basbaum, A. I., and Julius, D. (2011) A heteromeric Texas coral snake toxin targets acid-sensing ion channels to produce pain. *Nature* **479**, 410–414.
13. Sutherland, S. P., Benson, C. J., Adelman, J. P., and McCleskey, E. W. (2001) Acid-sensing ion channel 3 matches the acid-gated current in cardiac ischemia-sensing neurons. *Proc Natl Acad Sci USA* **98**, 711–716.
14. Benson, C. J., Eckert, S. P., and McCleskey, E. W. (1999) Acid-evoked currents in cardiac sensory neurons: a possible mediator of myocardial ischemic sensation. *Circ Res* **84**, 921–928.
15. Birdsong, W. T., Fierro, L., Williams, F. G., Spelta, V., Naves, L. A., Knowles, M., Marsh-Haffner, J., Adelman, J. P., Almers, W., Elde, R. P., and McCleskey, E. W. (2010) Sensing muscle ischemia: coincident detection of acid and ATP via interplay of two ion channels. *Neuron* **68**, 739–749.
16. Light, A. R., Hughen, R. W., Zhang, J., Rainier, J., Liu, Z., and Lee, J. (2008) Dorsal root ganglion neurons innervating skeletal muscle respond to physiological combinations of protons, ATP, and lactate mediated by ASIC, P2X, and TRPV1. *J Neurophysiol* **100**, 1184–1201.
17. Immke, D. C., and McCleskey, E. W. (2001) Lactate enhances the acid-sensing Na+ channel on ischemia-sensing neurons. *Nat Neurosci* **4**, 869–870.
18. Kraut, J. A., and Madias, N. E. (2010) Metabolic acidosis: pathophysiology, diagnosis and management. *Nat Rev Nephrol* **6**, 274–285.
19. Ai, M., Min, S., Grosjean, Y., Leblanc, C., Bell, R., Benton, R., and Suh, G. S. (2010) Acid sensing by the *Drosophila* olfactory system. *Nature* **468**, 691–695.
20. Suh, G. S., Wong, A. M., Hergarden, A. C., Wang, J. W., Simon, A. F., Benzer, S., Axel, R., and Anderson, D. J. (2004) A single population of olfactory sensory neurons mediates an innate avoidance behaviour in *Drosophila*. *Nature* **431**, 854–859.
21. Scott, K. (2011) Out of thin air: sensory detection of oxygen and carbon dioxide. *Neuron* **69**, 194–202.
22. Luo, M., Sun, L., and Hu, J. (2009) Neural detection of gases—carbon dioxide, oxygen—in vertebrates and invertebrates. *Curr Opin Neurobiol* **19**, 354–361.
23. Rassovsky, Y., and Kushner, M. G. (2003) Carbon dioxide in the study of panic disorder: issues of definition, methodology, and outcome. *J Anxiety Disord* **17**, 1–32.
24. Zandbergen, J., Pols, H., De Loof, C., Lousberg, H., and Griez, E. (1989) Effect of hypercapnia and other disturbances in the acid-base-balance on panic disorder. *Hillside J Clin Psychiatry* **11**, 185–197.

25. Smoller, J. W., Pollack, M. H., Otto, M. W., Rosenbaum, J. F., and Kradin, R. L. (1996) Panic anxiety, dyspnea, and respiratory disease. Theoretical and clinical considerations. *Am J Respir Crit Care Med* **154**, 6–17.
26. Ziemann, A. E., Allen, J. E., Dahdaleh, N. S., Drebot, II, Coryell, M. W., Wunsch, A. M., Lynch, C. M., Faraci, F. M., Howard, M. A., 3rd, Welsh, M. J., and Wemmie, J. A. (2009) The amygdala is a chemosensor that detects carbon dioxide and acidosis to elicit fear behavior. *Cell* **139**, 1012–1021.
27. Ziemann, A. E., Schnizler, M. K., Albert, G. W., Severson, M. A., Howard, M. A., 3rd, Welsh, M. J., and Wemmie, J. A. (2008) Seizure termination by acidosis depends on ASIC1a. *Nat Neurosci* **11**, 816–822.
28. Woo, Y. C., Park, S. S., Subieta, A. R., and Brennan, T. J. (2004) Changes in tissue pH and temperature after incision indicate acidosis may contribute to postoperative pain. *Anesthesiology* **101**, 468–475.
29. Deval, E., Noel, J., Gasull, X., Delaunay, A., Alloui, A., Friend, V., Eschalier, A., Lazdunski, M., and Lingueglia, E. (2011) Acid-sensing ion channels in postoperative pain. *J Neurosci* **31**, 6059–6066.
30. Tan, Z. Y., Lu, Y., Whiteis, C. A., Benson, C. J., Chapleau, M. W., and Abboud, F. M. (2007) Acid-sensing ion channels contribute to transduction of extracellular acidosis in rat carotid body glomus cells. *Circ Res* **101**, 1009–1019.
31. Jacobus, W. E., Taylor, G. J. t., Hollis, D. P., and Nunnally, R. L. (1977) Phosphorus nuclear magnetic resonance of perfused working rat hearts. *Nature* **265**, 756–758.
32. Cobbe, S. M., and Poole-Wilson, P. A. (1980) The time of onset and severity of acidosis in myocardial ischaemia. *J Mol Cell Cardiol* **12**, 745–760.
33. Immke, D. C., and McCleskey, E. W. (2003) Protons open acid-sensing ion channels by catalyzing relief of Ca^{2+} blockade. *Neuron* **37**, 75–84.
34. Schurr, A., Miller, J. J., Payne, R. S., and Rigor, B. M. (1999) An increase in lactate output by brain tissue serves to meet the energy needs of glutamate-activated neurons. *J Neurosci* **19**, 34–39.
35. Smith, D., Pernet, A., Hallett, W. A., Bingham, E., Marsden, P. K., and Amiel, S. A. (2003) Lactate: a preferred fuel for human brain metabolism in vivo. *J Cereb Blood Flow Metab* **23**, 658–664.
36. Gallagher, C. N., Carpenter, K. L., Grice, P., Howe, D. J., Mason, A., Timofeev, I., Menon, D. K., Kirkpatrick, P. J., Pickard, J. D., Sutherland, G. R., and Hutchinson, P. J. (2009) The human brain utilizes lactate via the tricarboxylic acid cycle: a 13C-labelled microdialysis and high-resolution nuclear magnetic resonance study. *Brain* **132**, 2839–2849.
37. Belanger, M., Allaman, I., and Magistretti, P. J. (2011) Brain energy metabolism: focus on astrocyte-neuron metabolic cooperation. *Cell Metab* **14**, 724–738.
38. Pellerin, L., Pellegri, G., Bittar, P. G., Charnay, Y., Bouras, C., Martin, J. L., Stella, N., and Magistretti, P. J. (1998) Evidence supporting the existence of an activity-dependent astrocyte-neuron lactate shuttle. *Dev Neurosci* **20**, 291–299.
39. Pellerin, L., and Magistretti, P. J. (1994) Glutamate uptake into astrocytes stimulates aerobic glycolysis: a mechanism coupling neuronal activity to glucose utilization. *Proc Natl Acad Sci USA* **91**, 10625–10629.
40. Suzuki, A., Stern, S. A., Bozdagi, O., Huntley, G. W., Walker, R. H., Magistretti, P. J., and Alberini, C. M. (2011) Astrocyte-neuron lactate transport is required for long-term memory formation. *Cell* **144**, 810–823.
41. Stewart, A. K., Boyd, C. A., and Vaughan-Jones, R. D. (1999) A novel role for carbonic anhydrase: cytoplasmic pH gradient dissipation in mouse small intestinal enterocytes. *J Physiol* **516 (Pt 1)**, 209–217.

42. Vaughan-Jones, R. D., Peercy, B. E., Keener, J. P., and Spitzer, K. W. (2002) Intrinsic H(+) ion mobility in the rabbit ventricular myocyte. *J Physiol* **541**, 139–158.

43. Pfeiffer, J., Johnson, D., and Nehrke, K. (2008) Oscillatory transepithelial H(+) flux regulates a rhythmic behavior in *C. elegans*. *Curr Biol* **18**, 297–302.

44. Beg, A. A., Ernstrom, G. G., Nix, P., Davis, M. W., and Jorgensen, E. M. (2008) Protons act as a transmitter for muscle contraction in *C-elegans*. *Cell* **132**, 149–160.

45. Du, J., Reznikov, L. R., Price, M. P., Zha, X. M., Lu, Y., Moninger, T. O., Wemmie, J. A., and Welsh, M. J. (2014) Protons are a neurotransmitter that regulates synaptic plasticity in the lateral amygdala. *Proc Natl Acad Sci USA* **111**, 8961–8966.

46. Gonzalez-Inchauspe, C., Urbano, F. J., Di Guilmi, M. N., and Uchitel, O. D. (2017) Acid-sensing ion channels activated by evoked released protons modulate synaptic transmission at the mouse calyx of held synapse. *J Neurosci* **37**, 2589–2599.

47. Li, H.S., Su, X.Y., Song, X.L., Qi, X., Li, Y., Wang, R.Q., Maximyuk, O., Krishtal, O., Wang, T., Fang, H., Liao, L., Cao, H., Zhang, Y.Q., Zhu, M.X., Liu, M.G., and Xu, T.L. (2019) Protein kinase C lambda mediates acid-sensing ion channel 1a-dependent cortical synaptic plasticity and pain Hypersensitivity. *J Neurosci* **39**, 5773–5793.

48. Casey, J. R., Grinstein, S., and Orlowski, J. (2010) Sensors and regulators of intracellular pH. *Nat Rev Mol Cell Biol* **11**, 50–61.

49. Ferenczi, E. A., Vierock, J., Atsuta-Tsunoda, K., Tsunoda, S. P., Ramakrishnan, C., Gorini, C., Thompson, K., Lee, S. Y., Berndt, A., Perry, C., Minniberger, S., Vogt, A., Mattis, J., Prakash, R., Delp, S., Deisseroth, K., and Hegemann, P. (2016) Optogenetic approaches addressing extracellular modulation of neural excitability. *Sci Rep* **6**, 23947.

50. Zeng, W. Z., Liu, D. S., Liu, L., She, L., Wu, L. J., and Xu, T. L. (2015) Activation of acid-sensing ion channels by localized proton transient reveals their role in proton signaling. *Sci Rep* **5**, 14125.

51. Orlowski, J., and Grinstein, S. (2004) Diversity of the mammalian sodium/proton exchanger SLC9 gene family. *Pflugers Arch* **447**, 549–565.

52. Slepkov, E. R., Rainey, J. K., Sykes, B. D., and Fliegel, L. (2007) Structural and functional analysis of the Na+/H+ exchanger. *Biochem J* **401**, 623–633.

53. Putney, L. K., Denker, S. P., and Barber, D. L. (2002) The changing face of the Na+/H+ exchanger, NHE1: structure, regulation, and cellular actions. *Annu Rev Pharmacol Toxicol* **42**, 527–552.

54. Denker, S. P., Huang, D. C., Orlowski, J., Furthmayr, H., and Barber, D. L. (2000) Direct binding of the Na–H exchanger NHE1 to ERM proteins regulates the cortical cytoskeleton and cell shape independently of H(+) translocation. *Mol Cell* **6**, 1425–1436.

55. Baumgartner, M., Patel, H., and Barber, D. L. (2004) Na(+)/H(+) exchanger NHE1 as plasma membrane scaffold in the assembly of signaling complexes. *Am J Physiol Cell Physiol* **287**, C844–C850.

56. Cardone, R. A., Casavola, V., and Reshkin, S. J. (2005) The role of disturbed pH dynamics and the Na+/H+ exchanger in metastasis. *Nat Rev Cancer* **5**, 786–795.

57. Gillies, R. J., Raghunand, N., Karczmar, G. S., and Bhujwalla, Z. M. (2002) MRI of the tumor microenvironment. *J Magn Reson Imaging* **16**, 430–450.

58. Reshkin, S. J., Bellizzi, A., Caldeira, S., Albarani, V., Malanchi, I., Poignee, M., Alunni-Fabbroni, M., Casavola, V., and Tommasino, M. (2000) Na+/H+ exchanger-dependent intracellular alkalinization is an early event in malignant transformation and plays an essential role in the development of subsequent transformation-associated phenotypes. *FASEB J* **14**, 2185–2197.

59. Luo, J., Chen, H., Kintner, D. B., Shull, G. E., and Sun, D. (2005) Decreased neuronal death in Na⁺/H⁺ exchanger isoform 1-null mice after in vitro and in vivo ischemia. *J Neurosci* **25**, 11256–11268.
60. Luo, J., Kintner, D. B., Shull, G. E., and Sun, D. (2007) ERK1/2-p90RSK-mediated phosphorylation of Na⁺/H⁺ exchanger isoform 1. A role in ischemic neuronal death. *J Biol Chem* **282**, 28274–28284.
61. DeCoursey, T. E., Morgan, D., and Cherny, V. V. (2003) The voltage dependence of NADPH oxidase reveals why phagocytes need proton channels. *Nature* **422**, 531–534.
62. Henderson, L. M., Chappell, J. B., and Jones, O. T. (1987) The superoxide-generating NADPH oxidase of human neutrophils is electrogenic and associated with an H⁺ channel. *Biochem J* **246**, 325–329.
63. Ramsey, I. S., Moran, M. M., Chong, J. A., and Clapham, D. E. (2006) A voltage-gated proton-selective channel lacking the pore domain. *Nature* **440**, 1213–1216.
64. Sasaki, M., Takagi, M., and Okamura, Y. (2006) A voltage sensor-domain protein is a voltage-gated proton channel. *Science* **312**, 589–592.
65. Decoursey, T. E. (2003) Voltage-gated proton channels and other proton transfer pathways. *Physiol Rev* **83**, 475–579.
66. Lishko, P. V., Botchkina, I. L., Fedorenko, A., and Kirichok, Y. (2010) Acid extrusion from human spermatozoa is mediated by flagellar voltage-gated proton channel. *Cell* **140**, 327–337.
67. Iovannisci, D., Illek, B., and Fischer, H. (2010) Function of the HVCN1 proton channel in airway epithelia and a naturally occurring mutation, M91T. *J Gen Physiol* **136**, 35–46.
68. Capasso, M., Bhamrah, M. K., Henley, T., Boyd, R. S., Langlais, C., Cain, K., Dinsdale, D., Pulford, K., Khan, M., Musset, B., Cherny, V. V., Morgan, D., Gascoyne, R. D., Vigorito, E., DeCoursey, T. E., MacLennan, I. C., and Dyer, M. J. (2010) HVCN1 modulates BCR signal strength via regulation of BCR-dependent generation of reactive oxygen species. *Nat Immunol* **11**, 265–272.
69. Musset, B., Morgan, D., Cherny, V. V., MacGlashan, D. W., Jr., Thomas, L. L., Rios, E., and DeCoursey, T. E. (2008) A pH-stabilizing role of voltage-gated proton channels in IgE-mediated activation of human basophils. *Proc Natl Acad Sci USA* **105**, 11020–11025.
70. Kettenmann, H., Hanisch, U. K., Noda, M., and Verkhratsky, A. (2011) Physiology of microglia. *Physiol Rev* **91**, 461–553.
71. De Simoni, A., Allen, N. J., and Attwell, D. (2008) Charge compensation for NADPH oxidase activity in microglia in rat brain slices does not involve a proton current. *Eur J Neurosci* **28**, 1146–1156.
72. Thomas, R. C., and Meech, R. W. (1982) Hydrogen ion currents and intracellular pH in depolarized voltage-clamped snail neurones. *Nature* **299**, 826–828.
73. Wu, L. J., Wu, G., Akhavan Sharif, M. R., Baker, A., Jia, Y., Fahey, F. H., Luo, H. R., Feener, E. P., and Clapham, D. E. (2012) The voltage-gated proton channel Hv1 enhances brain damage from ischemic stroke. *Nat Neurosci* **15**, 565–573.
74. Pastorekova, S., Parkkila, S., Pastorek, J., and Supuran, C. T. (2004) Carbonic anhydrases: current state of the art, therapeutic applications and future prospects. *J Enzyme Inhib Med Chem* **19**, 199–229.
75. Svichar, N., Waheed, A., Sly, W. S., Hennings, J. C., Hubner, C. A., and Chesler, M. (2009) Carbonic anhydrases CA4 and CA14 both enhance AE3-mediated Cl⁻–HCO₃⁻ exchange in hippocampal neurons. *J Neurosci* **29**, 3252–3258.
76. Shah, G. N., Ulmasov, B., Waheed, A., Becker, T., Makani, S., Svichar, N., Chesler, M., and Sly, W. S. (2005) Carbonic anhydrase IV and XIV knockout mice: roles of the

respective carbonic anhydrases in buffering the extracellular space in brain. *Proc Natl Acad Sci USA* **102**, 16771–16776.

77. Chandrashekar, J., Yarmolinsky, D., von Buchholtz, L., Oka, Y., Sly, W., Ryba, N. J., and Zuker, C. S. (2009) The taste of carbonation. *Science* **326**, 443–445.

78. Hu, H., Boisson-Dernier, A., Israelsson-Nordstrom, M., Bohmer, M., Xue, S., Ries, A., Godoski, J., Kuhn, J. M., and Schroeder, J. I. (2010) Carbonic anhydrases are upstream regulators of CO_2-controlled stomatal movements in guard cells. *Nat Cell Biol* **12**, 87–93; sup pp. 81–8.

79. Swietach, P., Patiar, S., Supuran, C. T., Harris, A. L., and Vaughan-Jones, R. D. (2009) The role of carbonic anhydrase 9 in regulating extracellular and intracellular ph in three-dimensional tumor cell growths. *J Biol Chem* **284**, 20299–20310.

80. Kwon, J. Y., Dahanukar, A., Weiss, L. A., and Carlson, J. R. (2007) The molecular basis of CO_2 reception in Drosophila. *Proc Natl Acad Sci USA* **104**, 3574–3578.

81. Jones, W. D., Cayirlioglu, P., Kadow, I. G., and Vosshall, L. B. (2007) Two chemo-sensory receptors together mediate carbon dioxide detection in Drosophila. *Nature* **445**, 86–90.

82. Fischler, W., Kong, P., Marella, S., and Scott, K. (2007) The detection of carbonation by the *Drosophila* gustatory system. *Nature* **448**, 1054–1057.

83. Yarmolinsky, D. A., Zuker, C. S., and Ryba, N. J. (2009) Common sense about taste: from mammals to insects. *Cell* **139**, 234–244.

84. Hu, J., Zhong, C., Ding, C., Chi, Q., Walz, A., Mombaerts, P., Matsunami, H., and Luo, M. (2007) Detection of near-atmospheric concentrations of CO_2 by an olfactory subsystem in the mouse. *Science* **317**, 953–957.

85. Wemmie, J. A., Taugher, R. J., and Kreple, C. J. (2013) Acid-sensing ion channels in pain and disease. *Nat Rev Neurosci* **14**, 461–471.

86. Wang, Y. Z., and Xu, T. L. (2011) Acidosis, acid-sensing ion channels, and neuronal cell death. *Mol Neurobiol* **44**, 350–358.

87. Xu, T. L., and Duan, B. (2009) Calcium-permeable acid-sensing ion channel in nociceptive plasticity: a new target for pain control. *Prog Neurobiol* **87**, 171–180.

88. Bartoi, T., Augustinowski, K., Polleichtner, G., Grunder, S., and Ulbrich, M. H. (2014) Acid-sensing ion channel (ASIC) 1a/2a heteromers have a flexible 2:1/1:2 stoichiometry. *Proc Natl Acad Sci USA* **111**, 8281–8286.

89. Zha, X. M., Costa, V., Harding, A. M., Reznikov, L., Benson, C. J., and Welsh, M. J. (2009) ASIC2 subunits target acid-sensing ion channels to the synapse via an association with PSD-95. *J Neurosci* **29**, 8438–8446.

90. Hattori, T., Chen, J., Harding, A. M., Price, M. P., Lu, Y., Abboud, F. M., and Benson, C. J. (2009) ASIC2a and ASIC3 heteromultimerize to form pH-sensitive channels in mouse cardiac dorsal root ganglia neurons. *Circ Res* **105**, 279–286.

91. Weng, J. Y., Lin, Y. C., and Lien, C. C. (2010) Cell type-specific expression of acid-sensing ion channels in hippocampal interneurons. *J Neurosci* **30**, 6548–6558.

92. Lin, J. H., Hung, C. H., Han, D. S., Chen, S. T., Lee, C. H., Sun, W. Z., and Chen, C. C. (2018) Sensing acidosis: nociception or sngception? *J Biomed Sci* **25**, 85.

93. Uchitel, O. D., Gonzalez Inchauspe, C., and Weissmann, C. (2019) Synaptic signals mediated by protons and acid-sensing ion channels. *Synapse* **73**, e22120.

94. Wemmie, J. A., Chen, J., Askwith, C. C., Hruska-Hageman, A. M., Price, M. P., Nolan, B. C., Yoder, P. G., Lamani, E., Hoshi, T., Freeman, J. H., Jr., and Welsh, M. J. (2002) The acid-activated ion channel ASIC contributes to synaptic plasticity, learning, and memory. *Neuron* **34**, 463–477.

95. Kreple, C. J., Lu, Y., Taugher, R. J., Schwager-Gutman, A. L., Du, J., Stump, M., Wang, Y., Ghobbeh, A., Fan, R., Cosme, C. V., Sowers, L. P., Welsh, M. J., Radley, J.

J., LaLumiere, R. T., and Wemmie, J. A. (2014) Acid-sensing ion channels contribute to synaptic transmission and inhibit cocaine-evoked plasticity. *Nat Neurosci* **17**, 1083–1091.

96. Hruska-Hageman, A. M., Benson, C. J., Leonard, A. S., Price, M. P., and Welsh, M. J. (2004) PSD-95 and Lin-7b interact with acid-sensing ion channel-3 and have opposite effects on H+-gated current. *J Biol Chem* **279**, 46962–46968.

97. Wang, Y. Z., Zeng, W. Z., Xiao, X., Huang, Y., Song, X. L., Yu, Z., Tang, D., Dong, X. P., Zhu, M. X., and Xu, T. L. (2013) Intracellular ASIC1a regulates mitochondrial permeability transition-dependent neuronal death. *Cell Death Differ.* **20**, 1359–1369.

98. Savic Azoulay, I., Liu, F., Hu, Q., Rozenfeld, M., Ben Kasus Nissim, T., Zhu, M. X., Sekler, I., and Xu, T. L. (2020) ASIC1a channels regulate mitochondrial ion signaling and energy homeostasis in neurons. *J Neurochem* **153**, 203–215.

99. Li, Z., Okamoto, K., Hayashi, Y., and Sheng, M. (2004) The importance of dendritic mitochondria in the morphogenesis and plasticity of spines and synapses. *Cell* **119**, 873–887.

100. Kann, O., and Kovacs, R. (2007) Mitochondria and neuronal activity. *Am J Physiol Cell Physiol* **292**, C641–C657.

101. Yermolaieva, O., Leonard, A. S., Schnizler, M. K., Abboud, F. M., and Welsh, M. J. (2004) Extracellular acidosis increases neuronal cell calcium by activating acid-sensing ion channel 1a. *Proc Natl Acad Sci USA* **101**, 6752–6757.

102. Zha, X. M., Wemmie, J. A., Green, S. H., and Welsh, M. J. (2006) Acid-sensing ion channel 1a is a postsynaptic proton receptor that affects the density of dendritic spines. *Proc Natl Acad Sci USA* **103**, 16556–16561.

103. Calabrese, B., Wilson, M. S., and Halpain, S. (2006) Development and regulation of dendritic spine synapses. *Physiology (Bethesda)* **21**, 38–47.

104. Sudhof, T. C. (2004) The synaptic vesicle cycle. *Annu Rev Neurosci* **27**, 509–547.

105. Ly, C. V., and Verstreken, P. (2006) Mitochondria at the synapse. *Neuroscientist* **12**, 291–299.

106. Calkins, M. J., and Reddy, P. H. (2011) Amyloid beta impairs mitochondrial anterograde transport and degenerates synapses in Alzheimer's disease neurons. *Biochim Biophys Acta* **1812**, 507–513.

107. Liu, Q. A., and Shio, H. (2008) Mitochondrial morphogenesis, dendrite development, and synapse formation in cerebellum require both Bcl-w and the glutamate receptor delta2. *PLoS Genet* **4**, e1000097.

108. Sung, J. Y., Engmann, O., Teylan, M. A., Nairn, A. C., Greengard, P., and Kim, Y. (2008) WAVE1 controls neuronal activity-induced mitochondrial distribution in dendritic spines. *Proc Natl Acad Sci USA* **105**, 3112–3116.

109. Liu, Z., Pei, H., Zhang, L., and Tian, Y. (2018) Mitochondria-targeted DNA nanoprobe for real-time imaging and simultaneous quantification of Ca(2+) and pH in Neurons. *ACS Nano* **12**, 12357–12368.

110. Miesenbock, G., De Angelis, D. A., and Rothman, J. E. (1998) Visualizing secretion and synaptic transmission with pH-sensitive green fluorescent proteins. *Nature* **394**, 192–195.

111. DeVries, S. H. (2001) Exocytosed protons feedback to suppress the Ca^{2+} current in mammalian cone photoreceptors. *Neuron* **32**, 1107–1117.

112. Pidoplichko, V. I., Aroniadou-Anderjaska, V., Prager, E. M., Figueiredo, T. H., Almeida-Suhett, C. P., Miller, S. L., and Braga, M. F. (2014) ASIC1a activation enhances inhibition in the basolateral amygdala and reduces anxiety. *J Neurosci* **34**, 3130–3141.

113. Li, W. G., Liu, M. G., Deng, S., Liu, Y. M., Shang, L., Ding, J., Hsu, T. T., Jiang, Q., Li, Y., Li, F., Zhu, M. X., and Xu, T. L. (2016) ASIC1a regulates insular long-term depression and is required for the extinction of conditioned taste aversion. *Nat Commun* **7**, 13770.

114. Cho, J. H., and Askwith, C. C. (2008) Presynaptic release probability is increased in hippocampal neurons from ASIC1 knockout mice. *J Neurophysiol* **99**, 426–441.

115. Wu, P. Y., Huang, Y. Y., Chen, C. C., Hsu, T. T., Lin, Y. C., Weng, J. Y., Chien, T. C., Cheng, I. H., and Lien, C. C. (2013) Acid-sensing ion channel-1a is not required for normal hippocampal LTP and spatial memory. *J Neurosci* **33**, 1828–1832.

116. Chiang, P. H., Chien, T. C., Chen, C. C., Yanagawa, Y., and Lien, C. C. (2015) ASIC-dependent LTP at multiple glutamatergic synapses in amygdala network is required for fear memory. *Sci Rep* **5**, 10143.

117. LeDoux, J. (2003) The emotional brain, fear, and the amygdala. *Cell Mol Neurobiol* **23**, 727–738.

118. Wemmie, J. A., Coryell, M. W., Askwith, C. C., Lamani, E., Leonard, A. S., Sigmund, C. D., and Welsh, M. J. (2004) Overexpression of acid-sensing ion channel 1a in transgenic mice increases acquired fear-related behavior. *Proc Natl Acad Sci USA* **101**, 3621–3626.

119. Coryell, M. W., Wunsch, A. M., Haenfler, J. M., Allen, J. E., McBride, J. L., Davidson, B. L., and Wemmie, J. A. (2008) Restoring acid-sensing ion channel-1a in the amygdala of knock-out mice rescues fear memory but not unconditioned fear responses. *J Neurosci* **28**, 13738–13741.

120. Coryell, M. W., Wunsch, A. M., Haenfler, J. M., Allen, J. E., Schnizler, M., Ziemann, A. E., Cook, M. N., Dunning, J. P., Price, M. P., Rainier, J. D., Liu, Z., Light, A. R., Langbehn, D. R., and Wemmie, J. A. (2009) Acid-sensing ion channel-1a in the amygdala, a novel therapeutic target in depression-related behavior. *J Neurosci* **29**, 5381–5388.

121. Chesler, M. (2003) Regulation and modulation of pH in the brain. *Physiol Rev* **83**, 1183–1221.

122. Magnotta, V. A., Heo, H. Y., Dlouhy, B. J., Dahdaleh, N. S., Follmer, R. L., Thedens, D. R., Welsh, M. J., and Wemmie, J. A. (2012) Detecting activity-evoked pH changes in human brain. *Proc Natl Acad Sci USA* **109**, 8270–8273.

123. Feinstein, J. S., Buzza, C., Hurlemann, R., Follmer, R. L., Dahdaleh, N. S., Coryell, W. H., Welsh, M. J., Tranel, D., and Wemmie, J. A. (2013) Fear and panic in humans with bilateral amygdala damage. *Nat Neurosci* **16**, 270–272.

124. Masland, R. H. (2012) The neuronal organization of the retina. *Neuron* **76**, 266–280.

125. Strauss, O. (2005) The retinal pigment epithelium in visual function. *Physiol Rev* **85**, 845–881.

126. Oakley, B., 2nd, and Wen, R. (1989) Extracellular pH in the isolated retina of the toad in darkness and during illumination. *J Physiol* **419**, 353–378.

127. Newman, E. A. (1996) Acid efflux from retinal glial cells generated by sodium bicarbonate cotransport. *J Neurosci* **16**, 159–168.

128. Ettaiche, M., Deval, E., Cougnon, M., Lazdunski, M., and Voilley, N. (2006) Silencing acid-sensing ion channel 1a alters cone-mediated retinal function. *J Neurosci* **26**, 5800–5809.

129. Brockway, L. M., Zhou, Z. H., Bubien, J. K., Jovov, B., Benos, D. J., and Keyser, K. T. (2002) Rabbit retinal neurons and glia express a variety of ENaC/DEG subunits. *Am J Physiol Cell Physiol* **283**, C126–C134.

130. Lilley, S., LeTissier, P., and Robbins, J. (2004) The discovery and characterization of a proton-gated sodium current in rat retinal ganglion cells. *J Neurosci* **24**, 1013–1022.

131. Liu, S., Wang, M. X., Mao, C. J., Cheng, X. Y., Wang, C. T., Huang, J., Zhong, Z. M., Hu, W. D., Wang, F., Hu, L. F., Wang, H., and Liu, C. F. (2014) Expression and functions of ASIC1 in the zebrafish retina. *Biochem Biophys Res Commun* **455**, 353–357.
132. Render, J. A., Howe, K. R., Wunsch, A. M., Guionaud, S., Cox, P. J., and Wemmie, J. A. (2010) Histologic examination of the eye of acid-sensing ion channel 1a knockout mice. *Int J Physiol Pathophysiol Pharmacol* **2**, 69–72.
133. Ettaiche, M., Guy, N., Hofman, P., Lazdunski, M., and Waldmann, R. (2004) Acid-sensing ion channel 2 is important for retinal function and protects against light-induced retinal degeneration. *J Neurosci* **24**, 1005–1012.
134. Ettaiche, M., Deval, E., Pagnotta, S., Lazdunski, M., and Lingueglia, E. (2009) Acid-sensing ion channel 3 in retinal function and survival. *Invest Ophthalmol Vis Sci* **50**, 2417–2426.
135. Lingueglia, E. (2007) Acid-sensing ion channels in sensory perception. *J Biol Chem* **282**, 17325–17329.
136. Osborne, N. N., Casson, R. J., Wood, J. P., Chidlow, G., Graham, M., and Melena, J. (2004) Retinal ischemia: mechanisms of damage and potential therapeutic strategies. *Prog Retin Eye Res* **23**, 91–147.
137. Molliver, D. C., Immke, D. C., Fierro, L., Pare, M., Rice, F. L., and McCleskey, E. W. (2005) ASIC3, an acid-sensing ion channel, is expressed in metaboreceptive sensory neurons. *Mol Pain* **1**, 35.
138. Price, M. P., McIlwrath, S. L., Xie, J., Cheng, C., Qiao, J., Tarr, D. E., Sluka, K. A., Brennan, T. J., Lewin, G. R., and Welsh, M. J. (2001) The DRASIC cation channel contributes to the detection of cutaneous touch and acid stimuli in mice. *Neuron* **32**, 1071–1083.
139. Diochot, S., Baron, A., Rash, L. D., Deval, E., Escoubas, P., Scarzello, S., Salinas, M., and Lazdunski, M. (2004) A new sea anemone peptide, APETx2, inhibits ASIC3, a major acid-sensitive channel in sensory neurons. *EMBO J* **23**, 1516–1525.
140. Wang, X., Li, W. G., Yu, Y., Xiao, X., Cheng, J., Zeng, W. Z., Peng, Z., Xi Zhu, M., and Xu, T. L. (2013) Serotonin facilitates peripheral pain sensitivity in a manner that depends on the nonproton ligand sensing domain of ASIC3 channel. *J Neurosci* **33**, 4265–4279.
141. Li, W. G., and Xu, T. L. (2011) ASIC3 channels in multimodal sensory perception. *ACS Chem Neurosci* **2**, 26–37.
142. Kang, S., Jang, J. H., Price, M. P., Gautam, M., Benson, C. J., Gong, H., Welsh, M. J., and Brennan, T. J. (2012) Simultaneous disruption of mouse ASIC1a, ASIC2 and ASIC3 genes enhances cutaneous mechanosensitivity. *PLoS One* **7**, e35225.
143. Mogil, J. S., Breese, N. M., Witty, M. F., Ritchie, J., Rainville, M. L., Ase, A., Abbadi, N., Stucky, C. L., and Seguela, P. (2005) Transgenic expression of a dominant-negative ASIC3 subunit leads to increased sensitivity to mechanical and inflammatory stimuli. *J Neurosci* **25**, 9893–9901.
144. Staniland, A. A., and McMahon, S. B. (2009) Mice lacking acid-sensing ion channels (ASIC) 1 or 2, but not ASIC3, show increased pain behaviour in the formalin test. *Eur J Pain* **13**, 554–563.
145. Page, A. J., Brierley, S. M., Martin, C. M., Martinez-Salgado, C., Wemmie, J. A., Brennan, T. J., Symonds, E., Omari, T., Lewin, G. R., Welsh, M. J., and Blackshaw, L. A. (2004) The ion channel ASIC1 contributes to visceral but not cutaneous mechanoreceptor function. *Gastroenterology* **127**, 1739–1747.
146. Lu, Y., Ma, X., Sabharwal, R., Snitsarev, V., Morgan, D., Rahmouni, K., Drummond, H. A., Whiteis, C. A., Costa, V., Price, M., Benson, C., Welsh, M. J., Chapleau, M. W.,

and Abboud, F. M. (2009) The ion channel ASIC2 is required for baroreceptor and autonomic control of the circulation. *Neuron* **64**, 885–897.

147. Abboud, F. M., and Benson, C. J. (2015) ASICs and cardiovascular homeostasis. *Neuropharmacology* **94**, 87–98.

148. Paukert, M., Sidi, S., Russell, C., Siba, M., Wilson, S. W., Nicolson, T., and Grunder, S. (2004) A family of acid-sensing ion channels from the zebrafish: widespread expression in the central nervous system suggests a conserved role in neuronal communication. *J Biol Chem* **279**, 18783–18791.

149. Rhoades, J. L., Nelson, J. C., Nwabudike, I., Yu, S. K., McLachlan, I. G., Madan, G. K., Abebe, E., Powers, J. R., Colon-Ramos, D. A., and Flavell, S. W. (2019) ASICs mediate food responses in an enteric serotonergic neuron that controls foraging behaviors. *Cell* **176**, 85–97 e14.

150. Kellenberger, S., and Schild, L. (2015) International union of basic and clinical pharmacology. XCI. structure, function, and pharmacology of acid-sensing ion channels and the epithelial Na^+ channel. *Pharmacol. Rev.* **67**, 1–35.

151. Price, M. P., Gong, H., Parsons, M. G., Kundert, J. R., Reznikov, L. R., Bernardinelli, L., Chaloner, K., Buchanan, G. F., Wemmie, J. A., Richerson, G. B., Cassell, M. D., and Welsh, M. J. (2014) Localization and behaviors in null mice suggest that ASIC1 and ASIC2 modulate responses to aversive stimuli. *Genes Brain Behav* **13**, 179–194.

152. Zeng, W. Z., Liu, D. S., and Xu, T. L. (2014) Acid-sensing ion channels: trafficking and pathophysiology. *Channels* **8**, 481–487.

153. Marra, S., Ferru-Clement, R., Breuil, V., Delaunay, A., Christin, M., Friend, V., Sebille, S., Cognard, C., Ferreira, T., Roux, C., Euller-Ziegler, L., Noel, J., Lingueglia, E., and Deval, E. (2016) Non-acidic activation of pain-related acid-sensing ion channel 3 by lipids. *EMBO J.* **35**, 414–428.

154. Deval, E., Noel, J., Lay, N., Alloui, A., Diochot, S., Friend, V., Jodar, M., Lazdunski, M., and Lingueglia, E. (2008) ASIC3, a sensor of acidic and primary inflammatory pain. *EMBO J* **27**, 3047–3055.

155. Walder, R. Y., Rasmussen, L. A., Rainier, J. D., Light, A. R., Wemmie, J. A., and Sluka, K. A. (2010) ASIC1 and ASIC3 play different roles in the development of hyperalgesia after inflammatory muscle injury. *J Pain* **11**, 210–218.

156. Duan, B., Liu, D. S., Huang, Y., Zeng, W. Z., Wang, X., Yu, H., Zhu, M. X., Chen, Z. Y., and Xu, T. L. (2012) PI3-kinase/Akt pathway-regulated membrane insertion of acid-sensing ion channel 1a underlies BDNF-induced pain hypersensitivity. *J Neurosci* **32**, 6351–6363.

157. Duan, B., Wu, L. J., Yu, Y. Q., Ding, Y., Jing, L., Xu, L., Chen, J., and Xu, T. L. (2007) Upregulation of acid-sensing ion channel ASIC1a in spinal dorsal horn neurons contributes to inflammatory pain hypersensitivity. *J Neurosci* **27**, 11139–11148.

158. Jones, N. G., Slater, R., Cadiou, H., McNaughton, P., and McMahon, S. B. (2004) Acid-induced pain and its modulation in humans. *J Neurosci* **24**, 10974–10979.

159. Sluka, K. A., Price, M. P., Breese, N. M., Stucky, C. L., Wemmie, J. A., and Welsh, M. J. (2003) Chronic hyperalgesia induced by repeated acid injections in muscle is abolished by the loss of ASIC3, but not ASIC1. *Pain* **106**, 229–239.

160. Ikeuchi, M., Kolker, S. J., and Sluka, K. A. (2009) Acid-sensing ion channel 3 expression in mouse knee joint afferents and effects of carrageenan-induced arthritis. *J Pain* **10**, 336–342.

161. Ikeuchi, M., Kolker, S. J., Burnes, L. A., Walder, R. Y., and Sluka, K. A. (2008) Role of ASIC3 in the primary and secondary hyperalgesia produced by joint inflammation in mice. *Pain* **137**, 662–669.

162. Gao, J., Duan, B., Wang, D. G., Deng, X. H., Zhang, G. Y., Xu, L., and Xu, T. L. (2005) Coupling between NMDA receptor and acid-sensing ion channel contributes to ischemic neuronal death. *Neuron* **48**, 635–646.

163. Song, N., Zhang, G., Geng, W., Liu, Z., Jin, W., Li, L., Cao, Y., Zhu, D., Yu, J., and Shen, L. (2012) Acid sensing ion channel 1 in lateral hypothalamus contributes to breathing control. *PLoS One* **7**, e39982.

164. Huda, R., Pollema-Mays, S. L., Chang, Z., Alheid, G. F., McCrimmon, D. R., and Martina, M. (2012) Acid-sensing ion channels contribute to chemosensitivity of breathing-related neurons of the nucleus of the solitary tract. *J Physiol* **590**, 4761–4775.

2 Acid-Sensing Ion Channels and Synaptic Plasticity: A Revisit

*Ming-Gang Liu, Michael X. Zhu,
and Tian-Le Xu*

CONTENTS

2.1 INTRODUCTION

Acid-sensing ion channels (ASICs) are proton-gated cation channels belonging to the degenerin/epithelial sodium channel family.[1,2] There are four genes encoding at least six different ASIC subunits (1a, 1b, 2a, 2b, 3 and 4).[3,4] Each ASIC subunit has two transmembrane domains. Functional ASIC channels can be assembled into trimers from either homomeric or heteromeric subunit combinations. Most ASICs can be activated by acidic pH to conduct Na^+ influx, although homomeric ASIC1a and ASIC1a/2b heteromers can also permeate Ca^{2+}[5–10]. ASIC1a, ASIC2a, ASIC2b, and ASIC4 are found in both the central and peripheral nervous systems, whereas ASIC1b and ASIC3 have been detected only in the peripheral nervous system[11–15]. Due to its ubiquitous expression and mediation of the largest fraction of proton-gated currents in the brain, the ASIC1a isoform has been most intensively investigated in the literature.

Since extracellular pH fluctuation is common in a variety of pathological conditions, ASICs, in particular the ASIC1a isoform, have been shown to play important roles in a wide spectrum of neurological diseases, such as chronic pain[4,16–20], ischemic stroke[21–26], multiple sclerosis[27,28], seizures[29–31], as well as emotional disorders[32–35]. Apart from the pathological functions of ASICs, interestingly, recent studies have begun to uncover the pivotal physiological role of ASIC1a in mediating/modulating synaptic transmission and plasticity in various brain regions[36,37]. These synaptic signals mediated by ASICs are of critical importance given the well-documented roles of synaptic plasticity in many higher brain functions and diseases[38–41]. In this chapter, we summarize and integrate previous reports on the synaptic distribution and function of ASICs, systematically discussing the results on protons and ASIC1a in synaptic transmission, as well as ASIC1a in multiple forms of functional synaptic plasticity and structural synaptic remodeling. We conclude this chapter by pointing out some unresolved issues and future research directions.

2.2 PROTONS AND ASICS IN SYNAPTIC TRANSMISSION

2.2.1 PROTONS ACT AS A NEUROTRANSMITTER IN SYNAPTIC SIGNALING

It is widely acknowledged that acidification can occur in the synaptic cleft during normal synaptic transmission[36]. It is generally agreed that vesicular pH is quite low as a result of H^+-ATPase activity, which generates an electrochemical gradient that drives transport of neurotransmitters into the vesicle[42,43]. Therefore, during exocytosis, protons are co-released with neurotransmitters, resulting in acidification of the synaptic cleft, which has been demonstrated by proton-induced blockade of voltage-gated calcium channels and NMDA receptors[44–48]. Using a pH-sensitive fluorescence protein, pHluorin, fused to the extracellular domain of a postsynaptic membrane protein, Du et al. (2014) directly measured the pH shift of synaptic cleft caused by synaptic stimulation in cultured amygdala slices. It was found that stimulating cortical inputs transiently reduced pH at spines and the neighboring dendrites, followed by a slower alkalinization, thus providing strong evidence for proton release during synaptic activity[49].

The first indication of protons acting as transmitters was found in the *C. elegans*, where the space between the intestine and the muscle is acidified during muscle contraction. The authors proposed that Na^+/H^+ ion exchanges or proton release from synaptic vesicles could be the source of acidification, implicating protons as neurotransmitters in the nematode[50]. Studying synapses formed by hair cells and the postsynaptic calyx, Highstein et al. (2014) showed in turtles that stimulus-evoked extrusion of protons from hair cells could acidify the synaptic cleft at hair cell calyx and induce a nonquantal excitatory postsynaptic current[51].

The intriguing hypothesis that protons act as neurotransmitters in synaptic signaling has also been investigated in mammalian neurons. Given a preferential somatodendritic distribution of ASIC1a, where it is primarily localized to dendritic spines with little expression in the axons[52–54], ASIC1a-mediated synaptic currents have been detected in a number of mouse brain regions, such as the amygdala[49], striatum[55,56] and anterior cingulate cortex (ACC)[20]. These are considered as strong evidence supporting the notion that ASIC1a is the postsynaptic receptor of the proton transmitter[15]. In the amygdala, Du et al. (2014) recorded excitatory post-synaptic glutamatergic currents (EPSCs) solely mediated by ASIC1/2 heteromeric channels after blocking glutamate receptors, which can be fully blocked by the ASIC blocker amiloride as well as by genetic deletion of ASIC1a[49]. In the nucleus accumbens (NAc), Kreple et al. (2014) identified a similar ASIC-mediated synaptic current that was sensitive to the pH buffer capacity of the bath solutions[56]. From this aspect, our previous work also demonstrated the existence of amiloride-sensitive, ASIC1a-mediated EPSCs in dorsal striatal medial spiny neurons[55] as well as pyramidal neurons of the ACC[20]. Finally, Gonzalez-Inchauspe et al. (2017) detected ASIC1a-mediated currents at the calyx of Held-medial nucleus of the trapezoid nucleus synapse after blocking AMPA and NMDA glutamate receptors, which were equally diminished in amplitude by amiloride and by an extracellular solution with increased pH buffer capacity[57]. Taken together, all these results strongly support the hypothesis of protons as neurotransmitters in synaptic signaling and indicate that presynaptic released protons contribute to synaptic transmission by activating the ASIC1a channel located in the postsynaptic site.

2.2.2 Modulation of Synaptic Transmission by ASICs

In addition to the above-described mediation of protonergic synaptic transmission by ASICs, previous reports also examined the modulatory effect of ASIC1a on basal synaptic communication. Using microisland cultures of hippocampal neurons, Cho and Askwith (2008) found that the AMPA/NMDA ratio and paired pulse ratio were reduced, and the frequency of spontaneous miniature EPSCs was increased in ASIC1 null mouse neurons compared with wild type (WT), indicating that ASIC1a might affect the probability of presynaptic neurotransmitter release[58]. Similarly, at the calyx of Held, genetic deletion or pharmacological inhibition of ASIC1a pro-duced an enhanced short-term depression of glutamatergic EPSCs during high-frequency stimulation, again suggesting the modulation of the presynaptic property by ASICs[57].

In the NAc, not only does ASIC1a directly contribute to synaptic transmission, but also it regulates glutamate receptor function by eliciting changes in AMPA/NMDA ratio and miniature EPSCs (mEPSCs). Specifically, the loss of ASIC1a causes an increase in GluA2-lacking AMPA receptors and AMPA/NMDA ratio, which is accompanied by an increased frequency of mEPSCs. All these alterations can be restored by overexpressing ASIC1a in the NAc, which also rescues the cocaine-conditioned place preference[56]. Consistent with the results in NAc, our recent work in the dorsal striatum also revealed an increased AMPA/NMDA ratio, mainly due to decreased NMDA receptor expression, in excitatory synaptic trans-mission by ASIC1a disruption, while re-introduction of ASIC1a in the knockout mice can rescue the defect in glutamate receptor expression and function[55]. Collectively, these observations add strength to the assertion that ASICs are the main postsynaptic proton receptors that can either mediate or modulate synaptic transmission in the brain.

2.3 ASIC1A IN SYNAPTIC PLASTICITY

Although the amplitude of the proton-activated EPSCs is rather small, ASIC1a-mediated Na^+ influx can induce postsynaptic cell depolarization and the consequent Ca^{2+} influx through either voltage-gated calcium channels or NMDA receptors. Furthermore, ASIC1a homomeric channels are Ca^{2+} permeable, which can mediate Ca^{2+} influx and in turn lead to intracellular Ca^{2+} rise when activated by synaptic stimulation. Thus, it has been naturally expected that ASIC1a could be important for certain forms of synaptic plasticity in brain regions where it exhibits high expression[36,37]. Here, we discuss the literature regarding the role of ASIC1a in synaptic plasticity, including long-term potentiation (LTP) and long-term potentiation (LTD) (Table 2.1).

2.3.1 ASIC1A IN LTP

2.3.1.1 Hippocampus

To date, controversial results have been reported about the role of ASIC1a in hippocampal LTP. In 2002, Wemmie and colleagues first demonstrated an

TABLE 2.1

Reported Roles of ASIC1a in Functional Synaptic Plasticity

Plasticity Type	Species	Brain Area	Approaches	Key Findings	Reference
LTP	Mouse	Hippocampus	Genetic knockout	Impaired hippocampal LTP by ASIC1a deletion	59
LTP	Rat and mouse	Hippocampus	Pharmacology and genetic knockout	No effect on LTP	60
LTP	Rat	Hippocampus	Pharmacology	Reduced hippocampal LTP	61
LTP and LTD	Mouse	Hippocampus	Pharmacology and genetic knockout	Impaired LTP but intact LTD	62
LTP	Mouse	Amygdala	Genetic knockout and pH buffering	Blocked LTP in lateral amygdala	49
LTP	Mouse	Amygdala	Genetic knockout	Impaired LTP at multiple glutamatergic synapses	68
LTP and LTD	Mouse	ACC	Pharmacology and genetic knockout	Impaired LTP but normal LTD	20
LTD	Mouse	Hippocampus	Pharmacology	Reduced mGluR-LTD in early adult but not juvenile mice	77
LTD	Mouse	Hippocampus	Pharmacology	Reduced NMDAR-LTD in both young and adult mice	78
LTP and LTD	Mouse	Insular cortex	Pharmacology and genetic knockout	Impaired LTD but normal LTP	82

attenuation of LTP induction in hippocampal slices of ASIC1a-null mice[59]. While these results were not confirmed in a subsequent study using both pharmacological and genetic approaches[60], another study showed that inhibiting ASIC1a channels with compound 5b, a novel potent antagonist of ASIC1a, indeed prevented high frequency stimulation (HFS)-induced LTP in the hippocampus[61]. These conflicting observations may be attributed, at least in part, to technical differences in a number of subtle experimental variables including electrophysiological approach, animal age and strain, different gene knockout strategies, drug dosage, etc.

In this context, our work employing a newly developed 64-channel multi-electrode array recording system to study synaptic plasticity in hippocampal slices from WT and ASIC1a KO mice may help reconcile this controversy[62]. Compared with other electrophysiological approaches, the multi-electrode array recording can detect the activity of neuronal networks in both space and time[63], with multiple sites recorded simultaneously and objectively[64,65]. Moreover, it enables a much longer time of LTP/LTD monitoring[66], thus providing a novel and unbiased approach to study the mechanisms of synaptic plasticity. Applying this system into the hippocampus, we first revealed a probabilistic nature of LTP induction at the CA3-CA1 synapses of normal mice. Second, we found that genetic ablation or pharmacological blockade of ASIC1a did not affect the baseline synaptic transmission, but strongly reduced, although not completely abolished, the probability of LTP induction by either HFS or theta burst stimulation (TBS) (Figure 2.1). Third, the unbiased recording and multisite analysis revealed two interesting insights into the specific function of ASIC1a in hippocampal LTP: (1) ASIC1a is not absolutely required for LTP induction at every site of the hippocampal slice; (2) the strength of LTP induction protocol (e.g. one vs four episodes of HFS) may influence the outcomes of single electrode recordings performed on ASIC1a KO animals, causing inconsistent conclusions in previous studies[59–61]. Fourth, our pharmacological rescue results indicate that both NMDAR-dependent and -independent mechanisms may be involved in ASIC1a regulation of hippocampal synaptic plasticity[62]. Taken together, these results reveal new insights into the role of ASIC1a in hippocampal LTP and may help reconcile some of the previous discrepancies concerning the role of ASIC1a in synaptic plasticity in the hippocampus.

2.3.1.2 Amygdala

The effect of genetic ablation of ASIC1a on synaptic plasticity has also been examined in the amygdala (Table 2.1), a brain area that can detect carbon dioxide and acidosis to drive fear behavior[67]. In the study of Du et al. (2014), eliminating ASIC1a or increasing pH buffer capacity reduced the magnitude of HFS-evoked LTP in the lateral amygdala pyramidal neurons, whereas exogenous application of protons directly induced LTP[49]. In addition, Chiang and colleagues reported that ASIC1a is also enriched in inhibitory interneurons in the basolateral and central amygdala, and selective deletion of ASIC1a in GABAergic neurons impaired LTP at intra-amygdala synapses and reduced fear memory[68]. Together with previous publications[13,69,70], it can be concluded that ASIC1a critically contributes to LTP induction at multiple glutamatergic synapses in the amygdala.

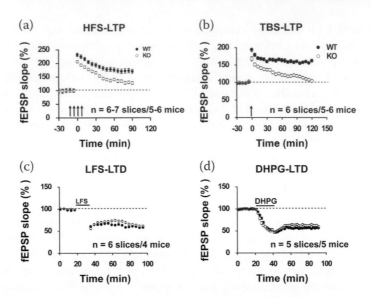

FIGURE 2.1   Genetic deletion of ASIC1a in the hippocampus impairs LTP induction without affecting LTD. a & b, Multi-channel electrophysiological data of HFS- (a) and TBS-evoked (b) LTP demonstrating a reduction in both forms of LTP in hippocampal slices from ASIC1a knockout mice compared to that from the wild-type mice. **c & d,** ASIC1s deletion had no effect on the induction of electrical (c, LFS-induced) or chemical (d, DHPG-induced) LTD in hippocampal slices. DHPG, (RS)-3,5-dihydroxyphenylglycine; HFS, high-frequency stimulation; LFS, low-frequency stimulation; LTD, long-term depression; LTP, long-term potentiation; TBS, theta-burst stimulation. (Adapted from Ref. 62).

2.3.1.3   ACC

ACC is an important forebrain structure involved in a variety of key brain functions such as pain perception, cognition, emotion, and social communication[71–74]. Of note, synaptic plasticity in the ACC is considered one of the most critical mechanisms underlying the pathogenesis of chronic pain[75]. We thus investigated the expression and function of ASIC1a in the ACC[20]. Western blot analysis and electrophysiological recordings show that ASIC1a is abundantly expressed in the ACC and contributes to excitatory synaptic transmission in cortical pyramidal neurons. ACC-specific genetic deletion or pharmacological blockade of ASIC1a reduced the probability of TBS-evoked LTP, without affecting low frequency stimulation (LFS)-induced LTD. Behaviorally, ASIC1a-dependent cingulate LTP correlated with the development and maintenance of inflammatory or neuropathic pain hypersensitivity (Figure 2.2). Mechanistically, we show that ASIC1a tunes pain-related cortical plasticity through protein kinase C λ-mediated increase of AMPA receptor trafficking to postsynaptic membranes in ACC excitatory neurons[20].

2.3.2   ASIC1A IN LTD

LTD is another form of synaptic plasticity in the central nervous system, which is critically involved in various physiological and pathological conditions[76]. A

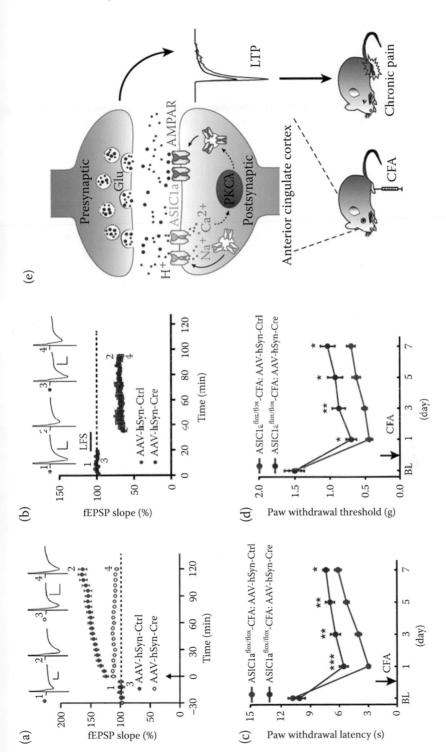

FIGURE 2.2 ASIC1a in the anterior cingulate cortex contributes to cortical LTP and pain hypersensitivity. a & b, Conditional deletion of ASIC1a in the anterior cingulate cortex (ACC) neurons reduced the induction probability of TBS-evoked LTP (a) without any influence on LFS-induced LTD (b). **c & d,** ACC-specific deletion of ASIC1a significantly attenuated CFA-evoked thermal hyperalgesia (c) and mechanical allodynia (d). **e,** A proposed model for ASIC1a regulation of cingulate LTP and its involvement in pain hypersensitivity. CFA, complete Freund's adjuvant; LFS, low-frequency stimulation; LTD, long-term depression; LTP, long-term potentiation; TBS, theta-burst stimulation. (Adapted from Ref. 20).

growing number of reports indicate that ASIC1a is required for LTD induction in some brain regions, although the complete picture for this regulation is still elusive (Table 2.1).

2.3.2.1 Hippocampus

There are two major types of LTD co-existing in the hippocampus: NMDA receptor-dependent LTD (NMDAR-LTD) and metabotropic glutamate receptor-dependent LTD (mGluR-LTD). In the work of Mango et al. (2017), the authors demonstrated an age-dependent involvement of ASIC1a in mGluR-LTD in the hippocampus[77]. Specifically, pharmacological inhibition of ASIC1a by psalmotoxin 1 resulted in an attenuation of LTD induced by application of the group I mGluR agonist (S)-3,5-Dihydroxyphenylglycine (DHPG) or paired-pulse LFS in postnatal days 30–40 (P30–40) animals. However, psalmotoxin 1 did not affect both forms of mGluR-LTD in P13–18 animals. With respect to NMDAR-LTD, the same group further reported that ASIC1a is required for induction of electrical LTD by LFS or chemical LTD by bath application of NMDA. Interestingly, the role of ASIC1a in NMDAR-LTD is age-independent[78]. Together, the results point toward the conclusion that ASIC1a is a crucial player in hippocampal LTD. In contrast, our recent work with the multi-electrode array recording system failed to detect any effect of ASIC1a gene deletion or pharmacological blockade on LFS- or DHPG-evoked LTD in the hippocampus[62]. The exact reasons for the discrepancies are not clear. However, the divergence might be explained by differences in experimental conditions, such as the animal age, recording method, recording temperature and drug concentration. Therefore, more elaborative work is needed to clarify this issue.

2.3.2.2 Insular Cortex

The insular cortex is an integrating forebrain structure involved in several sensory and cognitive functions, such as interoception[79], taste memory[80], and pain perception[81]. Notably, multiple forms of synaptic plasticity have been identified in insular slices of adult mice, including a long-lasting protein synthesis-dependent later-phase LTP, LFS-induced electrical LTD, and DHPG-evoked chemical LTD[64,66]. Of particular relevance, our recent work illustrated the function of ASIC1a in the two forms of insular LTD, without any contribution to the induction of LTP (Table 2.1). Moreover, ASIC1a-mediated LTD was shown to be important for the extinction of acquired taste aversion memory (Figure 2.3). Further experiments suggest that ASIC1a acts through activation of glycogen synthase kinase-3β signaling via an ion conductance-dependent mechanism[82].

Taken together, ASIC1a appears to play differential roles in synaptic plasticity in different brain regions (Table 2.1). The exact reasons for the region-specific regulations are not known. It might be due to the distribution of ASIC1a in distinct cell types in different brain regions, as well as the selective alterations in specific signaling pathways related to LTP or LTD resulting from ASIC1a inhibition or deletion. It could also stem from differences in ASIC subunit compositions and/or even binding partners in various brain regions[20,62,82]. Thus, the roles of ASIC1a in synaptic plasticity are rather complex, varying greatly depending on brain regions, cell types, and downstream signaling pathways.

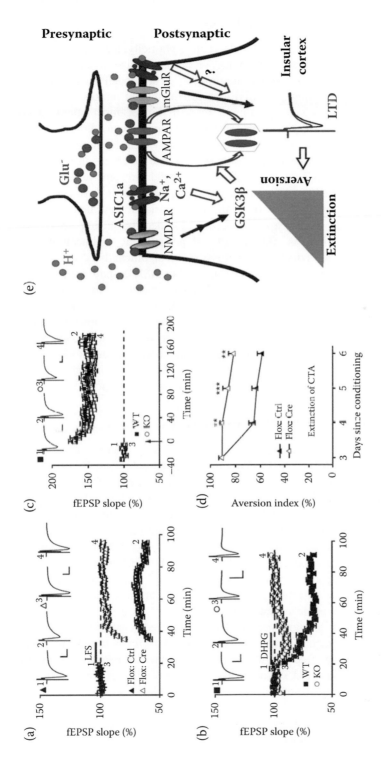

FIGURE 2.3 ASIC1a regulates LTD in the insular cortex and is required for the extinction of taste aversion memory. a–c, Conditional deletion of ASIC1a in the insular cortex impaired the induction of LFS- (a) and DHPG-evoked LTD (b) without any influence on TBS-induced LTP (c). **d,** Deletion of ASIC1a in the insular cortex significantly blocked the extinction of conditioned taste aversion. **e,** A proposed scheme for ASIC1a regulation of insular LTD and its involvement in taste aversion memory extinction. DHPG, (RS)-3,5-dihydroxyphenylglycine; LFS, low-frequency stimulation; LTD, long-term depression; LTP, long-term potentiation; TBS, theta-burst stimulation. (Adapted from Ref. 82).

2.4   ASICS IN SYNAPTIC REMODELING

In addition to the aforementioned functional synaptic plasticity, like LTP and LTD, synapses also exhibit structural remodeling in terms of size (enlargement or shrinkage) and density (formation or elimination)[83,84]. The subcellular localization of ASIC1a, coupled with its high sensitivity to pH fluctuation, makes it plausible that ASIC1a also regulates synaptic remodeling. In hippocampal slices, over-expression of ASIC1a increases the density of dendritic spines, while knockdown of ASIC1a reduces spine numbers[52]. Deleting ASIC2a, which facilitates ASIC1a targeting to spines, results in a similar reduction of dendritic spines in the hippo-campus[53]. By contrast, ASIC1a disruption increases dendritic spine density in the NAc, which is accompanied by altered glutamate receptor function[56]. In agreement with these findings, our previous work in the dorsal striatum revealed that gene deletion of ASIC1a leads to increased dendritic spine density but impaired spine morphology and maturation, as well as altered postsynaptic architecture. The NMDA receptor-mediated synaptic transmission is significantly decreased in medium spiny neurons of the dorsal striatum in the ASIC1a-null mice. These structural and functional changes underlie deficits in procedural motor learning, which can be rescued by striatal-specific overexpression of ASIC1a[55]. Compared to LTP and/or LTD, the role of ASICs in spine remodeling is less characterized and limited to only a very few brain areas. Future work is certainly warranted to in-vestigate the action of ASICs in spine dynamics in other brain regions. It would also be of great interest to test the possible involvement of ASICs in disease-induced alterations in synaptic remodeling under a wide range of pathological conditions.

2.5   SUMMARY AND FUTURE PERSPECTIVES

Research in recent years has demonstrated that protons act as neurotransmitters to regulate neuronal communication in synapses of several brain regions. It is generally accepted that neuronal activity triggers acidification in the synaptic cleft and activation of ASICs, especially ASIC1a, located in the postsynaptic compartment, which in turn produces a series of downstream signaling effects[36]. Depending on the different brain areas and synapses, ASIC1a, the most-studied ASIC family member, may directly mediate a component of excitatory synaptic transmission, or modulate the induction of LTP or LTD, or even spine remodeling[37]. All these synaptic functions endow ASIC1a with the ability to play significant roles in many physiological and pathological pro-cesses such as learning and memory, pain perception, emotional regulation, acidosis-induced neuronal injury, neurodegeneration, and addiction[1,85]. In spite of these exciting insights, many important questions remain to be answered regarding the synaptic function of ASICs, which require more exquisite studies in the future in order to paint a complete picture of synaptic roles of ASICs and their behavioral relevance. Here, we point out some of the unresolved and/or poorly characterized issues, hoping that they will stimulate more insightful studies in this important field.

1. As reviewed above, the role of ASIC1a in synaptic plasticity has been primarily examined in the hippocampus, amygdala, striatum, ACC, and insular cortex

(Table 2.1). It is also important to evaluate the involvement of ASIC1a in various forms of LTP or LTD in other brain regions of the central nervous system, where ASIC1a is highly expressed. As demonstrated by Wemmie et al. (2003)[13] and Price et al. (2014)[14], both ASIC1a and ASIC2 are widely distributed throughout the brain. It is thus interesting and important for future studies to examine the specific function of ASICs in those less characterized areas, such as lateral/medial habenula, inferior/superior colliculus, locus coeruleus, lateral dorsal tegmental nucleus, periaqueductal gray, etc. Notably, compared to LTP, much less is known about the role of ASICs in LTD. Therefore, special attention can be paid on dissecting the possible engagement of ASICs in induction, expression and maintenance of LTD in the aforementioned brain regions.

2. Synaptic plasticity is a generic term that applies to short- or long-lasting experience- or activity-dependent changes in the efficacy or connection of synaptic transmission in the brain. It can be classified into both functional and structural aspects of synaptic plasticity. For the former, except for LTP and LTD, it also includes short-term plasticity (like paired-pulse facilitation or depression), depotentiaion[86], metaplasticity (plasticity of synaptic plasticity)[87], and homeostatic plasticity[88] (scaling up or down of the synaptic strength in response to reduction or elevation of synaptic activity). For the latter, dendritic spines may undergo activity-dependent dynamic alterations in shape, size, density, or even composition during various behavioral tasks and/or synaptic stimulations[84]. To date, most studies have focused on the classical LTP, with much less emphasis placed on other forms of synaptic plasticity, although the possibilities that ASICs are equally important for depotentiaion (or metaplasticity) have not been fully excluded. It would be both necessary and exciting in future studies to test these possibilities.

3. Recent advances in circuit-based neuroscience research approaches, such as optogenetics/chemogenetics, fiber photometry, microendoscopic imaging, virus-mediated whole-brain input-output tracing, and labeling/manipulation of activated neural ensembles, are pushing our knowledge of how the brain works toward a new level[89,90]. More and more insightful findings have been obtained on the roles of specific cell types or subcircuits in execution and/or modulation of certain types of complicated behaviors. In this context, it would be of particular importance to extend the ion channel research into the circuit or network level. One inspiring starting point is to elucidate the function of ASICs in synaptic plasticity occurring in a defined projection circuit. Our recent work demonstrating ASIC1a in the ventral hippocampus as a critical molecular player in fear extinction memory provides a good example in this direction[91]. Electrophysiological analysis shows that extinction learning leads to adaptive changes in synaptic strength in the ventral hippocampus-prefrontal cortex circuit. Selective deletion of ASIC1a in the ventral hippocampus abolished fear extinction-triggered circuit plasticity[91]. It would be instructive to further explore the role of ASIC1a in behavior-driven synaptic plasticity in other projection circuits.

4. Synaptic strength is determined by both the amount of presynaptic neurotransmitter release and postsynaptic receptor abundance and responsiveness.

Current knowledge on the involvement of ASIC1a in synaptic plasticity mainly concentrates on the postsynaptically localized ASICs. However, it still remains obscure whether ASIC1a is also expressed in the presynaptic axonal terminals. If so, does it have any effect on the probability or dynamics of transmitter release? Almost all forms of synaptic plasticity (LTP, LTD, structural plasticity and homeostatic plasticity) have been reported to have a presynaptic component[92], but whether ASICs function in these forms of presynaptic plasticity remains unknown. Moreover, not only do excitatory synapses on pyramidal neurons exhibit activity-dependent plastic changes, but also excitatory synapses on inhibitory neurons and inhibitory synapses onto pyramidal neurons can equally undergo LTP or LTD[93]. Then, it would be tempting to ask whether and how ASIC1a contributes to synaptic plasticity at these synapses.

5. As introduced above, ASICs encompass a family of ion channels with several isoforms, among which ASIC1a, ASIC2a, ASIC2b and ASIC4 are all highly expressed in the central nervous system[1,36]. However, a large number of previous research on ASICs and synaptic plasticity have concentrated on the function of ASIC1a, with other isoforms being much less studied. Although ASIC2a has been shown to promote surface trafficking of ASIC1a[53], the exact roles of ASIC2a in different forms of synaptic plasticity are still not clear. Thus, future studies may be designed to evaluate the functions of homomeric or heteromeric ASIC2 isoforms in synaptic transmission and plasticity in various brain regions under a variety of physiological or pathological conditions.

ACKNOWLEDGEMENTS

The authors would like to express their apologies to those authors whose work is not cited here because of space limitation. Work in the authors' lab is supported by grants from the National Natural Science Foundation of China (81961128024, 81730095, and 31771157), Shanghai Natural Science Foundation (20ZR1430000), the Science and Technology Commission of Shanghai Municipality (18JC1420302), and the Shanghai Municipal Science and Technology Major Project (2018SHZDZX05).

REFERENCES

1. Kellenberger, S., & Schild, L. International union of basic and clinical pharmacology. XCI. structure, function, and pharmacology of acid-sensing ion channels and the epithelial Na+ channel. *Pharmacol Rev* **67**, 1–35 (2015).
2. Krishtal, O. The ASICs: Signaling molecules? Modulators? *Trends Neurosci* **26**, 477–483 (2003).
3. Wemmie, J.A., Price, M.P., & Welsh, M.J. Acid-sensing ion channels: advances, questions and therapeutic opportunities. *Trends Neurosci* **29**, 578–586 (2006).
4. Wemmie, J.A., Taugher, R.J., & Kreple, C.J. Acid-sensing ion channels in pain and disease. *Nat Rev Neurosci* **14**, 461–471 (2013).
5. Jasti, J., Furukawa, H., Gonzales, E.B., & Gouaux, E. Structure of acid-sensing ion channel 1 at 1.9 A resolution and low pH. *Nature* **449**, 316–323 (2007).

6. Bassler, E.L., Ngo-Anh, T.J., Geisler, H.S., Ruppersberg, J.P., & Grunder, S. Molecular and functional characterization of acid-sensing ion channel (ASIC) 1b. *J Biol Chem* **276**, 33782–33787 (2001).

7. Waldmann, R., Champigny, G., Bassilana, F., Heurteaux, C., & Lazdunski, M. A proton-gated cation channel involved in acid-sensing. *Nature* **386**, 173–177 (1997).

8. Sherwood, T.W., Lee, K.G., Gormley, M.G., & Askwith, C.C. Heteromeric acid-sensing ion channels (ASICs) composed of ASIC2b and ASIC1a display novel channel properties and contribute to acidosis-induced neuronal death. *J Neurosci* **31**, 9723–9734 (2011).

9. Sutherland, S.P., Benson, C.J., Adelman, J.P., & McCleskey, E.W. Acid-sensing ion channel 3 matches the acid-gated current in cardiac ischemia-sensing neurons. *Proc Natl Acad Sci USA* **98**, 711–716 (2001).

10. Yermolaieva, O., Leonard, A.S., Schnizler, M.K., Abboud, F.M., & Welsh, M.J. Extracellular acidosis increases neuronal cell calcium by activating acid-sensing ion channel 1a. *Proc Natl Acad Sci USA* **101**, 6752–6757 (2004).

11. Alvarez de la Rosa, D., Zhang, P., Shao, D., White, F., & Canessa, C.M. Functional implications of the localization and activity of acid-sensitive channels in rat peripheral nervous system. *Proc Natl Acad Sci USA* **99**, 2326–2331 (2002).

12. Alvarez de la Rosa, D. et al. Distribution, subcellular localization and ontogeny of ASIC1 in the mammalian central nervous system. *J Physiol* **546**, 77–87 (2003).

13. Wemmie, J.A. et al. Acid-sensing ion channel 1 is localized in brain regions with high synaptic density and contributes to fear conditioning. *J Neurosci* **23**, 5496–5502 (2003).

14. Price, M.P. et al. Localization and behaviors in null mice suggest that ASIC1 and ASIC2 modulate responses to aversive stimuli. *Genes Brain Behav* **13**, 179–194 (2014).

15. Zha, X.M. Acid-sensing ion channels: trafficking and synaptic function. *Mol Brain* **6**, 1 (2013).

16. Deval, E. et al. Acid-sensing ion channels (ASICs): pharmacology and implication in pain. *Pharmacol Ther* **128**, 549–558 (2010).

17. Deval, E., & Lingueglia, E. Acid-sensing ion channels and nociception in the peripheral and central nervous systems. *Neuropharmacology* (2015), **94**, 49–57.

18. Duan, B. et al. Upregulation of acid-sensing ion channel ASIC1a in spinal dorsal horn neurons contributes to inflammatory pain hypersensitivity. *J Neurosci* **27**, 11139–11148 (2007).

19. Duan, B. et al. PI3-kinase/Akt pathway-regulated membrane insertion of acid-sensing ion channel 1a underlies BDNF-induced pain hypersensitivity. *J Neurosci* **32**, 6351–6363 (2012).

20. Li, H.S. et al. Protein kinase c lambda mediates acid-sensing ion channel 1a-dependent cortical synaptic plasticity and pain hypersensitivity. *J Neurosci* **39**, 5773–5793 (2019).

21. Jiang, N. et al. Region specific contribution of ASIC2 to acidosis-and ischemia-induced neuronal injury. *J Cereb Blood Flow Metab* **37**, 528–540 (2017).

22. Wang, Y.Z., & Xu, T.L. Acidosis, acid-sensing ion channels, and neuronal cell death. *Mol Neurobiol* **44**, 350–358 (2011).

23. Xiong, Z.G. et al. Neuroprotection in ischemia: blocking calcium-permeable acid-sensing ion channels. *Cell* **118**, 687–698 (2004).

24. Gao, J. et al. Coupling between NMDA receptor and acid-sensing ion channel contributes to ischemic neuronal death. *Neuron* **48**, 635–646 (2005).

25. Wang, Y.Z. et al. Tissue acidosis induces neuronal necroptosis via ASIC1a channel independent of its ionic conduction. *Elife* **4** (2015), e05682.

26. Wang, J.J. et al. Disruption of auto-inhibition underlies conformational signaling of ASIC1a to induce neuronal necroptosis. *Nat Commun* **11**, 475 (2020).

27. Friese, M.A. et al. Acid-sensing ion channel-1 contributes to axonal degeneration in autoimmune inflammation of the central nervous system. *Nat Med* **13**, 1483–1489 (2007).
28. Vergo, S. et al. Acid-sensing ion channel 1 is involved in both axonal injury and demyelination in multiple sclerosis and its animal model. *Brain* **134**, 571–584 (2011).
29. Ziemann, A.E. et al. Seizure termination by acidosis depends on ASIC1a. *Nat Neurosci* **11**, 816–822 (2008).
30. Yang, F. et al. Astrocytic acid-sensing ion channel 1a contributes to the development of chronic epileptogenesis. *Sci Rep* **6**, 31581 (2016).
31. Chu, X.P., & Xiong, Z.G. Acid-sensing ion channels in pathological conditions. *Adv Exp Med Biol* **961**, 419–431 (2013).
32. Coryell, M.W. et al. Acid-sensing ion channel-1a in the amygdala, a novel therapeutic target in depression-related behavior. *J Neurosci* **29**, 5381–5388 (2009).
33. Wu, W.L., Lin, Y.W., Min, M.Y., & Chen, C.C. Mice lacking Asic3 show reduced anxiety-like behavior on the elevated plus maze and reduced aggression. *Genes Brain Behav* **9**, 603–614 (2010).
34. Lin, S.H., Sun, W.H., & Chen, C.C. Genetic exploration of the role of acid-sensing ion channels. *Neuropharmacology* **94**, 99–118 (2015).
35. Li, W.G., & Xu, T.L. Acid-sensing ion channels: a novel therapeutic target for pain and anxiety. *Curr Pharm Des* **21**, 885–894 (2015).
36. Uchitel, O.D., Gonzalez Inchauspe, C., & Weissmann, C. Synaptic signals mediated by protons and acid-sensing ion channels. *Synapse* **73**, e22120 (2019).
37. Huang, Y. et al. Two aspects of ASIC function: synaptic plasticity and neuronal injury. *Neuropharmacology* **94**, 42–48 (2015).
38. Duman, R.S., Aghajanian, G.K., Sanacora, G., & Krystal, J.H. Synaptic plasticity and depression: new insights from stress and rapid-acting antidepressants. *Nat Med* **22**, 238–249 (2016).
39. Kauer, J.A., & Malenka, R.C. Synaptic plasticity and addiction. *Nat Rev Neurosci* **8**, 844–858 (2007).
40. Luo, C., Kuner, T., & Kuner, R. Synaptic plasticity in pathological pain. *Trends Neurosci* **37**, 343–355 (2014).
41. Kessels, H.W., & Malinow, R. Synaptic AMPA receptor plasticity and behavior. *Neuron* **61**, 340–350 (2009).
42. Miesenbock, G., De Angelis, D.A., & Rothman, J.E. Visualizing secretion and synaptic transmission with pH-sensitive green fluorescent proteins. *Nature* **394**, 192–195 (1998).
43. Liu, Y., & Edwards, R.H. The role of vesicular transport proteins in synaptic transmission and neural degeneration. *Annu Rev Neurosci* **20**, 125–156 (1997).
44. Palmer, M.J., Hull, C., Vigh, J., & von Gersdorff, H. Synaptic cleft acidification and modulation of short-term depression by exocytosed protons in retinal bipolar cells. *J Neurosci* **23**, 11332–11341 (2003).
45. Cho, S., & von Gersdorff, H. Proton-mediated block of Ca^{2+} channels during multivesicular release regulates short-term plasticity at an auditory hair cell synapse. *J Neurosci* **34**, 15877–15887 (2014).
46. DeVries, S.H. Exocytosed protons feedback to suppress the Ca^{2+} current in mammalian cone photoreceptors. *Neuron* **32**, 1107–1117 (2001).
47. Vessey, J.P. et al. Proton-mediated feedback inhibition of presynaptic calcium channels at the cone photoreceptor synapse. *J Neurosci* **25**, 4108–4117 (2005).
48. Traynelis, S.F., & Cull-Candy, S.G. Proton inhibition of N-methyl-D-aspartate receptors in cerebellar neurons. *Nature* **345**, 347–350 (1990).
49. Du, J. et al. Protons are a neurotransmitter that regulates synaptic plasticity in the lateral amygdala. *Proc Natl Acad Sci USA* **111**, 8961–8966 (2014).

50. Beg, A.A., Ernstrom, G.G., Nix, P., Davis, M.W., & Jorgensen, E.M. Protons act as a transmitter for muscle contraction in *C. elegans. Cell* **132**, 149–160 (2008).
51. Highstein, S.M., Holstein, G.R., Mann, M.A., & Rabbitt, R.D. Evidence that protons act as neurotransmitters at vestibular hair cell-calyx afferent synapses. *Proc Natl Acad Sci USA* **111**, 5421–5426 (2014).
52. Zha, X.M., Wemmie, J.A., Green, S.H., & Welsh, M.J. Acid-sensing ion channel 1a is a postsynaptic proton receptor that affects the density of dendritic spines. *Proc Natl Acad Sci USA* **103**, 16556–16561 (2006).
53. Zha, X.M. et al. ASIC2 subunits target acid-sensing ion channels to the synapse via an association with PSD-95. *J Neurosci* **29**, 8438–8446 (2009).
54. Jing, L. et al. N-glycosylation of acid-sensing ion channel 1a regulates its trafficking and acidosis-induced spine remodeling. *J Neurosci* **32**, 4080–4091 (2012).
55. Yu, Z. et al. The acid-sensing ion channel ASIC1a mediates striatal synapse remodeling and procedural motor learning. *Sci Signal* **11** (2018), eaar4481.
56. Kreple, C.J. et al. Acid-sensing ion channels contribute to synaptic transmission and inhibit cocaine-evoked plasticity. *Nat Neurosci* **17**, 1083–1091 (2014).
57. Gonzalez-Inchauspe, C., Urbano, F.J., Di Guilmi, M.N., & Uchitel, O.D. Acid-sensing ion channels activated by evoked released protons modulate synaptic transmission at the mouse calyx of held synapse. *J Neurosci* **37**, 2589–2599 (2017).
58. Cho, J.H., & Askwith, C.C. Presynaptic release probability is increased in hippocampal neurons from ASIC1 knockout mice. *J Neurophysiol* **99**, 426–441 (2008).
59. Wemmie, J.A. et al. The acid-activated ion channel ASIC contributes to synaptic plasticity, learning, and memory. *Neuron* **34**, 463–477 (2002).
60. Wu, P.Y. et al. Acid-sensing ion channel-1a is not required for normal hippocampal LTP and spatial memory. *J Neurosci* **33**, 1828–1832 (2013).
61. Buta, A. et al. Novel potent orthosteric antagonist of ASIC1a prevents NMDAR-dependent LTP induction. *J Med Chem* **58**, 4449–4461 (2015).
62. Liu, M.G. et al. Acid-sensing ion channel 1a contributes to hippocampal LTP inducibility through multiple mechanisms. *Sci Rep* **6**, 23350 (2016).
63. Morin, F.O., Takumura, Y., & Tamiya, E. Investigating neuronal activity with planar microelectrode arrays: achievements and new perspectives. *J Biosci Bioeng* **100**, 131–143 (2005).
64. Liu, M.G. et al. Long-term depression of synaptic transmission in the adult mouse insular cortex in vitro. *Eur J Neurosci* **38**, 3128–3145 (2013).
65. Kang, S.J. et al. Plasticity of metabotropic glutamate receptor-dependent long-term depression in the anterior cingulate cortex after amputation. *J Neurosci* **32**, 11318–11329 (2012).
66. Liu, M.G. et al. Long-term potentiation of synaptic transmission in the adult mouse insular cortex: multielectrode array recordings. *J Neurophysiol* **110**, 505–521 (2013).
67. Ziemann, A.E. et al. The amygdala is a chemosensor that detects carbon dioxide and acidosis to elicit fear behavior. *Cell* **139**, 1012–1021 (2009).
68. Chiang, P.H., Chien, T.C., Chen, C.C., Yanagawa, Y., & Lien, C.C. ASIC-dependent LTP at multiple glutamatergic synapses in amygdala network is required for fear memory. *Sci Rep* **5**, 10143 (2015).
69. Wemmie, J.A. et al. Overexpression of acid-sensing ion channel 1a in transgenic mice increases acquired fear-related behavior. *Proc Natl Acad Sci USA* **101**, 3621–3626 (2004).
70. Coryell, M.W. et al. Restoring acid-sensing ion channel-1a in the amygdala of knockout mice rescues fear memory but not unconditioned fear responses. *J Neurosci* **28**, 13738–13741 (2008).
71. Chen, T. et al. Top-down descending facilitation of spinal sensory excitatory transmission from the anterior cingulate cortex. *Nat Commun* **9**, 1886 (2018).

72. Liu, M.G., Song, Q., & Zhuo, M. Loss of synaptic tagging in the anterior cingulate cortex after tail amputation in adult mice. *J Neurosci* **38**, 8060–8070 (2018).

73. Meda, K.S. et al. Microcircuit mechanisms through which mediodorsal thalamic input to anterior cingulate cortex exacerbates pain-related aversion. *Neuron* **102**, 944–959.e3 (2019).

74. Guo, B. et al. Anterior cingulate cortex dysfunction underlies social deficits in Shank3 mutant mice. *Nat Neurosci* **22**, 1223–1234 (2019).

75. Bliss, T.V., Collingridge, G.L., Kaang, B.K., & Zhuo, M. Synaptic plasticity in the anterior cingulate cortex in acute and chronic pain. *Nat Rev Neurosci* **17**, 485–496 (2016).

76. Collingridge, G.L., Peineau, S., Howland, J.G., & Wang, Y.T. Long-term depression in the CNS. *Nat Rev Neurosci* **11**, 459–473 (2010).

77. Mango, D. et al. Acid-sensing ion channel 1a is required for mGlu receptor dependent long-term depression in the hippocampus. *Pharmacol Res* **119**, 12–19 (2017).

78. Mango, D., & Nistico, R. Acid-sensing ion channel 1a is involved in N-methyl D-aspartate receptor-dependent long-term depression in the hippocampus. *Front Pharmacol* **10**, 555 (2019).

79. Craig, A.D. How do you feel – now? The anterior insula and human awareness. *Nat Rev Neurosci* **10**, 59–70 (2009).

80. Bermudez-Rattoni, F. Molecular mechanisms of taste-recognition memory. *Nat Rev Neurosci* **5**, 209–217 (2004).

81. Tan, L.L. et al. A pathway from midcingulate cortex to posterior insula gates nociceptive hypersensitivity. *Nat Neurosci* **20**, 1591–1601 (2017).

82. Li, W.G. et al. ASIC1a regulates insular long-term depression and is required for the extinction of conditioned taste aversion. *Nat Commun* **7**, 13770 (2016).

83. Holtmaat, A., & Svoboda, K. Experience-dependent structural synaptic plasticity in the mammalian brain. *Nat Rev Neurosci* **10**, 647–658 (2009).

84. Nishiyama, J., & Yasuda, R. Biochemical computation for spine structural plasticity. *Neuron* **87**, 63–75 (2015).

85. Chu, X.P., & Xiong, Z.G. Physiological and pathological functions of acid-sensing ion channels in the central nervous system. *Curr Drug Targets* **13**, 263–271 (2012).

86. Park, P. et al. Differential sensitivity of three forms of hippocampal synaptic potentiation to depotentiation. *Mol Brain* **12**, 30 (2019).

87. Abraham, W.C. Metaplasticity: tuning synapses and networks for plasticity. *Nat Rev Neurosci* **9**, 387 (2008).

88. Turrigiano, G. Too many cooks? Intrinsic and synaptic homeostatic mechanisms in cortical circuit refinement. *Annu Rev Neurosci* **34**, 89–103 (2011).

89. Lerner, T.N., Ye, L., & Deisseroth, K. Communication in neural circuits: tools, opportunities, and challenges. *Cell* **164**, 1136–1150 (2016).

90. Luo, L., Callaway, E.M., & Svoboda, K. Genetic dissection of neural circuits: a decade of progress. *Neuron* **98**, 865 (2018).

91. Wang, Q. et al. Fear extinction requires ASIC1a-dependent regulation of hippocampal-prefrontal correlates. *Sci Adv* **4**, eaau3075 (2018).

92. Monday, H.R., Younts, T.J., & Castillo, P.E. Long-term plasticity of neurotransmitter release: emerging mechanisms and contributions to brain function and disease. *Annu Rev Neurosci* **41**, 299–322 (2018).

93. Kullmann, D.M., & Lamsa, K.P. Long-term synaptic plasticity in hippocampal interneurons. *Nat Rev Neurosci* **8**, 687–699 (2007).

3 Trimeric Scaffold Ligand-Gated Ion Channels

Xiao-Na Yang, Si-Yu Wang, Jin Wang, and Ye Yu

CONTENTS

3.1 AN INTRODUCTION OF TRIMERIC SCAFFOLD OF LIGAND-GATED ION CHANNELS (TS-LGICS)

Ligand-gated ion channels (LGICs), as membrane receptors commonly referred to as ionotropic receptors, are a group of ion channels that open to allow ions such as Ca^{2+}, Na^+, K^+, and/or Cl^- to pass through the membrane in response to the binding of a ligand, such as a neurotransmitter or exogenous small molecule/peptide (1). Well-studied LGICs include members of the Cys-loop ion channel superfamily (2) (Figure 3.1A), such as the ionotropic acetylcholine receptor, γ-aminobutyric acid receptor ($GABA_A$), 5-HT_3 receptor, etc., which assemble as pentameric receptors. The members of tetrameric LGICs (3) have also been extensively studied in recent years (Figure 3.1B), for example, the α-Amino-3-hydroxy-5-methyl-4-isoxazole propionic acid receptor and ionotropic glutamate receptor. Recently, the trimeric scaffold ligand-gated ion channels (TS-LGICs, Figure 3.1C,D) have attracted widespread attention as novel drug targets for treatments of immunological, respiratory, cardiovascular, nervous, and genitourinary diseases, such as acid-

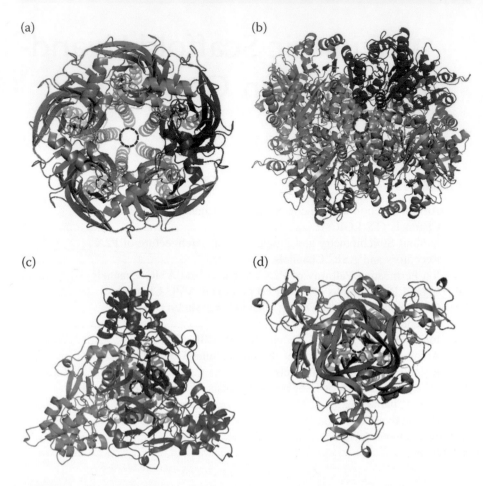

FIGURE 3.1 **Representative pentameric, tetrameric and trimeric ligand-gated ion channels**. **(A-D)** The three-dimensional (3D) architectures of pentameric (A, GABA$_A$ receptor, PDB ID: 4COF), tetrameric (B, GluA2/3 AMPA, PDB ID: 5IDE) and trimeric (C, ASIC1 channel, PDB ID: 6VTL; D, P2X4 receptor, PDB ID: 4DW0) ligand-gated ion channels viewed from the extracellular side. Each subunit is shown in different color for emphasis. Blue dashed circles indicate the boundary of the gate.

sensing ion channels (ASICs) (4), P2X receptors (5), FMRFamide peptide-gated sodium channels (FaNaC, only expression in mollusks) (6), and the epithelial sodium channel (ENaC, the channel with basic sustained activation that could also be directly activated by exogenous small molecules) (7).

The P2X receptor is one of TS-LGICs that could be directly activated by extracellular ATP, and seven different P2X subtypes have been identified so far, named as P2X1, P2X2, P2X3, P2X4, P2X5, P2X6, and P2X7 receptors (8). Wide distribution of these various P2X subtypes allows them to participate in diverse physiological processes, for instance, neural signaling, learning and memory, pain sensation, and immunological, respiratory, cardiovascular, and genitourinary functions regulations

(9–14). Disturbing channel functions mentioned above are considered to be relevant, also as important strategies in drug developments of many diseases like rheumatoid arthritis, mood disorder, stroke, neuropathic pain, and thrombogenesis (15–17). At present a series of P2X receptor-targeted small molecules have been in the main phase of clinical trials or pre-clinical stage aiming to treat rheumatoid arthritis, unexplained/refractory chronic cough, depression, and pain syndromes (18–21).

The ASIC channel is one of the degenerin/epithelial sodium channel (DEG/ENaC) superfamily, and the members of ASIC are capable of sensing extracellular mild and extreme acidosis. Slight acidification from pH 7.4 to 7.2–6.8 can directly elicit channel activation of ASIC3 and ASIC1a, while ASIC2a shows lower pH-sensitivity when compared to ASIC3 and ASIC1a (22–24). Each ASIC subtype distinguishes itself from the other subtypes by distinct distributions in various tissues. ASIC3 and ASIC1b are highly expressed in periphery nervous system and non-neural tissues, while ASIC1a, ASIC2a, and ASIC2b exist both in the central and peripheral nervous system (25,26). Although ASIC1a is abundant in both the central and peripheral nervous systems, its role in the brain and spinal cord has been studied more than that in peripheral system. This subtype has been implicated in signal transmission, learning and memory, hyperalgesia and allodynia, neuroinflammatory, ischemia-evoked neuronal death, fear memory, and so on (27–30). ASIC3 has been demonstrated as a chemical sensor and the mechanotransducer in the pain sensation and touch, respectively (31–34). ASIC4 is expressed in the pituitary gland and fails to form functional trimeric receptors, which might function together with other proteins independent of ion permeation (35).

ASICs and P2X receptors, two representative TS-LGICs having been extensively studied in recent years, contain diverse comparable characteristics, although their primary sequence identities are very low. Here, we present elucidation on some aspects of these two ion channels, including subunit stoichiometry/channel assembly, ion permeation pathway, ligand recognition, coordinated allostery during channel gating, ion selectivity, the architecture of the intracellular region, and the discovery of other endogenous/exogenous agonists, to advance our understanding of these unique architectures of nonclassical TS-LGICs.

3.2 SUBUNIT STOICHIOMETRY AND SINGLE SUBUNIT ARCHITECTURE OF P2X RECEPTORS AND ASIC CHANNELS

Only low sequence identity exists between the P2X receptors and ASIC channels (<30%), yet both of them assemble as trimeric ion channels. Current uncontroversial knowledge about the subunit stoichiometry of ASICs and P2X receptors is attributed to the tremendous advances in structural biology over the past decade.

Precedent studies on the subunit stoichiometry of ASICs never cease to be questioned by results that ASICs should be a tetramer (36), which is identical to those of FaNaC and ENaC (37,38). Nine subunits appear as another stoichiometric mode of the DEG/ENaC superfamily (39–41). It was commonly accepted that a functional ASIC is organized as a trimer until the crystal structure of chicken ASIC1 (cASIC1) at the desensitized state was determined by Dr. Gouaux and

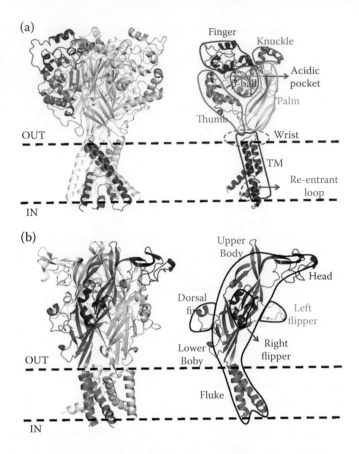

FIGURE 3.2   P2X receptors and ASIC channels. (A and B) The structures of cASIC1 and zfP2X4 receptor viewed parallel to the membrane. The ASIC1 subunit resembles a clenched hand. The P2X4 subunit has a dolphin-like shape. Distinctive body parts are shown in different colors.

colleagues in 2007 (42) (Figure 3.2A, *left*). This discrepancy also occurred on many membrane proteins; for example, the bacterial mechanosensitive ion channel MscL was firstly identified as a hexamer by chemical crosslinking (43), then was confirmed by X-ray crystallography as a pentamer (44). Similarly, although the majority of studies concerning P2X receptors concluded them as trimeric in formation, opinion by Kim et al. differed by presuming that the native assembling of P2X2 might be a tetramer (45), a prediction derived from the folding studies of the extracellular domain of P2X2. Ding et al. conjured the same result by evaluating the influence of Ca^{2+} on inactivation of P2X receptors (46). The crystal structure of zebra fish P2X4 (zfP2X4) determined by Gouaux et al. put this controversy to an end (47) (Figure 3.2B, *left*). The structure of cASIC and zfP2X4 indicate that three subunits of constituent formed a *pseudo* ternary symmetry by enclosing a central axis (48).

Pentameric classical LGICs possess both extracellular N- and C-terminal domains in one single subunit, which contains four transmembrane domains (TM). The tetrameric classical LGICs have an N-terminal extracellular domain and a C-terminal intracellular domain, accompanied by three TM domains. Notably, the individual subunit of TS-LGICs sharing semblable assembling is comprised of a large extracellular domain and two TM domains coupled with cytoplasmic N- and C-terminus. The extracellular region of one single cASIC1 subunit is a seemingly gripping "fist" named the thumb, finger, knuckle, palm, and β-ball domains. The linkage of the extracellular region and TM domain is a "wrist," and the TM segment is the vertical forearm (Figure 3.2A, *right*). The subunit of the P2X4 receptor looks like a dolphin springing from the sea with corresponding parts named the head, dorsal fin, left flipper, right fin, upper body, lower body, and tail domains (Figure 3.2B, *right*).

3.3   ION PERMEATION PATHWAY OF P2X RECEPTORS AND ASIC CHANNELS

ASIC channels and P2X receptors share similar three-dimensional (3D) architectures, particularly for their nearly identical ion permeation pathways (47,49) (Figure 3.3). Two possible ion permeation pathways were unveiled by the first resolved crystal structure of P2X receptors. *First*, cations enter the channel pore from the upper vestibule, passing the middle vestibule, extracellular vestibule, gating site, and intracellular vestibule at last into cytoplasm (Figure 3.3A). In reference to the P2X receptor, a too-narrow upper vestibule of the *apo*-state receptor leads to limited

(a)

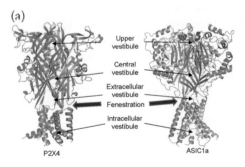

(b)

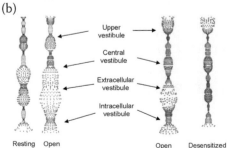

FIGURE 3.3   Various vestibules in P2X receptors and ASIC channels. **(A and B)** Location of the different vestibules (A) and pore-lining surfaces (B) in the zfP2X4 receptor and cASIC1 channel. *Left*, zfP2X4 receptor; *right*, cASIC1 channel. The structures are from PDB entry 4DW0 and 6VTL, respectively. The black dashed arrows indicate the ion influx pathway.

permeability of ions, whereas the bound agonist might elicit the expansion of the upper vestibule to facilitate the ion permeation (Figure 3.3B). *Second*, the subunit interface near the lipid bilayer contains a "fireplace"-like entrance (we call it "fenestration") that is directly exposed to the external solution. The ions move directly into the extracellular region, passing the gating site and intracellular vestibule eventually into the cells.

As revealed by the crystal structure of zfP2X4, Gd^{3+} inhibits ATP currents via changing the conformation of the central vestibule. The inhibition can be rescued by increasing concentration of ATP, implying that this blockade is not caused by the occupancy of Gd^{3+} in the ion-conducting pathway, and thus, the central vestibule might not be implicated in the ion-conducting pathway of P2X receptors (47). Other findings also indicate that it is effortless for monovalent cations to be transferred through the "fireplace"-like side window even for a channel at the *apo* state, yet a strong electrostatic barrier exists in the central path, which requires structural rearrangement to allow the ions to pass. Samways et al. introduced cysteine into the central path and side window, and found that residues along the central path couldn't be covalently modulated by MTSET, yet E56C and D58C located on the side window could be done (50). Moreover, the covalent modification-evoked significant inhibition can be reversed by reduction agent DTT. MTSET-mediated covalent modification is also observed in channels both with or without exposure to ATP, suggesting that the sulfhydryl reagent remains capable of accessing the side window even if P2X is closed. We also could find this "fireplace"-like side window in the open structure of zfP2X4. The central path in the open crystal structure of zfP2X4 remains too small to permeate the ions, yet the side window is open. Thus, the side window is the pathway for hydrated ions to be transferred through receptors. Once the ions pass through the side window, cations and anions could be attracted and repelled by the side window, respectively, while the concentrated cations will pass through the channel pore. Rokic et al. suggested that Cd^{2+} has modulatory effect on the gating by accessing extracellular and central vestibules through the side window (51). While residues on the TM1 domain accumulate the cations, three subunits' "TM2-TM2-TM2" residues chiefly respond to ATP-mediated channel gating and final permeation of ions. Thus, although the structure of P2X4 provides two possible ways for exogenous cations to enter the cell, most results deem that the pathway of the side window is the major accessing pathway for ions.

Similarly, as revealed by both the crystal structures and biochemical studies (52–54), the rehydrated cations accessing interiority of ASIC are also through the side window, forming cation-π interaction with the residue Y68 in the extracellular vestibule at the resting state. In circumstances of channel desensitization, Y68 interacts with R65 to form cation-π interaction to prevent ions trafficking, notwithstanding, this point requires further confirmation (54).

3.4 LIGAND RECOGNITIONS OF P2X RECEPTORS AND ASIC CHANNELS

In terms of specific extracellular regions for the ligand (ATP or proton) contained in ASICs and P2X receptors are the acidic pocket (proton-binding pocket) and ATP-binding site, respectively (Figure 3.4).

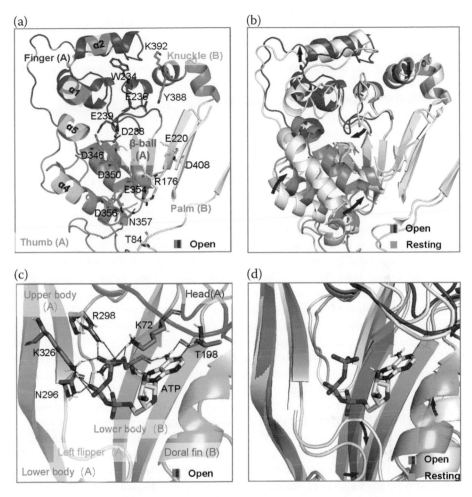

FIGURE 3.4 Agonist binding sites of P2X receptors and ASIC channels. (A) Acidic pocket of cASIC1 channel. Blue dashed lines indicate hydrogen bonds. (B) A superposition of the acidic pocket from the resting and open states. The blue arrows indicate the conformational changes during extracellular pH decreases. (C) The ATP-binding site of zfP2X4 at the open state. ATP molecules and key residues making contacts with ATP are shown as a stick for emphasis. (D) Superimposition of ATP-binding site at the resting and open states. The blue arrows indicate the conformational changes after ATP binding.

Enriching many acidic residues in the extracellular domain produces many protonation/deprotonation sites. Previous studies have figured out that four interaction pairs of residues participate in protonation through carboxyl-carboxyl interaction: D238-D350, E239-D346, and E220-D408 (cASIC1 numbering) that are located in the acidic pocket, as well as the E80-E417 pair in the lower palm domain (42,55). The acidic pocket is formed by residues from the thumb, finger, and β-ball, including several negative-charged amino acid side chains (42) (Figure 3.4A,B). As revealed by the desensitized ASIC crystal structures, these four pairs of amino acids are protonated under acidic conditions (55,56). Mutations carried out in these regions could significantly shift the curve of the pH-dependent response of ASIC (42,57,58). However, recent researches have shown that despite mutating residues of the acidic pocket being capable of altering pH dependence, ASIC still could be activated and desensitized regardless of studies from biochemical or computational approaches, suggesting that instead of an acidic pocket dispensable for acid activation, there are other protonation/deprotonation sites existing in the ASIC channels (59).

Combined previous studies show that, although the open conformation of cASIC1 is a MitTx toxin-binding state, it is sufficient to represent the conformation of the transient opening of cASIC1 elicited by acidosis (54,56,60). Compared with the resting state, the open structure unveiled that the thumb and knuckle domains shift to the center of the channel, the α2 helix of the finger domain moves outward, and the finger loop located centrally on the acidic pocket moves slightly to the thumb and the β-ball, which leads the entire acidic pocket to a state exhibiting a collapse/contraction/tightening conformation (60) (Figure 3.4B). Meanwhile, the acidic pocket at both the desensitization and open states share the same conformation (54). Quite a few electrophysiological and biochemical studies have been carried out to explore the relationship between the allostery and channel activation. For example, introducing a disulfide bond to narrow the distance between the finger α5 and the upper palm of the adjacent subunit (corresponding to E354-R176 of cASIC1) to dwindle the acidic pocket has led to a closing of the ASIC channel (61). The opposite, expanding it through narrowing the α5 and lower part of the palm domain, could significantly reduce ASIC desensitization (60). Moreover, the finger loop plays an essential role in acidification. In the desensitized cASIC1 structure, the distances of E239-D346 and D238-D350 on the loop are shortened by carboxyl-carboxylate interactions, corresponding to the loss-of-function in D238 and E239 mutants (57). Fluorescent patch clamp assay demonstrates that in the acidosis condition, E236 gathers with K392 and Y388 on the knuckle, yet W234 shifts outward from the finger (61). Furthermore, there is an affiliated interaction between the thumb and the finger regions during the processing of protonation and deprotonation of acidic residues, which leads to dislodging the thumb and the finger loop outward/downward of the center axis (62). The swing of the figure loop facilitates the development of the hydrogen bond (H-bond) between E236 and residues on the palm region of adjacent subunits to pull the knuckle domain inward (54). Meanwhile, the N357 of the α5 helix on the thumb interacts with R176 of the upper palm to stabilize the interface between adjacent subunits. These conformational changes in the afore-mentioned regions finally lead the whole acidic pocket to contract.

As to P2X receptors, three symmetrical ATP-binding sites exist in the extracellular domain of this trimeric ATP-gated receptor (49,63–65) (Figure 3.4C,D). The crystal structure shows that there is a large ATP-binding pocket between two adjacent subunits, comprised of the head and upper body of a subunit (chain A) and the lower body and dorsal fin of another subunit (chain B), containing the charged groups from K70, K72, R298, and K316 (zfP2X4 numbering) (49). Studied by Harrito et al., U-folded ATP binds to the zfP2X4 receptor. K70 (chain B) under the

U-shaped triphosphate interacts with the α, β, and γ-phosphate groups of ATP. The adenine group of ATP is buried into the ATP-binding pocket to form an H-bond with the carboxyl oxygen of K70 and T189 on the lower body, and make hydrophobic interaction with L191 on the lower body and I232 on the dorsal fin domain. Huang et al. suggested three models of ATP recognition, AR1, AR2, and AR3, corresponding to three parts of one ATP-recognition site, S1, S2, and S3 (66). In the AR1 model, the adenine ring contacts with L191, K70, K72, I232, and T186, parallel to Hattori and Gouaux's observation. The adenine ring interacts with I94, F297, Q97, and Y295 in AR2, a model being consistent with the study from Du et al. (67). W167, I173, L170, D145, and E171 function as a site to hold the adenine group of ATP in AR3, which is identical with the study of Jiang et al. (68). Molecular dynamics (MD) simulations show that AR1, rather than AR2 and AR3, is able to induce the head domain to move downward, and transit the conformational change of the extracellular domain to the channel pore, which facilitates the activation of P2X4.

3.5   COORDINATED ALLOSTERY DURING CHANNEL ACTIVATION OF P2X RECEPTORS AND ASIC CHANNELS

Under physiological circumstance, cASIC1 stays at the resting state including a dilating acidic pocket and a narrow extracellular central cavity (22). The conformational change occurs rapidly as extracellular pH decreases, where the α5 helix of the thumb domain rotates towards the central axis, and the thumb domain, the finger loop in one subunit, and the upper palm of the adjacent subunit close, which leads to a contracting acidic pocket (59,69,70). The finger loop also swings mildly to reduce the distance with the knuckle domain; meanwhile, the α2 helix of the finger domain separates with the β-ball. The contracted acidic pocket leads the upper palm to rotate and the lower part to move close to the membrane followed by alienating from the central axis to expand the extracellular central cavity. These alterations are sequentially coupled by the loop of the wrist region to translate and rotate the TM domain, leading to conformational changes of residues G432-Q437 and finally pore dilation (22,60,71). Sustained acid application would induce the channel to a rapid and complicated processing to desensitize in the AISC: it causes a mass of rearrangement in the β11-β12 loop of the palm domain, including the altered location of residues L414-N415, which detaches the upper extracellular domain from the lower part of the channel and facilitates the TM domain to regain the resting state-like conformation that is insensitive to acidosis (60). Removing extracellular acidosis allows the acidic pocket to relax and regain proton-sensitivity, in the case of which the β11-β12 loop moves back to the original position (Figure 3.5A,B).

Comparing the open state with the resting state, a conceivable model of P2X receptor activation is suggested as follows: first, ATP-binding elicits the head domain to shift downward and the dorsal fin domain upward; meanwhile, ATP pushes the left flipper domain to detach from it; then, the dorsal fin domain moves to the head, and the left flipper domain moves to the lower body, which expands the central cavity comprised of the three lower body domains; at the end, the

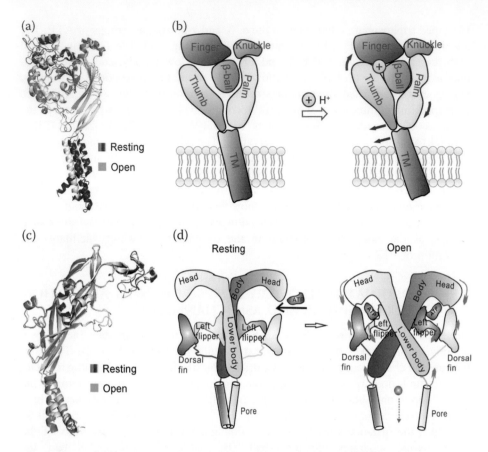

FIGURE 3.5 Agonist-induced allostery of P2X receptors and ASIC channels. (A) A superposition of a single ASIC1a subunit at resting and open states. (B) An illustration of the
allosteric changes during channel gating of the ASIC channel. First, at high concentration of
extracellular protons (acidosis), the acid pocket is in an obvious state of collapse, making the
thumb domain more close to the finger domain and the upper palm domain farther from the
neighboring subunit. Second, as the acid pocket collapses, the upper palm domain scaffold rotates
laterally, while the lower palm domain closes toward the membrane and away from three-fold
axis, which makes the extracellular fenestrations expand significantly. Finally, changes in the
extracellular domain drive shifts and rotations of two transmembrane (TM) domains, ultimately
creating an iris-like opening of the channel gate. (C) A superposition of a single P2X4 subunit at
resting and open states. (D) An illustration of the allosteric changes during channel gating of P2X
receptors. First, at the ATP-binding site, ATP promotes the closure of the jaw between the head
and dorsal fin domains, making the dorsal fin domain move upward toward the head domain to
accommodate ATP. Meanwhile, the bound ATP pushes the left flipper domain out of the ATP-
binding site. Second, because both the left flipper and dorsal fin domains are structurally coupled
with the lower body domain, the movement of those two domains leads to a concomitant outward
flexing of the lower body domains at the open state, which markedly expands the central vestibule. Finally, the lower body domains are directly coupled with two TM domains, TM1 and
TM2, and therefore their outward flexing can directly promote the opening of P2X by causing the
TM helices to expand in an iris-like motion.

expansion of the lower body domain drives the movement of TM1 and TM2 and finally pore opening (49,65,72) (Figure 3.5C,D). During this process, it is indispensable for the head domain to be pressed down (66,73–75). By combining voltage clamp fluorescence and electrophysiological studies, the involvements of residues N120 and G123 of the head domain in activation of P2X as well as P121 and I125 in the desensitization state have been found (76). Maintaining an ATP-binding state on rP2X2 via mutating L186 to cysteine and covalently linking NCS-ATP (a synthetic analog of ATP) on the site to promote the head domain downward could accelerate the opening of the channel (68). Suramin and its analog, NF499, are capable of recognizing the lysine K138 on the head to therefore prevent the channel from opening (77,78). Mutant S116H/T170H located in the ATP-binding pocket of rP2X2 facilitates the binding site jaw tightening, and will promote ATP-induced allostery by coordinating the binding of Zn^{2+} with D168 and H152 (73). The hydrophobic effects between the left flipper and the dorsal fin play an essential role in the opening of P2X receptors (74). ATP's binding alienates the left flipper from the ATP-binding pocket and hence drives the relative motions between the left flipper and dorsal fin domains. Introducing zinc-bridge and the disulfide bond between the left fin and dorsal fin restrains their movements, and altering hydrophobicity of K190, I205, L214, and V288 (number in rP2X4) could elicits disfunctions of channels (74). It has also been suggested by some studies that S275 on the left flipper domain of rP2X3 plays important roles in the ATP binding, desensitization, and recovery (79). A pair of salt bridges formed by E112 on the head and R198 on the dorsal fin domain coordinates the allosteric change of the extracellular domain, which can cause severe disfunctions by being caged by a disulfide bond.

3.6 IONIC SELECTIVITY OF P2X RECEPTORS AND ASIC CHANNELS

Ionic selectivity of ion channels is the relative high permeability to one or more kinds of ions yet impermeability to others. Although P2X receptors and ASICs are both TS-LGICs, selectivity of ions diverges significantly. The mechanism underlying ion selectivity correlates to complicated factors including pore size, state of charged residue, the state of ionic hydration, and the conformation of the TM domain.

ASIC is capable of permeating monovalent cations, exhibiting selectivity with the permeability sequence of $Na^+ > Li^+ > K^+ > Cs^+$; seepage of Na^+ is ten-fold higher than that of K^+, which is highly conserved in the DEG/ENaC superfamily of ion channels (22). Previous study considers the TM domain of AISC as a sequential two-segment helixes (TM1 and TM2) (42,55), recent structure shows that TM2 is discontinuous whenever in resting, opening, or desensitized state but cut off to two segments (TM2a and TM2b) by a horizontal pair of three residues "GAS", where nine dividing residues on three adjacent subunits beneath the gate are named the "GAS" belt (54,60). Generally considering, ionic selectivity of ASIC is relevant to "GAS"belt formed by the TM2 domain swapped. GAS is the narrowest position in the pore, and the main chain oxygen of G443 is inside-of-pore-oriented (80,81).

According to the research above, the pore is fit exactly for hydrated Na^+ to pass, but larger hydrated K^+ will be obstructed there, which should function as a selectivity filter (54,80,81). In a structure with a truncated N-terminal, the lower ion permeation pathway is completely formed by residues of TM2b, which constructs a new wide intracellular vestibule (54,56,60). Updated by very recent structure of SMA-cASIC1a, pre-TM1 forms a narrower passage beneath the "GAS" belt: the highly conserved motif "HG" located in N-terminal of pre-TM1 domain supports the "GAS" belt and the TM2 swap; the motif is beneath the "GAS" belt, facing the central axis and forming a smaller pore consisting of three H29 (56). However, in ENaCs, destruction of the "HG" motif attenuates the open possibility, which leads to Type 1 pseudo-hypoaldosteronism (80).

Differed from ASICs, the P2X receptor is a non-selective ion channel that permeates Na^+, K^+, and Ca^{2+}, one subtype of which, P2X5, is able to pass through Cl^- (82). Kawate et al. conjectured that, based on the crystal structure of zfP2X4, acidic residues D59 and D61 (zfP2X4 numbering) on the side window could not only directly bind cations, but also enrich cations and repel anions via long-range negative electrostatic potential (47). Meanwhile, the gate region determines whether the ions can pass as well as selectivity to cation and anion. After mutating T339 on the gate region of rP2X2 to lysine, the passage somewhat increases Cl^- permeability. The key discrepancy in ionic selectivity between P2X and ASIC lies on the permeability of calcium. The P2X receptor shows distinct preference to calcium-according subtypes, with P_{ca}/P_{Na} ranging from 1 to 13 (83). Xiong et al. found Ca^{2+} influx mediated by ASIC1a is essential for glutamic acid-independent neuronal damage; whereas ASIC3 cannot permeate Ca^{2+}, and Ca^{2+}-presence is able to block other ion permeation. Deprivation of extracellular calcium can directly activate ASIC3.

In general, ionic selectivity remains inherent and steady to each channel. Nonetheless, previous studies on P2X indicate that the effect of long-term application of agonists alters it. Referring to P2X2, P2X4 and P2X7, sustained activation by ATP will lead to altered reversal potential of P2X receptors (83). In short, ATP will evoke the channel to a rapid opening that permeates small cations at the beginning, such as Na^+, K^+, and Ca^{2+}. The pore diameter of P2X keeps dilating until large organic cation or dye molecules like NMDG and YO-PRO could pass through. In 2015 a new explanation suggested by Li et al., this phenomenon appears to be due to a drastic change in ion concentration, seen during an electrophysiology experiment: Na^+ decreased from 140 mM to 20 mM; on the contrary, NMDG increased from 0 mM to 200 mM (84).

3.7 ARCHITECTURE AND FUNCTION OF THE INTRACELLULAR DOMAIN OF P2X RECEPTORS AND ASIC CHANNELS

The N-terminal of the P2X receptor contains approximately 30 amino acids, yet the C-terminal is less conserved, ranging from 25 residues in P2X6 to 240 residues in P2X7. For acquisition of high-quality structure, N- and C-terminals were deleted in previous zfP2X4 crystal structures. Subsequent resolution of hP2X3 kept the relatively longer segment (85). Until recently, the structure of rat P2X7 determined by single particle electron cryo-microscopy (cryo-EM) has been based on the full length of P2X7 (86). Although the sequences of P2X3 and P2X7 vary significantly, the architecture of the intracellular domain they share exhibit similarity, for instance, crosslinking of N- and C-terminals is a sandwich-like manner. However, this assembling merely exists in the open state of P2X3; an order and stable folding remains whenever in resting or open state of rP2X7. Structures of P2X7 determined by cryo-EM reveal that the C-cys anchor as a cytoplasmic component could avoid desensitization by fixing on the membrane with a palmitoyl group; otherwise the intracellular domain of P2X7 also contains zinc- and guanosine-binding sites, which might contMoreover, conserved residues formingribute to P2X7 gating and other signaling that has not been clarified. Since truncation at this domain rarely influences the function of P2X7, we conjecture it might undertake some unclassical function as "metabolic gating (ion permeation-independent)."

Similarly, the cytoplasmic region of ASIC1 responds to acidosis-induced death of neurons (27,29). For a long time, ASIC1a-mediated cytoplasmic calcium overload has been considered as the only mechanism causing nerve damage (also known as the "ionic gating" death mechanism). In 2015, Wang et al. found protons not only evoke the death mechanism of "ionic gating," but also the entire allosteric transition of the channel, resulting in specific binding of RIPK1 and sequentially being phosphorylated at the C-terminal of ASIC1a to initiate apoptosis (29). This processing is independent from traditional ionic functions presented by ASIC1a. In addition, intracellular N- and C-terminals exhibit self-inhibitory behavior based on protein interaction; massive negative-charged glutamic acids (E6EEE9) in the N-terminal combine with positive-charged lysine (K468, K471, and K474) by electrostatic attraction, and the N- and C-terminals separate from each other after replacing glutamic acid with alanine (27).

The intracellular structures of most subtypes of P2X and ASIC are poorly identified, as well as the mechanism of how the interact between TM1 and TM2, or with other proteins that require structures with a high resolution of cytoplasmic domains.

3.8   OTHER ENDOGENOUS OR EXOGENOUS AGONISTS OF P2X RECEPTORS AND ASIC CHANNELS

Recently studies on P2X indicate that specifically activating or promoting each subtype in a particular circumstance is an available strategy for drug discovery (87). However, the binding site for endogenous agonist ATP is formed by five positive-charged basic amino acids and ones with a side chain of an aromatic ring, which requires compounds of high polarity to bind the site yet low suitability to be druggable. Moreover, conserved residues forming an orthosteric site cause the lack of selectivity for ATP to affect each subtype of P2X. On the other hand, the affinity with ATP that each subtype shows diverges intensively *in vitro*; especially P2X7 reaches a level of mM that will be hardly achieved under physiological conditions. Therefore, it is far-reaching to find other endogenous or exogenous non-ATP agonists. But it is regrettable that no relevant report on P2X-targeted non-ATP agonist has been published.

In this aspect, researches on ASICs agonists are in front of P2X receptors. Besides proton activation, there are many ways to activate the ASIC family in direct or indirect manner. A small molecule compound GMQ (2-guanidine-4-methylquinazoline), lipid, and neurotoxin could activate AISCs directly (54,88–90); Zn^{2+} indirectly activates AISC1a/2a heterodimer (91); neuropeptide and bile acid modulate the acidic activation of ASIC (92,93). GMQ is the first nonproton ligand of AISC that directly activates ASIC3 under the condition of pH 7.4, probably via acting on E79 on the palm domain of rASIC3 to promote the channel to maintain a constantly activated state (90,94). Toxins as a flashpoint have been identified in recent years, a few of which can bind to ASIC, including two direct agonists of ASIC, PcTx1 and MitTx, with effects on sites related to the acidic pocket (54,88). Functioning both as agonist and inhibitor, PcTx1 activates rASIC1 via binding the whole toxin to the α5 helix of the thumb domain by virtue of two arginines to expand to the acidic pocket, which also depends on the subtype and extracellular concentration of protons (88). MitTx activates ASIC in the form of dimer: its α subunit interacts with the lower palm and palm-finger loop, and β subunit binds on the thumb domain, presenting the whole molecule as a "bottle opener of triangular tip" attaching to the surface of the adjacent subunit. Lipids are the first class of ASIC endogenous agonists to be found (54). In the physiological condition, lipid signaling formed by lysophosphatidylcholine (LPC) and arachidonic acid directly activate ASIC3, which might be caused by decreasing the dependence of ASIC3 on acidosis (89). In addition to those substances, endogenous isoquinoline alkaloids (EIAs) can also activate ASIC3 in an acidic environment so that EIAs recover hASIC3 from desensitization and induce instantaneous current, but a high concentration of EIAs can activate rat ASIC3 both under physiological pH or alkaline condition (95). Other ligands have also been found to modulate maximum activation and desensitization of channels. Deoxycholic acid strengthens the acidic current in hASIC1a (96); the inhibitor of ASIC3, APETx2, increases that of ASIC1b and ASIC2a (97,98). FMRFamide, the FaNaC agonist, modulates the desensitization of ASIC1a and ASIC3 through increasing their sustained currents (93,99); FMRFamide-like neuropeptide FF and RPRFamide also modulate this type of current in ASIC3 channels (32,100).

To sum up, although many endogenous and exogenous ligands have been found, because proton-mediated activation exhibits properties of rapid desensitization, understanding of the mechanism underlying proton-sensing gating is incomplete to explain the long-term function of ASIC during physiological and pathological processes.

3.9 PERSPECTIVES

Since the first P2X receptor was cloned in 1994, and ASIC in 1997, TS-LGICs have been extensively studied, and many characteristics of TS-LGICs have been uncovered. In addition to the above summary, questions worth being well-studied include: the discovery of high subunit-specific agonists; other endogenous activation ways for ASIC and P2X; whether acidification and ATP signaling are coordinated in pathophysiological processes; the possibility for two families of TS-LGICs to mediate unnoticed functions in tissues with both P2X and ASIC expressions; whether the high degree of commonality shared by ASIC/P2X gating is compatible for other TS-LGICs, like FaNaC, etc.; is the AISC receptors a druggable target for pharmacological interventions for stroke, chronic pain, tumor, and so on; could candidate drugs targeting P2X pass proof-of-concept and come to market? In studies of TS-LGICs, the approaches and conclusions in one channel might edify another one. Acquired concepts on P2X receptors and ASIC channels studies can provide a new approach toward pharmaceutical discovery targeting this unique type of ligand-gating ion channel.

ACKNOWLEDGMENTS

This study was supported by grants from the National Natural Science Foundation of China (Nos. 31971146, 32000869, and 31900808), Open Project of State Key Laboratory of Natural Medicines (SKLNMZZRC201801), Innovation and Entrepreneurship Talent Program of Jiangsu Province, Natural Science Foundation of Jiangsu Province (BK20202002), Scientific Research and Practice Program for Graduate, Jiangsu Province (KYCX20_0682), Guangxi Funds for Distinguished Experts and State Key Laboratory of Utilization of Woody Oil Resource (2019XK2002).

REFERENCES

1. Alexander SPH, *et al.* (2019) THE CONCISE GUIDE TO PHARMACOLOGY 2019/ 20: Ion channels. *Br J Pharmacol* 176 (Suppl 1):S142–S228.
2. Sine SM & Engel AG (2006) Recent advances in Cys-loop receptor structure and function. *Nature* 440(7083):448 455.
3. Lodge D (2009) The history of the pharmacology and cloning of ionotropic glutamate receptors and the development of idiosyncratic nomenclature. *Neuropharmacology* 56(1):6 21.
4. Vullo S & Kellenberger S (2020) A molecular view of the function and pharmacology of acid-sensing ion channels. *Pharmacol Res* 154:104166.
5. Schmid R & Evans RJ (2019) ATP-gated P2X receptor channels: molecular insights into functional roles. *Annu Rev Physiol* 81:43–62.
6. Lingueglia E, Deval E, & Lazdunski M (2006) FMRFamide-gated sodium channel and ASIC channels: a new class of ionotropic receptors for FMRFamide and related peptides. *Peptides* 27(5):1138–1152.
7. Kleyman TR & Eaton DC (2020) Regulating ENaC's gate. *Am J Physiol Cell Physiol* 318(1):C150–C162.
8. Coddou C, Yan Z, Obsil T, Huidobro-Toro JP, & Stojilkovic SS (2011) Activation and regulation of purinergic P2X receptor channels. *Pharmacol Rev* 63(3):641–683.
9. Bele T & Fabbretti E (2015) P2X receptors, sensory neurons and pain. *Curr Med Chem* 22(7):845–850.
10. Bernier LP, Ase AR, & Seguela P (2018) P2X receptor channels in chronic pain pathways. *Br J Pharmacol* 175(12):2219–2230.
11. Franceschini A & Adinolfi E (2014) P2X receptors: new players in cancer pain. *World J Biol Chem* 5(4):429–436.
12. Burnstock G (2017) Purinergic signalling and neurological diseases: an update. *CNS Neurol Disord Drug Targets* 16(3):257–265.

13. Eltzschig HK, Sitkovsky MV, & Robson SC (2013) Purinergic signaling during inflammation. *N Engl J Med* 368(13):1260.

14. Yamamoto K, *et al.* (2006) Impaired flow-dependent control of vascular tone and remodeling in P2X4-deficient mice. *Nat Med* 12(1):133–137.

15. Burnstock G & Kennedy C (2011) P2X receptors in health and disease. *Adv Pharmacol* 61:333–372.

16. Hechler B & Gachet C (2015) Purinergic receptors in thrombosis and inflammation. *Arterioscler Thromb Vasc Biol* 35(11):2307–2315.

17. Khakh BS & North RA (2006) P2X receptors as cell-surface ATP sensors in health and disease. *Nature* 442(7102):527–532.

18. Smith JA, *et al.* (2020) Gefapixant, a P2X3 receptor antagonist, for the treatment of refractory or unexplained chronic cough: a randomised, double-blind, controlled, parallel-group, phase 2b trial. *Lancet Respir Med* 8(8):775–785.

19. Smith JA, *et al.* (2020) Gefapixant in two randomised dose-escalation studies in chronic cough. *Eur Respir J* 55(3), 1901615.

20. Abdulqawi R, *et al.* (2015) P2X3 receptor antagonist (AF-219) in refractory chronic cough: a randomised, double-blind, placebo-controlled phase 2 study. *Lancet* 385(9974):1198–1205.

21. Marucci G, *et al.* (2019) Update on novel purinergic P2X3 and P2X2/3 receptor antagonists and their potential therapeutic applications. *Expert Opin Ther Pat* 29(12): 943–963.

22. Kellenberger S & Schild L (2015) International Union of Basic and Clinical Pharmacology. XCI. Structure, function, and pharmacology of acid-sensing ion channels and the epithelial Na+ channel. *Pharmacol Rev* 67(1):1–35.

23. Kellenberger S & Schild L (2002) Epithelial sodium channel/degenerin family of ion channels: a variety of functions for a shared structure. *Physiol Rev* 82(3):735–767.

24. Chu XP, Papasian CJ, Wang JQ, & Xiong ZG (2011) Modulation of acid-sensing ion channels: molecular mechanisms and therapeutic potential. *Int J Physiol Pathophysiol Pharmacol* 3(4):288–309.

25. Papalampropoulou-Tsiridou M, Labrecque S, Godin AG, De Koninck Y, & Wang F (2020) Differential expression of acid-sensing ion channels in mouse primary afferents in naive and injured conditions. *Front Cell Neurosci* 14:103.

26. Wemmie JA, Taugher RJ, & Kreple CJ (2013) Acid-sensing ion channels in pain and disease. *Nat Rev Neurosci* 14(7):461–471.

27. Wang JJ, *et al.* (2020) Disruption of auto-inhibition underlies conformational signaling of ASIC1a to induce neuronal necroptosis. *Nat Commun* 11(1):475.

28. Qiang M, *et al.* (2018) Selection of an ASIC1a-blocking combinatorial antibody that protects cells from ischemic death. *Proc Natl Acad Sci USA* 115(32):E7469–E7477.

29. Wang YZ, *et al.* (2015) Tissue acidosis induces neuronal necroptosis via ASIC1a channel independent of its ionic conduction. *Elife* 4, e05682.

30. Gautam M & Benson CJ (2013) Acid-sensing ion channels (ASICs) in mouse skeletal muscle afferents are heteromers composed of ASIC1a, ASIC2, and ASIC3 subunits. *FASEB J* 27(2):793–802.

31. Stephan G, *et al.* (2018) The ASIC3/P2X3 cognate receptor is a pain-relevant and ligand-gated cationic channel. *Nat Commun* 9(1):1354.

32. Reimers C, *et al.* (2017) Identification of a cono-RFamide from the venom of Conus textile that targets ASIC3 and enhances muscle pain. *Proc Natl Acad Sci USA* 114(17):E3507–E3515.

33. Lin SH, *et al.* (2016) Evidence for the involvement of ASIC3 in sensory mechanotransduction in proprioceptors. *Nat Commun* 7:11460.

34. Peng Z, *et al.* (2015) ASIC3 mediates itch sensation in response to coincident stimulation by acid and nonproton ligand. *Cell Rep* 13(2):387–398.

35. Wang K, *et al.* (2020) Cell type-specific expression pattern of proton sensing receptors and channels in pituitary gland. *Biophys J*, 119(11): 2335–2348.
36. Cox DH, Cui J, & Aldrich RW (1997) Allosteric gating of a large conductance Ca-activated K+ channel. *J Gen Physiol* 110(3):257–281.
37. Poet M, *et al.* (2001) Exploration of the pore structure of a peptide-gated Na^+ channel. *EMBO J* 20(20):5595–5602.
38. Anantharam A & Palmer LG (2007) Determination of epithelial Na^+ channel subunit stoichiometry from single-channel conductances. *J Gen Physiol* 130(1):55–70.
39. Snyder PM, Cheng C, Prince LS, Rogers JC, & Welsh MJ (1998) Electrophysiological and biochemical evidence that DEG/ENaC cation channels are composed of nine subunits. *J Biol Chem* 273(2):681–684.
40. Staruschenko A, *et al.* (2004) Fluorescence resonance energy transfer analysis of subunit stoichiometry of the epithelial Na+ channel. *J Biol Chem* 279(26): 27729–27734.
41. Eskandari S, *et al.* (1999) Number of subunits comprising the epithelial sodium channel. *J Biol Chem* 274(38):27281–27286.
42. Jasti J, Furukawa H, Gonzales EB, & Gouaux E (2007) Structure of acid-sensing ion channel 1 at 1.9 A resolution and low pH. *Nature* 449(7160):316–323.
43. Hase CC, Minchin RF, Kloda A, & Martinac B (1997) Cross-linking studies and membrane localization and assembly of radiolabelled large mechanosensitive ion channel (MscL) of *Escherichia coli*. *Biochem Biophys Res Commun* 232(3):777–782.
44. Chang G, Spencer RH, Lee AT, Barclay MT, & Rees DC (1998) Structure of the MscL homolog from *Mycobacterium tuberculosis*: a gated mechanosensitive ion channel. *Science* 282(5397):2220–2226.
45. Kim M, Yoo OJ, & Choe S (1997) Molecular assembly of the extracellular domain of P2X2, an ATP-gated ion channel. *Biochem Biophys Res Commun* 240(3):618–622.
46. Ding S & Sachs F (2000) Inactivation of P2X2 purinoceptors by divalent cations. *J Physiol* 522 Pt 2:199–214.
47. Kawate T, Michel JC, Birdsong WT, & Gouaux E (2009) Crystal structure of the ATP-gated P2X(4) ion channel in the closed state. *Nature* 460(7255):592–598.
48. Kellenberger S & Grutter T (2015) Architectural and functional similarities between trimeric ATP-gated P2X receptors and acid-sensing ion channels. *J Mol Biol* 427(1): 54–66.
49. Hattori M & Gouaux E (2012) Molecular mechanism of ATP binding and ion channel activation in P2X receptors. *Nature* 485(7397):207–212.
50. Samways DS, Khakh BS, Dutertre S, & Egan TM (2011) Preferential use of unobstructed lateral portals as the access route to the pore of human ATP-gated ion channels (P2X receptors). *Proc Natl Acad Sci USA* 108(33):13800–13805.
51. Rokic MB, Stojilkovic SS, & Zemkova H (2014) Structural and functional properties of the rat P2X4 purinoreceptor extracellular vestibule during gating. *Front Cell Neurosci* 8:3.
52. Leng T, Shi Y, Xiong ZG, & Sun D (2014) Proton-sensitive cation channels and ion exchangers in ischemic brain injury: new therapeutic targets for stroke? *Prog Neurobiol* 115:189–209.
53. Grunder S & Chen X (2010) Structure, function, and pharmacology of acid-sensing ion channels (ASICs): focus on ASIC1a. *Int J Physiol Pathophysiol Pharmacol* 2(2): 73–94.
54. Baconguis I, Bohlen CJ, Goehring A, Julius D, & Gouaux E (2014) X-ray structure of acid-sensing ion channel 1-snake toxin complex reveals open state of a Na(+)-selective channel. *Cell* 156(4):717–729.

55. Gonzales EB, Kawate T, & Gouaux E (2009) Pore architecture and ion sites in acid-sensing ion channels and P2X receptors. *Nature* 460(7255):599–604.

56. Yoder N & Gouaux E (2020) The His-Gly motif of acid-sensing ion channels resides in a reentrant 'loop' implicated in gating and ion selectivity. *Elife* 9, e56527.

57. Krauson AJ, Rued AC, & Carattino MD (2013) Independent contribution of extracellular proton binding sites to ASIC1a activation. *J Biol Chem* 288(48):34375–34383.

58. Sherwood T, *et al.* (2009) Identification of protein domains that control proton and calcium sensitivity of ASIC1a. *J Biol Chem* 284(41):27899–27907.

59. Vullo S, *et al.* (2017) Conformational dynamics and role of the acidic pocket in ASIC pH-dependent gating. *Proc Natl Acad Sci USA* 114(14):3768–3773.

60. Yoder N, Yoshioka C, & Gouaux E (2018) Gating mechanisms of acid-sensing ion channels. *Nature* 555(7696):397–401.

61. Gwiazda K, Bonifacio G, Vullo S, & Kellenberger S (2015) Extracellular subunit interactions control transitions between functional states of acid-sensing ion channel 1a. *J Biol Chem* 290(29):17956–17966.

62. Liechti LA, *et al.* (2010) A combined computational and functional approach identifies new residues involved in pH-dependent gating of ASIC1a. *J Biol Chem* 285(21):16315–16329.

63. Wilkinson WJ, Jiang LH, Surprenant A, & North RA (2006) Role of ectodomain lysines in the subunits of the heteromeric P2X2/3 receptor. *Mol Pharmacol* 70(4):1159–1163.

64. Marquez-Klaka B, Rettinger J, & Nicke A (2009) Inter-subunit disulfide cross-linking in homomeric and heteromeric P2X receptors. *Eur Biophys J* 38(3):329–338.

65. Chataigneau T, Lemoine D, & Grutter T (2013) Exploring the ATP-binding site of P2X receptors. *Front Cell Neurosci* 7:273.

66. Huang LD, *et al.* (2014) Inherent dynamics of head domain correlates with ATP-recognition of P2X4 receptors: insights gained from molecular simulations. *PLoS One* 9(5):e97528.

67. Du J, Dong H, & Zhou HX (2012) Gating mechanism of a P2X4 receptor developed from normal mode analysis and molecular dynamics simulations. *Proc Natl Acad Sci U S A* 109(11):4140–4145.

68. Jiang R, *et al.* (2011) Agonist trapped in ATP-binding sites of the P2X2 receptor. *Proc Natl Acad Sci U S A* 108(22):9066–9071.

69. Bonifacio G, Lelli CI, & Kellenberger S (2014) Protonation controls ASIC1a activity via coordinated movements in multiple domains. *J Gen Physiol* 143(1):105–118.

70. Passero CJ, Okumura S, & Carattino MD (2009) Conformational changes associated with proton-dependent gating of ASIC1a. *J Biol Chem* 284(52):36473–36481.

71. Li T, Yang Y, & Canessa CM (2010) Asn415 in the beta11-beta12 linker decreases proton-dependent desensitization of ASIC1. *J Biol Chem* 285(41):31285–31291.

72. Habermacher C, Dunning K, Chataigneau T, & Grutter T (2016) Molecular structure and function of P2X receptors. *Neuropharmacology* 104:18–30.

73. Jiang R, *et al.* (2012) Tightening of the ATP-binding sites induces the opening of P2X receptor channels. *EMBO J* 31(9):2134–2143.

74. Zhao WS, *et al.* (2014) Relative motions between left flipper and dorsal fin domains favour P2X4 receptor activation. *Nat Commun* 5:4189.

75. Kowalski M, *et al.* (2014) Conformational flexibility of the agonist binding jaw of the human P2X3 receptor is a prerequisite for channel opening. *Br J Pharmacol* 171(22):5093–5112.

76. Lorinczi E, *et al.* (2012) Involvement of the cysteine-rich head domain in activation and desensitization of the P2X1 receptor. *Proc Natl Acad Sci USA* 109(28):11396–11401.

77. Sim JA, Broomhead HE, & North RA (2008) Ectodomain lysines and suramin block of P2X1 receptors. *J Biol Chem* 283(44):29841–29846.
78. Braun K, *et al.* (2001) NF449: a subnanomolar potency antagonist at recombinant rat P2X(1) receptors. *Naunyn Schmiedebergs Arch Pharmacol* 364(3):285–290.
79. Petrenko N, Khafizov K, Tvrdonova V, Skorinkin A, & Giniatullin R (2011) Role of the ectodomain serine 275 in shaping the binding pocket of the ATP-gated P2X3 receptor. *Biochemistry* 50(39):8427 8436.
80. Lynagh T, *et al.* (2017) A selectivity filter at the intracellular end of the acid-sensing ion channel pore. *Elife* 6, e24630.
81. Li T, Yang Y, & Canessa CM (2011) Outlines of the pore in open and closed conformations describe the gating mechanism of ASIC1. *Nat Commun* 2:399.
82. Ruppelt A, Ma W, Borchardt K, Silberberg SD, & Soto F (2001) Genomic structure, developmental distribution and functional properties of the chicken P2X(5) receptor. *J Neurochem* 77(5):1256–1265.
83. Khakh BS & North RA (2012) Neuromodulation by extracellular ATP and P2X receptors in the CNS. *Neuron* 76(1):51 69.
84. Li M, Toombes GE, Silberberg SD, & Swartz KJ (2015) Physical basis of apparent pore dilation of ATP-activated P2X receptor channels. *Nat Neurosci* 18(11):1577–1583.
85. Mansoor SE, *et al.* (2016) X-ray structures define human P2X(3) receptor gating cycle and antagonist action. *Nature* 538(7623):66–71.
86. McCarthy AE, Yoshioka C, & Mansoor SE (2019) Full-length P2X7 structures reveal how palmitoylation prevents channel desensitization. *Cell* 179(3):659–670.e613.
87. Stokes L, Bidula S, Bibic L, & Allum E (2020) To iinhibit or enhance? Is there a benefit to positive allosteric modulation of P2X receptors? *Front Pharmacol* 11:627.
88. Baconguis I & Gouaux E (2012) Structural plasticity and dynamic selectivity of acid-sensing ion channel-spider toxin complexes. *Nature* 489(7416):400–405.
89. Marra S, *et al.* (2016) Non-acidic activation of pain-related Acid-Sensing Ion Channel 3 by lipids. *EMBO J* 35(4):414–428.
90. Alijevic O & Kellenberger S (2012) Subtype-specific modulation of acid-sensing ion channel (ASIC) function by 2-guanidine-4-methylquinazoline. *J Biol Chem* 287(43):36059–36070.
91. Joeres N, Augustinowski K, Neuhof A, Assmann M, & Grunder S (2016) Functional and pharmacological characterization of two different ASIC1a/2a heteromers reveals their sensitivity to the spider toxin PcTx1. *Sci Rep* 6:27647.
92. Younger MA, Muller M, Tong A, Pym EC, & Davis GW (2013) A presynaptic ENaC channel drives homeostatic plasticity. *Neuron* 79(6):1183–1196.
93. Xie J, Price MP, Wemmie JA, Askwith CC, & Welsh MJ (2003) ASIC3 and ASIC1 mediate FMRFamide-related peptide enhancement of H+-gated currents in cultured dorsal root ganglion neurons. *J Neurophysiol* 89(5):2459–2465.
94. Yu Y, *et al.* (2010) A nonproton ligand sensor in the acid-sensing ion channel. *Neuron* 68(1):61–72.
95. Osmakov DI, Koshelev SG, Andreev YA, & Kozlov SA (2017) Endogenous iso-quinoline alkaloids agonists of acid-sensing ion channel type 3. *Front Mol Neurosci* 10:282.
96. Ilyaskin AV, Diakov A, Korbmacher C, & Haerteis S (2017) Bile acids potentiate proton-activated currents in *Xenopus laevis* oocytes expressing human acid-sensing ion channel (ASIC1a). *Physiol Rep* 5(3), e13132.
97. Lee JYP, *et al.* (2018) Inhibition of acid-sensing ion channels by diminazene and APETx2 evoke partial and highly variable antihyperalgesia in a rat model of inflammatory pain. *Br J Pharmacol* 175(12):2204–2218.

98. Kozlov SA, *et al.* (2012) [Polypeptide toxin from sea anemone inhibiting proton-sensitive channel ASIC3]. *Bioorg Khim* 38(6):653–659.

99. Niu R, *et al.* (2019) ASIC1a promotes synovial invasion of rheumatoid arthritis via Ca(2+)/Rac1 pathway. *Int Immunopharmacol* 79:106089.

100. Askwith CC, *et al.* (2000) Neuropeptide FF and FMRFamide potentiate acid-evoked currents from sensory neurons and proton-gated DEG/ENaC channels. *Neuron* 26(1): 133–141.

4 Eukaryotic Mechanosensitive Ion Channels

Wei-Zheng Zeng and Bailong Xiao

CONTENTS

4.1 INTRODUCTION

Cells communicate via electrical and chemical signals. Many ion channels have been identified that are dedicated to this process. Cells are also subjected to the mechanical environment and can translate mechanical forces into biological signals, known as mechanotransduction. Mechanotransduction is critical for a wide range of physiological processes in all organisms. For instance, the senses of touch, proprioception, mechanical pain, hearing, and balance depend on mechanically activated (MA) ion channels [1]. Besides sensory systems, mechanotransduction is involved in diverse physiological functions, including cardiovascular tone and blood flow regulation, bone and muscle homeostasis, and stretch sensation of internal organs [1]. Many membrane proteins are involved in mechanotransduction, including ion channels, G-protein-coupled receptors, specialized cytoskeletal proteins, cell junction molecules, and kinases [2]. Among all the mechanosensitive

65

molecules, MA ion channels convert mechanical forces into electrical signals within mini-seconds [2], which is particularly suitable for the fast signaling that occurs in the sensory process involved in touch, hearing, cardiopulmonary regulation, and internal organ stretch sensing.

Mechanosensing ion channels are divided into two categories, the mechanically activated (MA) ion channels and mechanosensitive (MS) ion channels. They present in the plasma membranes of organisms from three domains of life: bacteria, archaea, and eukaryote [3]. Given the fundamental physiological implications, a search for these ion channels across species attracts enormous interests. The major difference between MS and MA ion channels is that MA ion channels, like Piezo1 and Piezo2, are bona fide ion channels directly gated by mechanical force, while MS ion channels are primarily activated by a non-mechanical stimulus and are modulated by mechanical force to their gating domains, such as Kv1.1, Nav1.5, ASICs, and TRP channels [3]. A gold standard for characterizing intrinsic MA channel activities is by heterologous expression of candidate channels in naive cells, which do not contain any endogenous MA currents.

In the first part of the chapter, we will focus on discussing MS channels with long-standing mechanotransduction roles, especially those involved in both chemo- and mechano-somatosensation, like ASICs and TRP channels. Then we will cover the recent discoveries of newly identified genes that might encode MA channels, including TMEM63 (OSCA), TMEM120a (TACAN), and TMEM87a (Elkin1). In the second part, we will dive deep to discuss the first bona fide MA cation channels in mammals, Piezo1 and Piezo2. We will discuss the physiology and activation mechanisms of Piezos given the recent breakthroughs on both fronts.

4.1.1 MS AND NOVEL MA ION CHANNELS—CONCEPTS AND PHYSIOLOGY

4.1.1.1 ASICs

Acid-sensing ion channels (ASICs) are activated by extracellular acidosis and belong to the degenerin/epithelial sodium channel protein family, a group of cation channels expressing in the nervous system and types of epithelial and immune cells [4]. As the major proton sensors of the cell, ASICs detect tissue acidosis occurring from tissue injury, inflammation, ischemia, stroke, and tumors as well as transmit pain signals to the brain in the peripheral sensory neurons [5]. Though ASICs have been recognized as a chemosensor, accumulating evidence shows that ASICs are gated by mechanical force in a tether model, in which the extracellular matrix or cytoplasmic cytoskeletons act like a gating-spring to tether and transmit the force to the channels. Accordingly, genetic knock-out of ASICs in mice suggests their roles in proprioceptors, mechanoreceptors, and nociceptors to monitor the homoeostatic status of muscle contraction, blood volume, and blood pressure as well as pain sensation [4]. Although the murine phenotypes show that ASICs are involved in many mechanotransduction systems, there is no evidence to reconstitute the ASIC-mediated MA current in a heterologous expression system. On the other hand, a recent work by Fronius et al. found that shear stress modulates the ASICs channel activity at low pH or in the presence of non-proton ligands in Xenopus oocytes, while it doesn't gate ASCIs at neutral pH when the channels are closed.

The finding supports the notion that ASICs can be modulated but not directly gated by mechanical force [6].

4.1.1.2 TRP Channels

Various transient receptor potential (TRP) channels are involved in mechanosensation in sensory neurons, including TRPV1, TRPC5, and TRPV4. **TRPV1** has recently been implicated in controlling the stretch response of bladder urothelium, since the knock-out mice showed defects in voiding. However, no mechano-currents have been recorded from urothelial cells, suggesting that TRPV1 may detect the presence of molecules produced by stretch and may act downstream of the real MA channel [7]. **TRPC5** has been shown opened by hypo-osmolality and plays a role in baroreceptor mechanosensing [8]. However, since a hypo-osmotic stimulus contains both chemical (decreased ionic strength) and mechanical (cell swell) components, more specific mechanical activation should be taken into consideration. A recent study provides clear evidence that membrane stretch failed to activate TRPC5 in a heterologous expression system [9]. These studies suggest that both TRPV1 and TRPC5 serve the downstream of mechanosensation as MS channels.

Emerging evidence has implicated the polymodal and non-selective **TRPV4** cation channel in a wide variety of mechanosensory processes in different cells and tissues. Although the most studied mechanosensory role of TRPV4 in the mammalian nervous system is in osmolarity sensing, it remains unclear whether TRPV4 is directly gated by mechanical stimuli or is activated downstream [7]. Interestingly, a recent elegant study found that TRPV4 does not respond to membrane stretch and indentation (poking) but can be efficiently gated by substrate deflections in a heterologous system, suggesting a direct gating of TRPV4 by mechanical force as an MA channel [10].

4.1.1.3 TMEM150C

TMEM150c/Tentonin3 is a transmembrane protein proposed to be an MA channel mediating slowly inactivating MA current in proprioceptive neurons in mouse DRG [11] and detect blood pressure in baroreceptors [12]. However, data from different groups failed to reproduce any TMEM150c related MA currents in transfected Piezo1-deficient HEK cells [13,14], suggesting that using conventional HEK cell line as a heterologous expression system for studying MA currents can be misleading due to endogenously expressed PIEZO1 channels (to be discussed later). Intriguingly, by co-expressing TMEM150c with bona fide MA channels (Piezo1, Piezo2, and Trek1) in HEK cells without endogenous *PIEZO1*, Anderson et al. found that TMEM150c prolongs the duration of mechano-currents of these channels, indicating that TMEM150c is a common modulator for mechano-gated ion channels in a variety of cells and tissues [13]. The modulation mechanism remains to be elucidated.

4.1.1.4 TMEM63 (OSCA)

Recent studies by different groups identified various members of the OSCA and TMEM63 family of proteins from plants, flies, and mammals as MA channels in transfected Piezo1-deficient HEK cells [15–18]. By using various electrophysiological approaches, Murthy et al. show that AtOSCAs, together with human and mouse homologs TMEM63a,b,c, evoke high-threshold mechano-activated currents ($P_{0.5}$ at 60–80 mm Hg) [16]. A detailed characterization of plant OSCA channels found that both AtOSCA1.1 and AtOSCA1.2 from *Arabidopsis thaliana* are inherently mechanosensitive, pore-forming ion channels [16]. OSCAs are cationic and non-selective, with a slight permeability for chloride. Reconstitution of purified AtOSCA1.2 into artificial bilayers induces robust stretch-activated currents, demonstrating that it is intrinsically mechanosensitive [16]. Notably, only AtOSCA1.1 and AtOSCA1.2 exhibit MA responses to both cell in-dentation and membrane stretch stimuli, while the TMEM63 family members only response to membrane stretch. The stretch-activated single-channel currents in-duced by TMEM63s are too small to be resolved, as with AtOSCA2.3 [16]. These data suggest that MA currents induced by the TMEM63 family members had un-ique gating properties with slow activation and inactivation kinetics compared to Piezo channels. Future efforts will be needed to address which tissues have en-dogenous TMEM63 channel activities, and the specific pathophysiological role of each TMEM63a,b,c isoforms in humans.

4.1.1.5 TMEM120A (TACAN)

The search for MA ion channels mediating the mechanical pain in nociceptors continues. Starting with a list of over 70 candidates involved in mechan-otransduction of smooth muscle cells, Sharif-Naeini et al. identified TMEM120a, a novel MA ion channel involved in sensing mechanical pain [19]. They rename it TACAN, for "movement" in Farsi. TMEM120a has six putative transmembrane domains with both intracellular amino and carboxyl termini. Notably, while het-erologous expression of TMEM120a in Piezo1-deficient cells evokes tiny stretch-induced currents in the single digital pA range, purification and reconstitution of TMEM120a in synthetic lipids unexpectedly generates large spontaneous single-channel currents. The reason for such paradoxical results remains explained. TMEM120a has a broad expression, highly expressive in the heart, kidneys, colon, and sensory neurons of the dorsal root ganglia (DRG). Given the somatosensation role of DRG, they generate a nociceptor-specific inducible knockout of TMEM120a to evaluate its function. Interestingly, knockdown of TMEM120a by siRNA in nociceptors caused a significant decrease of neurons displaying the ultra-slow adapting current, which kinetics distinguished from Piezo2 (to be discussed later) [19]. Supporting this finding, the conditional knock-out mice show decreased behavioral responses to painful mechanical stimuli but not to thermal or touch stimuli [19]. The novel discovery supports the notion that TMEM120a might be one of the mysterious candidates that mediate mechanical pain. Structural determination will validate whether TACAN is indeed a bona fide MA ion channel.

4.1.1.6 TMEM87A (Elkin1)

Mechanotransduction happens both outside and inside the cells. However, the molecules and mechanisms sensing cellular internal forces are less understood. Emerging works suggest that MA channels, including Piezo1, sense the cellular internal forces generated by changes in the extracellular matrix, cell adhesions, or the interaction of MA channels with cytoskeleton [20]. However, Piezo channels do not account for all mammalian MA channel activity in tissues where they don't exist. To identify additional MA channels and characterize their function represent outstanding challenges in the field. Recently, Poole et al. identified that the presence of MA currents in skin cancer cells (melanoma) are dependent on TMEM87a [21]. They renamed it Elkin1, from the Greek word "elko", meaning "to pull." Importantly, heterologous expression of TMEM87a in *Piezo1*-deficient cells is sufficient to reconstitute mechanically activated currents [21]. Another notable finding is that TMEM87a is only gated by substrate deflection and cell indentation stimuli, not by membrane stretch generated by a high-speed pressure pump [21], suggesting a refined mechanogating mechanism like Piezo2. Given that mechanotransduction is also involved in tumor development and metastasis, they tested the role of TMEM87a in melanoma cells. Deleting TMEM87a caused melanoma cells to move more slowly and dissociate more easily from tumor-like clusters of cells [21]. More studies are needed to further understand the TMEM87a activation mechanism and its interaction with surrounding matrix.

4.1.2   Piezo Channels—Physiology and Activation Insights

4.1.2.1   Piezo Channels Are Bona fide Mechanically Activated Ion Channels

To date, only a few classes of ion channels satisfy the criteria for bona fide MA ion channels—namely Msc, DEG/ENaC, Trpn, and Piezo channels [1,3]. The well-characterized Msc originating from bacteria—the MscM, MscS, and MscL channels (mechanosensitive channels of mini, small, and large conductance) regulating osmotic pressure in cells by releasing intracellular fluid when they are stretched—were identified decades ago [3]. A variety of large-scale genetic screen studies have successfully demonstrated that the Trpn1 channel—also called TRP-4 in *Caenorhabditis elegans* (*C. elegans*) and NompC in *Drosophila*—is an MA ion channel in invertebrate species. *C. elegans* TRPN1 is involved in the response that worms reduce their locomotion speed upon mechanical contact with a food source. The homolog of Trpn1, NompC in *Drosophila*, is involved in the gentle touch sensation in *Drosophila* larvae [1]. However, the bacteria and invertebrate MA ion channels either are not conserved in vertebrates or lost their mechanotransduction properties during vertebrate evolution. Although MA cation currents present in vertebrate sensory neurons are relatively well described, the identities for vertebrate MA ion channels remain a long-standing mystery.

The discovery of Piezo channels as the first mammalian MA cation channels in 2010 has enormous impacts on the field, boosting the understanding of mechanotransduction in a wide range of research areas [22]. Coste et al. had applied a reductionist approach to uncover Piezo channels. To identify mammalian MA ion channels, they sought a mammalian cell line that expresses an MA current like those recorded from primary sensory neurons. By applying force to the cell surface via a piezo-electrically driven glass probe while patch-clamp recording in the whole-cell configuration with another pipette, they screened several mouse and rat cell lines, and identified that the Neuro2A (N2A) mouse neuroblastoma cell line expressed the most consistent MA currents. Coste et al. used RNA interference to decrease the expression of candidate genes systematically and identified the *Fam38a* gene, renaming it as *Piezo1* after the Greek "pressure." They found that the *Piezo1* gene is required for mechanically stimulated cation conductance in N2A cells. Further sequence homology analysis led to the identification of the *Piezo2*, a *Piezo1* homolog gene with 47% identity. Piezo2 is required for rapid MA current in naïve dorsal root ganglion neurons. Interestingly, Piezo1 and Piezo2 show no similarity to any other protein and contain no known protein domains. Genomic analysis predicts single Piezo orthologs in a range of species from protozoa, *Caenorhabditis elegans*, plants to vertebrates. Interestingly, no homologs have been identified in bacteria or yeast, suggesting the evolution of Piezo channels is independent of bacteria MA channels in higher organisms [2,22]. Subsequent studies have purified the mouse Piezo1 proteins and showed that they formed oligomers and mediated spontaneous single-channel activities when reconstituted into lipid layers, providing strong evidence that Piezo1 is the pore-forming subunit of a bona fide class of MA cation channel [23].

Evidence from different independent investigators demonstrated that Piezo channels are specialized force transducers activated by any physiological membrane tension generated by diverse mechanical forces, such as shear stress (blood flow), stiffness (bone) and stretch (lung) [3,24]. The most common in vitro mechanical stimulations are "stretch" and "poke" in combination with patch-clamp electrophysiology. In the "stretch" model, the membrane is stimulated using a high-speed pressure clamp, which results in highly reproducible pressure–response relationships. In the "poke" model, the membrane is indented with a blunt glass probe, leading to larger current amplitudes [24]. Piezo1, but not Piezo2, can also be activated by chemical activators, including Yoda1 [25], Jedi1/Jedi2 [26], and intrinsically by single strand RNAs [27]. Piezo channels can be inhibited by nonspecific blockers, including ruthenium red, gadolinium ion, and the spider toxin GsMTx4 [22,28,29].

Interestingly, Piezo1 and Piezo2 have distinct channel properties [3,22]. First, while both of them show voltage-dependent inactivation, the time constant for inactivation kinetics (τ_{inac}) of Piezo2 (<10 ms) in heterologous systems is relatively faster than that of Piezo1 (>15 ms) [22]. Secondly, Piezo1 can be effectively activated by stretch with a P_{50} of ~ −30 mmHg, while Piezo2 responds poorly to stretch stimulation [24]. Third, physiological forces, like blood flow-induced shear stress, can initiate a Ca^{2+} response by Piezo1 [30], while similar responses by Piezo2 have not yet been reported. Fourth, Piezo1 can be activated by chemical activators, while Piezo2 has not been reported to respond to chemical activation [28]. These lines of evidence strongly suggest that Piezo1 might serve as a polymodal sensor of diverse forms of mechanical forces, whereas Piezo2 could be more narrowly tuned to detect specialized mechanical forces.

4.1.2.2 Structures and Activation Mechanisms of Piezo Channels

Piezo1 and Piezo2 form novel Ca^{2+}-permeable nonselective cationic channels that do not resemble any known cation channels [22]. They are large membrane proteins

(2521 and 2752 amino acids for human PIEZO1 and human PIEZO2, respectively) and form homotrimers with a molecular weight near 1 MDa [3]. The determination of mouse Piezo1 structures by several groups has revealed that the trimeric channel complex consists of 114 transmembrane helices (TMs) in total (38 TMs per each subunit, in which the first 12 TMs in the N-terminal region were not structurally resolved), marking the Piezo channel membrane protein complex with the highest number of TMs among all known membrane proteins. The channel assembles as a three-bladed, propeller shape with a central ion-conducting pore topped with an extracellular cap-like domain and surrounded by three peripheral blades, each of which contains nine repetitive 4-TM-constituted transmembrane helical units (THUs) or Piezo repeats and a 9 nm-long intracellular beam domain. Remarkably, the three TM blades are highly curved outwardly to form an inverted dome or nano-bowl configuration, which might locally deform the membrane into a nano-bowl shape [31–34].

The full-length mouse Piezo2 structure has been determined as well, in which the 38 TMs in each subunit were completely resolved (Figure 4.1a-c) [35]. Despite distinct biophysical properties, Piezo2 adopts an overall similar three-bladed propeller-like structure as that of Piezo1 (Figure 4.1a) [35]. Based on the complete structure of Piezo2, the three highly curved blades shape a nano-bowl area of a mid-plane opening diameter of 24 nm diameter and a depth of 10 nm, producing a mid-plane dome surface area of 700 nm^2 and a projected in-plane area of 450 nm^2 (Figure 4.1a, left panel) [35]. The transmembrane pore of Piezo2 is fully closed, while that of Piezo1 is dilated, revealing distinct conformational states and the upper and lower transmembrane gates (TM gates) located in the pore-lining TM38, also termed inner helix (IH) (Figure 4.1d) [35].

Based on the unique structure of Piezo1/2, it has been proposed that tension-induced expansion of the curved Piezo-membrane area might provide the gating energy for tension-induced channel opening [31,32]. Furthermore, Piezo channels might deform the membrane area outside of their perimeter into a large curved membrane footprint, which might further amplify the mechanosensitivity of the Piezo channels [31]. In line with the unusually curved Piezo channel-membrane system, it has been shown that reconstituted mouse Piezo1 protein in droplet lipid bilayers (containing no other cellular components) opens under asymmetric bilayer-induced membrane curvature, strongly suggesting that Piezo1 is intrinsically mechanosensitive [36]. Thus, on the basis of the signature bowl-shaped feature of the Piezo-membrane system and the electrophysiological characterizations, Piezo channels might adopt a force-from-lipids gating mechanism to enable a versatile response to changes in local curvature and membrane tension.

Structural analysis of Piezo1, Piezo2, and a Piezo1 splicing variant (Piezo1.1) has led to the identification of the transmembrane gates located in the pore-lining inner helix (IH, TM38) [35] and the cytosolic lateral plug gates that physically block the three lateral ion-conducting portals following the transmembrane pore (Figure 4.2a) [37]. Intrinsically, the lateral plug gate has been identified to undergo alternative splicing in both Piezo1 and Piezo2 [37,38], accounting for the sensitized mechanosensitivity, enlarged single-channel conductance, and altered ion selectivity of the resulting splicing variant Piezo1.1 (Figure 4.2c) [37]. It has been proposed that Piezo channels might utilize a dual-gating mechanism, in which the transmembrane gate is dominantly controlled by the top extracellular cap [35], while the lateral plug gates are controlled by the peripheral blade-beam apparatus

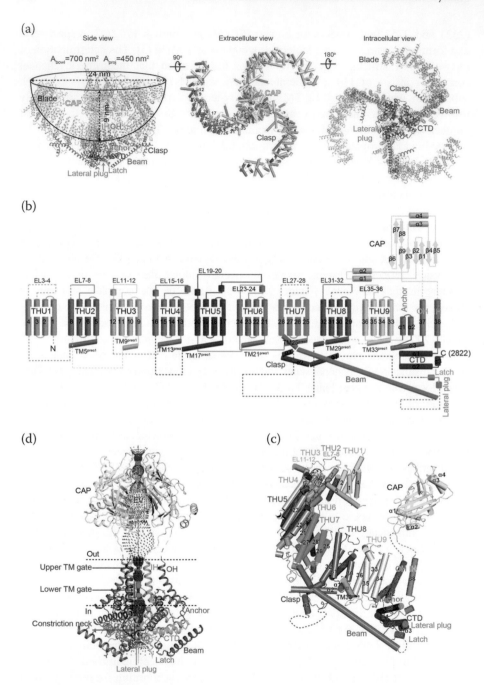

FIGURE 4.1 **The homo-trimeric structure of mouse Piezo2 and its 38-TM topological organization** [35]. **(a)** The indicated view of cartoon models of Piezo2 with the major structural domains colored and labeled. In the left panel, the mid-plane opening diameter, depth, surface area (A_{dome}) and projection area (A_{proj}) of the illustrated nano-bowl configuration shaped by the highly curved transmembrane regions of the three blades are labeled.

via an elegant plug-and-latch mechanism (Figure 4.2a,b) [26,33,35,37,39]. The cap domain is embedded in the center of the nano-bowl shaped by the curved Piezo1-membrane system and sits right on the top of the transmembrane pore of Piezo1. Structural analysis has revealed that motion of the cap domain is strictly coupled to the transmembrane gate residing in the pore-lining IH [35]. Furthermore, either deleting [35] or crosslinking the cap [40] completely abolished mechanical activation of Piezo1. Remarkably, it has been shown that the cap domain of Piezo1 interacts with the extracellular ectodomain of the adhesion molecule E-cadherin, which is linked to the F-actin cytoskeleton via the β-catenin-vinculin mechanotransduction complex [41]. Thus, a physical interaction of E-cadherin with Piezo1 might allow a direct focus of cytoskeleton-transmitted force on the extracellular cap domain, providing a structural basis for a tether model or force-from-filament model for mechanogating of Piezo1 [41].

The peripheral blade-beam apparatus might undergo force-induced conformational change to gate the intracellular lateral plug gates (Figure 4.2). In line with this, it has been elegantly demonstrated that the highly curved blade can undergo reversible flattening at biologically relevant pressures [32]. Furthermore, a serial of key mechanotransduction sites along the blade-beam-lateral plug gate [26,33,37,42] have been identified and proposed to form an intramolecular lever-like mechanotransduction pathway for converting long-range conformational changes from the distal blades to the cytosolic lateral plug gates (Figure 4.2) [26,28,33,37,42]. Interestingly, Piezo1 activators, including Yoda1 and Jedi1/2, might utilize the transduction pathway to allosterically activate Piezo1 [26]. Collectively, these studies suggest that the blade-beam apparatus might serve as the molecular basis for a force-from-lipids model to gate the cytosolic lateral plug gates via an elegant plug-and-latch gating mechanism (Figure 4.2a–e). Notably, the lateral plug gates might also be subjected to the E-cadherin-β-catenin-vinculin-F-actin-mediated tether-gating model as the intracellular domains in close proximity to the lateral plug gates interact with the cytosolic domain of E-cadherin [41].

Given the existence of the structurally and functionally distinct molecular bases for the force-from-filament and force-from-lipids models, Piezo channels might incorporate these two distinct yet non-exclusive gating models to serve as a versatile mechanotransducer [41]. Using the dual mechanogating models, Piezo channels are not only able to constantly monitor changes in local membrane curvature and tension, but also overcome the obstacle of limited membrane tension

---

FIGURE 4.1 (Continued). **(b)** The 38-TM topological model of a Piezo2 protomer. Dashed lines indicate unresolved regions. **(c)** The 38-TM topological model of a Piezo2 protomer. **(d)** Ribbon diagram showing the OH-Cap-IH-CTD-constituted pore module together with the Anchor, Lateral plug, Latch and Beam domains. The central solvent-accessible pathway is marked with a dotted mesh generated by the program HOLE (pore radius: red < 1 Å < green < 2 Å < purple). The extracellular vestibule (EV), membrane vestibule (MV), and intracellular vestibule (IV) are labeled. The constricted sites formed by IHs serve as upper TM gate and lower TM gate. The cytosolic constriction neck might not serve as major ion-conducting pathway. (Adapted from Wang et al., Nature, 2019).

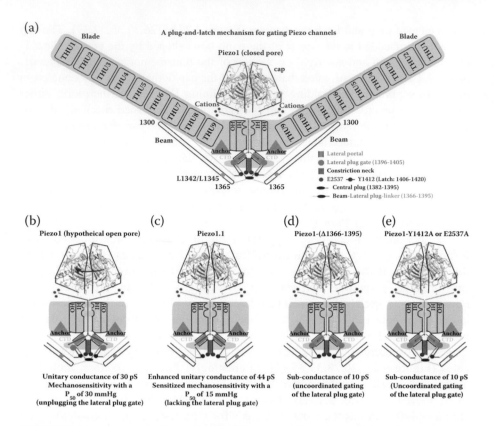

FIGURE 4.2 A proposed plug-and-latch mechanism for gating the lateral ion-conducting portals [37]. **(a)** A model depicting the key structural elements of Piezo1, which are conserved in Piezo2. For simplicity, only two of the three subunits are illustrated. The extracellular cap is shown in yellow structure and the membrane region is shown in gray. The lateral plug gates of the lateral ion-conducting portals are latched onto the central axis through the central plug and the Y1412-E2537 interactions. The gray dot on the beam represents L1342 and L1342 in Piezo1, which have been identified to be critical for mechanical activation of Piezo1 [26,33]. Red dashed arrows indicate the closed ion-permeating pathways. THU: transmembrane helical unit; TM gate (in yellow): transmembrane gate; CTD: C-terminal domain. **(b)** A hypothetical model of Piezo1 showing opening of the transmembrane gate by motion of the top Cap and partial unplugging of the lateral plug gates via flattening the curved blade-membrane system. **(c–e)** Illustration of the altered gating modes of the lateral portals of the indicated Piezo1.1 isoform **(c)** or Piezo1 mutants **(d, e)** upon either removal or uncoordinated gating of the lateral plug gates. Solid and dashed red arrows indicate open and closed ion-conducting pathway, respectively (adapted from Geng et al., Neuron, 2020).

propagation within an intact cellular membrane to effectively detect long-range mechanical perturbation across a cell or between cells. Indeed, Piezo1 can mediate both localized and whole-cell mechanical responses regardless of whether the mechanical stimuli are either exogenously applied or endogenously originated [22,43–47].

4.1.2.3 Physiology of Piezo Channels

The two Piezo channels are complementarily expressed in a wide range of mechanically sensitive cells. Piezo1 is widely expressed in non-sensory tissues exposed to stretch and stress (abundant in blood vessels, kidney, lung, bladder, etc.), whereas Piezo2 is predominantly found in sensory tissues such as dorsal root ganglia (DRG) sensory neurons and Merkel cells in skin exposed to light touch [3,22,24]. Interestingly, a few cell types express both Piezo1 and Piezo2 (articular cartilage chondrocytes [48] and baroreceptor sensory neurons [49]. A synergy model in which two distinct mechanosensors cooperate to receive a broader range of mechanical forces may exist [50]. For example, a recent exciting finding identified Piezo1 and Piezo2 together as the long-sought-after baroreceptor mechanosensors for acute sensing of blood pressure changes and initiating baroreflex, a homeostatic mechanism that helps to keep blood pressure stable [49].

Piezo1 has been shown to sense cell volume in red blood cells and shear stress in vascular endothelial cells, and mediate stretch-activated currents in flow-sensitive cells, including renal epithelial and bladder urothelial cells [2,3,24,24,24,30]. Piezo1 also detects forces from across cells (e.g. niche or substrate stiffness) in neuron oligodendrocyte progenitor cells, and other cell lines, thereby promoting downstream changes in cell cycle, lineage choice, and migration [51–53]. A large amount of evidence has shown that Piezo2 in DRG mediates much of the organism's response to light, but not noxious mechanical stimulations—specifically in physiological processes like touch and proprioception—and contributes to mechanical allodynia in the pathological conditions [2,3,24,24,24,54]. Interestingly, Piezo2 has also been shown in internal organ sensory neurons (nodose-jugular ganglia) in sensing airway stretch of lung tissue [55]. Compelling evidence by Knight et al. showed that intestinal mechanosensory afferents control food intake, suggesting a potential role of Piezo2 in sensing volume changes of the gastrointestinal tract [56].

Over 50 human mutations in *PIEZO1/2* have been linked to multiple disorders. Although most PIEZO mutations have not been characterized by mechanoelectrical study, few of these are identified as gain- or loss-of-function mutations by their channel activity and inactivation rate. For example, *PIEZO1* gain-of-function mutations that show slow inactivation rate are associated with dehydrated hereditary xerocytosis [24]. Surprisingly, one common *PIEZO1* gain-of-function allele (E756del) exists in one third of the African population. RBCs from E756del carriers are dehydrated and show reduced susceptibility to malaria [57]. This is an intriguing example of how deeply the mechanotransduction mediated by PIEZO1 affects human health and natural selection in human evolution. Several *PIEZO1* loss-of-function mutations occur in patients with congenital lymphatic dysplasia [24]. Given a strict expression pattern of PIEZO2 in sensory neurons, *PIEZO2* human mutations lead to a unique syndrome during development and adulthood. *PIEZO2* gain-of-function mutations are associated with several arthrogryposis disorders, mostly distal arthrogryposis subtype 5, which is characterized by multiple congenital contractures of limbs without primary neurologic and/or muscle disease [58]. In line with touch and proprioception phenotypes revealed by the mouse model, *PIEZO2* loss-of-function carriers have a selective loss of discriminative touch perception but responded to other gentle mechanical stimulation on hairy skin [59]. The patients also had profoundly decreased proprioception leading to ataxia and dysmetria that were markedly worse without visual cues [59].

4.2   CONCLUSION AND FUTURE DIRECTIONS

Over the past ten years, our understanding of mechanotransduction has been greatly expanded due to the groundbreaking discoveries of several bona fide mechanically activated ion channels. Moreover, the physiological and structural/functional studies of these channels are growing rapidly. Yet, there remain some outstanding questions in the field, including a few "holy grails." First, the molecular identity of the hair cell mechanotransduction channel in the tip link that initiates our sense of hearing is still elusive. Although Piezo2 is found in the hair cells, it is at the base of the stereocilia and its opening is caused by negative pressure near the apical surface and reverse polarity, independently of the tip link [60]. Accumulating evidence indicates the tip link channel as multiple transmembrane protein complexes other than a single MA channel, given that several deafness-related candidates such as TMC1/2, TMIE, and TMHS alone failed to reproduce MA currents in the heterologous system [61,62]. Another long-standing mystery is the molecular identity of the noxious mechanoreceptor detecting the high-threshold forces in sensory neurons, where Piezo2's role is limited [63]. Much effort has been focused on screening novel MA channels. As a promising future direction, the high-throughput approach offers many advantages relative to conventional hypothesis-driven screening. First, high-throughput approaches enable a full spectrum genomic, non-biased screen for the candidates that are not limited to ion channels. In contrast, conventional screening generates a candidate list based on preferred expression profile analysis or bioinformatics prediction [22], which may miss potential hits. Second, a high-throughput assay such as calcium imaging is time effective, while the conventional electrophysiology recordings require the overexpression or knock-down of candidate genes one by one [22]. As a proof-of-concept, Xu et al. creatively designed a 384-well screening system that applies physiological shear stress on cultured cells, and identified a mechanosensitive GPCR, GPR68, by a focused RNAi library [64]. Future efforts on high-throughput approaches will pave the way to the next breakthrough discoveries. Last but not least, in light of the observation that different classes of MA channels appear to have distinct structural organizations, how diverse groups of MA channels utilize their distinct structural bases to fulfill their designated cellular mechanotransduction function represents a fascinating question to be addressed.

REFERENCES

1. Delmas, P. and B. Coste, Mechano-gated ion channels in sensory systems. *Cell*, 2013. **155**(2): p. 278–284.
2. Parpaite, T. and B. Coste, Piezo channels. *Curr Biol*, 2017. **27**(7): p. R250–R252.

3. Murthy, S.E., A.E. Dubin, and A. Patapoutian, Piezos thrive under pressure: mechanically activated ion channels in health and disease. *Nat Rev Mol Cell Biol*, 2017. **18**(12): p. 771–783.
4. Cheng, Y.R., B.Y. Jiang, and C.C. Chen, Acid-sensing ion channels: dual function proteins for chemo-sensing and mechano-sensing. *J Biomed Sci*, 2018. **25**(1): p. 46.
5. Wemmie, J.A., R.J. Taugher, and C.J. Kreple, Acid-sensing ion channels in pain and disease. *Nat Rev Neurosci*, 2013. **14**(7): p. 461–471.
6. Barth, D. and M. Fronius, Shear force modulates the activity of acid-sensing ion channels at low pH or in the presence of non-proton ligands. *Sci Rep*, 2019. **9**(1): p. 6781.
7. Christensen, A.P. and D.P. Corey, TRP channels in mechanosensation: direct or indirect activation? *Nat Rev Neurosci*, 2007. **8**(7): p. 510–521.
8. Lau, O.C., et al., TRPC5 channels participate in pressure-sensing in aortic baroreceptors. *Nat Commun*, 2016. **7**: p. 11947.
9. Nikolaev, Y.A., et al., Mammalian TRP ion channels are insensitive to membrane stretch. *J Cell Sci*, 2019. **132**(23), jcs238360.
10. Servin-Vences, M.R., et al., Direct measurement of TRPV4 and PIEZO1 activity reveals multiple mechanotransduction pathways in chondrocytes. *Elife*, 2017. **6**, e21074.
11. Hong, G.S., et al., Tentonin 3/TMEM150c confers distinct mechanosensitive currents in dorsal-root ganglion neurons with proprioceptive function. *Neuron*, 2016. **91**(3): p. 708–710.
12. Lu, H.J., et al., Tentonin 3/TMEM150C senses blood pressure changes in the aortic arch. *J Clin Invest*, 2020. **130**(7): p. 3671–3683.
13. Anderson, E.O., et al., TMEM150C/Tentonin3 is a regulator of mechano-gated ion channels. *Cell Rep*, 2018. **23**(3): p. 701–708.
14. Dubin, A.E., et al., Endogenous Piezo1 can confound mechanically activated channel identification and characterization. *Neuron*, 2017. **94**(2): p. 266–270.e3.
15. Zhang, M., et al., Structure of the mechanosensitive OSCA channels. *Nat Struct Mol Biol*, 2018. **25**(9): p. 850–858.
16. Murthy, S.E., et al., OSCA/TMEM63 are an evolutionarily conserved family of mechanically activated ion channels. *Elife*, 2018. **7**, e41844.
17. Liu, X., J. Wang, and L. Sun, Structure of the hyperosmolality-gated calcium-permeable channel OSCA1.2. *Nat Commun*, 2018. **9**(1): p. 50–60.
18. Jojoa-Cruz, S., et al., Cryo-EM structure of the mechanically activated ion channel OSCA1.2. *Elife*, 2018. **7**, e41845.
19. Beaulieu-Laroche, L., et al., TACAN is an ion channel involved in sensing mechanical pain. *Cell*, 2020. **180**(5): p. 956–967.e17.
20. Nourse, J.L. and M.M. Pathak, How cells channel their stress: interplay between Piezo1 and the cytoskeleton. *Semin Cell Dev Biol*, 2017. **71**: p. 3–12.
21. Patkunarajah, A., et al., TMEM87a/Elkin1, a component of a novel mechanoelectrical transduction pathway, modulates melanoma adhesion and migration. *Elife*, 2020. **9**, e53308.
22. Coste, B., et al., Piezo1 and Piezo2 are essential components of distinct mechanically activated cation channels. *Science*, 2010. **330**(6000): p. 55–60.
23. Coste, B., et al., Piezo proteins are pore-forming subunits of mechanically activated channels. *Nature*, 2012. **483**(7388): p. 176–181.
24. Wu, J., A.H. Lewis, and J. Grandl, Touch, tension, and transduction – the function and regulation of Piezo ion channels. *Trends Biochem Sci*, 2017. **42**(1): p. 57–71.
25. Syeda, R., et al., Chemical activation of the mechanotransduction channel Piezo1. *Elife*, 2015. **4**, e07369.

26. Wang, Y., et al., A lever-like transduction pathway for long-distance chemical- and mechano-gating of the mechanosensitive Piezo1 channel. *Nat Commun*, 2018. **9**(1): p. 1300.

27. Sugisawa, E., et al., RNA sensing by gut Piezo1 is essential for systemic serotonin synthesis. *Cell*, 2020. **182**(3): p. 609–624.e21.

28. Xiao, B., Levering mechanically activated Piezo channels for potential pharmacological intervention. *Annu Rev Pharmacol Toxicol*, 2020, **60**:195–218.

29. Bae, C., F. Sachs, and P.A. Gottlieb, The mechanosensitive ion channel Piezo1 is inhibited by the peptide GsMTx4. *Biochemistry*, 2011. **50**(29): p. 6295–6300.

30. Beech, D.J. and A.C. Kalli, Force sensing by Piezo channels in cardiovascular health and disease. *Arterioscler Thromb Vasc Biol*, 2019. **39**(11): p. 2228–2239.

31. Haselwandter, C.A. and R. MacKinnon, Piezo's membrane footprint and its contribution to mechanosensitivity. *Elife*, 2018. **7**, e41968.

32. Lin, Y.C., et al., Force-induced conformational changes in PIEZO1. *Nature*, 2019. **573**(7773): p. 230–234.

33. Zhao, Q., et al., Structure and mechanogating mechanism of the Piezo1 channel. *Nature*, 2018. **554**(7693): p. 487–492.

34. Saotome, K., et al., Structure of the mechanically activated ion channel Piezo1. *Nature*, 2018. **554**(7693): p. 481–486.

35. Wang, L., et al., Structure and mechanogating of the mammalian tactile channel PIEZO2. *Nature*, 2019. **573**(7773): p. 225–229.

36. Syeda, R., et al., Piezo1 channels are inherently mechanosensitive. *Cell Rep*, 2016. **17**(7): p. 1739–1746.

37. Geng, J., et al., A plug-and-latch mechanism for gating the mechanosensitive Piezo channel. *Neuron*, 2020. **106**(3): p. 438–451.e6.

38. Szczot, M., et al., Cell-type-specific splicing of Piezo2 regulates mechanotransduction. *Cell Rep*, 2017. **21**(10): p. 2760–2771.

39. Zhao, Q., et al., The mechanosensitive Piezo1 channel: a three-bladed propeller-like structure and a lever-like mechanogating mechanism. *FEBS J*, 2019, **286**(13): 2461–2470.

40. Lewis, A.H. and J. Grandl, Inactivation kinetics and mechanical gating of Piezo1 ion channels depend on subdomains within the Cap. *Cell Rep*, 2020. **30**(3): p. 870–880.e2.

41. Wang, J.J.J., X. Yang, L. Wang, & B. Xiao, Tethering Piezo channels to the actin cytoskeleton for mechanogating via the E-cadherin-β-catenin mechanotransduction complex. *bioRxiv*, 2020.

42. Geng, J., et al., A plug-and-latch mechanism for gating the mechanosensitive Piezo channel. *Neuron*, 2020, **106**(3): 438–451.e6.

43. Lewis, A.H. and J. Grandl, Mechanical sensitivity of Piezo1 ion channels can be tuned by cellular membrane tension. *Elife*, 2015. **4**, e12088.

44. Cox, C.D., et al., Removal of the mechanoprotective influence of the cytoskeleton reveals PIEZO1 is gated by bilayer tension. *Nat Commun*, 2016. **7**: p. 10366.

45. Bavi, N., et al., PIEZO1-mediated currents are modulated by substrate mechanics. *ACS Nano*, 2019. **13**(11): p. 13545–13559.

46. Ellefsen, K.L., et al., Myosin-II mediated traction forces evoke localized Piezo1-dependent Ca(2+) flickers. *Commun Biol*, 2019. **2**: p. 298.

47. Shi, Z., et al., Cell membranes resist flow. *Cell*, 2018. **175**(7): p. 1769–1779.e13.

48. Lee, W., et al., Synergy between Piezo1 and Piezo2 channels confers high-strain mechanosensitivity to articular cartilage. *Proc Natl Acad Sci USA*, 2014. **111**(47): p. E5114–E5122.

49. Zeng, W.Z., et al., PIEZOs mediate neuronal sensing of blood pressure and the baroreceptor reflex. *Science*, 2018. **362**(6413): p. 464–467.

50. Gnanasambandam, R., et al., Functional analyses of heteromeric human PIEZO1 channels. *PLoS One*, 2018. **13**(11): p. e0207309.
51. Gudipaty, S.A., et al., Mechanical stretch triggers rapid epithelial cell division through Piezo1. *Nature*, 2017. **543**(7643): p. 118–121.
52. Pathak, M.M., et al., Stretch-activated ion channel Piezo1 directs lineage choice in human neural stem cells. *Proc Natl Acad Sci USA*, 2014. **111**(45): p. 16148–16153.
53. Segel, M., et al., Niche stiffness underlies the ageing of central nervous system progenitor cells. *Nature*, 2019. **573**(7772): p. 130–134.
54. Murthy, S.E., et al., The mechanosensitive ion channel Piezo2 mediates sensitivity to mechanical pain in mice. *Sci Transl Med*, 2018. **10**(462).
55. Nonomura, K., et al., Piezo2 senses airway stretch and mediates lung inflation-induced apnoea. *Nature*, 2017. **541**(7636): p. 176–181.
56. Bai, L., et al., Genetic identification of vagal sensory neurons that control feeding. *Cell*, 2019. **179**(5): p. 1129–1143.e23.
57. Ma, S., et al., Common PIEZO1 allele in African populations causes RBC dehydration and attenuates Plasmodium infection. *Cell*, 2018. **173**(2): p. 443–455.e12.
58. Coste, B., et al., Gain-of-function mutations in the mechanically activated ion channel PIEZO2 cause a subtype of distal arthrogryposis. *Proc Natl Acad Sci USA*, 2013. **110**(12): p. 4667–4672.
59. Chesler, A.T., et al., The role of PIEZO2 in human mechanosensation. *N Engl J Med*, 2016. **375**(14): p. 1355–1364.
60. Wu, Z., et al., Mechanosensory hair cells express two molecularly distinct mechanotransduction channels. *Nat Neurosci*, 2017. **20**(1): p. 24–33.
61. Qiu, X. and U. Muller, Mechanically gated ion channels in mammalian hair cells. *Front Cell Neurosci*, 2018. **12**: p. 100.
62. Wu, Z. and U. Muller, Molecular identity of the mechanotransduction channel in hair cells: not quiet there yet. *J Neurosci*, 2016. **36**(43): p. 10927–10934.
63. Wood, J.N. and N. Eijkelkamp, Noxious mechanosensation – molecules and circuits. *Curr Opin Pharmacol*, 2012. **12**(1): p. 4–8.
64. Xu, J., et al., GPR68 senses flow and is essential for vascular physiology. *Cell*, 2018. **173**(3): p. 762–775.e16.

5 Ion Channels in Human Pluripotent Stem Cells and Their Neural Derivatives

Ritika Raghavan, Robert Juniewicz,
Maharaib Syed, Michael Lin, and Peng Jiang

CONTENTS

5.1 INTRODUCTION

Though mouse and other animal models have provided valuable insight into brain development, functions, and diseases, we cannot neglect the differences between human and other organisms in gene expression patterns and pathophysiological properties. The scarce availability of live and functional human brain tissue has limited our ability to understand mechanisms underlying neurological diseases. The ability of human pluripotent stem cells (PSCs), comprised of human induced pluripotent stem cells (hiPSCs) and human embryonic stem cells (hESCs), to self-renew and differentiate into any types of cells in our body provides a novel and powerful tool to understand human brain development and pathology. Human ESCs are pluripotent cells derived from the inner cell mass of blastocyst embryos. Human iPSCs are generated by the reprogramming of patients' somatic cells via the expression of defined transcriptional factors to an embryonic stem cell-like state (1–3). Both hiPSCs and hESCs are able to differentiate into neural lineage cells, including neural progenitor cells (NPCs), different types of neurons, macroglial cells (i.e. oligodendroglia and astroglia), as well as microglia, the brain-resident innate immune cells (Figure 5.1). The generation of neural lineage cells from hPSCs can be achieved through a differentiation approach that is built upon fundamental developmental principles learned from animal models, or through an accelerated induction approach using transcriptional factors. In addition, human neural lineage cells can also be generated by somatic cell reprogramming, in which somatic cells, such as skin fibroblasts, are directly converted to neural lineage cells without reverting back to pluripotent state (4). With human stem cell technology, previous studies have established in vitro 2-dimensional (2D) human neural cell culture models and 3-dimensional (3D) brain organoid models, as well as in vivo transplantation and human-mouse chimeric brain models for understanding biological properties of human neural cells and revealing mechanisms of a variety of neurological disorders (5,6). Ion channels have been also examined in human neural lineage cells in both in vitro and in vivo hPSC-based modeling systems. Both classical and nonclassical ion channels have been observed across all different types of models. Classical ion channels such as voltage gated sodium (Na_V), potassium (K_V), and calcium channels (Ca_V) have been identified in various types of functional neurons along with measurements of various electrophysiological parameters. Studies on nonclassical ion channels such as Piezo (a member of the stretch activated channel (SAC) family), transient receptor potential channel (TRP), hyperpolarization-activated cyclic nucleotide–gated channel (HCN), and acid sensing ion channels (ASICs) have also been conducted. Furthermore, many of the models have been used to study ion channelopathies that have a significant role in many neuropathologies and diseases (Figure 5.1).

5.1.1 DERIVATION OF NEURONS AND GLIA FROM hPSCs

The generation of functionally mature neurons and glia from hPSCs is essential to understanding their neurophysiology and neuropathology. Overall, there are three approaches to generating human neurons and glia, including cell differentiation approach based on canonical neural developmental principle, accelerated differentiation approach using neural lineage-restricted transcription factors, and direct reprogramming of somatic cells (such as fibroblasts). In the first cell differentiation

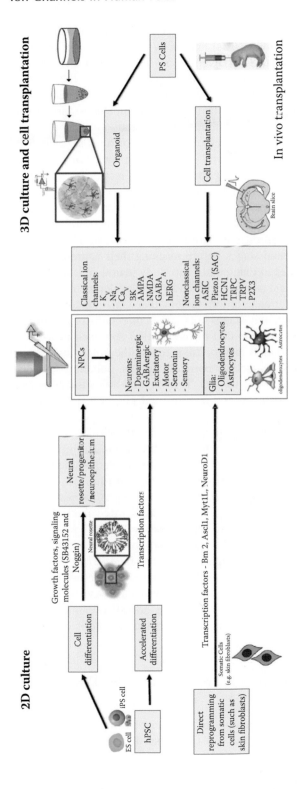

FIGURE 5.1 Various methods to obtain human neural lineage cells, including the cell differentiation method involving the formation of neuroepithelial cells, accelerated differentiation method using transcription factors to generate neural cells from hPSCs, and direct reprogramming method using transcription factors to convert somatic cells directly to mature neurons and glia. Previous studies have examined ion channel properties in 2D cultures, 3D cerebral organoids, and in vivo engrafted human neural cells and have discovered a variety of classical and nonclassical ion channels to be present in these cells.

approach, hPSCs are usually deprived of their pluripotency by forming multicellular aggregates known as embryoid bodies (EBs) and/or by using inhibitors of TGFβ and BMP signaling pathways (dual SMAD inhibition) (7,8). These cells are re-plated to allow differentiation of primitive neuroepithelial cells (NE; neural rosettes—radial arrangements of elongated cells that resemble early neural tube morphology), which are neural progenitor cells (NPCs) and are able to further give rise to mature neurons and macroglia (9). Although the formation of neural lineage cells has long been considered one of the default pathways of hPSCs, the process can be facilitated at both the EB and neural rosette stage by adding different combinations of growth factors and signaling molecules to direct differentiation to various types of neurons or macroglia (10). In addition, region-specific patterning of the CNS is controlled by the concentration and gradient of various morphogens which can also be applied to NE cells to derive region-specific neural subtypes, such as interneurons (11). This cell differentiation approach allows hPSCs to un-dergo various neural developmental stages before reaching their neuronal or glial fates, which provide opportunities to manipulate and study neural cells at different developmental stages to understand neurogenesis and neurodevelopmental diseases.

Studies have also identified transcription factors that can induce hPSCs to bypass the sequential developmental stages and drive direct differentiation to mature neural cell types (4,12). Accelerated differentiation approach utilizes the forced expression of transcriptional factors to mediate the differentiation of hPSCs to various induced neurons, such as cortical neurons, dopaminergic neu-rons, motor neurons, and interneurons (13–17). This method holds the advantage of differentiating human PSCs to functional neurons and glia in a fast, efficient, and highly targeted manner (12).

With advances in cellular reprogramming research, the direct reprogram-ming approach aims to use transcription factors, small molecules, microRNAs, and epigenetic modifiers or a combination of them to convert somatic cells, such as skin fibroblasts, directly to functional neurons and glia (18–20). The original study that first reported the direct reprogramming of terminally dif-ferentiated mouse fibroblasts into induced neurons used three neuronal tran-scription factors: Brn2, Ascl1, and Myt1L, known as the "BAM" factors. The fibroblasts were observed to drastically change in morphology into and become neuronal cells without cell division through a transient stage (21). Along with the addition of NeuroD1, these BAM factors were also able to convert human fibroblasts to functional neurons (22). This method retains the advantages of the accelerated differentiation approach along with the added advantage of not having to use PSCs as the starting cells in culture as somatic cells are readily available (23).

5.1.2 Models to Study Ion Channel Properties of hPSC Derived Neurons and Glia

The ability to acquire, maintain, and manipulate stem cells has led to the formation of various models to study them and the cell types that differentiate from them.

Both in vitro and in vivo models have been successfully used in studying ion channel properties in PSCs and PSC neural derivatives. The classical 2D cell culture system, and the recently developed 3D organoid/spheroid culture system are the two most widely used in vitro models. In addition, the transplantation model in which hPSC neural derivatives are engrafted into neonatal or adult mouse brains provides an in vivo system to examine neurophysiology of human neural cells in the intact CNS.

By utilizing a flat, monolayer surface with substrates that support an extracellular matrix (ECM) environment and cell behavior, the 2D culture model has allowed researchers to differentiate and grow cells in an efficient and cost-effective way. Basic aspects of brain development and disease, such as cell differentiation, migration, and neuron-glia interactions, can be examined using hPSC-based in vitro 2D cell culture models. However, proper cell growth and differentiation is dependent on the cell-cell and cell-ECM interactions which surround the cell and act on it in a three-dimensional manner. The need for tissue and organ level model structures that can recapitulate an in vivo environment to recapitulate brain development and disease mechanisms urges the development of 3D organoid models (24). 3D cerebral organoids are derived from human PSCs, based upon the remarkable self-organizing abilities of NE cells and their ability to differentiate into functional neural cell types in vitro. Organoids have a great potential in modeling the complex organization, composition, and connectivity of the human brain through their ability to recapitulate tissue and organ level function (25). The hPSC-derived cerebral organoids are usually embedded in ECM gel systems, such as Matrigel or hydrogel, that support neural progenitor cell sheets and neuroepithelial bud proliferation. The buds then progress into various brain regions with progenitor and neuronal/glial organization. This results in cerebral organoids that possess complex 3D cytoarchitecture, cellular interactions, and cell morphology closely resembling the in vivo human brain environment and composition (26,27). Engraftment of hPSC-derived neural derivatives into mouse brains has allowed researchers to study human neurodevelopment and disease pathology using human neural cells in vivo. Human PSCs can now be transplanted into the adult mouse brain to create a transplantation model with local distribution of human neural cells or can be transplanted into neonatal mouse brains to allow a wide disperse of hPSC-derived neural cells across multiple brain regions in the adult mouse brain, thus creating a human-mouse chimeric brain model (28–32). The cell transplantation model has proven to be a promising approach to study the in vivo behavior and function of patient-derived neural cells.

5.2   ION CHANNELS IN hPSCs

Human iPSCs and ESCs are morphologically identical and largely share electrophysiological properties at the pluripotent state. They appear to both be electrophysiologically homogenous, with both cell types exhibiting depolarization-activated delayed rectifier K^+ currents (I_{KDR}) and 4-aminopyradine (4-AP) sensitive currents but not voltage-gated sodium channel currents (I_{Na}), voltage-gated calcium currents (I_{Ca})and Ca^{2+} activated K^+ currents (I_{KCa}) (33). At transcriptomic level, the expression of transcripts encoding ion channels could not be distinguished in hESCs and hiPSCs. KCNQ2 transcript has the highest expression in both hiPSCs and hESCs, which encodes the non-inactivating, slowly deactivating M-current (33,34). Also expressed in both hiPSC and hESCs are CNC4, which encodes delayed rectifier voltage-gated potassium channels ($K_v3.4$), and KCNS3 which encodes the silent modulatory α-subunit of I_{KDR} channels (33). The detection of I_{KDR} was at a high current density (47.5 ± 7.9 pA/pF at +40 mV) for both hESCs and hiPSCs and these I_{KDR} are believed to have a role in proliferation (33,34). Tetraethylammonium (TEA), a blocker of I_{KDR}, dose-dependently inhibited hESC proliferation with an EC_{50} of 11.6 ± 2.0mM and hiPSC proliferation with an EC_{50} of 7.8 ± 1.2 mM (33,34).

Interestingly, human ether-a-go-go (hERG) channels (voltage-dependent K^+ channels) were only found to be in hESCs, but not in hiPSCs (33,34). In addition, hyperpolarization-activated pacemaker currents (I_f) were detected in hESCs, but were absent in hiPSCs. hERG channel-mediated K^+ currents and I_f are known to be important for cardiomyocyte development and functions (35,36). The differences in the expression of these ion channels in hESCs and hiPSCs may affect their differentiation to cardiac cell lineages (37,38).

5.3 ION CHANNELS IN HPSC NEURAL DERIVATIVESIN VITRO AND IN VIVO

5.3.1 HUMAN PSC-DERIVED MULTIPOTENT NEURAL PROGENITOR CELLS (NPCs)

Multipotent NPCs have three possible fates of differentiation: induction into a neuron, astrocyte, or oligodendrocyte. The ion channels, Piezo1, and TRP channels have been detected in hPSC-derived NPCs, and the former has also been shown to greatly influence the fate of differentiation of these NPCs. Piezo1 is a nonclassical ion channel belonging to the stretch-activated channel (SAC) family, which act as nonspecific cationic channels that can be activated by mechanical force (39,40). It was observed that the Piezo1 ion receptor was able to dictate whether neural progenitor cells differentiated into neurons or astrocytes depending on the stiffness of the experienced mechanical force. The fate of these Piezo1-expressing NPCs was determined by the stiffness of their local environment, such as the force exerted by the ECM or adjacent cells (41). The stiffness regulated the elicited inward calcium transient which directed the fate of NPCs. Increasing the stiffness led to an increase in the calcium transient which directed the NPCs toward neurogenesis, while a decrease in stiffness caused a lower calcium influx resulting in an astrocytic fate (41). The canonical transient receptor potential (TRPC) channel was also found to be present in NPCs and its function was implicated in transient calcium influx which controls cell proliferation in NE cells. Transient calcium influx was reduced significantly in hPSC-derived NE cells upon application of TRPC antagonists (42). These findings have implications in developing NPC-based stem cell regenerative medicine, in which NPCs can be transplanted to promote neural repair.

Manipulation of Piezo1 channel in NPCs may control their fate after implantation to achieve optimal regenerative effects under disease conditions, while the manipulation of TRPC channels can be used to study their functions in neurodevelopment.

5.3.2 HUMAN PSC-DERIVED EXCITATORY CORTICAL NEURONS

Cortical excitatory neurons can be generated through differentiation from hPSCs following fundamental developmental principles (10,11), accelerated differentiation, and the direct reprogramming method (22,23,43). Through patch-clamp recordings, ion channel properties have been examined in human excitatory neurons. These ion channels include the K_V and Na_V channels, α-amino-3-hydroxy-5-methyl-4-isoxazolepropionic acid (AMPA) receptors, and TRPC channels. The human excitatory neurons exhibit gradual maturation along with the culture time as indicated by the hyperpolarized resting membrane potential (-70 mV) while also increasing the amplitude and decreasing the duration of action potentials (43). Neurons differentiated from hPSCs also developed synapses and exhibited synaptic activities (44). The development of functional synapses can also be accelerated by co-culture with astrocytes (45). Activation of TRPC produces calcium transients, which play a key role in governing many neurodevelopmental processes such as proliferation, migration, and neurite growth (46–48). The TRPC1 and TRPC4 ion channels were shown to be expressed in hPSC-derived excitatory neurons through q-PCR. Using calcium imaging, it was also observed that calcium transients decreased significantly upon the application of TRPC antagonist: SKF96365. Furthermore, TRPC channel inhibition resulted in reduction of neurite outgrowth and neurite extension (42).

5.3.3 HUMAN PSC-DERIVED GABAERGIC INTERNEURONS

GABAergic interneurons originate from the ganglionic eminences of the early embryonic brain. Thus, this specific type of neuron can be derived from hPSCs by directing their differentiation to medial ganglionic eminence (MGE) neural progenitors. Using Sonic Hedgehog (SHH) molecule and agonist of SHH signaling, such as purmorphamine, hPSCs can be induced to become MGE progenitors which further give rise to functional GABAergic interneurons (38,49,50). Whole-cell patch-clamp recordings showed that these neurons robustly fired action potentials, indicating the presence of K_V and Na_V channels. Moreover, another study observed spontaneous synaptic activities which were almost completely eliminated by $GABA_A$ antagonist, bicuculline, indicating primarily GABAergic inhibitory neurotransmission inputs in these cultures (38). It is worth mentioning that functional maturation of these hPSC-derived GABAergic interneurons requires an extended timeline in culture. At week 8 post-differentiation, these neurons were comparatively immature and exhibited broad action potential (AP) half width, small afterhyperpolarization (AHP), and the inability to fire repetitively upon current injection. By week 30 post-differentiation, these cells exhibited mature properties as shown by AP firing properties with faster and consistent AP velocity near threshold firing, smaller AP half-width, and larger AHPs as well as high-frequency repetitive AP firing upon current injection. Spontaneous postsynaptic currents were detected as early as week 8 post-differentiation (49).

GABAergic interneurons can also be generated using the direct reprogramming method with high efficiency. By expressing the transcription factors Foxg1, Sox2, Ascl1, Dlx5, and Lhx6 in in human fibroblasts, researchers were able to produce functional GABAergic interneurons in vitro (51). The electrophysiological activity of these neurons was measured using patch clamp analysis. As early as week four post-differentiation, these neurons acquired the ability to fire APs repetitively at high frequencies, typical of the interneurons. GABAergic inhibitory synaptic neurotransmission of the fibroblast-derived GABAergic interneurons was also detected (51). Other studies have shown that using transcription factors, such as Ascl1, Dlx2, NKX2.1, and Lhx6, hPSCs can be rapidly induced to differentiate to GABAergic interneurons. GABAergic interneurons generated using this method produce mature APs and exhibit synapse activities within 5 to 8 weeks (13,15).

5.3.4   Human PSC-Derived Dopaminergic Neurons

Human PSC- and fibroblast-derived dopaminergic (DA) neurons have been successfully generated through accelerated differentiation approach and direct reprogramming approach respectively (52–54). Patch clamp experiments showed that human DA neurons exhibit normal resting membrane potential (RMP), I_{Na}, firing of APs, prominent I_K and afterspike hyperpolarization (ASH) (54). Upon electrical stimulation, depolarization of these cells caused the release of dopamine. Furthermore, high-performance liquid chromatography measurements revealed that these human DA neurons also contain high levels of intracellular dopamine. In another study, calcium imaging experiments also revealed the presence of Ca_V channels (52). In the early stage of maturation (~weeks 4), the neurons exhibited slow and weak Ca^{2+} transients which gradually became sharper and larger by week 6. At week 8, these spontaneous Ca^{2+} signals were most likely dependent on Ca^{2+} influx through the plasma membrane and at week 10, spontaneous oscillatory Ca^{2+} signals were also observed. Furthermore, these Ca^{2+} signals were eliminated by removal of extracellular Ca^{2+}, strongly indicating that these signals are initiated through the plasma membrane, most likely via Ca_V channels. Postsynaptic spontaneous currents have also been observed in hiPSC-derived DA cells which are mainly glutaminergic and GABAergic inputs (53).

5.3.5   Human PSC-Derived Motor Neurons

Motor neurons have been mainly derived using cell differentiation method that goes through NPC stages. The differentiated cells exhibited motor neuron gene expression, specifically the expression of *HB9* and *ChAT* genes (55–58). The electrophysiological activity of these neurons was tested at both immature and mature stages. At day 5, the cells had weak APs, low spike magnitude and rate, and did not exhibit bursting activity. At day 21, the cells had faster rising spikes, higher spiking frequency, some bursting behavior, and K_V- and Na_V-mediated currents (55). Calcium imaging studies confirmed the presence of calcium channels along with spontaneous oscillations of intracellular calcium concentrations. Excitatory postsynaptic potentials (EPSPs) were reduced and inhibited in human motor neurons by the application (2R)-amino-5-phosphonovaleric acid; (2R)-amino-5-phosphonopentanoate (D-AP5) and 6-Cyano-2, 3-dihydroxy-7-nitro-quinoxaline (CNQX), indicating the formation of functional excitatory synapses (56). Inhibitory postsynaptic potentials (IPSPs) could be blocked through the application of strychnine and bicuculline, indicating the expression of functional glycine and GABA$_A$ receptors in human motor neurons (56). The expression of Acetylcholine (ACh) channels were tested in hPSC-derived motor neurons co-cultured along with C2C12 myocytes. This showed that hPSC motor neurons can form neuromuscular junctions capable of releasing Ach to initiate muscle contraction (58).

5.3.6 HUMAN PSC-DERIVED SENSORY NEURONS

Sensory neurons, specifically dorsal root ganglia nociceptors, have been successfully developed through the cell differentiation method using hESC cells. These neurons expressed nonclassical ion channels including ASIC, ATP-gated P2X$_3$ channel, HCN1, and TRPV1 as well as various classical ion channels. ASICs expressed in sensory neurons respond to acidic extracellular conditions which achieved optimal activation in the 6-6.5 pH range as shown by patch clamp analysis. Transient inward currents (I$_{pH}$) were induced when the extracellular pH decreased rapidly (59). Additionally, these channels remained inactive for approximately 200 seconds after activation before they were able to respond to another stimulus. The HCN1 receptor belongs to a family of nonselective cation channels found in cardiac cells and neurons that regulate pace making (60). The HCN1 channel in the hPSC sensory neurons was found to regulate their excitability through hyperpolarization activated currents (59). The genes encoding P2X$_3$ and TRPV1 receptors were also expressed in hPSC-derived sensory neurons through qPCR analysis. hPSC derived sensory neurons, further differentiated into nociceptors, showed classical channels such as GABA$_A$ receptors, Na$_V$, Ca$_V$, and K$_V$ channels (61). All of the classical ion channels were shown to be functional through the use of whole cell patch-clamp experiments. Creating functional human sensory neuron models have great potential in the study of pain.

5.3.7 HUMAN PSC-DERIVED SEROTONIN NEURONS

Serotonin-releasing neurons, possessing characteristics of neurons found within the raphe nuclei in the brain, were differentiated from hPSCs by activation of the WNT pathway (62). These cells have been shown to exhibit classical ion channels, such as Na$_V$ and K$_V$ channels, through whole cell patch-clamp experiments. The channels were functional and enable the firing of APs as well as maintenance of RMP. In addition, these cells were shown to have active synapses capable of responding to the elicited APs, resulting in the release of serotonin. Upon application of commonly prescribed serotonin reuptake inhibitors (SSRI), such as Lexapro, as well as serotonin releasing drugs, these neurons released serotonin in increasing concentrations. Both drug types had their intended effect of increasing the concentration of serotonin in the synaptic cleft, indicating that the hPSC derived neurons were successfully differentiated into serotonin releasing neurons which were capable of synthesizing and releasing the neurotransmitter (62).

5.3.8   HUMAN PSC-DERIVED GLIAL CELLS

Human PSCs have been efficiently differentiated into astrocytes and oligoden-drocyte progenitor cells (OPCs). Astrocytes derived from hESCs were shown to be heterogenous when compared to astrocytes differentiated from hESC-derived subpopulations of different NPCs. A previous study showed that astrocytes differentiated from both Olig2-expressing (Olig2$^+$) and Olig2-negative (Olig2$^-$) NPCs exhibited K_V-mediated currents. Interestingly, Olig2$^+$ NPC-derived astrocytes, but not Olig2$^-$ NPC-derived astrocytes showed Tetrodotoxin (TTX)-sensitive Na_V-mediated currents (63). A subpopulation of OPCs derived from hESC cells were shown to express K_V-mediated currents and large Na_V-mediated currents, which enable them to fire APs (64). Studies in a mouse ESC differentiation model demonstrated that the expression of Na_V in OPCs might promote their maturation to oligodendrocytes and the formation myelin (63).

5.3.9   HUMAN PSC-DERIVED CELLS IN TRANSPLANTATION MODELS

Transplantation of hPSC neural derivatives into rodent models has been used to study the function and integration of engrafted human neural cells in vivo. Human PSC-derived cortical excitatory neurons, GABAergic neurons, and dopaminergic have been shown to successfully integrate and express classical ion channels in vivo.

Human PSC-derived cortical neurons have been observed to integrate within mouse neural circuitry after transplantation. Specifically, hESC-derived pyramidal cortical neurons transplanted into mouse brain are able to successful integrate and communicate with host neurons within the mouse brain (65). Xenotransplanted human cortical neurons took 11 months to mature, which is notably longer than the few weeks mESC-derived cortical neurons typically take to mature, but they showed characteristics of mature APs, indicating the presence of NaV and KV channels (31). These neurons developing over the 11-month period in the mouse brain gradually displayed signs of maturation including membrane potential hyperpolarization, decreased input resistance, increase in maximum sodium currents, and shortening of AP half-width. Furthermore, these neurons showed simultaneous synaptic maturation as indicated by an increase in incoming synaptic signals. Calcium transients were also detected in those neurons via calcium imaging. Excitatory cortical neurons differentiated from Alzheimer's disease patient-derived hiPSC were transplanted into mouse brains (66). Seven months post transplantation, the brain tissue was sampled, and the electrophysiology of the transplanted neurons was assessed. Whole cell patch clamp recordings indicated that the neurons were capable of firing AP and receiving spontaneous postsynaptic currents, indicating the presence of functional Na_V and K_V channels in vivo. In addition, when N-methyl-D-aspartate receptor (NMDA), AMPA, and GABA$_A$ antagonists were applied, all postsynaptic events could be blocked, indicating the presence of functional synapses containing those respective ion channels. Patch-clamp recordings on hPSC-derived GABA neurons engrafted into mouse brains showed that these human GABA neurons were able to fire APs, indicating the presence of Na_V and K_V channels (49). The GABA neurons successfully integrated into the host's CNS as shown by receiving both bicuculine-sensitive GABAergic and CNQX-sensitive glutamatergic neurotransmissions. GABA neurons derived from Down Syndrome (DS) patient hiPSCs were transplanted into mouse brains to examine the abnormal development of DS GABA interneurons in vivo. Patch clamp analysis demonstrated that transplanted GABA neurons in the chimeric mouse brain exhibited both spontaneous postsynaptic potentials and currents (67). Human iPSC-derived DA neurons have been shown to integrate and relieve symptoms of Parkinson's disease (PD) when transplanted in rodent models of PD (68). These human DA neurons were shown to be able to fire mature APs and release significant amounts of dopamine and DOPAC, a metabolite of dopamine.

5.3.10 TWO-DIMENSIONAL HUMAN CELL CULTURE AND CELL TRANSPLANTATION AS MODELS TO STUDY CHANNELOPATHY

The use of hPSCs, specifically patient-derived hPSCs to model complex nervous system disorders is a new and exciting approach to studying pathology, understanding disease mechanisms, and drug discovery. Researchers have been able to use this technology to study specific channelopathies and devise therapeutic strategies for various Autism Spectrum Disorders (ASD) and motor neuron diseases. Caused by mutations in the X-linked gene encoding the MeCP2 protein, ASD Rett Syndrome (RTT) is a progressive neurological disorder involving impaired motor function, hypotonia, seizures, and autistic behavior from as early as 18 months in individuals (69). Given the importance of intracellular calcium levels in the activation of signaling pathways in neural development, it was found that calcium oscillations were significantly decreased in RTT patient-derived iPSCs (70). This result was consistent with a previous study on MeCP2 KO mouse model and suggests a deficiency in neuronal network connectivity and activity dynamics (70,71). Furthermore, it was revealed that RTT neurons have a significant decrease in frequency and amplitude of spontaneous postsynaptic currents as compared to neurons differentiated from normal individual-derived hiPSCs, further reiterating an altered neuronal network in RTT patients. Another ASD, Timothy syndrome (TS), is caused by a dominant mutation in an alternatively spliced exon of the *CACNA1C* gene which encodes for the a subunit of the $Ca_V1.2$ channel (72). The Ca_V channel plays an important role in neuronal development by promoting dendritic growth and arborization (73). In TS patient hiPSC-derived neurons, researchers observed a lack of dendrite growth (coupled with dendrite retraction upon depolarization) thus reinforcing the connection between the Ca_V channel mutation in the disease and calcium's role in neuron development (74). It was also found that TS hiPSC-derived neurons displayed wider APs, increased calcium signals, and abnormal levels of norepinephrine and dopamine (75). Furthermore, these phenotypes were corrected by the application of roscovitine, which is known to restore calcium and electrical signaling. Dravet syndrome, displaying symptoms of developmental delays in language, motor function, and social skills, is another type of childhood epilepsy and ASD that is caused by the loss-of-function mutation in the SCN1A gene that encodes the $Na_V1.1$ channel (76–78). Using whole-cell patch clamp recording, it was found that the I_{Na} amplitude was significantly reduced in Dravet syndrome hiPSC-derived telencephalic inhibitory neurons (79). The maximal firing frequency was also reduced by 40% in patient hiPSC-derived inhibitory neurons. These phenotypes were recapitulated in control neurons where $Na_V1.1$ expression was suppressed using shRNAs, and again rescued in patient hiPSC-derived neurons by forced expression of $Na_V1.1$ expression, highlighting the importance of $Na_V1.1$ in controlling the excitability of inhibitory neurons.

Amyotrophic lateral sclerosis (ALS) is a devastating neurodegenerative disease characterized by a progressive degeneration of motor nerve cells in the brain (upper motor neurons) and spinal cord (lower motor neurons). About 20% of familial ALS cases involve a mutation in the superoxide dismutase 1 (SOD1) gene (80). Different theories posit either the involvement of excessive glutamatergic neurotransmission, leading to calcium overload, and cell death, or the involvement of increased axonal membrane excitability, which could be caused by either increased persistent sodium or reduced delayed-rectifier potassium currents (81–84). In a recent study which conducted multi electrode array (MEA) and patch-clamp recordings on ALS hiPSC-derived neurons, it was observed that these neurons fired significantly more spontaneous APs relative to neurons derived from their isogenic control lines (85). By using gene targeting and homologous recombination techniques, the SOD1 mutation was corrected in the iPSC lines and the electrophysiological properties were recorded. The corrected neurons showed spiking rates similar to that of the isogenic control lines. Upon the application of retigabine, a specific activator of subthreshold K_V7 (KCNQ) currents and a known anticonvulsant, the hyperexcitability phenotype of ALS motor neurons was rescued.

5.4   ION CHANNELS IN HPSC-DERIVED BRAIN ORGANOIDS

5.4.1   ION CHANNELS IN BRAIN ORGANOIDS

Human iPSC-derived 3D brain organoids provide new opportunities to better understand human brain development and neuropathology at varied levels (27,86). Initial electrophysiological studies of cells developed in brain organoids revealed the presence of electrically active mature neurons (87). Dissociated neurons from mature organoids replated in monolayer displayed spontaneous calcium spikes in addition to I_{Na} after depolarization. Using patch-clamp recordings, researchers observed that the I_{Na} could be blocked by the sodium channel blocker, TTX, and was followed by the activation of a more sustained I_K. Depolarizing current injections also stimulate action potential firing. Additionally, neurons in these organoids underwent synaptogenesis and formed functional synapses, as indicated by the expression of the NMDA receptor subunit NR2B. Synaptic currents mediated by glutamate receptors AMPA and NMDA were also recorded and abolished upon application of AMPA and NMDA antagonists. To test the functionality of these neurons in the intact 3D organoids, whole-cell recordings were conducted on slices of these organoids. In response to electrical stimulation, the neurons present in intact organoid slices also displayed large amplitude EPSPs indicative of an intricate and functional neuronal network. In a separate study that also conducted whole-cell recording of organoid slices, the following classical ion channels were identified in brain organoids: Na$_V$ and ligand-gated receptor ion channels, including AMPA, NMDA, and GABA$_A$ receptors (24). Notably, the gene expression of various Na$_V$ subunits increased significantly over time, indicating neuronal maturation in these organoids. The presence of functional ligand-gated ion channels was verified by the ionic currents recorded upon application of glutamate, NMDA, and GABA. Additionally, upon application of propofol, an anesthetic drug known to be the agonist of the GABA$_A$ receptor, the GABA current was prolonged.

One recent study aiming to recapitulate midbrain dopaminergic neurons (mDA) in organoids found that these mDA neurons were electrically mature (53). Whole-cell patch recordings showed the presence of I$_{Na}$ and I$_K$ which increased in amplitude over time along with a decrease in membrane resistance (Rm) and increase in membrane capacitance (Cm); changes corresponding to the functional maturation of these neurons. Upon depolarization in current clamp mode, APs were generated. Large-amplitude EPSPs were also generated indicating the presence of a functional neuronal network. Electrical properties more specific to mDA neurons were also observed by more than 20% of the total neuronal population in these organoids. These properties include rhythmic discharges and rebound depolarizations resulting in AP generation after short hyperpolarization. The presence of mDA neurons was confirmed by high performance liquid chromatography (HPLC) measuring the DA content which was found to be comparable to native mDA neurons.

5.4.2 Ion Channels in Sensory Organ-Specific Organoids

It is worth mentioning that PSCs can be directed to form sensory organ-specific tissues. A previous study examined ion channel activity on mechanosensitive hair cells of the inner ear using a mouse ESC-derived organoid model (88). It was found that hair cells in the organoids had the mechanosensitivity and the intrinsic electrical properties that resemble native hair cells. Upon physical stimulation, the ES derived hair cells displayed mechanotransduction current amplitudes similar to native hair cells. The hair cells also displayed voltage dependent currents, namely, large outward I$_K$, fast inward rectifying I$_K$, transiently expressed small I$_{Na}$, and I$_h$. The amplitude of delayed rectifier I$_K$ and I$_h$ current increased with time in culture, and the I$_{Na}$ peaked in amplitude followed by a decrease as the culture matured. These findings were consistent with the developmental timeline of native perinatal utricle hair cells. The presence of the HCN1 channel, which mediates I$_h$, was also confirmed by RT-PCR. Organoid hair cells were also found to have normal voltage responses by performing recordings in current-clamp mode to measure their resting potential and by measuring the voltage excursions upon step current injections. The membrane voltage followed sinusoidal currents which mimic oscillating stimuli such as walking, vibration or acoustic stimuli. Ca^{2+} currents were also measured in organoid hair cells and were found to be sufficient to stimulate synaptic release of neurotransmitters. Lastly, although the majority of organoid hair cells analyzed showed properties of type II hair cells, some cells also showed electrophysiological properties of type I hair cells (a large outward rectifier K+ conductance with a negative activation range referred to as $G_{K,L}$).

5.4.3 ORGANOIDS AS A MODEL TO STUDY CHANNELOPATHIES

In addition to studying the basic properties of ion channels, a recent study employed the brain organoid model to examine the dysfunction of ion channels in Angelman's syndrome (AS) (89). AS is a neurodevelopmental disorder characterized by delayed development, intellectual disability, and seizures. Approximately 90% of AS cases are caused by a loss-of-function mutation of the ubiquitin protein ligase E3A (UBE3A) gene, which encodes HECT E3 ubiquitin ligase. The loss of the UBE3A protein potentially results in the accumulation of relevant substrates, thus causing the disease (90). By using CRISPR-Cas9 technology, UBE3A knockout (KO) neurons were generated from hESCs. It was found that the fast components of after-hyperpolarization (fAHPs) were significantly increased in UBE3A KO neurons. fAHPs are primarily mediated by calcium- and voltage-dependent big potassium channels (BK). Consistently, UBE3A KO neurons exhibited larger BK currents along with increased levels of BK protein expression. Furthermore, by using im-munoprecipitation, hemagglutinin (HA) tagging of BK, and in vitro ubiquitination assays, it was found that UBE3A indeed mediated the ubiquitination and protea-somal degradation of BK, leading to increased fAHPs and neuronal excitability. These results were also reproduced in brain organoids generated from AS patient-derived hiPSCs. Organoids cultured from Alzheimer's Disease (AD) patient hiPSCs have been reported to display a significant increase in AP firing rate as compared to control neurons differentiated from normal individual-derived hiPSCs, when plated in a MEA recording chamber (91). In monolayer cultures of these AD hiPSC-derived neurons, increased sodium current density was detected. In line with these findings, enhanced Nav1.6 sodium channel expression was demonstrated in AD mice (92).

5.5 CONCLUDING REMARKS

Building upon stem cell technologies, generation of functional human neurons and glial cells provides new opportunities to study ion channel properties and dysfunction of ion channels in a variety of neurological disorders. Notably, recent studies have reported successful derivation of microglia cells from hPSCs (93–97). These hPSC-derived microglia been incorporated into cerebral organoids (93,98) or used to develop human microglia mouse chimeras (30,99–102). Ion channels are one of the important determinants of microglial functions under homeostatic and disease conditions (103). Given the fact that rodent microglia are not able to fully mirror the properties of human microglia (104,105), these hPSC-based in vitro and in vivo microglia models will provide new opportunities to understand the roles of ion channels in microglia. Brain organoids have proven to be valuable in modeling human development and disease pathology. Due to the fact that hPSC-derived brain organoids consist of purely human cells and possess complex 3D brain cytoarchitecture, they provide a powerful tool to study channelopathies. Furthermore, brain organoids can be generated at high throughput, which could facilitate drug screening and developing therapeutic strategies targeting ion channels. As issues such as a dearth of functional glial cells, a lack of vascularization, and limited functional cell maturation are overcome, brain organoids will become an increasingly effective model for studying the pathology behind not only complex neurodevelopmental disorders, but also neurodegenerative disorders, such as Alzheimer's disease (AD) and amyotrophic lateral sclerosis (ALS) (26). Studies have demonstrated survival, migration, and functional integration of engrafted human neural cells into mouse brains. Generation of in vivo transplantation and human-mouse chimeric brain models will complement organoid models and facilitate studying ion channel properties and functions using human neural cells developed in an intact brain environment. Neural circuits are the underlying functional units of the human brain that govern complex behavior and higher-order cognitive processes. Further development of cerebral organoid and human-mouse chimeric brain models may allow the modeling of neural circuits formed by human neural cells and investigation of the role of classical and nonclassical ion channels in the brain circuitry that underlies specific brain functions and is implicated in neurological disorders.

Abbreviations:

2D	2-Dimensional
3D	3-Dimensional
AD	Alzheimer's disease
AHP	After hyper polarization
ALS	Amyotrophic lateral sclerosis
AMPA	α-amino-3-hydroxy-5-methyl-4-isoxazolepropionic acid
AP	Action potential
AS	Angelman's syndrome
ASD	Autism spectrum disorders
ASH	afterspike hyperpolarization
ASIC	acid sensing ion channels
BK	Big potassium channels
Ca_V	Voltage gated Ca channel
CNQX	6-Cyano-2, 3-dihydroxy-7-nitro-quinoxaline
CNS	Central nervous system
D-AP5	(2R)-amino-5-phosphonovaleric acid; (2R)-amino-5-phosphonopentanoate
DA	Dopaminergic
EB	Embryoid bodies
ECM	Extracellular matrix
EPSP	Excitatory postsynaptic potentials
fAHP	Fast components of after-hyperpolarization
GABA	Gamma-aminobutyric acid

$GABA_A$	GABA A-type
HA	Hemagglutinin
HCN	Hyperpolarization-activated cyclic nucleotide–gated channel
hERG	Human ether-a-go-go
hESC	Human embryonic stem cell
hiPSC	Human induced pluripotent stem cell
HPLC	High performance liquid chromatography
I_{Ca}	Voltage-gated calcium currents
I_f	Pacemaker currents
I_h	Hyperpolarization currents
I_K	Voltage-gated potassium currents
I_{KDR}	Delayed rectifier K^+ currents
I_{Na}	Voltage-gated sodium channel currents
IPSP	Inhibitory postsynaptic potentials
KO	Knockout
K_V	Voltage-gated K channel
mDA	Midbrain dopaminergic
MEA	Multielectrode array
MGE	Medial ganglionic eminence
MS	Multiple sclerosis
Na_V	Voltage-gated Na channel
NE	Neuroepithelial
NMDA	N-methyl D-aspartate receptor
NPC	Neural progenitor cell
OPC	Oligodendrocyte progenitor cells
PD	Parkinson's disease
PNS	Peripheral nervous system
PSC	Pluripotent stem cell
q-PCR	Quantitative polymerase chain reaction
RMP	Resting membrane potential
RT-PCR	Reverse transcription polymerase chain reaction
RTT	Rett syndrome
SAC	Stretch activated channel
SHH	sonic hedgehog
SOD	Superoxide dismutase
SSRI	Serotonin reuptake inhibitors
TEA	Tetraethylammonium
TRP	Transient receptor potential
TRPC	Canonical transient receptor potential
TS	Timothy syndrome
TTX	Tetrodotoxin
UBE3A	Ubiquitin protein ligase E3A

REFERENCES

1. K. Takahashi *et al.*, Induction of pluripotent stem cells from adult human fibroblasts by defined factors. *Cell* **131**, 861–872 (2007).
2. J. Yu *et al.*, Induced pluripotent stem cell lines derived from human somatic cells. *Science* **318**, 1917–1920 (2007).
3. I. H. Park *et al.*, Reprogramming of human somatic cells to pluripotency with defined factors. *Nature* **451**, 141–146 (2008).
4. J. Mertens, M. C. Marchetto, C. Bardy, F. H. Gage, Evaluating cell reprogramming, differentiation and conversion technologies in neuroscience. *Nat Rev Neurosci* **17**, 424–437 (2016).
5. K. Ardhanareeswaran, J. Mariani, G. Coppola, A. Abyzov, F. M. Vaccarino, Human induced pluripotent stem cells for modelling neurodevelopmental disorders. *Nat Rev Neurol* **13**, 265–278 (2017).
6. A. Sharma, S. Sances, M. J. Workman, C. N. Svendsen, Multi-lineage human iPSC-derived platforms for disease modeling and drug discovery. *Cell Stem Cell* **26**, 309–329 (2020).
7. S. M. Chambers *et al.*, Highly efficient neural conversion of human ES and iPS cells by dual inhibition of SMAD signaling. *Nat Biotechnol* **27**, 275–280 (2009).
8. B. Bhattacharya *et al.*, Comparison of the gene expression profile of undifferentiated human embryonic stem cell lines and differentiating embryoid bodies. *BMC Dev Biol* **5**, 22 (2005).
9. M. T. Pankratz *et al.*, Directed neural differentiation of human embryonic stem cells via an obligated primitive anterior stage. *Stem Cells* **25**, 1511–1520 (2007).
10. P. H. Schwartz, D. J. Brick, A. E. Stover, J. F. Loring, F. J. Muller, Differentiation of neural lineage cells from human pluripotent stem cells. *Methods* **45**, 142–158 (2008).
11. Y. Tao, S. C. Zhang, Neural subtype specification from human pluripotent stem cells. *Cell Stem Cell* **19**, 573–586 (2016).
12. Y. Oh, J. Jang, Directed differentiation of pluripotent stem cells by transcription factors. *Mol Cells* **42**, 200–209 (2019).
13. A. X. Sun *et al.*, Direct induction and functional maturation of forebrain GABAergic neurons from human pluripotent stem cells. *Cell Rep* **16**, 1942–1953 (2016).
14. I. Theka *et al.*, Rapid generation of functional dopaminergic neurons from human induced pluripotent stem cells through a single-step procedure using cell lineage transcription factors. *Stem Cells Transl Med* **2**, 473–479 (2013).
15. N. Yang *et al.*, Generation of pure GABAergic neurons by transcription factor programming. *Nat Methods* **14**, 621–628 (2017).
16. G. Miskinyte *et al.*, Transcription factor programming of human ES cells generates functional neurons expressing both upper and deep layer cortical markers. *PLoS One* **13**, e0204688 (2018).
17. M. E. Hester *et al.*, Rapid and efficient generation of functional motor neurons from human pluripotent stem cells using gene delivered transcription factor codes. *Mol Ther* **19**, 1905–1912 (2011).
18. S. Chanda *et al.*, Generation of induced neuronal cells by the single reprogramming factor ASCL1. *Stem Cell Reports* **3**, 282–296 (2014).
19. W. Hu *et al.*, Direct conversion of normal and Alzheimer's disease human fibroblasts into neuronal cells by small molecules. *Cell Stem Cell* **17**, 204–212 (2015).
20. R. Ambasudhan *et al.*, Direct reprogramming of adult human fibroblasts to functional neurons under defined conditions. *Cell Stem Cell* **9**, 113–118 (2011).
21. T. Vierbuchen *et al.*, Direct conversion of fibroblasts to functional neurons by defined factors. *Nature* **463**, 1035–1041 (2010).

22. Z. P. Pang *et al.*, Induction of human neuronal cells by defined transcription factors. *Nature* **476**, 220–223 (2011).
23. S. Chen, J. Zhang, D. Zhang, J. Jiao, Acquisition of functional neurons by direct conversion: Switching the developmental clock directly. *J Genet Genomics* **46**, 459–465 (2019).
24. S. Logan *et al.*, Dynamic characterization of structural, molecular, and electro-physiological phenotypes of human-induced pluripotent stem cell-derived cerebral organoids, and comparison with fetal and adult gene profiles. *Cells* **9**, (2020).
25. C. T. Lee, R. M. Bendriem, W. W. Wu, R. F. Shen, 3D brain organoids derived from pluripotent stem cells: Promising experimental models for brain development and neurodegenerative disorders. *J Biomed Sci* **24**, 59 (2017).
26. A. T. Brawner, R. Xu, D. Liu, P. Jiang, Generating CNS organoids from human induced pluripotent stem cells for modeling neurological disorders. *Int J Physiol Pathophysiol Pharmacol* **9**, 101–111 (2017).
27. M. A. Lancaster *et al.*, Cerebral organoids model human brain development and microcephaly. *Nature* **501**, 373–379 (2013).
28. C. Chen, W. Y. Kim, P. Jiang, Humanized neuronal chimeric mouse brain generated by neonatally engrafted human iPSC-derived primitive neural progenitor cells. *JCI Insight* **1**, e88632 (2016).
29. S. A. Goldman, M. Nedergaard, M. S. Windrem, Modeling cognition and disease using human glial chimeric mice. *Glia* **63**, 1483–1493 (2015).
30. P. Jiang, L. Turkalj, R. Xu, High-fidelity modeling of human microglia with pluripotent stem cells. *Cell Stem Cell* **26**, 629–631 (2020).
31. D. Linaro *et al.*, Xenotransplanted human cortical neurons reveal species-specific development and functional integration into mouse visual circuits. *Neuron* **104**, 972–986 e976 (2019).
32. A. T. Crane, J. P. Voth, F. X. Shen, W. C. Low, Concise review: Human-animal neurological chimeras: Humanized animals or human cells in an animal? *Stem Cells* **37**, 444–452 (2019).
33. P. Jiang *et al.*, Electrophysiological properties of human induced pluripotent stem cells. *Am J Physiol Cell Physiol* **298**, C486–C495 (2010).
34. K. Wang *et al.*, Electrophysiological properties of pluripotent human and mouse embryonic stem cells. *Stem Cells* **23**, 1526–1534 (2005).
35. A. O. Verkerk, R. Wilders, Hyperpolarization-activated current, If, in mathematical models of rabbit sinoatrial node pacemaker cells. *Biomed Res Int* **2013**, 872454 (2013).
36. M. Perry, M. Sanguinetti, J. Mitcheson, Revealing the structural basis of action of hERG potassium channel activators and blockers. *J Physiol* **588**, 3157–3167 (2010).
37. L. Sartiani *et al.*, Developmental changes in cardiomyocytes differentiated from human embryonic stem cells: A molecular and electrophysiological approach. *Stem Cells* **25**, 1136–1144 (2007).
38. J. Ma *et al.*, High purity human-induced pluripotent stem cell-derived cardiomyocytes: Electrophysiological properties of action potentials and ionic currents. *Am J Physiol Heart Circ Physiol* **301**, H2006–H2017 (2011).
39. I. Borbiro, T. Rohacs, Regulation of piezo channels by cellular signaling pathways. *Curr Top Membr* **79**, 245–261 (2017).
40. B. Coste *et al.*, Piezo1 and Piezo2 are essential components of distinct mechanically activated cation channels. *Science* **330**, 55–60 (2010).
41. M. M. Pathak *et al.*, Stretch-activated ion channel Piezo1 directs lineage choice in human neural stem cells. *Proc Natl Acad Sci USA* **111**, 16148–16153 (2014).
42. J. P. Weick, M. Austin Johnson, S. C. Zhang, Developmental regulation of human embryonic stem cell-derived neurons by calcium entry via transient receptor potential channels. *Stem Cells* **27**, 2906–2916 (2009).

43. Y. Zhang *et al.*, Rapid single-step induction of functional neurons from human pluripotent stem cells. *Neuron* **78**, 785–798 (2013).
44. Y. Shi, P. Kirwan, J. Smith, H. P. Robinson, F. J. Livesey, Human cerebral cortex development from pluripotent stem cells to functional excitatory synapses. *Nat Neurosci* **15**, 477–486, S471 (2012).
45. M. A. Johnson, J. P. Weick, R. A. Pearce, S. C. Zhang, Functional neural development from human embryonic stem cells: Accelerated synaptic activity via astrocyte coculture. *J Neurosci* **27**, 3069–3077 (2007).
46. B. T. Jacques-Fricke, Y. Seow, P. A. Gottlieb, F. Sachs, T. M. Gomez, Ca^{2+} influx through mechanosensitive channels inhibits neurite outgrowth in opposition to other influx pathways and release from intracellular stores. *J Neurosci* **26**, 5656–5664 (2006).
47. H. Komuro, T. Kumada, Ca^{2+} transients control CNS neuronal migration. *Cell Calcium* **37**, 387–393 (2005).
48. T. A. Weissman, P. A. Riquelme, L. Ivic, A. C. Flint, A. R. Kriegstein, Calcium waves propagate through radial glial cells and modulate proliferation in the developing neocortex. *Neuron* **43**, 647–661 (2004).
49. C. R. Nicholas *et al.*, Functional maturation of hPSC-derived forebrain interneurons requires an extended timeline and mimics human neural development. *Cell Stem Cell* **12**, 573–586 (2013).
50. Y. Liu *et al.*, Directed differentiation of forebrain GABA interneurons from human pluripotent stem cells. *Nat Protoc* **8**, 1670–1679 (2013).
51. G. Colasante *et al.*, Rapid conversion of fibroblasts into functional forebrain GABAergic interneurons by direct genetic reprogramming. *Cell Stem Cell* **17**, 719–734 (2015).
52. E. M. Hartfield *et al.*, Physiological characterisation of human iPS-derived dopaminergic neurons. *PLoS One* **9**, e87388 (2014).
53. J. Xi *et al.*, Specification of midbrain dopamine neurons from primate pluripotent stem cells. *Stem Cells* **30**, 1655–1663 (2012).
54. M. Caiazzo *et al.*, Direct generation of functional dopaminergic neurons from mouse and human fibroblasts. *Nature* **476**, 224–227 (2011).
55. F. Bianchi *et al.*, Rapid and efficient differentiation of functional motor neurons from human iPSC for neural injury modelling. *Stem Cell Res* **32**, 126–134 (2018).
56. X. J. Li *et al.*, Specification of motoneurons from human embryonic stem cells. *Nat Biotechnol* **23**, 215–221 (2005).
57. Y. Liu, P. Jiang, W. Deng, OLIG gene targeting in human pluripotent stem cells for motor neuron and oligodendrocyte differentiation. *Nat Protoc* **6**, 640–655 (2011).
58. B. Y. Hu, S. C. Zhang, Differentiation of spinal motor neurons from pluripotent human stem cells. *Nat Protoc* **4**, 1295–1304 (2009).
59. G. T. Young *et al.*, Characterizing human stem cell-derived sensory neurons at the single-cell level reveals their ion channel expression and utility in pain research. *Mol Ther* **22**, 1530–1543 (2014).
60. O. Postea, M. Biel, Exploring HCN channels as novel drug targets. *Nat Rev Drug Discov* **10**, 903–914 (2011).
61. S. M. Chambers *et al.*, Combined small-molecule inhibition accelerates developmental timing and converts human pluripotent stem cells into nociceptors. *Nat Biotechnol* **30**, 715–720 (2012).
62. J. Lu *et al.*, Generation of serotonin neurons from human pluripotent stem cells. *Nat Biotechnol* **34**, 89–94 (2016).
63. P. Jiang *et al.*, hESC-derived $Olig^{2+}$ progenitors generate a subtype of astroglia with protective effects against ischaemic brain injury. *Nat Commun* **4**, 2196 (2013).

64. S. R. Stacpoole *et al.*, High yields of oligodendrocyte lineage cells from human embryonic stem cells at physiological oxygen tensions for evaluation of translational biology. *Stem Cell Reports* **1**, 437–450 (2013).
65. I. Espuny-Camacho *et al.*, Pyramidal neurons derived from human pluripotent stem cells integrate efficiently into mouse brain circuits in vivo. *Neuron* **77**, 440–456 (2013).
66. R. Najm *et al.*, In vivo chimeric Alzheimer's disease modeling of apolipoprotein E4 toxicity in human neurons. *Cell Rep* **32**, 107962 (2020).
67. R. Xu *et al.*, OLIG2 drives abnormal neurodevelopmental phenotypes in human iPSC-based organoid and chimeric mouse models of down syndrome. *Cell Stem Cell* **24**, 908–926 e908 (2019).
68. D. Doi *et al.*, Isolation of human induced pluripotent stem cell-derived dopaminergic progenitors by cell sorting for successful transplantation. *Stem Cell Reports* **2**, 337–350 (2014).
69. R. E. Amir *et al.*, Rett syndrome is caused by mutations in X-linked MECP2, encoding methyl-CpG-binding protein 2. *Nat Genet* **23**, 185–188 (1999).
70. M. C. Marchetto *et al.*, A model for neural development and treatment of Rett syndrome using human induced pluripotent stem cells. *Cell* **143**, 527–539 (2010).
71. S. L. Mironov *et al.*, Remodelling of the respiratory network in a mouse model of Rett syndrome depends on brain-derived neurotrophic factor regulated slow calcium buffering. *J Physiol* **587**, 2473–2485 (2009).
72. I. Splawski *et al.*, Ca(V)1.2 calcium channel dysfunction causes a multisystem disorder including arrhythmia and autism. *Cell* **119**, 19–31 (2004).
73. R. O. Wong, A. Ghosh, Activity-dependent regulation of dendritic growth and patterning. *Nat Rev Neurosci* **3**, 803–812 (2002).
74. J. F. Krey *et al.*, Timothy syndrome is associated with activity-dependent dendritic retraction in rodent and human neurons. *Nat Neurosci* **16**, 201–209 (2013).
75. S. P. Pasca *et al.*, Using iPSC-derived neurons to uncover cellular phenotypes associated with Timothy syndrome. *Nat Med* **17**, 1657–1662 (2011).
76. W. A. Catterall, F. Kalume, J. C. Oakley, NaV1.1 channels and epilepsy. *J Physiol* **588**, 1849–1859 (2010).
77. C. Dravet, M. Bureau, B. Dalla Bernardina, R. Guerrini, Severe myoclonic epilepsy in infancy (Dravet syndrome) 30 years later. *Epilepsia* **52 Suppl 2**, 1–2 (2011).
78. C. Dravet, The core Dravet syndrome phenotype. *Epilepsia* **52 Suppl 2**, 3–9 (2011).
79. Y. Sun *et al.*, A deleterious Nav1.1 mutation selectively impairs telencephalic inhibitory neurons derived from Dravet syndrome patients. *Elife* **5**, (2016).
80. J. J. Kuo, T. Siddique, R. Fu, C. J. Heckman, Increased persistent Na(+) current and its effect on excitability in motoneurones cultured from mutant SOD1 mice. *J Physiol* **563**, 843–854 (2005).
81. D. W. Cleveland, J. D. Rothstein, From Charcot to Lou Gehrig: Deciphering selective motor neuron death in ALS. *Nat Rev Neurosci* **2**, 806–819 (2001).
82. P. Pasinelli, R. H. Brown, Molecular biology of amyotrophic lateral sclerosis: Insights from genetics. *Nat Rev Neurosci* **7**, 710–723 (2006).
83. H. Bostock, M. K. Sharief, G. Reid, N. M. Murray, Axonal ion channel dysfunction in amyotrophic lateral sclerosis. *Brain* **118 (Pt 1)**, 217–225 (1995).
84. M. Nakata *et al.*, Distal excitability changes in motor axons in amyotrophic lateral sclerosis. *Clin Neurophysiol* **117**, 1444–1448 (2006).
85. B. J. Wainger *et al.*, Intrinsic membrane hyperexcitability of amyotrophic lateral sclerosis patient-derived motor neurons. *Cell Rep* **7**, 1–11 (2014).
86. X. Qian, H. Song, G. L. Ming, Brain organoids: Advances, applications and challenges. *Development* **146**, dev166074 (2019).
87. A. M. Pasca *et al.*, Functional cortical neurons and astrocytes from human pluripotent stem cells in 3D culture. *Nat Methods* **12**, 671–678 (2015).

88. X. P. Liu, K. R. Koehler, A. M. Mikosz, E. Hashino, J. R. Holt, Functional development of mechanosensitive hair cells in stem cell-derived organoids parallels native vestibular hair cells. *Nat Commun* **7**, 11508 (2016).
89. A. X. Sun *et al.*, Potassium channel dysfunction in human neuronal models of Angelman syndrome. *Science* **366**, 1486–1492 (2019).
90. T. Kishino, M. Lalande, J. Wagstaff, UBE3A/E6-AP mutations cause Angelman syndrome. *Nat Genet* **15**, 70–73 (1997).
91. S. Ghatak *et al.*, Mechanisms of hyperexcitability in Alzheimer's disease hiPSC-derived neurons and cerebral organoids vs isogenic controls. *Elife* **8**, e50333 (2019).
92. C. Liu, F. C. Tan, Z. C. Xiao, G. S. Dawe, Amyloid precursor protein enhances Nav1.6 sodium channel cell surface expression. *J Biol Chem* **290**, 12048–12057 (2015).
93. E. M. Abud *et al.*, iPSC-derived human microglia-like cells to study neurological diseases. *Neuron* **94**, 278–293 e279 (2017).
94. H. Pandya *et al.*, Differentiation of human and murine induced pluripotent stem cells to microglia-like cells. *Nat Neurosci* **20**, 753–759 (2017).
95. J. Muffat *et al.*, Efficient derivation of microglia-like cells from human pluripotent stem cells. *Nat Med* **22**, 1358–1367 (2016).
96. P. Douvaras *et al.*, Directed differentiation of human pluripotent stem cells to microglia. *Stem Cell Reports* **8**, 1516–1524 (2017).
97. W. Haenseler *et al.*, A highly efficient human pluripotent stem cell microglia model displays a neuronal-co-culture-specific expression profile and inflammatory response. *Stem Cell Reports* **8**, 1727–1742 (2017).
98. P. R. Ormel *et al.*, Microglia innately develop within cerebral organoids. *Nat Commun* **9**, 4167 (2018).
99. J. Hasselmann *et al.*, Development of a chimeric model to study and manipulate human microglia in vivo. *Neuron* **103**, 1016–1033 e1010 (2019).
100. R. Mancuso *et al.*, Stem-cell-derived human microglia transplanted in mouse brain to study human disease. *Nat Neurosci* **22**, 2111–2116 (2019).
101. D. S. Svoboda *et al.*, Human iPSC-derived microglia assume a primary microglia-like state after transplantation into the neonatal mouse brain. *Proc Natl Acad Sci USA* **116**, 25293–25303 (2019).
102. R. Xu *et al.*, Human iPSC-derived mature microglia retain their identity and functionally integrate in the chimeric mouse brain. *Nat Commun* **11**, 1577 (2020).
103. P. Izquierdo, D. Attwell, C. Madry, Ion channels and receptors as determinants of microglial function. *Trends Neurosci* **42**, 278–292 (2019).
104. D. Gosselin *et al.*, An environment-dependent transcriptional network specifies human microglia identity. *Science* **356**, (2017).
105. T. F. Galatro *et al.*, Transcriptomic analysis of purified human cortical microglia reveals age-associated changes. *Nat Neurosci* **20**, 1162–1171 (2017).

6 Exocytosis of Nonclassical Neurotransmitters

Xiao Su, Vincent R. Mirabella,
Kenneth G. Paradiso, and Zhiping P. Pang

CONTENTS

6.1 INTRODUCTION

Brain function relies on the regulated release of chemical substances known as neurotransmitters at specialized junctions between neurons called synapses. Synaptic transmission is mainly mediated by classical neurotransmitters such as glutamate and γ-amino butyric acid (GABA), which normally transduce fast information flow in the brain. This process is further regulated by released neuromodulators including monoamines and neuropeptides. Dysfunctional neurotransmission is a major component of many neurological disorders and neuropsychiatric disorders, which include schizophrenia, depression, bipolar disorders, and eating disorders, as well as neurodevelopmental disorders such as autism spectrum disorders (ASDs). Since the discovery of synapses over 100 years ago, scientists have made major breakthroughs in determining the mechanisms and function of synaptic transmission. While it is impossible to fully comprehend all that we now know about synaptic transmission, we will review many of the fundamental discoveries and our current knowledge on neurotransmitter release. It is indisputable that understanding the mechanism and regulation of neurotransmitter release is a fundamental area of investigation in unraveling how the brain works.

The existence of some forms of communication between neurons was established by Cajal's careful work in the late 1800s, clearly showing how neurons in the brain are discrete entities: individual cells that form close contacts but are separate. This resulted in several major disputes on how neurons communicate, including arguments from Camillo Golgi against the finding that neurons were in fact separate. However, Cajal's work convinced the vast majority of scientists to accept that neurons are separate. The points of contact were the logical choice for being the location where communication occurred, and these points of contact became known as synapses after Sherrington coined the word "synapse" from syn (together), haptein (to clasp) and apse (a domed or vaulted recesses or terminal portion of a structure), and used the term in his seminars and teaching lectures from 1897. Given the existence of the synapse, this meant that neurons must communicate by an electrical method that crossed the synapse, or by chemical transmission. The argument against chemical transmission was daunting: it was beyond rational belief that chemical release could occur with the speed and precision necessary to allow rapid neuronal communication.

Despite the arguments against it, evidence for the existence of chemical release from neurons was first presented by Thomas Elliott who, in a series of papers, provided strong evidence for the release of epinephrine from sympathetic neurons. Following this, Sir Henri Dale discovered and isolated acetylcholine from fungal extracts and tested its effects, and Otto Loewi discovered its release and its effects after stimulation from the vagus nerve. Sherrington continued to make additional contributions, among them the concept that inhibitory neurotransmitters existed and that neuronal inhibition was not simply a reduction in excitatory activity but an active suppression of activity. Soon after, John Eccles and others used electrophysiology to define the functional evidence of chemical neurotransmission in the 1950s. How chemical neurotransmitters are released from the synapse thus became a focus of modern neuroscience. Classic work from Bernard Katz and his colleagues in 1950–1960 elucidated the essential principles of the quantal release of neurotransmitter at the synapse and the fundamental role of Ca^{2+} in regulating the release of neurotransmitters, and many studies that continue today have focused on investigating how Ca^{2+} regulates neurotransmitter release. Motivated by the findings from Bernard Katz, and his theory that presynaptic vesicles contain neurotransmitter, John Heuser and Thomas Reese used electron microscopy (EM) combined with rapid freezing techniques to effectively produce snapshots of presynaptic activity to show vesicles fusing with membranes, and vesicle recycling following neuronal stimulation. In the 1980–1990s, Thomas Südhof and others conducted a series of biochemical, genetic, structural, and functional studies to determine the molecular identities of Ca^{2+}-sensing and -triggering of synaptic vesicle exocytosis. With the exceptions of the discoveries made by Thomas Elliot, and by Heuser and Reese, each of these studies won Nobel prizes (Cajal 1906; Sherrington 1932; Dale and Loewi 1936; Eccles 1963; Katz 1970, and Südhof 2013). It's also important to note that numerous other discoveries have been made in parallel to the ones listed above, and they serve to strengthen and in some cases dispute our understanding of neurotransmitter release and its effects in the nervous system.

To fully appreciate the effects of neurotransmitter release, it is important to note that two different modes of neurotransmitter transmission have been proposed: point-to-point transmission that is typically localized to individual synaptic contacts (classical neurotransmitters) and volume transmission that can affect a much larger area (nonclassical neurotransmitters including dopamine, norepinephrine, and neuropeptides, which are suggested to spill over from the synaptic cleft and reach the extra-synaptic receptors). Here, we give a brief introduction about the universal machinery involved in Ca^{2+}-triggered exocytosis of classical neurotransmitters and then focus on Ca^{2+}-triggered exocytosis for nonclassical neurotransmitters such as dopamine, and neuropeptides such as oxytocin and neuropeptide Y. In addition, we will also describe how Ca^{2+}-triggered exocytosis works in endocrine cells among other types of cells.

6.2 Ca^{2+} TRIGGERING OF CLASSICAL NEUROTRANSMITTER (GLUTAMATE AND Γ-AMINOBUTYRIC ACID, GABA) VESICLE EXOCYTOSIS AT SYNAPSES

It has been well understood that neurotransmitter release depends on extracellular Ca^{2+}, which was initially established by electrophysiological recordings demonstrating that neurotransmitter release fails when calcium is removed from the extracellular medium, but release can be restored by the focal and appropriately timed application of calcium (Katz and Miledi, 1965). Earlier work done by Bernard Katz revealed the quantal nature of neurotransmitter release events, suggesting that neurotransmitter release is mediated through exocytosis from discrete "packets" of cellular contents, later recognized as synaptic vesicles (SVs) (Katz, 1971). Physiological recordings provided the foundation of and continue to contribute to our understanding of how calcium affects neurotransmitter release. However, physiological measurements alone can only define the nature and characteristics of neurotransmitter release, not how it actually occurs.

A major and more recent contribution to our understanding of how calcium triggers neurotransmitter release comes from identifying the specific proteins that are required for, or involved in, the calcium-triggered fusion of synaptic vesicles. Our understanding of the molecular mechanism underlying Ca^{2+}-dependent SV release began in the early 1990s with the emerging techniques of molecular cloning, mouse genetics as well as biochemistry. Perin et al. identified a new synaptic vesicle protein synaptotagmin (Syt) and suggested its potential role in Ca^{2+} regulation and SV exocytosis due to the presence of a repeated sequence in its cytosolic region that is homologous to the regulatory C2-domain of protein kinase C (PKC) (Perin et al., 1990, 1991). The structure and properties of Syt suggest its potential role as a Ca^{2+} sensor involved in SV exocytosis (Perin et al., 1990). In the past three decades, numerous studies had characterized the essential role of Syt as a Ca^{2+} sensor involved in Ca^{2+}-triggered SV exocytosis in presynaptic terminal, large dense-core vesicle (LDCV, often containing neuromodulators) exocytosis including those in neuroendocrine system (Schonn et al., 2008) and endocrine cells (Gustavsson et al., 2009; Li et al., 2007), and cytokine exocytosis in mast cells (Baram et al., 2001). In general, the universal soluble N-ethylmaleimide-sensitive factor attachment protein receptor (SNARE) proteins complex (formed by SNARE α-helical motifs from synaptobrevin, SNAP25, and syntaxin) mediates the membrane fusion process (Chen and Scheller, 2001). In addition, synaptic vesicle exocytosis also requires active presynaptic proteins to dock and prime the SV for exocytosis, recruit voltage-gated Ca^{2+} channels to the site of exocytosis, and position the active zone exactly opposite to post-synaptic specializations via transsynaptic cell-adhesion molecules. The presynaptic proteins allow precise spatial and temporal signal transmission. The presynaptic active zones are composed of an evolutionarily conserved protein complex that includes RIM, Munc13, RIM-BP, α-liprin, and ELKS proteins, which mediate short-term and long-term synaptic plasticity (Südhof, 2012). This work indicates that there are some universally shared mechanisms involved in Ca^{2+}-induced vesicle exocytosis. Not surprisingly, it has also shown some differences in different types of cells.

The finding that synaptotagmin-1 (Syt 1) functions as the major Ca^{2+} sensor for synchronous (time-locked) neurotransmitter release represents a fundamental contribution to understanding of vesicle release (Geppert, 1994; Fernandez-Chacon, 2001). This work was done using innovations in electrophysiological techniques that enabled intracellular recordings using the high-resistance whole-cell patch clamp technique (Neher, 1992), as well biochemical analyses combined with mouse genome engineering technology (Südhof, 2014) (Figure 6.1). Importantly, the identification of Syt 1 allowed functional studies to be done. Mutations were used to identify the Ca^{2+}-binding domains of Syt 1 based on their effects to increase or decrease SV release (Pang et al., 2006b). This work was significant in demonstrating the critical role for Ca^{2+}-dependent Syt1/SNARE binding in classical neurotransmitter exocytosis. Next, transgenic mouse models were used to explore the contribution of other Syts. Mutation or deletion of synaptotagmin-2 (Syt 2) established that Syt 2 is a fast Ca^{2+} sensor for synaptic vesicle release (Pang et al., 2006a, 2006c; Sun et al., 2007). Later, a dual Ca^{2+} sensor model precisely quantified the kinetics of Ca^{2+}-triggered synaptic vesicle release (Sun et al., 2007), as well as the role of calmodulin and Doc2 family proteins in the regulation of synaptic release properties through the activation of CaMK II (Pang et al., 2010b) and synaptic protein expression (Pang et al., 2010a). Although a great deal of molecular mechanistic insights into SV fusion have been attained over three decades of study, critical questions remain regarding the ultrafast nature of SV fusion (compared to other vesicles), as well as the molecular mechanisms mediating delayed asynchronous SV release and the molecular mechanisms encoding of release mode heterogeneity at distinct synapses. Physiological recordings have shown that

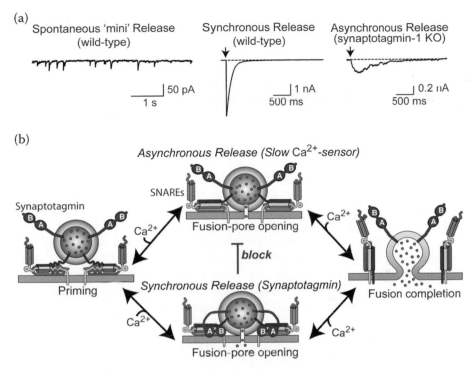

FIGURE 6.1 Ca^{2+}-triggering of SV exocytosis. (a) Three modes of synaptic vesicle release at the synapse. (b) Model of the molecular steps that lead to SV exocytosis. Synaptotagmin (Syt) mediates fast synchronous release and the slow Ca^{2+} sensor(s) mediates asynchronous release. Syt 1 is believed to compete with the slow sensor for exocytosis.

there are three different modes for classical neurotransmitter (glutamate and GABA) release: synchronous, asynchronous and spontaneous "mini" release. The contribution of specific Syts to these different types of release is still an area of intense interest.

6.3 SYNAPTOTAGMINS AS Ca^{2+} SENSORS FOR EXOCYTOSIS

As of now, 17 mammalian Syt isoforms have been identified, 8 of which bind Ca^{2+} (Syt 1–3, 5–7, 9, 10). There are 16 genes encoding the canonical Syt present in mammals, as well as a Syt-related gene occasionally called Syt 17, which encodes a similar protein (B/K protein) but notably lacks a transmembrane region. Syts are evolutionarily conserved and all invertebrates express multiple Syt isoforms. However, plants and unicellular eukaryotes do not express Syt, suggesting that they emerged at the same times with animals during evolution. Even though Syt protein families contain structurally diverse Syt isoforms, they play s similar role in regulating exocytosis of vesicle cargos.

Syt1 was the first of the Syt isoforms to be identified and cloned in mammals (Perin et al., 1990, 1991). Syt 1 was isolated from synaptic vesicles and was found to contain a single N-terminal non-cytosolic region followed by a transmembrane region and two protein kinase C-like C2 domains, which function as an autonomous Ca^{2+}-binding domain. These two C2 domains contain an eight-stranded anti-parallel β-sandwich structure, where two and three Ca^{2+} ions bind the loop regions formed at the top of the C2A and C2B β-sandwich structure, respectively. In addition, the C2B domain contains an additional α-helix between the β7 and β8 strands. The work done on Syt 1 has served as a template for understanding the activity and molecular components of the other Syt proteins.

Ca^2 binding to Syt triggers fusion by binding to SNARE complexes and phospholipids (Figure 6.1). Interestingly, the C2A and C2B domains contribute unequally to regulate vesicle exocytosis. Blocking Ca^{2+} from binding to the C2B domain blocks over 95% of synchronous exocytosis (Mackler et al., 2002) while blocking Ca^{2+} from binding to the C2A domain decreases around 40% of exocytosis and also caused a significant decrease in apparent Ca^{2+} cooperativity of release (Shin et al., 2009). Although the C2A domain is not required for release, Ca^{2+} binding to the C2A domain significantly contributes to the regulation of overall Ca^{2+} cooperativity and the triggering of neurotransmitter release. Biochemically, Ca^{2+} binding to the C2B domain primarily determines binding to phospholipids and displacing of complexin from SNARE trans complexes.

6.4 THREE MODES OF SYNAPTIC RELEASE AND THEIR RESPECTIVE CALCIUM SENSOR(S)

6.4.1 SYNCHRONOUS RELEASE

Evoked synchronous release initiates within a millisecond after an action potential induces Ca^{2+} influx into a presynaptic terminal. Fast synchronous release measured as the postsynaptic response can be fitted by a double exponential function, and thus can be arbitrarily subdivided into a fast and a slow phase (Pang et al., 2006c). Syt 1, 2, and 9 are known as Ca^{2+} sensors for operating fast synchronous neurotransmitter release but exhibit distinct expression patterns and properties (Pang et al., 2006a, 2006b; Xu et al., 2007). Syt 1 was initially proposed to be the Ca^{2+} sensor by Thomas Südhof based on the special domain structure of Syt1 (double C2 domain), its ability to bind to Ca^{2+} and phospholipids in the presence of Ca^{2+}, and the fact that deletion of Syt 1 eliminates the fast synchronous release (Perin et al, 1990; Geppert et al., 1994). Moreover, Syt 1 mutants with reduced Ca^{2+}- and phospholipid-binding also reduced synaptic transmission (Fernandez-Chacon et al., 2001). Now it is generally accepted that Syt 1 and Syt 2 are two major Ca^{2+} sensors in charge of fast synchronous neurotransmitter release, with Syt 1 predominating the forebrain region and Syt 2 predominating the caudal region of the brain including spinal cord, cerebellum, and brainstem. Syt 2 is also selectively expressed in forebrain neurons including most striatal neurons, and some hypothalamic, cortical, and hippocampal neurons (Pang et al., 2006a). Overexpression of Syt1 and Syt 2 in Syt 1-null cells indicates that Syt 1 and Syt 2 are alternative, but not identical, calcium sensors for readily releasable pool (RRP) fusion (Nagy et al., 2006). The different structure in the C2A domain caused a slightly lower Ca^{2+} affinity and phospholipid binding in Syt 2 than in Syt 1 (Pang et al., 2006a). Syt 9 is primarily expressed in the limbic system and striatum, which is involved in reward pathways (Xu et al., 2007). Acute deletion of Syt 9 in striatal neurons severely impaired fast synchronous without affecting the RRP size (Xu et al., 2007).

6.4.2 ASYNCHRONOUS RELEASE

Evoked asynchronous release sets in with a delay after an action potential and is normally negligible, either because it is outcompeted by the synchronous release mechanism or because the synchronous release mechanism suppresses asynchronous release. However, asynchronous release becomes a dominant form of release in some synapses during high-frequency trains of action potentials, particularly in inhibitory synapses. Asynchronous release is primarily activated during and following high-frequency stimulus trains.

Different from three Syts (1, 2 & 9) mentioned above, Syt 7 was suggested to mediate asynchronous neurotransmitter release (Bacaj et al., 2013). Loss of function of Syt 7 in Syt 1-deficient excitatory and inhibitory neurons suppressed asynchronous release. Even in Syt 1 containing neurons, ablation of Syt 7 impaired partially asynchronous release induced by extended high-frequency stimulus trains. However, no major phenotype was detected upon ablation of Syt 7 in wild-type neurons, suggesting Syt 7 function is normally occluded by Syt 1 (Bacaj et al., 2013). Syt 7 exhibits its function selectively via C2A domains whereas Syt 1 requires the C2B domain to exhibit its function. Similarly, Syt 7 KO in calyx synapses has no effect on altering fast release and short-term plasticity in calyx of Held synapses. Syt 7-dependent asynchronous release induces a steady-state, asynchronous postsynaptic current during prolonged Ca^{2+} influx, sustaining high-fidelity neurotransmission driven by synchronous release during high-frequency stimulus trains (Luo and Südhof, 2017).

6.4.3 SPONTANEOUS RELEASE

Finally, spontaneous "mini" release represents the exocytosis of single vesicles that is independent of action potentials, but is nevertheless largely Ca^{2+}-dependent (Xu et al., 2009). Spontaneous release is induced by resting Ca^{2+} concentrations, or may be stimulated by stochastic Ca^{2+}-channel openings and/or Ca^{2+} sparks via Ca^{2+} influx from internal Ca^{2+} stores (Llano et al., 2000).

It has been suggested that the double C2-domain-containing proteins, e.g. Doc2, are alternative Ca^{2+} sensors for spontaneous synaptic release. Docs (including Doc2A, Doc2B, Doc2G) were identified over two decades ago (Orita et al., 1995; Sakaguchi et al., 1995), and rabphilin shares a similar domain structure. Doc2 proteins contain two C2 domains separated by a linker region, a spacer region, and an N terminal Munc 13-interacting domains, which lacks the transmembrane region found in the C2-domain containing Syt. The Doc2 proteins translocate from the cytosol to the plasma membrane with high Ca^{2+} sensitivity and can bind to SNARE complexes in competition with Syt 1 (Groffen et al., 2004, 2006). Doc2 proteins have been implicated in spontaneous and asynchronous neurotransmitter release (Groffen et al., 2010; Yao et al., 2011). Several studies used genetic editing to investigate the role of Doc2 proteins involved in synaptic release. A Doc2A and Doc2B double knockout (DKO) mouse model showed a reduction in spontaneous release frequency (Groffen et al., 2010), suggesting their role as Ca^{2+} sensors for spontaneous release. However, a quadruple knockdown model of the Doc2 protein family (Doc2A, Doc2B, Dob2G, and rabphilin) in mouse cortical neurons caused over a 60% reduction in mini release (Pang et al., 2011). This was confirmed by another group independently (Ramirez et al., 2017). Interestingly, the impaired spontaneous release was resued by expressing a Ca^{2+}-binding-deficient mutant of Doc2B (Pang et al., 2011), implying Doc2B functions in spontaneous release in a Ca^{2+}-independent manner and again emphasizing the primary role of Syt 1 as a Ca^{2+} sensor. In support of this, one recent study suggests that Doc2 proteins may function differently in different synapses by showing no influence on transmission at Purkinje cell to deep nuclei synapses in Doc2B KO mice (Khan and Regehr, 2020).

In addition, another study using a chimeric construct strategy in a triple loss-of-functionmodel of Syt1, Doc2A, and Doc2B suggested that the C2AB domains of the Doc2 protein and Syt 1 function in synchronous and spontaneous release by separate mechanisms (Diez-Arazola et al., 2020). The study about short-term synaptic plasticity indicates that Doc2-mediated superpriming supports synaptic augmentation. Doc2 also plays a role in neuroendocrine and endocrine cells. For example, Doc2 isoforms Doc2A and Doc2B are shown to have dual roles in insulin secretion and insulin-stimulated glucose uptake using Doc2A/Doc2B knockout model (Li et al., 2014).

6.5   NONCLASSICAL NEUROTRANSMITTERS AND NEUROPEPTIDE RELEASE

Proper brain function relies not only on tightly regulated synaptic transmission mediated by fast-acting classical neurotransmitters and SV exocytosis, but also nonclassical neurotransmitters (e.g. neuropeptides, monoamines, and other small molecules) that are also released by neurons through vesicle fusion and critically modulate synaptic transmission and neuronal activity (van den Pol, 2012). In particular, neuropeptide secretion and the functions of peptidergic neuronal circuits remain enigmatic (Südhof, 2017Südhof 2017), largely due to a lack of reliable real-time detection methods (van den Pol, 2012). Unlike classical neurotransmitters, neuropeptides are released from large dense core vesicles (LDCVs) and bind G-protein coupled receptors (GPCRs) (Figure 6.2), which activate second messenger signaling cascades to induce delayed cellular responses. Thus, time-locked neuropeptide release and the study of LDCV exocytosis cannot be reliably detected by conventional synaptic physiology. Moreover, unlike monoamine or other small molecule-based nonclassical neurotransmitters, such as dopamine and norepinephrine, neuropeptides are typically chemically inert (non-oxidizable) and thus cannot be studied with oxidizable chemoelectric probes through amperometry or voltammetry. In general, it is accepted that LDCV exocytosis, like SV exocytosis, depends on neuronal activity and an increase in intracellular Ca^{2+} concentration. SV release is tightly regulated by Ca^{2+} influx acting through Syts within the active zone of presynaptic terminals (Südhof, 2004). Neuropeptide-containing vesicles, however, are normally not tightly coupled to Ca^{2+} channels, and the release sites in the nervous system remain largely unknown. Moreover, the molecular identities of

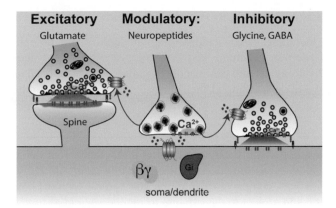

FIGURE 6.2 Neuromodulator release from the nerve terminal.

the Ca^{2+} sensors that mediate neuropeptide exocytosis remain to be unambiguously established. This represents a critical gap in our basic molecular understanding of neurotransmitter release and key elementary processes underlying brain function.

Volume transmission has been proposed to mediate the modulatory functions for most of neuromodulators, i.e. nonclassical neurotransmitters. Volume transmission is defined as a form of communication mediated by extracellular diffusion of neuroactive substances through the extracellular space to mediate effects in many cells over a large area or a long distance (Agnati et al., 1995; Caille et al., 1996; Fuxe et al., 2007; Liu et al., 2018). The term "extrasynaptic communication" was developed to define volume transmission in response to transmitter liberation from extrasynaptic sites, such as in the soma, dendrites, and axons (De-Miguel and Nicholls, 2015). To be noted, extrasynaptic release is suggested to be unrelated to postsynaptic counterparts. Synaptically released transmitters such as dopamine and norepinephrine can spill over from the synaptic cleft and reach extrasynaptic receptors, therefore contributing to extrasynaptic communication (Courtney and Ford, 2014; Rice and Cragg, 2008). Most low molecular weight transmitters such as serotonin and different peptides including oxytocin and vasopressin are released extrasynaptically (Trueta and De-Miguel, 2012). Serotonin, oxytocin, and vasopressin are packed in LDCVs that rest at a distance from the plasma membrane (Coggeshall, 1972; Schimchowitsch et al., 1983). Rapid trains of impulses or large depolarizations instead of single action potentials, trigger a massive exocytosis that lasts for hundreds of seconds (Ludwig and Stern, 2015; Trueta et al., 2003). Unlike the classical neurotransmitters, e.g. glutamate, GABA, and glycine, that can elicit postsynaptic current responses which can be detected by using electrophysiology in high temporal resolution, methods for studying the release of neuromodulators, especially neuropeptides, are very limited. Therefore, the how and when of nonclassical neurotransmitters remain enigmatic.

6.6 MONOAMINE NEUROTRANSMITTERS

Monoamines including catecholamines (e.g. dopamine and norepinephrine) and other amines such as histamine and serotonins are important neuromodulators. While their physiological and pharmacological functions have been appreciated for a long time, the understanding of their release mechanisms are limited.

6.6.1 DOPAMINE

Dopamine plays important functions including motor control, decision making, and reward. The most well-studied dopamine-secreting neurons are located in the substantia nigra pars compacta (SNpc) and the ventral tegmental area (VTA) in midbrain region and project to the striatum to regulate essential functions including movement, motivation, and reward-mediated learning (Howe and Dombeck, 2016).

Different from fast transmitters that form classical point-to-point transmission, dopamine was suggested to be released via a volume transmission mode, where neuromodulators diffuse to mediate effects in many cells over a large area or a longer distance (Agnati et al., 1995; Caille et al., 1996; Liu et al., 2018). Dopamine can also be released from soma and dendrites (Geffen et al., 1976). A large amount of morphological and functional evidence including immunohistochemistry, amperometry, and whole-cell voltage clamp support the belief that most of dopamine transmission is mediated by vesicular exocytosis (Caille et al., 1996; Kress et al., 2014; Staal et al., 2004; Uchigashima et al., 2016; Yung et al., 1995). The ablation of vesicular monoamine transporter type 2 (VMAT2) which is a major vesicular transporter for dopamine eliminates dopamine transmission (Fon et al., 1997). Here, we mainly focus on axonal dopamine release and related machinery involved in dopamine secretion.

Vesicular exocytosis of dopamine also relies on the SNARE complex to mediate membrane fusion which consists of syntaxin, SNAP25, and synaptobrevin. Dopamine release is shown to be repressed partially by botulinum toxin A and B which are proteases for SNAP25 and synaptobrevin-2 respectively while another synaptobrevin-2 protease, tetanus toxin, has no effect on dopamine release (Bergquist et al., 2002; Fortin et al., 2006). This indirect evidence suggests that SNARE proteins play a role in regulating dopamine release. Nevertheless, future studies need to identify the relevant SNARE proteins. For fast synaptic transmission, the presynaptic active zone proteins dock and prime synaptic vesicles, recruit voltage-gated Ca^{2+} channels, align fusion locations with postsynaptic receptors, and mediate short- and long-term presynaptic plasticity (Südhof, 2012). However, the nature of volume transmission does not require precise localization and timing of secretion. This raises the question of how much the active zone machinery contributes to the release of dopamine. One recent study has shown that ~30% of dopamine varicosities in striatal dopamine neurons contain the presynaptic active zone consisting of bassoon, RIM, and ELKS (Liu et al., 2018). Knockout of RIM disrupted the presynaptic active zone scaffolding and impaired dopamine release while knockout of ELKS has no effect on dopamine release (Liu et al., 2018). The dopamine release was fast with a very high initial release probability with the assistance of protein scaffolds for coupling Ca^{2+} influx to vesicle fusion. In summary, the regulation of sparse specialized active zone-like release sites supports precise spatial and temporal secretion of dopamine (Liu et al., 2018).

Similar to glutamate and GABA, dopamine release is triggered by Ca^{2+} influx and several studies have shown that voltage-gated calcium channels (VGCCs) operate dynamically on dopaminergic axons including N-, Q-, T- and L-VGCCs with some local differences in striatum, which is different from classic synapses that rely on N- and P-type VGCCs (Brimblecombe et al., 2015; Takahashi and Momiyama, 1993). In addition, Syt is also involved in regulating dopamine release. It is interesting to note that some evidence suggests that differential Ca^{2+} requirement of terminal and somatodendritic dopamine release requires different Syt isoforms. Down-regulation of the dendritically expressed Syt 4 and Syt 7 reduces dopamine release severely whereas terminal release requires Syt 1 (Mendez et al., 2011). It has also been suggested that Syt 11 in Substantia Nigra pars compacta dopaminergic neurons is required for Parkinson's Disease-like neurotoxicity induced by parkin dysfunction (Wang et al., 2018). A recent elegant study from the Kaeser laboratory utilized high-resolution imaging, amperometry, and mouse genetics and provided very strong evidence that Syt 1 mediates the fast release of dopamine from the nerve terminal, while the loss of Syt 1 does not affect strong depolarization induced by high K^+ caused by the release of dopamine (Banerjee et al., 2020).

6.6.2 NOREPINEPHRINE

The release of norepinephrine throughout the brain is important for modulating attention, arousal, and cognition during many behaviors (Schwarz and Luo, 2015). Norepinephrine released in the brain is produced by a very small, bilateral nucleus in the brainstem called locus coeruleus (LC). LC is involved in modulating numerous behaviors, including sleep/wake states, attention and memory during cognitive tasks and stress response (Berridge and Waterhouse, 2003). LC neurons have been reported to exhibit synchronous firing patterns and project throughout the brain and spinal cord to release norepinephrine (Aston-Jones and Bloom, 1981; Dahlstroem and Fuxe, 1964; Grzanna and Molliver, 1980; Ishimatsu and Williams, 1996; Swanson and Hartman, 1975). Although several lines of evidence support the idea that the locus coeruleus-norepinephrine (LC-NE) system is largely homogeneous, it has some distinct characteristics that may provide heterogeneity to its function with at least two types of cells distinguished by distinct morphology (Schwarz and Luo, 2015; Swanson, 1976).

6.6.3 SEROTONIN

Serotonin, a phylogenetically ancient signaling molecule, is the most widely distributed neurotransmitter in the brain, although the CNS content is less than 5% of the whole body content of serotonin (Dahlstroem and Fuxe, 1964; Steinbusch, 1981). Serotoninergic neurons are spatially clustered in the raphe nucleus (Dahlstrom and Fuxe, 1964) including the dorsal raphe (DR) and median raphe (MR). The DR and MR serotonin systems are linked to many processes including anxiety, mood, aggression, learning, reward, and social interaction (Brimblecombe et al., 2015). Serotonergic neurons showed diversities in their projections, connectivity, electrophysiological properties, and the behaviors they affect (Alonso et al., 2013; Calizo et al., 2011; Okaty et al., 2019). L-type calcium channels mediate the somatic exocytosis of serotonin whereas blockers of N-, P-, Q- or invertebrate calcium channels do not affect somatic exocytosis (Trueta et al., 2003).

Although, the release machinery, the Ca^{2+}-dependency, and the release sites of norepinephrine and serotonin in the nervous system remain a mystery, the recent development of genetically encoded optical sensors for norepinephrine (Wan et al., 2020) and serotonin (Feng et al., 2019) provide important tools to delineate these questions.

6.7 NEUROPEPTIDE EXOCYTOSIS

Over 100 known neuropeptides, such as neuropeptide Y (NPY) and oxytocin, have been identified in the nervous system, and most of them function as neuromodulators to regulate brain function. They function unlike classical neuronal signal transmission which is mediated by vesicle release of neurotransmitter from the presynaptic terminals to postsynaptic neuron, with the neurotransmitter action being largely confined to the pre- and post-synaptic area. Instead, neuropeptides can spill over to have extrasynaptic actions and sometimes can be released in the soma and dendrites (Tobin et al., 2012). In particular, neuropeptides are unlikely to be confined to synapses because they are packaged in large dense-cored vesicles (LDCVs), containing much more cargos than conventional synaptic vesicles (Leng and Ludwig, 2008). Also, they have higher affinities for their receptors than conventional neurotransmitters with longer half-lives and broader presence throughout the brain (Leng and Ludwig, 2008). Due to direct synaptic transmission and extrasynaptic transmission, neuropeptides are believed to act directly on postsynaptic neurons, and at neurons further away from the sites of release given that neuropeptides diffuse a long distance (van den Pol, 2012). In some cases of nonsynaptic transmission, neuropeptides can directly act on neighboring postsynaptic neurons (van den Pol, 2012). Although classic synaptic secretion is typically mediated by the activation of P-, Q- and N-type Ca^{2+} channels (Reuter, 1996), the release of hormones from excitable endocrine cells depends mostly on L-type Ca^{2+} channels (Kits and Mansvelder, 2000).

The mechanism by which neuropeptides are released in the the nervous system remains largely unknown. Elegant biochemical and functional studies revealed that neuropeptide release, such as that of substance P (Iversen et al., 1976) and oxytocin (Ludwig et al., 2002), requires depolarization and likely Ca^{2+}-influx. Here we highlight recent advances in studying oxytocin and NPY. release

6.7.1 OXYTOCIN

Oxytocin is a peptide hormone synthesized in the supraoptic and paraventricular nuclei of the hypothalamus and released via the posterior pituitary gland to peripheral targets (Pittman et al., 1981; Swaab et al., 1975). Animal studies emphasize the importance of oxytocin in parturition, milk let-down, protective aggression, social behaviors, and pair bonding between mothers and infants and in mating pairs

(Insel and Young, 2001), which is also supported by human studies (Macdonald and Macdonald, 2010).

The release of oxytocin from the nervous system is a topic of considerable interest. It has been reported that oxytocin can be released from dendrites and the process is dependent on the entry of Ca^{2+} into the cell, similar to nerve terminal release, but it is less tightly coupled to the action potential (Leng and Ludwig, 2008). Interestingly, the axonal and dendritic release of oxytocin can be regulated independently, which is supported by studies with α-MSH. It is known that α-MSH can act on oxytocin cells that express MC4 receptors and trigger intracellular Ca^{2+} release in oxytocin cells, evoke dendritic release of oxytocin, and induce expression of the immediate-early gene *c-fos*. However, α-MSH inhibits the electrical activity of oxytocin neurons and thus reduces the secretion of oxytocin into the blood. The dendritic release of oxytocin might happen with or without an increase in electrical activity at the cell bodies, which is dependent on the nature of the afferent or physiological stimulus to the oxytocin cells (Ludwig and Leng, 2006).

Oxytocin release from supraoptic neuron dendrites was dependent primarily on N-type Ca^{2+} channels, and to a lesser extent the P/Q- type Ca^{2+}; other Ca^{2+} calcium channels also contribute to mature oxytocin neurons (Hirasawa et al., 2001; Tobin et al., 2011). Ca^{2+} channel-independent mechanisms differ between axons and the cell body: the endoplasmic reticulum plays a role in calcium regulation/peptide release from the oxytocin cell body but does not affect the axon terminal. Both intracellular Ca^{2+} stores and VGCCs, especially N-type Ca^{2+} channels, appear to be important for dendritic release of oxytocin (Tobin et al., 2011).

Although Syt 4 does not bind to Ca^{2+}, a study has reported that Syt 4 negatively regulates oxytocin release in hypothalamic paraventricular nucleus and the inhibition of Syt 4 prevents dietary obesity by normalizing oxytocin release and energy balance under chronic nutritional excess (Zhang et al., 2011). Moreover, although the conventional SNARE complex involved in dendritic release was reviewed (Ovsepian and Dolly, 2011), other results dispute this idea by showing a lack of some core proteins of the SNARE complex in dendrites or no colocalization with oxytocin (Deleuze et al., 2005; Tobin et al., 2012).

6.7.2 NEUROPEPTIDE Y (NPY)

Neuropeptide Y (NPY), the most abundant peptide in the hypothalamus of the mammalian central nervous system, plays a key role in regulating energy homeostasis including food intake, storage of energy, and stimulating insulin secretion (Billington et al., 1991; Stanley et al., 1986, 1992). NPY contributes a large degree to inhibiting neurotransmitter release from excitatory CA3 neurons in the hippocampus and inhibits excitatory neurotransmission presynaptically at the striatum radiatum-CA1 synapse by reducing calcium influx into the axon terminal (Colmers et al., 1988). NPY shares the same SNARE complex-mediated vesicular exocytosis as oxytocin. However, it can be packaged into dense core vesicles with a smaller volume than LDCV in magnocellular neurons (van den Pol, 2012). Using an optical imaging reporter (i.e. NPY-pHluorin), Dr. Matthijs Verhage's laboratory conducted a series of interesting studies that revealed that NPY release from the DCV requires active zone proteins such as Munc13, as well as the RAB3-Rim pathway. Their study proposed that RIMs and Munc13 work as mammalian alternatives to the yeast exocyst complex RAB3/SEC4 and regulates DCV fusion sites by positioning Munc13 and recruiting NPY-DCVs via RAB3. By applying genetic techniques to knockout different components of active zone proteins, they verified the essential function of each component. For example, NPY-DCV fusion was reduced by over 90% in RAB3 quadruple knockout neurons. In RIM-conditional knockout neurons, DCV release was completely lost and fully restored by expressing the N-terminal RAB3- and Munc13-interacting domains of RIM in a RAB3-dependent manner. Overexpression of Munc13-2 and N-terminal truncated Munc13-2 also rescued the DCV fusion in RIM-deficient neurons (Persoon et al., 2019). In the Munc13-1 and -2 double deletion model, synaptic DCV fusion was reduced but not abolished with loss of synaptic preference, and the remaining fusion required prolonged stimulation, similar to extrasynaptic fusion in WT neurons. However, the overexpression of Munc13-1 promoted extrasynaptic DCV release without prolonged stimulation. Munc13-1/2 facilitated DCV fusion in a different way for synaptic vesicles, which is not essential for DCV release, and overexpression of Munc13 can promote efficient DCV release extrasynaptically (van de Bospoort et al., 2012).

6.8   NEURO-ENDOCRINE EXOCYTOSIS

It has been proposed that endocrine exocytosis shares similar mechanisms as synaptic vesicle exocytosis. For example, chromaffin granule exocytosis in chromaffin cells requires Munc18. The study of Munc18-1 lacking mouse chromaffin cells showed 10-fold reduction of Ca^{2+}-dependent LDCV exocytosis and morphologically docked LDCVs without changing the kinetic properties of the remaining release, suggesting that Munc18-1 functions upstream of SNARE-complex formation and promotes LDCV docking (Voets et al., 2001). Moreover, nearly all of Ca^{2+}-triggering is mediated equally by Syt 1 and Syt 7 (Schonn et al., 2008). The deletion of Syt 1 reduced the overall secretion only 20%, largely due to the persistence of slow exocytosis mediated by Syt 7. Syt 7 knockout or knock in of mutant Syt 7 decreased the Ca^{2+}-triggered exocytosis dramatically. Double knockout of Syt 1 and Syt 7 results in exocytosis levels that are 30% of WT exocytosis levels, and this remaining level of exocytosis is due to a very slow release component that persists (Schonn et al., 2008).

Syts are also reported to be involved in the vesicular exocytosis of other endocrine cells such as pancreatic islet α and β cells and adipocytes (Dolai et al., 2016; Gao et al., 2000; Gauthier et al., 2008; Gustavsson et al., 2009). In pancreatic insulin and glucagon-secreting cells, Syt 7 is the dominant Ca^{2+} sensor (Dolai et al., 2016; Gao et al., 2000; Gauthier et al., 2008; Gustavsson et al., 2009). Ca^{2+}-induced exocytosis in mast cells is also regulated by Syt 1 and Syt 2, where Syt 2 negatively regulates exocytosis of lysosomes (Baram et al., 1998, 1999). Ca^{2+}-triggered exocytosis in fibroblasts and sperms is also regulated by Syt (Feany and Buckley, 1993; Hutt et al., 2002).

Insulin is another highly important hormone, due to its regulation of peripheral glucose metabolism. It is released into the circulation by pancreatic β cells in response to changes in blood glucose. The mechanism of release is initiated by glucose entry into the β cell through the glucose transporter GLUT2. Intracellular glucose is phosphorylated by the enzyme glucokinase and subsequently metabolized to generate ATP. The subsequent increase in ATP:ADP ratio inhibits the ATP-gated potassium channel (K_{ATP}), resulting in membrane depolarization, activation of VGCC, and increased intracellular Ca^{2+} influx. The rise in Ca^{2+} influx then stimulates the exocytosis of insulin-containing vesicles. Once in circulation, insulin binds to membrane receptors in the liver to decrease glucose production, and in fat and skeletal muscle to increase glucose uptake. The transport of glucose into these cells is mediated by GLUT4, which is stored in intracellular vesicles that are translocated to the plasma membrane after insulin-receptor activation. Insulin may also activate GLUT4 present on cell surface (Michelle Furtado et al., 2003). Syt 7 deficient mice are glucose tolerant, due to abnormalities not in insulin secretion but also in GLUT4 trafficking (Li et al., 2007). In addition, adipocyte-secreted leptin has been proved to be secreted in membrane-bound vesicles and undergo Ca^{2+}-triggered secretory granule exocytosis (Wang et al., 2014). Although Ca^{2+} alone is not sufficient to induce leptin secretion, it is required for insulin-stimulated Akt phosphorylation, which stimulates leptin secretion (Wang et al., 2014).

6.9   SUMMARY AND OUTLOOK

Ca^{2+} mediates many activity-dependent cellular signaling events. The molecular mechanisms mediating classical neurotransmitter vesicle exocytosis, including release machinery (SNAREs and active zone proteins) and Ca^{2+} sensor(s), as well as Ca^{2+} dependency, have been well-characterized. However, the mechanism underlying nonclassical neurotransmitter (i.e. LDCVs) release is not known, mainly because currently there are no effective methods that can be used to detect the release of endogenous neuropeptides. In general, although it is believed that LDCVs are not tightly coupled to Ca^{2+} channels, their release is clearly dependent on Ca^{2+} (Pang and Südhof, 2010). Their release sites also remain unknown.

- It is not known whether LDCV exocytosis relies on specific Ca^{2+} sensor proteins or what the molecular identity of Ca^{2+}-binding proteins triggering neuropeptide release from neurons is. Ca^{2+}-binding proteins in the mouse genome can be classified into three categories based on analysis of their protein domain structures using protein the SMART database: C2 domain containing proteins, EF-hand proteins, annexins, and others. C2 domain proteins include all 17 members of Syts, Doc2s, MCTP1–2, Ferlins, Esyt1–3, Copine1–9, SytL1–5, Rab11fip1–5, CC2d1–2, PLCs, PKCs, Munc13s; E-F hand proteins include CaM1–3, NeCab1–3, NCS1, Efcab1–5, Calbinidin1–2, Casbp22; Annexin 1–11; Others include Plscls1–4; rolling balckout1–2; and syntaxin 1. Conclusively identifying the Ca^{2+} sensor(s) and release mode of neuropeptides/DCV fusion will represent a significant advance in our fundamental knowledge of brain function. As mentioned, Syts form a large protein family composed of 17 members. Eight Syts (Syt 1–3, 5–7, 9, and 10) have unambiguous Ca^{2+}-binding sites in their C2 domains. Therefore, it will be a significant effort to systematically address the possible involvement of

Syts in LDCV release. It is well established that Syt 1 and 2 are the primary Ca^{2+} sensors for fast, synchronous SV release, as described in the introductory material (Pang and Südhof, 2010). It has been shown that Syt 1 and 7 are also involved in insulin release (Gustavsson et al., 2008) and Syt 10 is involved in IGF release (Cao et al., 2011). Thus, there is a strong basis for the hypothesis that some of Syts may be involved in neuropeptide release. Nevertheless, an unambiguous role for Syts in mediating Ca^{2+}-dependent neuropeptide release in the brain is not clearly established.

- The Ca^{2+}-dependency of LDCV exocytosis is not known. Moreover, the strict dependence of peptide release on action potentials should be confirmed using Na^+ channel blockers and K^+ channel blockers such as 4-aminopyridine. The inclusion of intracellular free calcium buffers with distinct buffering kinetics (EGTA vs BAPTA) will provide additional insight into the localization of release and dynamic dependence on calcium for peptide release. These experiments will lay the framework for the molecular dissection of the peptide release machinery and new insights into LDCV release mode and recycling.
- Neuropeptide release can be from the somatodendritic areas or axonal compartment of a neuron, but the exact release sites are not known. It is likely that neuropeptides with diverse regulatory effects will have diverse release mechanisms.

Addressing these unknowns requires future technological development, similar to patch clamp electrophysiology for classical neurotransmitters. Current development in optical sensors for neurotransmitter and neuromodulators likely will provide novel tools for detecting endogenous neuropeptide release. These novel technologies combined with mouse genetics and behavioral assays will provide us with exciting insight into harnessing the power of endogenous neuromodulators, especially neuropeptides, in addressing the role of neuromodulation in health and disease.

REFERENCES

Agnati, L.F., Zoli, M., Stromberg, I., and Fuxe, K. (1995). Intercellular communication in the brain: wiring versus volume transmission. Neuroscience *69*, 711–726.

Alonso, A., Merchan, P., Sandoval, J.E., Sanchez-Arrones, L., Garcia-Cazorla, A., Artuch, R., Ferran, J.L., Martinez-de-la-Torre, M., and Puelles, L. (2013). Development of the serotonergic cells in murine raphe nuclei and their relations with rhombomeric domains. Brain Struct Funct *218*, 1229–1277.

Aston-Jones, G., and Bloom, F.E. (1981). Activity of norepinephrine-containing locus coeruleus neurons in behaving rats anticipates fluctuations in the sleep-waking cycle. J Neurosci *1*, 876–886.

Bacaj, T., Wu, D., Yang, X., Morishita, W., Zhou, P., Xu, W., Malenka, R.C., and Südhof, T.C. (2013). Synaptotagmin-1 and synaptotagmin-7 trigger synchronous and asynchronous phases of neurotransmitter release. Neuron *80*, 947–959.

Banerjee, A., Lee, J., Nemcova, P., Liu, C., and Kaeser, P.S. (2020). Synaptotagmin-1 is the Ca(2+) sensor for fast striatal dopamine release. Elife *9*.

Baram, D., Adachi, R., Medalia, O., Tuvim, M., Dickey, B.F., Mekori, Y.A., and Sagi-Eisenberg, R. (1999). Synaptotagmin II negatively regulates Ca^{2+}-triggered exocytosis of lysosomes in mast cells. J Exp Med *189*, 1649–1658.

Baram, D., Linial, M., Mekori, Y.A., and Sagi-Eisenberg, R. (1998). Ca^{2+}-dependent exocytosis in mast cells is stimulated by the Ca^{2+} sensor, synaptotagmin I. J Immunol *161*, 5120–5123.

Baram, D., Mekori, Y.A., and Sagi-Eisenberg, R. (2001). Synaptotagmin regulates mast cell functions. Immunol Rev *179*, 25–34.

Bergquist, F., Niazi, H.S., and Nissbrandt, H. (2002). Evidence for different exocytosis pathways in dendritic and terminal dopamine release in vivo. Brain Res *950*, 245–253.

Berridge, C.W., and Waterhouse, B.D. (2003). The locus coeruleus-noradrenergic system: modulation of behavioral state and state-dependent cognitive processes. Brain Res Brain Res Rev *42*, 33–84.

Billington, C.J., Briggs, J.E., Grace, M., and Levine, A.S. (1991). Effects of intracerebroventricular injection of neuropeptide Y on energy metabolism. Am J Physiol *260*, R321–R327.

Brimblecombe, K.R., Gracie, C.J., Platt, N.J., and Cragg, S.J. (2015). Gating of dopamine transmission by calcium and axonal N-, Q-, T- and L-type voltage-gated calcium channels differs between striatal domains. J Physiol *593*, 929–946.

Caille, I., Dumartin, B., and Bloch, B. (1996). Ultrastructural localization of D1 dopamine receptor immunoreactivity in rat striatonigral neurons and its relation with dopaminergic innervation. Brain Res *730*, 17–31.

Calizo, L.H., Akanwa, A., Ma, X., Pan, Y.Z., Lemos, J.C., Craige, C., Heemstra, L.A., and Beck, S.G. (2011). Raphe serotonin neurons are not homogenous: electrophysiological, morphological and neurochemical evidence. Neuropharmacology *61*, 524–543.

Cao, P., Maximov, A., and Südhof, T.C. (2011). Activity-dependent IGF-1 exocytosis is controlled by the Ca(2+)-sensor synaptotagmin-10. Cell *145*, 300–311.

Chen, Y.A., and Scheller, R.H. (2001). SNARE-mediated membrane fusion. Nat Rev Mol Cell Biol *2*, 98–106.

Coggeshall, R.E. (1972). Autoradiographic and chemical localization of 5-hydroxytryptamine in identified neurons in the leech. Anat Rec *172*, 489–498.

Colmers, W.F., Lukowiak, K., and Pittman, Q.J. (1988). Neuropeptide Y action in the rat hippocampal slice: site and mechanism of presynaptic inhibition. J Neurosci *8*, 3827–3837.

Courtney, N.A., and Ford, C.P. (2014). The timing of dopamine- and noradrenaline-mediated transmission reflects underlying differences in the extent of spillover and pooling. J Neurosci *34*, 7645–7656.

Dahlstroem, A., and Fuxe, K. (1964). Evidence for the existence of monoamine-containing neurons in the central nervous system. I. Demonstration of monoamines in the cell bodies of brain stem neurons. Acta Physiol Scand Suppl, SUPPL 232, 1–55.

Dahlstrom, A., and Fuxe, K. (1964). Localization of monoamines in the lower brain stem. Experientia *20*, 398–399.

De-Miguel, F.F., and Nicholls, J.G. (2015). Release of chemical transmitters from cell bodies and dendrites of nerve cells. Philos Trans R Soc Lond B Biol Sci *370*.

Deleuze, C., Alonso, G., Lefevre, I.A., Duvoid-Guillou, A., and Hussy, N. (2005). Extrasynaptic localization of glycine receptors in the rat supraoptic nucleus: further evidence for their involvement in glia-to-neuron communication. Neuroscience *133*, 175–183.

Diez-Arazola, R., Meijer, M., Bourgeois-Jaarsma, Q., Cornelisse, L.N., Verhage, M., and Groffen, A.J. (2020). Doc2 proteins are not required for the increased spontaneous release rate in synaptotagmin-1-deficient neurons. J Neurosci *40*, 2606–2617.

Dolai, S., Xie, L., Zhu, D., Liang, T., Qin, T., Xie, H., Kang, Y., Chapman, E.R., and Gaisano, H.Y. (2016). Synaptotagmin-7 functions to replenish insulin granules for exocytosis in human islet beta-cells. Diabetes 65, 1962–1976.

Feany, M.B., and Buckley, K.M. (1993). The synaptic vesicle protein synaptotagmin promotes formation of filopodia in fibroblasts. Nature 364, 537–540.

Feng, J., Zhang, C., Lischinsky, J.E., Jing, M., Zhou, J., Wang, H., Zhang, Y., Dong, A., Wu, Z., Wu, H., et al. (2019). A genetically encoded fluorescent sensor for rapid and specific in vivo detection of norepinephrine. Neuron 102, 745–761 e748.

Fernandez-Chacon, R., Konigstorfer, A., Gerber, S.H., Garcia, J., Matos, M.F., Stevens, C.F., Brose, N., Rizo, J., Rosenmund, C., and Südhof, T.C. (2001). Synaptotagmin I functions as a calcium regulator of release probability. Nature 410, 41–49.

Fon, E.A., Pothos, E.N., Sun, B.C., Killeen, N., Sulzer, D., and Edwards, R.H. (1997). Vesicular transport regulates monoamine storage and release but is not essential for amphetamine action. Neuron 19, 1271–1283.

Fortin, G.D., Desrosiers, C.C., Yamaguchi, N., and Trudeau, L.E. (2006). Basal somatodendritic dopamine release requires snare proteins. J Neurochem 96, 1740–1749.

Fuxe, K., Dahlstrom, A., Hoistad, M., Marcellino, D., Jansson, A., Rivera, A., Diaz-Cabiale, Z., Jacobsen, K., Tinner-Staines, B., Hagman, B., et al. (2007). From the Golgi-Cajal mapping to the transmitter-based characterization of the neuronal networks leading to two modes of brain communication: wiring and volume transmission. Brain Res Rev 55, 17–54.

Gao, Z., Reavey-Cantwell, J., Young, R.A., Jegier, P., and Wolf, B.A. (2000). Synaptotagmin III/VII isoforms mediate Ca^{2+}-induced insulin secretion in pancreatic islet beta-cells. J Biol Chem 275, 36079–36085.

Gauthier, B.R., Duhamel, D.L., Iezzi, M., Theander, S., Saltel, F., Fukuda, M., Wehrle-Haller, B., and Wollheim, C.B. (2008). Synaptotagmin VII splice variants alpha, beta, and delta are expressed in pancreatic beta-cells and regulate insulin exocytosis. FASEB J 22, 194–206.

Geffen, L.B., Jessell, T.M., Cuello, A.C., and Iversen, L.L. (1976). Release of dopamine from dendrites in rat substantia nigra. Nature 260, 258–260.

Geppert, M., Goda, Y., Hammer, R.E., Li, C., Rosahl, T.W., Stevens, C.F., and Südhof, T.C. (1994). Synaptotagmin I: a major Ca^{2+} sensor for transmitter release at a central synapse. Cell 79, 717–727.

Groffen, A.J., Brian, E.C., Dudok, J.J., Kampmeijer, J., Toonen, R.F., and Verhage, M. (2004). Ca(2+)-induced recruitment of the secretory vesicle protein DOC2B to the target membrane. J Biol Chem 279, 23740–23747.

Groffen, A.J., Friedrich, R., Brian, E.C., Ashery, U., and Verhage, M. (2006). DOC2A and DOC2B are sensors for neuronal activity with unique calcium-dependent and kinetic properties. J Neurochem 97, 818–833.

Groffen, A.J., Martens, S., Diez Arazola, R., Cornelisse, L.N., Lozovaya, N., de Jong, A.P., Goriounova, N.A., Habets, R.L., Takai, Y., Borst, J.G., et al. (2010). Doc2b is a high-affinity Ca2+ sensor for spontaneous neurotransmitter release. Science 327, 1614–1618.

Grzanna, R., and Molliver, M.E. (1980). The locus coeruleus in the rat: an immunohistochemical delineation. Neuroscience 5, 21–40.

Gustavsson, N., Lao, Y., Maximov, A., Chuang, J.C., Kostromina, E., Repa, J.J., Li, C., Radda, G.K., Südhof, T.C., and Han, W. (2008). Impaired insulin secretion and glucose intolerance in synaptotagmin-7 null mutant mice. Proc Natl Acad Sci USA 105, 3992–3997.

Gustavsson, N., Wei, S.H., Hoang, D.N., Lao, Y., Zhang, Q., Radda, G.K., Rorsman, P., Südhof, T.C., and Han, W. (2009). Synaptotagmin-7 is a principal Ca^{2+} sensor for Ca^{2+}-induced glucagon exocytosis in pancreas. J Physiol 587, 1169–1178.

Hirasawa, M., Kombian, S.B., and Pittman, Q.J. (2001). Oxytocin retrogradely inhibits evoked, but not miniature, EPSCs in the rat supraoptic nucleus: role of N- and P/Q-type calcium channels. J Physiol 532, 595–607.

Howe, M.W., and Dombeck, D.A. (2016). Rapid signalling in distinct dopaminergic axons during locomotion and reward. Nature 535, 505–510.

Hutt, D.M., Cardullo, R.A., Baltz, J.M., and Ngsee, J.K. (2002). Synaptotagmin VIII is localized to the mouse sperm head and may function in acrosomal exocytosis. Biol Reprod 66, 50–56.

Insel, T.R., and Young, L.J. (2001). The neurobiology of attachment. Nat Rev Neurosci 2, 129–136.

Ishimatsu, M., and Williams, J.T. (1996). Synchronous activity in locus coeruleus results from dendritic interactions in pericoerulear regions. J Neurosci 16, 5196–5204.

Iversen, L.L., Jessell, T., and Kanazawa, I. (1976). Release and metabolism of substance P in rat hypothalamus. Nature 264, 81–83.

Katz, B. (1971). Quantal mechanism of neural transmitter release. Science 173, 123–126.

Katz, B., and Miledi, R. (1965). The effect of calcium on acetylcholine release from motor nerve terminals. Proc R Soc Lond B Biol Sci 161, 496–503.

Khan, M.M., and Regehr, W.G. (2020). Loss of Doc2b does not influence transmission at Purkinje cell to deep nuclei synapses under physiological conditions. Elife 9.

Kits, K.S., and Mansvelder, H.D. (2000). Regulation of exocytosis in neuroendocrine cells: spatial organization of channels and vesicles, stimulus-secretion coupling, calcium buffers and modulation. Brain Res Brain Res Rev 33, 78–94.

Kress, G.J., Shu, H.J., Yu, A., Taylor, A., Benz, A., Harmon, S., and Mennerick, S. (2014). Fast phasic release properties of dopamine studied with a channel biosensor. J Neurosci 34, 11792–11802.

Leng, G., and Ludwig, M. (2008). Neurotransmitters and peptides: whispered secrets and public announcements. J Physiol 586, 5625–5632.

Li, J., Cantley, J., Burchfield, J.G., Meoli, C.C., Stockli, J., Whitworth, P.T., Pant, H., Chaudhuri, R., Groffen, A.J., Verhage, M., et al (2014). DOC2 isoforms play dual roles in insulin secretion and insulin-stimulated glucose uptake. Diabetologia 57, 2173–2182.

Li, Y., Wang, P., Xu, J., Gorelick, F., Yamazaki, H., Andrews, N., and Desir, G.V. (2007). Regulation of insulin secretion and GLUT4 trafficking by the calcium sensor synaptotagmin VII. Biochem Biophys Res Commun 362, 658–664.

Liu, C., Kershberg, L., Wang, J., Schneeberger, S., and Kaeser, P.S. (2018). Dopamine secretion is mediated by sparse active zone-like release sites. Cell 172, 706–718 e715.

Llano, I., Gonzalez, J., Caputo, C., Lai, F.A., Blayney, L.M., Tan, Y.P., and Marty, A. (2000). Presynaptic calcium stores underlie large-amplitude miniature IPSCs and spontaneous calcium transients. Nat Neurosci 3, 1256–1265.

Ludwig, M., and Leng, G. (2006). Dendritic peptide release and peptide-dependent behaviours. Nat Rev Neurosci 7, 126–136.

Ludwig, M., Sabatier, N., Bull, P.M., Landgraf, R., Dayanithi, G., and Leng, G. (2002). Intracellular calcium stores regulate activity-dependent neuropeptide release from dendrites. Nature 418, 85–89.

Ludwig, M., and Stern, J. (2015). Multiple signalling modalities mediated by dendritic exocytosis of oxytocin and vasopressin. Philos Trans R Soc Lond B Biol Sci 370.

Luo, F., and Südhof, T.C. (2017). Synaptotagmin-7-mediated asynchronous release boosts high-fidelity synchronous transmission at a central synapse. Neuron 94, 826–839 e823.

Macdonald, K., and Macdonald, T.M. (2010). The peptide that binds: a systematic review of oxytocin and its prosocial effects in humans. Harv Rev Psychiatry 18, 1–21.

Mackler, J.M., Drummond, J.A., Loewen, C.A., Robinson, I.M., and Reist, N.E. (2002). The C(2)B Ca(2+)-binding motif of synaptotagmin is required for synaptic transmission in vivo. Nature 418, 340–344.

Mendez, J.A., Bourque, M.J., Fasano, C., Kortleven, C., and Trudeau, L.E. (2011). Somatodendritic dopamine release requires synaptotagmin 4 and 7 and the participation of voltage-gated calcium channels. J Biol Chem 286, 23928–23937.

Michelle Furtado, L., Poon, V., and Klip, A. (2003). GLUT4 activation: thoughts on possible mechanisms. Acta Physiol Scand 178, 287–296.

Nagy, G., Kim, J.H., Pang, Z.P., Matti, U., Rettig, J., Südhof, T.C., and Sorensen, J.B. (2006). Different effects on fast exocytosis induced by synaptotagmin 1 and 2 isoforms and abundance but not by phosphorylation. J Neurosci 26, 632–643.

Neher, E. (1992). Nobel lecture. Ion channels for communication between and within cells. EMBO J 11, 1672–1679.

Okaty, B.W., Commons, K.G., and Dymecki, S.M. (2019). Embracing diversity in the 5-HT neuronal system. Nat Rev Neurosci 20, 397–424.

Orita, S., Sasaki, T., Naito, A., Komuro, R., Ohtsuka, T., Maeda, M., Suzuki, H., Igarashi, H., and Takai, Y. (1995). Doc2: a novel brain protein having two repeated C2-like domains. Biochem Biophys Res Commun 206, 439–448.

Ovsepian, S.V., and Dolly, J.O. (2011). Dendritic SNAREs add a new twist to the old neuron theory. Proc Natl Acad Sci USA 108, 19113–19120.

Pang, Z.P., Bacaj, T., Yang, X., Zhou, P., Xu, W., and Südhof, T.C. (2011). Doc2 supports spontaneous synaptic transmission by a Ca(2+)-independent mechanism. Neuron 70, 244–251.

Pang, Z.P., Cao, P., Xu, W., and Südhof, T.C. (2010a). Calmodulin controls synaptic strength via presynaptic activation of calmodulin kinase II. J Neurosci 30, 4132–4142.

Pang, Z.P., Melicoff, E., Padgett, D., Liu, Y., Teich, A.F., Dickey, B.F., Lin, W., Adachi, R., and Südhof, T.C. (2006a). Synaptotagmin-2 is essential for survival and contributes to Ca^{2+} triggering of neurotransmitter release in central and neuromuscular synapses. J Neurosci 26, 13493–13504.

Pang, Z.P., Shin, O.H., Meyer, A.C., Rosenmund, C., and Sudhof, T.C. (2006b). A gain-of-function mutation in synaptotagmin-1 reveals a critical role of Ca^{2+}-dependent soluble N-ethylmaleimide-sensitive factor attachment protein receptor complex binding in synaptic exocytosis. J Neurosci 26, 12556–12565.

Pang, Z.P., and Südhof, T.C. (2010). Cell biology of Ca^{2+}-triggered exocytosis. Curr Opin Cell Biol 22, 496–505.

Pang, Z.P., Sun, J., Rizo, J., Maximov, A., and Südhof, T.C. (2006c). Genetic analysis of synaptotagmin 2 in spontaneous and Ca^{2+}-triggered neurotransmitter release. EMBO J 25, 2039–2050.

Pang, Z.P., Xu, W., Cao, P., and Südhof, T.C. (2010b). Calmodulin suppresses synaptotagmin-2 transcription in cortical neurons. J Biol Chem 285, 33930–33939.

Perin, M.S., Fried, V.A., Mignery, G.A., Jahn, R., and Südhof, T.C. (1990). Phospholipid binding by a synaptic vesicle protein homologous to the regulatory region of protein kinase C. Nature 345, 260–263.

Perin, M.S., Johnston, P.A., Ozcelik, T., Jahn, R., Francke, U., and Südhof, T.C. (1991). Structural and functional conservation of synaptotagmin (p65) in Drosophila and humans. J Biol Chem 266, 615–622.

Persoon, C.M., Hoogstraaten, R.I., Nassal, J.P., van Weering, J.R.T., Kaeser, P.S., Toonen, R.F., and Verhage, M. (2019). The RAB3-RIM pathway is essential for the release of neuromodulators. Neuron 104, 1065–1080.e1012.

Pittman, Q.J., Blume, H.W., and Renaud, L.P. (1981). Connections of the hypothalamic paraventricular nucleus with the neurohypophysis, median eminence, amygdala, lateral septum and midbrain periaqueductal gray: an electrophysiological study in the rat. Brain Res 215, 15–28.

Ramirez, D.M.O., Crawford, D.C., Chanaday, N.L., Trauterman, B., Monteggia, L.M., and Kavalali, E.T. (2017). Loss of Doc2-dependent spontaneous neurotransmission augments glutamatergic synaptic strength. J Neurosci *37*, 6224–6230.

Reuter, H. (1996). Diversity and function of presynaptic calcium channels in the brain. Curr Opin Neurobiol *6*, 331–337.

Rice, M.E., and Cragg, S.J. (2008). Dopamine spillover after quantal release: rethinking dopamine transmission in the nigrostriatal pathway. Brain Res Rev *58*, 303–313.

Sakaguchi, G., Orita, S., Maeda, M., Igarashi, H., and Takai, Y. (1995). Molecular cloning of an isoform of Doc2 having two C2-like domains. Biochem Biophys Res Commun *217*, 1053–1061.

Schimchowitsch, S., Stoeckel, M.E., Klein, M.J., Garaud, J.C., Schmitt, G., and Porte, A. (1983). Oxytocin-immunoreactive nerve fibers in the pars intermedia of the pituitary in the rabbit and hare. Cell Tissue Res *228*, 255–263.

Schonn, J.S., es> (2017). Molecular neuroscience in the 21(st) century: a personal perspective. Neuron *96*, 536–541.

Sun, J., Pang, Z.P., Qin, D., Fahim, A.T., Adachi, R., and Südhof, T.C. (2007). A dual-Ca^{2+}-sensor model for neurotransmitter release in a central synapse. Nature *450*, 676–682.

Swaab, D.F., Pool, C.W., and Nijveldt, F. (1975). Immunofluorescence of vasopressin and oxytocin in the rat hypothalamo-neurohypophypopseal system. J Neural Transm *36*, 195–215.

Swanson, L.W. (1976). The locus coeruleus: a cytoarchitectonic, Golgi and immunohistochemical study in the albino rat. Brain Res *110*, 39–56.

Swanson, L.W., and Hartman, B.K. (1975). The central adrenergic system. An immunofluorescence study of the location of cell bodies and their efferent connections in the rat utilizing dopamine-beta-hydroxylase as a marker. J Comp Neurol *163*, 467–505.

Takahashi, T., and Momiyama, A. (1993). Different types of calcium channels mediate central synaptic transmission. Nature *366*, 156–158.

Tobin, V., Leng, G., and Ludwig, M. (2012). The involvement of actin, calcium channels and exocytosis proteins in somato-dendritic oxytocin and vasopressin release. Front Physiol *3*, 261.

Tobin, V.A., Douglas, A.J., Leng, G., and Ludwig, M. (2011). The involvement of voltage-operated calcium channels in somato-dendritic oxytocin release. PLoS One *6*, e25366.

Trueta, C., and De-Miguel, F.F. (2012). Extrasynaptic exocytosis and its mechanisms: a source of molecules mediating volume transmission in the nervous system. Front Physiol *3*, 319.

Trueta, C., Mendez, B., and De-Miguel, F.F. (2003). Somatic exocytosis of serotonin mediated by L-type calcium channels in cultured leech neurones. J Physiol *547*, 405–416.

Uchigashima, M., Ohtsuka, T., Kobayashi, K., and Watanabe, M. (2016). Dopamine synapse is a neuroligin-2-mediated contact between dopaminergic presynaptic and GABAergic postsynaptic structures. Proc Natl Acad Sci USA *113*, 4206–4211.

van de Bospoort, R., Farina, M., Schmitz, S.K., de Jong, A., de Wit, H., Verhage, M., and Toonen, R.F. (2012). Munc13 controls the location and efficiency of dense-core vesicle release in neurons. J Cell Biol *199*, 883–891.

van den Pol, A.N. (2012). Neuropeptide transmission in brain circuits. Neuron *76*, 98–115.

Voets, T., Toonen, R.F., Brian, E.C., de Wit, H., Moser, T., Rettig, J., Sudhof, T.C., Neher, E., and Verhage, M. (2001). Munc18-1 promotes large dense-core vesicle docking. Neuron *31*, 581–591.

Wan, J., Peng, W., Li, X., Qian, T., Song, K., Zeng, J., Deng, F., Hao, S., Feng, J., Zhagn, P., et al. (2020). A genetically encoded GRAB sensor for measuring serotonin dynamics in vivo. *bioRxiv*.

Wang, C., Kang, X., Zhou, L., Chai, Z., Wu, Q., Huang, R., Xu, H., Hu, M., Sun, X., Sun, S., et al. (2018). Synaptotagmin-11 is a critical mediator of parkin-linked neurotoxicity and Parkinson's disease-like pathology. Nat Commun 9, 81.

Wang, Y., Ali, Y., Lim, C.Y., Hong, W., Pang, Z.P., and Han, W. (2014). Insulin-stimulated leptin secretion requires calcium and PI3K/Akt activation. Biochem J *458*, 491–498.

Xu, J., Mashimo, T., and Südhof, T.C. (2007). Synaptotagmin-1, -2, and -9: Ca(2+) sensors for fast release that specify distinct presynaptic properties in subsets of neurons. Neuron *54*, 567–581.

Xu, J., Pang, Z.P., Shin, O.H., and Südhof, T.C. (2009). Synaptotagmin-1 functions as a Ca^{2+} sensor for spontaneous release. Nat Neurosci *12*, 759–766.

Yao, J., Gaffaney, J.D., Kwon, S.E., and Chapman, E.R. (2011). Doc2 is a Ca2+ sensor required for asynchronous neurotransmitter release. Cell *147*, 666–677.

Yung, K.K., Bolam, J.P., Smith, A.D., Hersch, S.M., Ciliax, B.J., and Levey, A.I. (1995). Immunocytochemical localization of D1 and D2 dopamine receptors in the basal ganglia of the rat: light and electron microscopy. Neuroscience *65*, 709–730.

Zhang, G., Bai, H., Zhang, H., Dean, C., Wu, Q., Li, J., Guariglia, S., Meng, Q., and Cai, D. (2011). Neuropeptide exocytosis involving synaptotagmin-4 and oxytocin in hypothalamic programming of body weight and energy balance. Neuron *69*, 523–535.

7 Nonclassical Ion Channels in Learning and Memory

Ze-Jie Lin, Xue Gu, Tian-Le Xu, and Wei-Guang Li

CONTENTS

7.1 INTRODUCTION

Learning and memory are the basic functions of the brain through which both humans and animals can adjust their behaviors to survive in ever-changing internal and external environments. Memory is a dynamic process wherein a higher living organism (e.g. humans) recognizes, maintains, recurs, or reconsiders an experience and thus is the basis for thinking, imagination, and other higher cognitive activities (Figure 7.1A). The mechanisms of learning and memory remain one of the great unanswered questions in the field of biology and neuroscience. A combination of top-down and bottom-up approaches has been taken to elucidate the neurobiology of learning and memory, with the former focusing on animal behaviors associated with memory acquisition, retrieval, consolidation, reconsolidation, and extinction, as well as the brain regions where these processes occur. On the other hand, the bottom-up approach explores the cellular and circuit mechanisms of memory encoding and storage by examining the efficacy of neuronal discharge patterns and synaptic transmission (Poo et al. 2016). In order to build a bridge between these two lines of research, the concept of memory engram with four defining criteria—persistence, ecphory, content, and dormancy—has been proposed to define it as a persistent offline representation of past experience (Figure 7.1A),

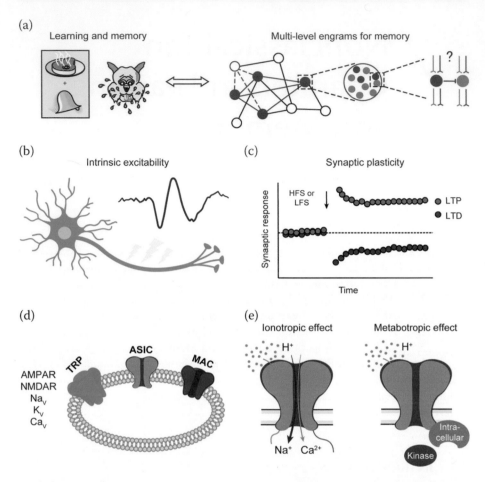

FIGURE 7.1 Schematic diagram of the basic principles of ion channel involvement in learning and memory. (A) Relationship between a memory and its engram in the brain. (*Left*) Memory is the capacity of an organism to acquire, store and recover information based on experience. Shown is an example of Pavlovian reflexes as the definition of memory. Initially, placing meat powder in a dog's mouth elicited salivation (unconditioned stimulus, US), whereas the presence of a neutral stimulus such as a bell did not. The conditioning process involved pairing the innocuous stimuli (i.e. the bell, conditioned stimulus, CS) with the food, such that they preceded food delivery. The fascinating result was that after a number of these pairings, the mere presentation of the paired stimulus (i.e. the bell) elicited salivation. (*Right*) The memory engram (i.e. the physical trace of a memory in the brain) has been proposed as a persistent offline representation that involves strengthening connections between different brain regions, as well as between sets of neurons (neuronal ensemble) that are active (red) during an event. Notably, an engram is not yet a memory but provides the necessary physical conditions for a memory to emerge. **(B, C)** Both intrinsic excitability **(B)** and synaptic plasticity [**C**, including long-term potentiation (LTP) and long-term depression (LTD), that are typically induced by high-frequency stimulation (HFS) and low-frequency stimulation (LFS), respectively] are the two necessary components for the formation and regulation of memory engrams. **(D)** Typical examples of classical and nonclassical ion

although its exact neurobiological nature in the nervous system is still under active investigation (Josselyn, Kohler, and Frankland 2015; Josselyn and Tonegawa 2020). Advances in memory engram research suggest that both increased intrinsic excitability (Figure 7.1B) and synaptic plasticity (Figure 7.1C) work hand in hand to form memory engrams, namely subsets of neurons (likely also including synapses or other unidentified units) that are recruited during learning and memory retrieval, and that provide the necessary physical conditions for a memory to emerge, and these mechanisms are also implicated in memory consolidation and retrieval processes. While the control of intrinsic excitability is eventually accomplished by the voltage-gated ion channels permeable to either sodium (Na_V) or potassium (K_V) (Isacoff, Jan, and Minor 2013; Pignatelli et al. 2019), control of synaptic plasticity is considered to be mainly mediated by the ligand-gated ion channels, typified by the glutamatergic N-methyl-D-aspartic acid receptors (NMDARs) and α-amino-3-hydroxy-5-methyl-4-isoxazolepropionic acid receptors (AMPARs) (Figure 7.1D) (Bliss and Collingridge 1993; Hong et al. 2013; Nabavi et al. 2014; Wright et al. 2020). The Ca^{2+} entry through NMDARs (Lee et al. 2009) in addition to voltage-gated Ca^{2+} channels (Ca_V) (Wheeler et al. 2012) results in the activation of Ca^{2+}/calmodulin-dependent kinase II (CaMKII) (Sanhueza and Lisman 2013), which initiates a series of molecular events to drive cytoskeleton rearrangements and gene expression for memory storage. Mechanistically, for a particular subtype of ion channel, both the ionotropic and metabotropic mechanisms could be responsible for its involvement in learning and memory processes. Ion channels allow ions to cross membranes, leading to changes in membrane excitability as well as specific ion-mediated signal transduction processes with a focus on the Ca^{2+} ion (Li, Tadross, and Tsien 2016; Yap and Greenberg 2018), both of which are referred to be the ionotropic signal mechanism. Furthermore, ion channels possess a largely unidentified metabotropic cascade, i.e. a conformational

---

FIGURE 7.1 (Continued). channels described in the present chapter. While the control of intrinsic excitability for the formation and regulation of memory engram is eventually accomplished by the voltage-gated ion channels permeable to either sodium (Na_V) or potassium (K_V) that drive action potential generation and propagation, that of synaptic plasticity for memory engram is considered to be centrally mediated by the ligand-gated ion channels from fast chemical synapses, typified by the glutamatergic N-methyl-D-aspartic acid receptors (NMDARs) and α-amino-3-hydroxy-5-methyl-4-isoxazolepropionic acid receptor (AMPARs). Beyond these classical voltage- and ligand-gated ion channels, several recently discovered ion channels, including acid-sensing ion channels (ASICs), transient receptor potential (TRP) channels, and mechanically activated channels (MACs), emerge to be important for various aspects of synaptic plasticity and thus for learning and memory. (E) For a particular ion channel, both the ionotropic and metabotropic mechanisms are thought to be responsible for its involvement in learning and memory processes. Ion channels allow ions to cross membranes, leading to changes in membrane excitability as well as specific ion-mediated signal transduction processes with a focus on the Ca^{2+} ion, both of which are referred to be the ionotropic signal mechanism. Furthermore, ion channels possess a largely unidentified metabotropic cascade, i.e. a conformational change downstream of the channel (signaling complexes formation), through which some protein kinases are usually directly or indirectly activated.

change downstream of the channel (signaling complexes formation), through which some protein kinases are usually directly or indirectly activated (Figure 7.1E) (Li, Tadross, and Tsien 2016; Valbuena and Lerma 2016; Petrovic et al. 2017). Overall, the involvement of these two classical classes of ion channels, i.e. voltage- and ligand-gated ion channels, in learning and memory has been well documented and comprehensively reviewed (Malinow and Malenka 2002; Lynch 2004; Nakazawa et al. 2004; Huganir and Nicoll 2013; Herring and Nicoll 2016; Iacobucci and Popescu 2017; Trimmer and Rhodes 2004; Bean 2007; Adelman, Maylie, and Sah 2012; Biel et al. 2009; He et al. 2014).

Besides classical voltage-gated channels and ionic receptors, the less-characterized ion channels, such as acid-sensing ion channels (ASICs) (Waldmann et al. 1997), transient receptor potential (TRP) channels (Clapham, Runnels, and Strubing 2001; Minke and Cook 2002; Ramsey, Delling, and Clapham 2006; Nilius et al. 2007), and mechanically activated channels (MACs) (Coste et al. 2012; Kim et al. 2012; Yan et al. 2013; Ranade, Syeda, and Patapoutian 2015; Murthy, Dubin, and Patapoutian 2017), have been shown to be equally important for synaptic plasticity and memory (Figure 7.1D). Understanding such parts of nonclassical ion channels in learning and memory should inform our ability to connect the operational rules and channel activities with the fulfillment of the memory engrams. Focusing on this integrative aspect of channel function, it will be essential to uncover how the complex intracellular signaling network of a neuron shapes the dynamics of electrical activity and synaptic signaling for learning and memory. In this chapter, we review the biologically salient features of nonclassical ion channels, in particular ASICs as the paradigmatic examples, and highlight the more specialized activities necessary for channel function with an emphasis on the particular cellular contexts under the behavioral tasks of learning and memory, hopefully advancing the mechanistic understanding of cellular and molecular substrates in the memory field.

7.2 MAIN TEXT

7.2.1 DISTRIBUTION AND ACTIVATION OF ASICS

ASICs are members of the sodium-selective cation channels belonging to the degenerin/epithelial sodium channel (DEG/ENaC) family, which can be activated by extracellular acidosis and are widely expressed in the mammalian nervous system (Waldmann et al. 1997). To date, at least six ASIC subunits (ASIC1a, ASIC1b, ASIC2a, ASIC2b, ASIC3, and ASIC4; where a and b refer to splice variants) have been identified to be encoded by four genes (*Asic1*, *Asic2*, *Asic3*, and *Asic4*) (Kellenberger and Schild 2002, 2015; Wemmie, Price, and Welsh 2006; Wemmie, Taugher, and Kreple 2013). In mammals, there is a relative specificity of the distribution of various ASIC subtypes between the central and peripheral nervous systems (Krishtal 2003). While ASIC1a, ASIC2a, and ASIC2b are expressed in both peripheral and central nervous systems, ASIC1b and ASIC3 are primarily expressed in sensory neurons and non-neuronal tissues. As the most sensitive ion channel subtypes to extracellular acidosis, ASIC1a and ASIC3 have been proposed as ideal acid sensors in the central and peripheral nervous systems (Wemmie, Price, and Welsh 2006). ASIC3 is well positioned to be involved in multimodal sensory perception (Li and Xu 2011), especially in the acidosis-related chemosensation (Yagi et al. 2006; Birdsong et al. 2010; Yu et al. 2010; Peng et al. 2015; Marra et al. 2016; Stephan et al. 2018) and nociception (Chen et al. 2002; Sluka et al. 2003, 2007; Ikeuchi et al. 2008; Deval et al. 2008, 2011; Wang, Li, et al. 2013; Sluka and Gregory 2015) as well as mechanosensation (Price et al. 2001; Fromy et al. 2012; Jalalvand et al. 2016; Lin et al. 2016). By contrast, ASIC1a has been pathophysiologically implicated in ischemic neuronal death (Xiong et al. 2004; Gao et al. 2005; Pignataro, Simon, and Xiong 2007; Duan et al. 2011; Zeng et al. 2013; Wang, Zeng, et al. 2013; Wang et al. 2015, 2020) and chronic pain (Wu et al. 2004; Mazzuca et al. 2007; Duan et al. 2007, 2012; Bohlen et al. 2011; Diochot et al. 2012; Li et al. 2019). Of note, ASIC1a is also proposed as a novel synaptic receptor to cleft acidification and actively engaged in synaptic transmission and diverse types of synaptic plasticity (Kreple et al., 2014; Gonzalez-Inchauspe et al. 2017; Hill and Ben-Shahar 2018; Uchitel, Gonzalez Inchauspe, and Weissmann 2019; Mango and Nistico 2020). Due to the behavioral outputs associated with its synaptic action, ASIC1a has been shown to be of great importance for multiple forms of learning and memory. Here, we summarize these typical examples of ASIC1a involvement in learning and memory, which may advance our understanding of the physiological roles of this type of nonclassical ion channels.

7.2.2 ASIC1A IN SPATIAL LEARNING AND MEMORY

The Morris water maze is one of the most widely used behavioral paradigms for studying the neurobiological mechanisms of spatial learning and memory. During this behavioral task, animals (usually rats or mice) are placed in a large circular pool of water and required to escape from water onto a hidden platform whose location can normally be identified only through spatial memory. There are no local cues indicating where the platform is located (Morris 1981). Conceptually, this task requires place cells, which are neurons in the hippocampus that recognize or represent points in space in the environment (O'Keefe 1976). Cellularly, long-term potentiation (LTP) of synaptic transmission in the hippocampal CA1 region serves as the primary experimental model for investigating the synaptic basis of spatial learning and memory (Bliss and Collingridge 1993; Tsien, Huerta, and Tonegawa 1996).

ASIC1a is believed to be the dominant isoform in the central nervous system, since loss of ASIC1a largely abolished acid-evoked currents. Whether ASIC1a participates in the hippocampal LTP and thus in spatial learning and memory remains controversial. Initially, Wemmie et al. (2002) identified that ASIC1a was expressed in hippocampal synaptosomes and in dendrites localized at synapses. Global knockout of *Asic1a* in mice led to a reduction in excitatory postsynaptic potentials and impaired NMDAR activation during high frequency stimulation (HFS) for LTP induction, thereby modestly impaired this form of synaptic plasticity in hippocampal slices (Wemmie et al. 2002). As a result, the *Asic1a*-null mice showed a mild deficit in spatial learning that could be overcome by intensive training (Wemmie et al. 2002). However, another research group (Wu et al. 2013) took advantages of a newly generated floxed ASIC1a mice as the conditioned *Asic1a*-null mice but failed to observe the LTP deficit. Consistently, their behavioral analysis showed that mice lacking ASIC1a had normal performance in hippocampus-dependent spatial memory (Wu et al. 2013). Additional investigation of the exact roles of ASIC1a in hippocampal synaptic plasticity (Buta et al. 2015; Quintana et al. 2015; Liu et al. 2016) tends to support its modest but significant contribution to LTP, although their behavioral relevance in spatial learning and memory needs to be re-examined in the future.

Recently, ASIC1a has been reported to be critical for both metabotropic glutamate receptor (mGluR)-dependent long-term depression (LTD) (Mango et al. 2017) and NMDAR-dependent LTD (Mango and Nistico 2019) in the hippocampus. It is suggested that ASIC1a may act by weakening previously encoded memory traces when new information is learned and is thus required for behavioral flexibility in both the Morris water maze and a delayed nonmatch to place T-maze task (Nicholls et al. 2008). Therefore, a more comprehensive study on the roles of ASIC1a in different aspects of spatial learning and memory, including reversal learning, is desirable in the future. Together, these results identify ASIC1a as a key component of hippocampal synaptic plasticity and spatial learning and memory.

7.2.3   ASIC1a in Fear Learning and Memory

In addition to the hippocampus, ASICs are also abundantly distributed in several brain areas with strong excitatory synaptic inputs such as the olfactory bulb, whisker barrel cortex, cingulate cortex, striatum, nucleus accumbens, amygdala, and cerebellar cortex (Wemmie et al. 2003). In particular, the level of ASIC1a is higher in the amygdala than hippocampal neurons and disrupting the *Asic1a* gene eliminates acid-evoked currents in the amygdala. Behaviorally, consistent with the well-recognized roles of amygdala circuits in fear memory (LeDoux 2000), the deficiency of ASIC1a impairs the performance of mice in cue and context fear conditioning, but does not affect the baseline fear of mice in the elevated maze (Wemmie et al. 2003). Moreover, mice lacking ASIC1a also exhibit deficits in unconditioned fear behaviors (Coryell et al. 2007). Conversely, overexpression of ASIC1a throughout the brain in transgenic mice enhances neuronal acid-evoked cation currents and consequently increases acquired fear-related behaviors (Wemmie et al. 2004). These results suggest that ASIC1a indeed acts as a major modulator for bidirectional control of fear learning and memory.

To probe a specific location of ASIC1a action in fear memory, researchers selectively expressed ASIC1a in the basolateral amygdala (BLA) of $Asic1a^{-/-}$ mice using viral vector-mediated gene transfer, and found that it successfully rescued the context-dependent fear memory, but not the freezing deficit during training or the unconditioned fear response to predator odor. These data pinpoint the BLA as one crucial site where ASIC1a contributes to fear memory (Coryell et al. 2008). Interestingly, ASIC1a in BLA can also be activated by carbon dioxide (CO_2)-induced acidosis (Ziemann et al. 2009), which provides a molecular and neuronal explanation for how rising CO_2 concentrations can elicit intense fear and provide a foundation for dissecting the bases of anxiety and panic disorders. The CO_2 itself causes a reduction in brain pH, which triggers significant freezing response and reduces the explorative activity in the open field test. CO_2 does not act as an unconditioned stimulus but could enhance fear memory when coupled with footshocks as the unconditioned stimulus during contextual fear conditioning. All of these effects of CO_2 are dependent on the presence of ASIC1a in BLA (Ziemann et al. 2009). Finally, beyond the amygdala (Feinstein et al. 2013), ASIC1a in the bed nucleus of the stria terminalis (BNST) also plays an important role in the CO_2-evoked fear-related behaviors (Taugher et al. 2014).

LTP at lateral amygdala neurons has long been considered to be crucial for cued fear learning and it represents a candidate mechanism through which subsets of neurons (i.e. engram cells) are recruited during fear learning and memory retrieval (Nabavi et al. 2014; Bocchio, Nabavi, and Capogna 2017). Despite the observation that stimulation of presynaptic terminals can increase the proton concentration in synapses, the direct evidence that synaptically released protons serve as a neurotransmitter and activate postsynaptic ASIC1a has always been lacking. To address this issue, Du et al. (2014) identified that presynaptic stimulation transiently reduced extracellular pH in the amygdala. They also reported the excitatory postsynaptic currents generated by the protons-activated ASIC1a in lateral amygdala pyramidal neurons. Notably, both protons and ASIC1a are required for LTP in lateral amygdala neurons. Collectively, these results identify protons as a neurotransmitter and establish ASIC1a as a postsynaptic receptor involved in LTP in the lateral amygdala critical for amygdala-dependent fear learning and memory.

Since the amygdala comprises a heterogeneous collection of nuclei, including the BLA, the central amygdala (CeA), and intercalated cell mass (ICM), it is not clear whether the distribution of ASIC1a varies among different cell types within the amygdala circuits, or whether synaptic plasticity in other cell types depends on ASIC1a and such plasticity also plays a role in fear learning. In this context, Chiang et al. (2015) compared the ASIC currents in different types of amygdala neurons in acute brain slices and found that ASIC was differentially expressed in different cells and regions of the amygdala. Notably, the level of ASIC expression in postsynaptic neurons affected the degree of LTP in different glutamatergic synapses. Furthermore, selective deficiency of ASIC1a in amygdala output neurons could eliminate LTP in these cells and hence reduce the same degree of fear learning. In summary, these findings demonstrate that ASIC1a is expressed differentially in diverse cell types within the amygdala network, and fear learning requires ASIC1a-dependent LTP at multiple amygdala synapses.

The current understanding on dynamic memory processes posits that after retrieval of a memory, the corresponding engram is reactivated and subsequently undergoes a destabilization process thereby necessitating reconsolidation to restabilize the reactivated/destabilized memory (Nader, Schafe, and Le Doux 2000; Lee, Nader, and Schiller 2017; Yan et al. 2020). As a result, the reconsolidation process enables memories to be updated with new information. Moreover, disruption of reconsolidation by administration of amnestic agents or behavioral interference shortly after memory reactivation typically causes long-lasting memory impairment in a variety of experimental paradigms including fear memory in both laboratory animals (Nader, Schafe, and Le Doux 2000) and human subjects (Schiller et al. 2010). Accordingly, increasing lability after retrieval might make the memory more amenable to modification by reconsolidation-update procedure, whilst augmenting synaptic signaling during retrieval would increase memory lability (Clem and Huganir 2010; Hong et al. 2013; Wright et al. 2020). Taking advantage of the role of ASIC1a in adjusting synaptic transmission and plasticity, Du et al. (2017) subjected mice to CO_2 inhalation to activate ASIC1a in amygdala neurons, which together with retrieval increased activation of amygdala neurons bearing the memory trace and increased the synaptic exchange from Ca^{2+}-impermeable to Ca^{2+}-permeable AMPA receptors. As a result, such transient acidification increased the retrieval-induced lability of an aversive memory, which could then be weakened by an extinction protocol or strengthened by reconditioning. These results suggest that transient acidosis during retrieval by enhancing synaptic signal in an ASIC1a-dependent manner renders the memory of an aversive event more labile through enhancing synaptic signaling in an ASIC1a-dependent manner.

7.2.4   ASIC1A IN EXTINCTION LEARNING AND MEMORY

Extinction learning, a critical form of learning and memory, is necessary for adaptation of the organism to the constantly changing environment. Numerous brain diseases, especially anxiety disorder and post-traumatic stress disorder, are intimately connected with deficiencies in extinction learning (Milad and Quirk 2012; Dunsmoor et al. 2015; Bocchio, Nabavi, and Capogna 2017; Lebois et al. 2019). Fear extinction learning and memory requires dynamic involvement of the interconnected circuits including ventral hippocampus (vHPC), medial prefrontal cortex (mPFC), and basolateral amygdala (BLA), but the core molecular players that regulate these circuits to achieve fear extinction are still unclear. Exploring the circuit-specific function of ASIC1a, our previous results identified that this channel in vHPC, but not dorsal hippocampus, mPFC, or BLA, as a crucial molecular regulator of fear extinction (Figure 7.2) (Wang et al. 2018). In stark contrast, ASIC1a expression in BLA is critical for acquisition of cued fear memory. To uncover the synaptic correlates underlying the fear extinction learning, we further found that ASIC1a in vHPC drives distinct synaptic adaptations at the neural projections from vHPC to the infralimbic and prelimbic subdivisions of mPFC (termed as IL/mPFC and PL/mPFC), respectively. Namely, extinction learning caused a significant enhancement of both presynaptic neurotransmitter release and postsynaptic NMDAR function at the projection circuit from vHPC to IL/mPFC, but decreased these indexes at the circuit from vHPC to PL/mPFC, all of which were diminished in the vHPC *Asic1a* conditional knockout mice. Several neuronal activity-regulated and memory-related genes, including *Fos*, *Npas4*, and *Bdnf* as the potential mediators of ASIC1a regulation of fear extinction, were identified by gene expression profiling analysis and validating experiments. Notably, either genetic overexpression of BDNF in vHPC or supplement of BDNF protein in mPFC rescued the deficiency in fear extinction as well as extinction-driven maladaptive alterations of hippocampal-prefrontal correlates caused by the *Asic1a* gene

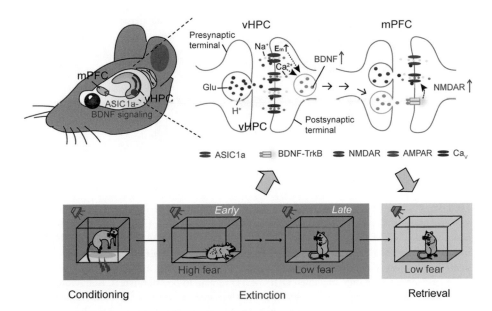

FIGURE 7.2   A working model of vHPC ASIC1a-dependent regulation of fear extinction. vHPC ASIC1a activity drives BDNF expression in vHPC in response to fear extinction training and subsequently alters the synaptic efficacy of projections from vHPC to mPFC via forward BDNF signaling. Adapted from Wang et al. (2018) with permission. Please see text for more details.

inactivation in vHPC. Together, these findings emphasize ASIC1a as a critical constituent in fear extinction circuits and thus a promising target for managing adaptive behaviors related to fear memory.

Incapacity of aversive memory extinction is one of the common causes for brain diseases such as anxiety disorder and anorexia nervosa. For instance, if the conditioned memory of taste aversion cannot be extinguished effectively, it will easily lead to refractory feeding difficulties during the growth and development of children, which will irreversibly have adverse effects on the physical and intellectual welfare of young children. The insular cortex is well known for its critical role in encoding taste learning (Rosenblum et al. 1997) and processing aversively motivated learning tasks, such as conditioned taste aversion (CTA) (Garcia, Kimeldorf, and Koelling 1955; Chambers 1990; Elkobi et al. 2008; Adaikkan and Rosenblum 2015), a form of associative learning where the subject associates a novel taste with a subsequent transient visceral illness. At the synaptic level, LTP in insular cortex is demonstrated to contribute to the acquisition of CTA (Escobar, Alcocer, and Chao 1998; Jones et al. 1999). However, the molecular and synaptic mechanisms underlying CTA memory extinction (Berman and Dudai 2001; Eisenberg et al. 2003) remain undetermined. To answer this question, we reported the involvement of ASIC1a in mediating long-term depression (LTD) through a mechanism that requires glycogen synthase kinase-3β (GSK3β) at mouse insular synapses and extinction of CTA memory (Figure 7.3) (Li et al. 2016). Genetic ablation or pharmacological inhibition of ASIC1a reduces the induction probability of LTD by

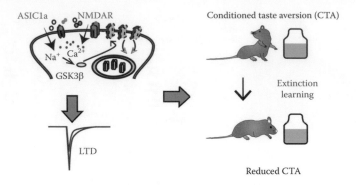

FIGURE 7.3 A proposed scheme for ASIC1a regulation of insular LTD and its involvement in CTA extinction. ASIC1a activation leads to a rise in postsynaptic Ca^{2+} concentration, which in turn activates GSK3β (at least for NMDAR-dependent LTD). This triggers a series of downstream signaling cascades including AMPAR endocytosis, and ultimately leads to a decrease in insular synaptic efficacy. The molecular mechanisms of ASIC1a-dependent mGluR-mediated DHPG-LTD remain to be established. Behaviorally, the ASIC1a-dependent synaptic plasticity at insular synapses is critical for extinction of CTA (adapted from Li et al. (2016) with permission. Please see text for more details).

either low-frequency stimulation (LFS) or bath application of the group I metabotropic glutamate receptor agonist, 3,5-dihydroxyphenylglycine (DHPG), but without affecting the LTP induction in the insular cortex. Interestingly, aversive taste extinction learning leads to reduced synaptic efficacy in the insular cortex, which precludes further LTD induction, implicating the eligibility of LTP and LTD at insular synapses for bidirectional modulation of the CTA memory (Rodriguez-Duran, Martinez-Moreno, and Escobar 2017). Behaviorally, disruption of ASIC1a also attenuates the extinction of established taste aversion memory without altering the initial associative taste learning or its long-term retention. Moreover, the impaired insular LTD and extinction learning in *Asic1a*-null mice can be restored by virus-mediated expression of wild-type ASIC1a, but not its ion-impermeable mutant, in the insular cortices. These data thus demonstrate the involvement of an ASIC1a-mediated insular synaptic depression mechanism in CTA memory extinction.

7.2.5 ASIC1A IN APPETITIVE LEARNING AND MEMORY

Both appetitiveness and aversiveness, as two opposing motivational states, drive the formation of associative memories. Substance use disorders are associated with dysfunction in the brain's motivational and reward processing, therefore, understanding how the reward system works is the key to looking for new therapies. The most appreciable reward center in the brain is the mesolimbic dopamine system. In particular, synaptic plasticity or adaptations occurring at the nucleus accumbens (NAc) are essential central cellular mechanisms underlying the drug abuse. For example, in the NAc medium spiny neurons (MSNs), cocaine exposure changes dendritic spine density and morphology, alters glutamate receptor composition and function, and thus increases the probability to subsequent cocaine exposure (Luscher and Malenka 2011). However, much remains to be learned regarding the mechanisms underlying these addiction-related synaptic abnormalities. It might be of interest to note that ASIC1a is enriched in the NAc, which has been shown to exert an unexpected role in inhibiting excitatory synaptic transmission, thereby restricting cocaine-induced plasticity (Kreple et al., 2014). While selective disruption of ASIC1a in the mouse NAc increased cocaine-induced place preference, overexpression of ASIC1a in the rat NAc reduced cocaine self-administration. To investigate the underlying mechanisms, researchers further identified a previously unknown postsynaptic current mediated by ASIC1a-containing channels and revealed changes in dendritic spine density and glutamate receptor function following *Asic1a* deletion in the NAc. Overall, these data suggest that ASIC1a inhibits the plasticity underlying addiction-related behavior and raise the possibility of targeting ASIC-dependent neurotransmission to develop new treatment strategies for drug addiction.

Apart from cocaine addiction, ASIC1a also plays a vital role in Pavlovian conditioning to nondrug reward (Ghobbeh et al. 2019). Interestingly, the effects of ASIC1a on reward conditioning depend on the temporal relationship between the CS and the US. When CS signaled upcoming reward before US, the *Asic1a*-null mice displayed a strong deficit in conditioning. Nevertheless, when CS and US appeared at the same time signaling an immediate reward, the *Asic1a*-null mice showed an increase in conditioned responses. Moreover, when CS and US partially overlapped in time, or CS shortened and coexisted with US, the *Asic1a*-null mice had no significant effect on conditioned responses. Although the precise circuit and synaptic mechanisms regulated by ASIC1a need to be identified in the future, these results indicate a critical role of ASIC1a in Pavlovian reward conditioning.

7.2.6 ASIC1A IN PROCEDURAL LEARNING AND MEMORY

Procedural learning and memory, also known as skill memory, is a memory of how to do something, including the memory of perceptual skills, cognitive skills, and motor skills, in which stimulus-response associations or habits are incrementally acquired (Packard and Knowlton 2002). Typical examples of procedural memories include swimming, bicycling, playing the piano, and language learning. Pathologically, the impairment of procedural memory is closely associated with many brain diseases such as autism, Parkinson's disease, and stroke. The formation and storage of procedural memory are mainly governed by neural circuits involving the cerebellum and basal ganglia. The dorsal striatum is a major component of the basal ganglia, which receives glutamatergic and dopaminergic projections and transmits them to other parts of the basal ganglia (Hikosaka 1991; Graybiel et al. 1994). It is thus a critical brain region in procedural learning and memory. Striatum-related motor control and procedural memory is accompanied by remarkable synaptic remodeling (Kreitzer and Malenka 2008), but the key molecular determinants that mediate these processes are still unclear. It is well recognized that ASIC1a is an essential factor for regulating synaptic structure and function (Zha 2013). Our

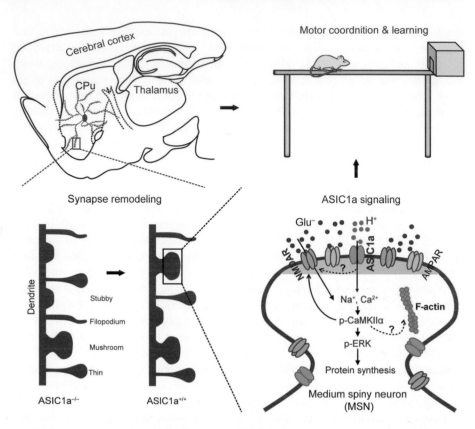

FIGURE 7.4   A proposed mechanism by which ASIC1a regulates striatal synaptic remodeling and procedural motor learning. Postsynaptic ASIC1a is activated by decreased pH in the synaptic cleft associated with striatal synaptic activity, leading to an influx of cations (i.e., Na^+, Ca^{2+}). Elevated intracellular Ca^{2+} activates the downstream CaMKII-ERK signaling pathway. CaMKII contributes to actin dynamics promoting structural remodeling of dendritic spines and regulating postsynaptic distribution and function of NMDAR, whereas ERK signaling is involved in activity-dependent transcriptional regulation of a set of neuronal proteins that in turn drive long-term synaptic plasticity. Together, the ASIC1a-CaMKII-ERK signaling cascade represents a novel molecular mechanism that promoting synaptic remodeling in the striatum, which is important for procedural motor learning (e.g. the balance beam test). Adapted from Yu et al. (2018) with permission. Please see text for more details.

recent work revealed that ASIC1a in the dorsal striatum promoted maturation of excitatory synapse structure and function, thereby improving the efficiency of procedural motor learning efficiency (Figure 7.4) (Yu et al. 2018). We found that ASIC1a was highly enriched in the postsynaptic density (PSD) fractions of the excitatory synapse at the mouse striatum. Importantly, ASIC1a can be activated by synaptic transmission in the cortico-striatal pathway. *Asic1a* knockout results in an increase in the number of dendritic spines characterized as excitatory synapses in MSNs of the dorsal striatum, but an increase in the proportion of immature types and a decrease in the proportion of mature dendritic spines. Consistently, the thickness and length of PSDs were significantly reduced, and the composition and function of postsynaptic glutamate receptors were significantly impaired. These structural and functional changes are largely due to reductions in the phosphorylation of CaMKII and extracellular regulated protein kinases (ERKs). Behaviorally, *Asic1a*-null mice exhibit poor performance in multiple motor tasks, which can be corrected by striatal specific expression of either ASIC1a or CaMKII in adult mutant animals. Collectively, our findings present a novel mechanism mediated by ASIC1a that regulates excitatory synaptic function and striatum-related procedural learning and memory. Pathologically, synaptic dysfunction in striatum and its related circuits are associated with neurological and psychological disorders (Kreitzer 2009; Smith, Villalba, and Raju 2009) including Huntington's disease, Parkinson's disease, and obsessive-compulsive disorder. Intriguingly, previous studies also implicated ASIC1a was involved in the development of these neurodegenerative diseases (Arias et al. 2008; Wong et al. 2008). Therefore, our elucidation of the pivotal role of ASIC1a in striatal synaptic remodeling raises an exciting possibility that targeting ASIC1a in the striatum might represent a feasible therapeutic strategy to treat these striatum-related diseases in the future.

7.3 SUMMARY

This chapter has summarized the role of ASIC1a in various forms of learning and memory, including spatial, fear, extinction, reward, and procedural learning and memory. This function is mainly determined by the dynamic functional modification and structural remodeling of relevant circuits and synapses driven by the channel activity. It is of great significance to consider that ASICs may be potential drug targets for the treatment of memory-related brain diseases. Importantly, these studies not only lay the foundation for further exploration of neurobiological mechanisms of learning and memory involving in the coordination of different channel complexes with downstream signaling cascades, but also provide additional insights into the physiological operation rules for more nonclassical ion channel subtypes.

ACKNOWLEDGMENTS

We thank Dr. Ming-Gang Liu for carefully proofreading the manuscript. Studies by authors for this chapter were supported by grants from the National Natural Science Foundation of China (31930050, 81961128024, 81730095, 81771214, and 32071023), the Science and Technology Commission of Shanghai Municipality (18JC1420302 and 18QA1402500), the Shanghai Municipal Science and Technology Major Project (2018SHZDZX05), and innovative research team of high-level local universities in Shanghai.

REFERENCES

Adaikkan, C., and K. Rosenblum. 2015. "A molecular mechanism underlying gustatory memory trace for an association in the insular cortex." *Elife* 4:e07582. doi: 10.7554/eLife.07582.

Adelman, J. P., J. Maylie, and P. Sah. 2012. "Small-conductance Ca^{2+}-activated K+ chan-
    nels: form and function." *Annu Rev Physiol* 74:245–269. doi: 10.1146/annurev-
    physiol-020911-153336.
Arias, R. L., M. L. Sung, D. Vasylyev, M. Y. Zhang, K. Albinson, K. Kubek, N. Kagan, C.
    Beyer, Q. Lin, J. M. Dwyer, M. M. Zaleska, M. R. Bowlby, J. Dunlop, and M.
    Monaghan. 2008. "Amiloride is neuroprotective in an MPTP model of Parkinson's
    disease." *Neurobiol Dis* 31 (3):334–341. doi: 10.1016/j.nbd.2008.05.008.
Bean, B. P. 2007. "The action potential in mammalian central neurons." *Nat Rev Neurosci* 8
    (6):451–465. doi: 10.1038/nrn2148.
Berman, D. E., and Y. Dudai. 2001. "Memory extinction, learning anew, and learning the
    new: dissociations in the molecular machinery of learning in cortex." *Science* 291
    (5512):2417–2419. doi: 10.1126/science.1058165.
Biel, M., C. Wahl-Schott, S. Michalakis, and X. Zong. 2009. "Hyperpolarization-activated
    cation channels: from genes to function." *Physiol Rev* 89 (3):847–885. doi: 10.1152/
    physrev.00029.2008.
Birdsong, W. T., L. Fierro, F. G. Williams, V. Spelta, L. A. Naves, M. Knowles, J. Marsh-
    Haffner, J. P. Adelman, W. Almers, R. P. Elde, and E. W. McCleskey. 2010. "Sensing
    muscle ischemia: coincident detection of acid and ATP via interplay of two ion
    channels." *Neuron* 68 (4):739–749. doi: 10.1016/j.neuron.2010.09.029.
Bliss, T. V., and G. L. Collingridge. 1993. "A synaptic model of memory: long-term
    potentiation in the hippocampus." *Nature* 361 (6407):31–39. doi: 10.1038/361031a0.
Bocchio, M., S. Nabavi, and M. Capogna. 2017. "Synaptic plasticity, engrams, and network
    oscillations in amygdala circuits for storage and retrieval of emotional memories."
    Neuron 94 (4):731–743. doi: 10.1016/j.neuron.2017.03.022.
Bohlen, C. J., A. T. Chesler, R. Sharif-Naeini, K. F. Medzihradszky, S. Zhou, D. King, E. E.
    Sanchez, A. L. Burlingame, A. I. Basbaum, and D. Julius. 2011. "A heteromeric Texas
    coral snake toxin targets acid-sensing ion channels to produce pain." *Nature* 479
    (7373):410–414. doi: 10.1038/nature10607.
Buta, A., O. Maximyuk, D. Kovalskyy, V. Sukach, M. Vovk, O. Ievglevskyi, E. Isaeva, D.
    Isaev, A. Savotchenko, and O. Krishtal. 2015. "Novel potent orthosteric antagonist of
    ASIC1a prevents NMDAR-dependent LTP induction." *J Med Chem* 58 (11):4449–4461.
    doi: 10.1021/jm5017329.
Chambers, K. C. 1990. "A neural model for conditioned taste aversions." *Annu Rev Neurosci*
    13:373–385. doi: 10.1146/annurev.ne.13.030190.002105.
Chen, C. C., A. Zimmer, W. H. Sun, J. Hall, M. J. Brownstein, and A. Zimmer. 2002. "A role
    for ASIC3 in the modulation of high-intensity pain stimuli." *Proc Natl Acad Sci USA*
    99 (13):8992–8997. doi: 10.1073/pnas.122245999.
Chiang, P. H., T. C. Chien, C. C. Chen, Y. Yanagawa, and C. C. Lien. 2015. "ASIC-
    dependent LTP at multiple glutamatergic synapses in amygdala network is required for
    fear memory." *Sci Rep* 5:10143. doi: 10.1038/srep10143.
Clapham, D. E., L. W. Runnels, and C. Strubing. 2001. "The TRP ion channel family." *Nat
    Rev Neurosci* 2 (6):387–396. doi: 10.1038/35077544.
Clem, R. L., and R. L. Huganir. 2010. "Calcium-permeable AMPA receptor dynamics
    mediate fear memory erasure." *Science* 330 (6007):1108–1112. doi: 10.1126/
    science.1195298.
Coryell, M. W., A. M. Wunsch, J. M. Haenfler, J. E. Allen, J. L. McBride, B. L. Davidson,
    and J. A. Wemmie. 2008. "Restoring acid-sensing ion channel-1a in the amygdala of
    knock-out mice rescues fear memory but not unconditioned fear responses." *J Neurosci*
    28 (51):13738–13741. doi: 10.1523/jneurosci.3907-08.2008.
Coryell, M. W., A. E. Ziemann, P. J. Westmoreland, J. M. Haenfler, Z. Kurjakovic, X. M.
    Zha, M. Price, M. K. Schnizler, and J. A. Wemmie. 2007. "Targeting ASIC1a reduces

innate fear and alters neuronal activity in the fear circuit." *Biol Psychiatry* 62 (10):1140–1148. doi: 10.1016/j.biopsych.2007.05.008.

Coste, B., B. Xiao, J. S. Santos, R. Syeda, J. Grandl, K. S. Spencer, S. E. Kim, M. Schmidt, J. Mathur, A. E. Dubin, M. Montal, and A. Patapoutian. 2012. "Piezo proteins are poreforming subunits of mechanically activated channels." *Nature* 483 (7388):176–181. doi: 10.1038/nature10812.

Deval, E., J. Noel, X. Gasull, A. Delaunay, A. Alloui, V. Friend, A. Eschalier, M. Lazdunski, and E. Lingueglia. 2011. "Acid-sensing ion channels in postoperative pain." *J Neurosci* 31 (16):6059–6066. doi: 10.1523/JNEUROSCI.5266-10.2011.

Deval, E., J. Noel, N. Lay, A. Alloui, S. Diochot, V. Friend, M. Jodar, M. Lazdunski, and E. Lingueglia. 2008. "ASIC3, a sensor of acidic and primary inflammatory pain." *EMBO J* 27 (22):3047–3055. doi: 10.1038/emboj.2008.213.

Diochot, S., A. Baron, M. Salinas, D. Douguet, S. Scarzello, A. S. Dabert-Gay, D. Debayle, V. Friend, A. Alloui, M. Lazdunski, and E. Lingueglia. 2012. "Black mamba venom peptides target acid-sensing ion channels to abolish pain." *Nature* 490 (7421):552–555. doi: 10.1038/nature11494.

Du, J., M. P. Price, R. J. Taugher, D. Grigsby, J. J. Ash, A. C. Stark, M. Z. Hossain Saad, K. Singh, J. Mandal, J. A. Wemmie, and M. J. Welsh. 2017. "Transient acidosis while retrieving a fear-related memory enhances its lability." *Elife* 6. doi: 10.7554/eLife.22564.

Du, J., L. R. Reznikov, M. P. Price, X. M. Zha, Y. Lu, T. O. Moninger, J. A. Wemmie, and M. J. Welsh. 2014. "Protons are a neurotransmitter that regulates synaptic plasticity in the lateral amygdala." *Proc Natl Acad Sci USA* 111 (24):8961–8966. doi: 10.1073/pnas.1407018111.

Duan, B., D. S. Liu, Y. Huang, W. Z. Zeng, X. Wang, H. Yu, M. X. Zhu, Z. Y. Chen, and T. L. Xu. 2012. "PI3-kinase/Akt pathway-regulated membrane insertion of acid-sensing ion channel 1a underlies BDNF-induced pain hypersensitivity." *J Neurosci* 32 (18):6351–6363. doi: 10.1523/JNEUROSCI.4479-11.2012.

Duan, B., Y. Z. Wang, T. Yang, X. P. Chu, Y. Yu, Y. Huang, H. Cao, J. Hansen, R. P. Simon, M. X. Zhu, Z. G. Xiong, and T. L. Xu. 2011. "Extracellular spermine exacerbates ischemic neuronal injury through sensitization of ASIC1a channels to extracellular acidosis." *J Neurosci* 31 (6):2101–2112. doi: 10.1523/JNEUROSCI.4351-10.2011.

Duan, B., L. J. Wu, Y. Q. Yu, Y. Ding, L. Jing, L. Xu, J. Chen, and T. L. Xu. 2007. "Upregulation of acid-sensing ion channel ASIC1a in spinal dorsal horn neurons contributes to inflammatory pain hypersensitivity." *J Neurosci* 27 (41):11139–11148. doi: 10.1523/JNEUROSCI.3364-07.2007.

Dunsmoor, J. E., Y. Niv, N. Daw, and E. A. Phelps. 2015. "Rethinking extinction." *Neuron* 88 (1):47–63. doi: 10.1016/j.neuron.2015.09.028.

Eisenberg, M., T. Kobilo, D. E. Berman, and Y. Dudai. 2003. "Stability of retrieved memory: inverse correlation with trace dominance." *Science* 301 (5636):1102–1104. doi: 10.1126/science.1086881.

Elkobi, A., I. Ehrlich, K. Belelovsky, L. Barki-Harrington, and K. Rosenblum. 2008. "ERK-dependent PSD-95 induction in the gustatory cortex is necessary for taste learning, but not retrieval." *Nat Neurosci* 11 (10):1149–1151. doi: 10.1038/nn.2190.

Escobar, M. L., I. Alcocer, and V. Chao. 1998. "The NMDA receptor antagonist CPP impairs conditioned taste aversion and insular cortex long-term potentiation in vivo." *Brain Res* 812 (1-2):246–251.

Feinstein, J. S., C. Buzza, R. Hurlemann, R. L. Follmer, N. S. Dahdaleh, W. H. Coryell, M. J. Welsh, D. Tranel, and J. A. Wemmie. 2013. "Fear and panic in humans with bilateral amygdala damage." *Nat Neurosci* 16 (3):270–272. doi: 10.1038/nn.3323.

Fromy, B., E. Lingueglia, D. Sigaudo-Roussel, J. L. Saumet, and M. Lazdunski. 2012. "Asic3 is a neuronal mechanosensor for pressure-induced vasodilation that protects against pressure ulcers." *Nat Med* 18 (8):1205–1207. doi: 10.1038/nm.2844.

Gao, J., B. Duan, D. G. Wang, X. H. Deng, G. Y. Zhang, L. Xu, and T. L. Xu. 2005. "Coupling between NMDA receptor and acid-sensing ion channel contributes to ischemic neuronal death." *Neuron* 48 (4):635–646. doi: 10.1016/j.neuron.2005.10.011.

Garcia, J., D. J. Kimeldorf, and R. A. Koelling. 1955. "Conditioned aversion to saccharin resulting from exposure to gamma radiation." *Science* 122 (3160):157–158.

Ghobbeh, A., R. J. Taugher, S. M. Alam, R. Fan, R. T. LaLumiere, and J. A. Wemmie. 2019. "A novel role for acid-sensing ion channels in Pavlovian reward conditioning." *Genes Brain Behav* 18 (7):e12531. doi: 10.1111/gbb.12531.

Gonzalez-Inchauspe, C., F. J. Urbano, M. N. Di Guilmi, and O. D. Uchitel. 2017. "Acid-sensing ion channels activated by evoked released protons modulate synaptic transmission at the mouse calyx of held synapse." *J Neurosci* 37 (10):2589–2599. doi: 10.1523/JNEUROSCI.2566-16.2017.

Graybiel, A. M., T. Aosaki, A. W. Flaherty, and M. Kimura. 1994. "The basal ganglia and adaptive motor control." *Science* 265 (5180):1826–1831.

He, C., F. Chen, B. Li, and Z. Hu. 2014. "Neurophysiology of HCN channels: from cellular functions to multiple regulations." *Prog Neurobiol* 112:1–23. doi: 10.1016/j.pneurobio.2013.10.001.

Herring, B. E., and R. A. Nicoll. 2016. "Long-term potentiation: from CaMKII to AMPA receptor trafficking." *Annu Rev Physiol* 78:351–365. doi: 10.1146/annurev-physiol-021014-071753.

Hikosaka, O. 1991. "Basal ganglia—possible role in motor coordination and learning." *Curr Opin Neurobiol* 1 (4):638–643.

Hill, A. S., and Y. Ben-Shahar. 2018. "The synaptic action of degenerin/epithelial sodium channels." *Channels (Austin)* 12 (1):262–275. doi: 10.1080/19336950.2018.1495006.

Hong, I., J. Kim, J. Kim, S. Lee, H. G. Ko, K. Nader, B. K. Kaang, R. W. Tsien, and S. Choi. 2013. "AMPA receptor exchange underlies transient memory destabilization on retrieval." *Proc Natl Acad Sci USA* 110 (20):8218–8223. doi: 10.1073/pnas.1305235110.

Huganir, R. L., and R. A. Nicoll. 2013. "AMPARs and synaptic plasticity: the last 25 years." *Neuron* 80 (3):704–717. doi: 10.1016/j.neuron.2013.10.025.

Iacobucci, G. J., and G. K. Popescu. 2017. "NMDA receptors: linking physiological output to biophysical operation." *Nat Rev Neurosci* 18 (4):236–249. doi: 10.1038/nrn.2017.24.

Ikeuchi, M., S. J. Kolker, L. A. Burnes, R. Y. Walder, and K. A. Sluka. 2008. "Role of ASIC3 in the primary and secondary hyperalgesia produced by joint inflammation in mice." *Pain* 137 (3):662–669. doi: 10.1016/j.pain.2008.01.020.

Isacoff, E. Y., L. Y. Jan, and D. L. Minor, Jr. 2013. "Conduits of life's spark: a perspective on ion channel research since the birth of neuron." *Neuron* 80 (3):658–674. doi: 10.1016/j.neuron.2013.10.040.

Jalalvand, E., B. Robertson, P. Wallen, and S. Grillner. 2016. "Ciliated neurons lining the central canal sense both fluid movement and pH through ASIC3." *Nat Commun* 7:10002. doi: 10.1038/ncomms10002.

Jones, M. W., P. J. French, T. V. Bliss, and K. Rosenblum. 1999. "Molecular mechanisms of long-term potentiation in the insular cortex in vivo." *J Neurosci* 19 (21):RC36.

Josselyn, S. A., S. Kohler, and P. W. Frankland. 2015. "Finding the engram." *Nat Rev Neurosci* 16 (9):521–534. doi: 10.1038/nrn4000.

Josselyn, S. A., and S. Tonegawa. 2020. "Memory engrams: recalling the past and imagining the future." *Science* 367 (6473). doi: 10.1126/science.aaw4325.

Kellenberger, S., and L. Schild. 2002. "Epithelial sodium channel/degenerin family of ion channels: a variety of functions for a shared structure." *Physiol Rev* 82 (3):735–767. doi: 10.1152/physrev.00007.2002.

Kellenberger, S., and L. Schild. 2015. "International Union of Basic and Clinical Pharmacology. XCI. structure, function, and pharmacology of acid-sensing ion channels and the epithelial Na+ channel." *Pharmacol Rev* 67 (1):1–35. doi: 10.1124/pr.114.009225.

Kim, S. E., B. Coste, A. Chadha, B. Cook, and A. Patapoutian. 2012. "The role of Drosophila Piezo in mechanical nociception." *Nature* 483 (7388):209–212. doi: 10.1038/nature10801.

Kreitzer, A. C. 2009. "Physiology and pharmacology of striatal neurons." *Annu Rev Neurosci* 32:127–147. doi: 10.1146/annurev.neuro.051508.135422.

Kreitzer, A. C., and R. C. Malenka. 2008. "Striatal plasticity and basal ganglia circuit function." *Neuron* 60 (4):543–554. doi: 10.1016/j.neuron.2008.11.005.

Kreple, C. J., Y. Lu, R. T. LaLumiere, and J. A. Wemmie. 2014. "Drug abuse and the simplest neurotransmitter." *ACS Chem Neurosci* 5 (9):746–748. doi: 10.1021/cn500154w.

Kreple, C. J., Y. Lu, R. J. Taugher, A. L. Schwager-Gutman, J. Du, M. Stump, Y. Wang, A. Ghobbeh, R. Fan, C. V. Cosme, L. P. Sowers, M. J. Welsh, J. J. Radley, R. T. LaLumiere, and J. A. Wemmie. 2014. "Acid-sensing ion channels contribute to synaptic transmission and inhibit cocaine-evoked plasticity." *Nat Neurosci* 17 (8):1083–1091. doi: 10.1038/nn.3750.

Krishtal, O. 2003. "The ASICs: signaling molecules? Modulators?" *Trends Neurosci* 26 (9):477–483. doi: 10.1016/s0166-2236(03)00210-8.

Lebois, L. A. M., A. V. Seligowski, J. D. Wolff, S. B. Hill, and K. J. Ressler. 2019. "Augmentation of extinction and inhibitory learning in anxiety and trauma-related disorders." *Annu Rev Clin Psychol* 15:257–284. doi: 10.1146/annurev-clinpsy-050718-095634.

LeDoux, J. E. 2000. "Emotion circuits in the brain." *Annu Rev Neurosci* 23:155–184. doi: 10.1146/annurev.neuro.23.1.155.

Lee, J. L. C., K. Nader, and D. Schiller. 2017. "An update on memory reconsolidation updating." *Trends Cogn Sci* 21 (7):531–545. doi: 10.1016/j.tics.2017.04.006.

Lee, S. J., Y. Escobedo-Lozoya, E. M. Szatmari, and R. Yasuda. 2009. "Activation of CaMKII in single dendritic spines during long-term potentiation." *Nature* 458 (7236):299–304. doi: 10.1038/nature07842.

Li, B., M. R. Tadross, and R. W. Tsien. 2016. "Sequential ionic and conformational signaling by calcium channels drives neuronal gene expression." *Science* 351 (6275):863–867. doi: 10.1126/science.aad3647.

Li, H. S., X. Y. Su, X. L. Song, X. Qi, Y. Li, R. Q. Wang, O. Maximyuk, O. Krishtal, T. Wang, H. Fang, L. Liao, H. Cao, Y. Q. Zhang, M. X. Zhu, M. G. Liu, and T. L. Xu. 2019. "Protein kinase C lambda mediates acid-sensing ion channel 1a-dependent cortical synaptic plasticity and pain hypersensitivity." *J Neurosci* 39 (29):5773–5793. doi: 10.1523/JNEUROSCI.0213-19.2019.

Li, W. G., M. G. Liu, S. Deng, Y. M. Liu, L. Shang, J. Ding, T. T. Hsu, Q. Jiang, Y. Li, F. Li, M. X. Zhu, and T. L. Xu. 2016. "ASIC1a regulates insular long-term depression and is required for the extinction of conditioned taste aversion." *Nat Commun* 7:13770. doi: 10.1038/ncomms13770.

Li, W. G., and T. L. Xu. 2011. "ASIC3 channels in multimodal sensory perception." *ACS Chem Neurosci* 2 (1):26–37. doi: 10.1021/cn100094b.

Lin, S. H., Y. R. Cheng, R. W. Banks, M. Y. Min, G. S. Bewick, and C. C. Chen. 2016. "Evidence for the involvement of ASIC3 in sensory mechanotransduction in proprioceptors." *Nat Commun* 7:11460. doi: 10.1038/ncomms11460.

Liu, M. G., H. S. Li, W. G. Li, Y. J. Wu, S. N. Deng, C. Huang, O. Maximyuk, V. Sukach, O. Krishtal, M. X. Zhu, and T. L. Xu. 2016. "Acid-sensing ion channel 1a contributes to hippocampal LTP inducibility through multiple mechanisms." *Sci Rep* 6:23350. doi: 10.1038/srep23350.

Luscher, C., and R. C. Malenka. 2011. "Drug-evoked synaptic plasticity in addiction: from molecular changes to circuit remodeling." *Neuron* 69 (4):650–663. doi: 10.1016/j.neuron.2011.01.017.

Lynch, M. A. 2004. "Long-term potentiation and memory." *Physiol Rev* 84 (1):87–136. doi: 10.1152/physrev.00014.2003.

Malinow, R., and R. C. Malenka. 2002. "AMPA receptor trafficking and synaptic plasticity." *Annu Rev Neurosci* 25:103–126. doi: 10.1146/annurev.neuro.25.112701.142758.

Mango, D., E. Braksator, G. Battaglia, S. Marcelli, N. B. Mercuri, M. Feligioni, F. Nicoletti, Z. I. Bashir, and R. Nistico. 2017. "Acid-sensing ion channel 1a is required for mGlu receptor dependent long-term depression in the hippocampus." *Pharmacol Res* 119:12–19. doi: 10.1016/j.phrs.2017.01.028.

Mango, D., and R. Nistico. 2019. "Acid-sensing ion channel 1a is involved in N-methyl D-aspartate receptor-dependent long-term depression in the hippocampus." *Front Pharmacol* 10:555. doi: 10.3389/fphar.2019.00555.

Mango, D., and R. Nistico. 2020. "Role of ASIC1a in normal and pathological synaptic plasticity." *Rev Physiol Biochem Pharmacol* 177: 83–100. doi: 10.1007/112_2020_45.

Marra, S., R. Ferru-Clement, V. Breuil, A. Delaunay, M. Christin, V. Friend, S. Sebille, C. Cognard, T. Ferreira, C. Roux, L. Euller-Ziegler, J. Noel, E. Lingueglia, and E. Deval. 2016. "Non-acidic activation of pain-related acid-sensing ion channel 3 by lipids." *EMBO J* 35 (4):414–428. doi: 10.15252/embj.201592335.

Mazzuca, M., C. Heurteaux, A. Alloui, S. Diochot, A. Baron, N. Voilley, N. Blondeau, P. Escoubas, A. Gelot, A. Cupo, A. Zimmer, A. M. Zimmer, A. Eschalier, and M. Lazdunski. 2007. "A tarantula peptide against pain via ASIC1a channels and opioid mechanisms." *Nat Neurosci* 10 (8):943–945. doi: 10.1038/nn1940.

Milad, M. R., and G. J. Quirk. 2012. "Fear extinction as a model for translational neuroscience: ten years of progress." *Annu Rev Psychol* 63:129–151. doi: 10.1146/annurev.psych.121208.131631.

Minke, B., and B. Cook. 2002. "TRP channel proteins and signal transduction." *Physiol Rev* 82 (2):429–472. doi: 10.1152/physrev.00001.2002.

Morris, R. G. 1981. "Spatial localization does not require the presence of local cues." *Learn Motiv* 12:239–260. doi: 10.1016/0023-9690(81)90020-5.

Murthy, S. E., A. E. Dubin, and A. Patapoutian. 2017. "Piezos thrive under pressure: mechanically activated ion channels in health and disease." *Nat Rev Mol Cell Biol* 18 (12):771–783. doi: 10.1038/nrm.2017.92.

Nabavi, S., R. Fox, C. D. Proulx, J. Y. Lin, R. Y. Tsien, and R. Malinow. 2014. "Engineering a memory with LTD and LTP." *Nature* 511 (7509):348–352. doi: 10.1038/nature13294.

Nader, K., G. E. Schafe, and J. E. Le Doux. 2000. "Fear memories require protein synthesis in the amygdala for reconsolidation after retrieval." *Nature* 406 (6797):722–726. doi: 10.1038/35021052.

Nakazawa, K., T. J. McHugh, M. A. Wilson, and S. Tonegawa. 2004. "NMDA receptors, place cells and hippocampal spatial memory." *Nat Rev Neurosci* 5 (5):361–372. doi: 10.1038/nrn1385.

Nicholls, R. E., J. M. Alarcon, G. Malleret, R. C. Carroll, M. Grody, S. Vronskaya, and E. R. Kandel. 2008. "Transgenic mice lacking NMDAR-dependent LTD exhibit deficits in behavioral flexibility." *Neuron* 58 (1):104–117. doi: 10.1016/j.neuron.2008.01.039.

Nilius, B., G. Owsianik, T. Voets, and J. A. Peters. 2007. "Transient receptor potential cation channels in disease." *Physiol Rev* 87 (1):165–217. doi: 10.1152/physrev.00021.2006.

O'Keefe, J. 1976. "Place units in the hippocampus of the freely moving rat." *Exp Neurol* 51 (1):78–109. doi: 10.1016/0014-4886(76)90055-8.

Packard, M. G., and B. J. Knowlton. 2002. "Learning and memory functions of the Basal Ganglia." *Annu Rev Neurosci* 25:563–593. doi: 10.1146/annurev.neuro.25.112701.142 937.

Peng, Z., W. G. Li, C. Huang, Y. M. Jiang, X. Wang, M. X. Zhu, X. Cheng, and T. L. Xu. 2015. "ASIC3 mediates itch sensation in response to coincident stimulation by acid and nonproton ligand." *Cell Rep* 13 (2):387–398. doi: 10.1016/j.celrep.2015.09.002.

Petrovic, M. M., S. Viana da Silva, J. P. Clement, L. Vyklicky, C. Mulle, I. M. Gonzalez-Gonzalez, and J. M. Henley. 2017. "Metabotropic action of postsynaptic kainate receptors triggers hippocampal long-term potentiation." *Nat Neurosci* 20 (4):529–539. doi: 10.1038/nn.4505.

Pignataro, G., R. P. Simon, and Z. G. Xiong. 2007. "Prolonged activation of ASIC1a and the time window for neuroprotection in cerebral ischaemia." *Brain* 130 (Pt 1):151–158. doi: 10.1093/brain/awl325.

Pignatelli, M., T. J. Ryan, D. S. Roy, C. Lovett, L. M. Smith, S. Muralidhar, and S. Tonegawa. 2019. "Engram cell excitability state determines the efficacy of memory retrieval." *Neuron* 101 (2):274–284.e5. doi: 10.1016/j.neuron.2018.11.029.

Poo, M. M., M. Pignatelli, T. J. Ryan, S. Tonegawa, T. Bonhoeffer, K. C. Martin, A. Rudenko, L. H. Tsai, R. W. Tsien, G. Fishell, C. Mullins, J. T. Goncalves, M. Shtrahman, S. T. Johnston, F. H. Gage, Y. Dan, J. Long, G. Buzsaki, and C. Stevens. 2016. "What is memory? The present state of the engram." *BMC Biol* 14:40. doi: 10.11 86/s12915-016-0261-6.

Price, M. P., S. L. McIlwrath, J. Xie, C. Cheng, J. Qiao, D. E. Tarr, K. A. Sluka, T. J. Brennan, G. R. Lewin, and M. J. Welsh. 2001. "The DRASIC cation channel contributes to the detection of cutaneous touch and acid stimuli in mice." *Neuron* 32 (6):1071–1083. doi: 10.1016/s0896-6273(01)00547-5.

Quintana, P., D. Soto, O. Poirot, M. Zonouzi, S. Kellenberger, D. Muller, R. Chrast, and S. G. Cull-Candy. 2015. "Acid-sensing ion channel 1a drives AMPA receptor plasticity following ischaemia and acidosis in hippocampal CA1 neurons." *J Physiol* 593 (19):4373–4386. doi: 10.1113/JP270701.

Ramsey, I. S., M. Delling, and D. E. Clapham, 2006. "An Introduction to TRP channels." *Annu Rev Physiol* 68:619–647. doi: 10.1146/annurev.physiol.68.040204.100431.

Ranade, S. S., R. Syeda, and A. Patapoutian. 2015. "Mechanically activated ion channels." *Neuron* 87 (6):1162–1179. doi: 10.1016/j.neuron.2015.08.032.

Rodriguez-Duran, L. F., A. Martinez-Moreno, and M. L. Escobar. 2017. "Bidirectional modulation of taste aversion extinction by insular cortex LTP and LTD." *Neurobiol Learn Mem* 142 (Pt A):85–90. doi: 10.1016/j.nlm.2016.12.014.

Rosenblum, K., D. E. Berman, S. Hazvi, R. Lamprecht, and Y. Dudai. 1997. "NMDA receptor and the tyrosine phosphorylation of its 2B subunit in taste learning in the rat insular cortex." *J Neurosci* 17 (13):5129–5135.

Sanhueza, M., and J. Lisman. 2013. "The CaMKII/NMDAR complex as a molecular memory." *Mol Brain* 6:10. doi: 10.1186/1756-6606-6-10.

Schiller, D., M. H. Monfils, C. M. Raio, D. C. Johnson, J. E. Ledoux, and E. A. Phelps. 2010. "Preventing the return of fear in humans using reconsolidation update mechanisms." *Nature* 463 (7277):49–53. doi: 10.1038/nature08637.

Sluka, K. A., and N. S. Gregory. 2015. "The dichotomized role for acid sensing ion channels in musculoskeletal pain and inflammation." *Neuropharmacology* 94:58–63. doi: 10.1 016/j.neuropharm.2014.12.013.

Sluka, K. A., M. P. Price, N. M. Breese, C. L. Stucky, J. A. Wemmie, and M. J. Welsh. 2003. "Chronic hyperalgesia induced by repeated acid injections in muscle is abolished by the loss of ASIC3, but not ASIC1." *Pain* 106 (3):229–239. doi: 10.1016/s0304-395 9(03)00269-0.

Sluka, K. A., R. Radhakrishnan, C. J. Benson, J. O. Eshcol, M. P. Price, K. Babinski, K. M. Audette, D. C. Yeomans, and S. P. Wilson. 2007. "ASIC3 in muscle mediates mechanical, but not heat, hyperalgesia associated with muscle inflammation." *Pain* 129 (1-2):102–112. doi: 10.1016/j.pain.2006.09.038.

Smith, Y., R. M. Villalba, and D. V. Raju. 2009. "Striatal spine plasticity in Parkinson's disease: pathological or not?" *Parkinsonism Relat Disord* 15 Suppl 3:S156–S161. doi: 10.1016/S1353-8020(09)70805-3.

Stephan, G., L. Huang, Y. Tang, S. Vilotti, E. Fabbretti, Y. Yu, W. Norenberg, H. Franke, F. Goloncser, B. Sperlagh, A. Dopychai, R. Hausmann, G. Schmalzing, P. Rubini, and P. Illes. 2018. "The ASIC3/P2X3 cognate receptor is a pain-relevant and ligand-gated cationic channel." *Nat Commun* 9 (1):1354. doi: 10.1038/s41467-018-03728-5.

Taugher, R. J., Y. Lu, Y. Wang, C. J. Kreple, A. Ghobbeh, R. Fan, L. P. Sowers, and J. A. Wemmie. 2014. "The bed nucleus of the stria terminalis is critical for anxiety-related behavior evoked by CO_2 and acidosis." *J Neurosci* 34 (31):10247–10255. doi: 10.1523/JNEUROSCI.1680-14.2014.

Trimmer, J. S., and K. J. Rhodes. 2004. "Localization of voltage-gated ion channels in mammalian brain." *Annu Rev Physiol* 66:477–519. doi: 10.1146/annurev.physiol.66.032102.113328.

Tsien, J. Z., P. T. Huerta, and S. Tonegawa. 1996. "The essential role of hippocampal CA1 NMDA receptor-dependent synaptic plasticity in spatial memory." *Cell* 87 (7):1327–1338. doi: 10.1016/s0092-8674(00)81827-9.

Uchitel, O. D., C. Gonzalez Inchauspe, and C. Weissmann. 2019. "Synaptic signals mediated by protons and acid-sensing ion channels." *Synapse* 73 (10):e22120. doi: 10.1002/syn.22120.

Valbuena, S., and J. Lerma. 2016. "Non-canonical signaling, the hidden life of ligand-gated ion channels." *Neuron* 92 (2):316–329. doi: 10.1016/j.neuron.2016.10.016.

Waldmann, R., G. Champigny, F. Bassilana, C. Heurteaux, and M. Lazdunski. 1997. "A proton-gated cation channel involved in acid-sensing." *Nature* 386 (6621):173–177.

Wang, J. J., F. Liu, F. Yang, Y. Z. Wang, X. Qi, Y. Li, Q. Hu, M. X. Zhu, and T. L. Xu. 2020. "Disruption of auto-inhibition underlies conformational signaling of ASIC1a to induce neuronal necroptosis." *Nat Commun* 11 (1):475. doi: 10.1038/s41467-019-13873-0.

Wang, Q., Q. Wang, X. L. Song, Q. Jiang, Y. J. Wu, Y. Li, T. F. Yuan, S. Zhang, N. J. Xu, M. X. Zhu, W. G. Li, and T. L. Xu. 2018. "Fear extinction requires ASIC1a-dependent regulation of hippocampal-prefrontal correlates." *Sci Adv* 4 (10):eaau3075. doi: 10.112 6/sciadv.aau3075.

Wang, X., W. G. Li, Y. Yu, X. Xiao, J. Cheng, W. Z. Zeng, Z. Peng, M. Xi Zhu, and T. L. Xu. 2013. "Serotonin facilitates peripheral pain sensitivity in a manner that depends on the nonproton ligand sensing domain of ASIC3 channel." *J Neurosci* 33 (10):4265–4279. doi: 10.1523/JNEUROSCI.3376-12.2013.

Wang, Y. Z., J. J. Wang, Y. Huang, F. Liu, W. Z. Zeng, Y. Li, Z. G. Xiong, M. X. Zhu, and T. L. Xu. 2015. "Tissue acidosis induces neuronal necroptosis via ASIC1a channel independent of its ionic conduction." *Elife* 4. doi: 10.7554/eLife.05682.

Wang, Y. Z., W. Z. Zeng, X. Xiao, Y. Huang, X. L. Song, Z. Yu, D. Tang, X. P. Dong, M. X. Zhu, and T. L. Xu. 2013. "Intracellular ASIC1a regulates mitochondrial permeability transition-dependent neuronal death." *Cell Death Differ* 20 (10):1359–1369. doi: 10.1 038/cdd.2013.90.

Wemmie, J. A., C. C. Askwith, E. Lamani, M. D. Cassell, J. H. Freeman, Jr., and M. J. Welsh. 2003. "Acid-sensing ion channel 1 is localized in brain regions with high synaptic density and contributes to fear conditioning." *J Neurosci* 23 (13):5496–5502. doi: 10.1523/jneurosci.23-13-05496.2003.

Wemmie, J. A., J. Chen, C. C. Askwith, A. M. Hruska-Hageman, M. P. Price, B. C. Nolan, P. G. Yoder, E. Lamani, T. Hoshi, J. H. Freeman, Jr., and M. J. Welsh. 2002. "The acid-activated ion channel ASIC contributes to synaptic plasticity, learning, and memory." *Neuron* 34 (3):463–477. doi: 10.1016/s0896-6273(02)00661-x.

Wemmie, J. A., M. W. Coryell, C. C. Askwith, E. Lamani, A. S. Leonard, C. D. Sigmund, and M. J. Welsh. 2004. "Overexpression of acid-sensing ion channel 1a in transgenic mice increases acquired fear-related behavior." *Proc Natl Acad Sci USA* 101 (10):3621–3626. doi: 10.1073/pnas.0308753101.

Wemmie, J. A., M. P. Price, and M. J. Welsh. 2006. "Acid-sensing ion channels: advances, questions and therapeutic opportunities." *Trends Neurosci* 29 (10):578–586. doi: 10.1 016/j.tins.2006.06.014.

Wemmie, J. A., R. J. Taugher, and C. J. Kreple. 2013. "Acid-sensing ion channels in pain and disease." *Nat Rev Neurosci* 14 (7):461–471. doi: 10.1038/nrn3529.

Wheeler, D. G., R. D. Groth, H. Ma, C. F. Barrett, S. F. Owen, P. Safa, and R. W. Tsien. 2012. "Ca(V)1 and Ca(V)2 channels engage distinct modes of Ca(2+) signaling to control CREB-dependent gene expression." *Cell* 149 (5):1112–1124. doi: 10.1016/j.cell.2012.03.041.

Wong, H. K., P. O. Bauer, M. Kurosawa, A. Goswami, C. Washizu, Y. Machida, A. Tosaki, M. Yamada, T. Knopfel, T. Nakamura, and N. Nukina. 2008. "Blocking acid-sensing ion channel 1 alleviates Huntington's disease pathology via an ubiquitin-proteasome system-dependent mechanism." *Hum Mol Genet* 17 (20):3223–3235. doi: 10.1093/hmg/ddn218.

Wright, W. J., N. M. Graziane, P. A. Neumann, P. J. Hamilton, H. M. Cates, L. Fuerst, A. Spenceley, N. MacKinnon-Booth, K. Iyer, Y. H. Huang, Y. Shaham, O. M. Schluter, E. J. Nestler, and Y. Dong. 2020. "Silent synapses dictate cocaine memory destabilization and reconsolidation." *Nat Neurosci* 23 (1):32–46. doi: 10.1038/s41593-019-0537-6.

Wu, L. J., B. Duan, Y. D. Mei, J. Gao, J. G. Chen, M. Zhuo, L. Xu, M. Wu, and T. L. Xu. 2004. "Characterization of acid-sensing ion channels in dorsal horn neurons of rat spinal cord." *J Biol Chem* 279 (42):43716–43724. doi: 10.1074/jhc.M403557200.

Wu, P. Y., Y. Y. Huang, C. C. Chen, T. T. Hsu, Y. C. Lin, J. Y. Weng, T. C. Chien, I. H. Cheng, and C. C. Lien. 2013. "Acid-sensing ion channel-1a is not required for normal hippocampal LTP and spatial memory." *J Neurosci* 33 (5):1828–1832. doi: 10.1523/jneurosci.4132-12.2013.

Xiong, Z. G., X. M. Zhu, X. P. Chu, M. Minami, J. Hey, W. L. Wei, J. F. MacDonald, J. A. Wemmie, M. P. Price, M. J. Welsh, and R. P. Simon. 2004. "Neuroprotection in ischemia: blocking calcium-permeable acid-sensing ion channels." *Cell* 118 (6):687–698. doi: 10.1016/j.cell.2004.08.026.

Yagi, J., H. N. Wenk, L. A. Naves, and E. W. McCleskey. 2006. "Sustained currents through ASIC3 ion channels at the modest pH changes that occur during myocardial ischemia." *Circ Res* 99 (5):501–509. doi: 10.1161/01.RES.0000238388.79295.4c.

Yan, Y., L. Zhang, T. Zhu, S. Deng, B. Ma, H. Lv, X. Shan, H. Cheng, K. Jiang, T. Zhang, B. Meng, B. Mei, W. G. Li, and F. Li. 2020. "Reconsolidation of a post-ingestive nutrient memory requires mTOR in the central amygdala." *Mol Psychiatry*. doi: 10.1038/s413 80-020-00874-5.

Yan, Z., W. Zhang, Y. He, D. Gorczyca, Y. Xiang, L. E. Cheng, S. Meltzer, L. Y. Jan, and Y. N. Jan. 2013. "Drosophila NOMPC is a mechanotransduction channel subunit for gentle-touch sensation." *Nature* 493 (7431):221–225. doi: 10.1038/nature11685.

Yap, E. L., and M. E. Greenberg. 2018. "Activity-regulated transcription: bridging the gap between neural activity and behavior." *Neuron* 100 (2):330–348. doi: 10.1016/j.neuron.2018.10.013.

Yu, Y., Z. Chen, W. G. Li, H. Cao, E. G. Feng, F. Yu, H. Liu, H. Jiang, and T. L. Xu. 2010. "A nonproton ligand sensor in the acid-sensing ion channel." *Neuron* 68 (1):61–72. doi: 10.1016/j.neuron.2010.09.001.

Yu, Z., Y. J. Wu, Y. Z. Wang, D. S. Liu, X. L. Song, Q. Jiang, Y. Li, S. Zhang, N. J. Xu, M. X. Zhu, W. G. Li, and T. L. Xu. 2018. "The acid-sensing ion channel ASIC1a mediates striatal synapse remodeling and procedural motor learning." *Sci Signal* 11 (542). doi: 10.1126/scisignal.aar4481.

Zeng, W. Z., D. S. Liu, B. Duan, X. L. Song, X. Wang, D. Wei, W. Jiang, M. X. Zhu, Y. Li, and T. L. Xu. 2013. "Molecular mechanism of constitutive endocytosis of Acid-sensing ion channel 1a and its protective function in acidosis-induced neuronal death." *J Neurosci* 33 (16):7066–7078. doi: 10.1523/JNEUROSCI.5206-12.2013.

Zha, X. M. 2013. "Acid-sensing ion channels: trafficking and synaptic function." *Mol Brain* 6:1. doi: 10.1186/1756-6606-6-1.

Ziemann, A. E., J. E. Allen, N. S. Dahdaleh, Drebot, II, M. W. Coryell, A. M. Wunsch, C. M. Lynch, F. M. Faraci, M. A. Howard, 3rd, M. J. Welsh, and J. A. Wemmie. 2009. "The amygdala is a chemosensor that detects carbon dioxide and acidosis to elicit fear behavior." *Cell* 139 (5):1012–1021. doi: 10.1016/j.cell.2009.10.029.

8 Neuropeptide Regulation of Ion Channels and Food Intake

Xiaobing Zhang

CONTENTS

Food intake is well controlled by the central nervous system that responds to circulating signals of energy homeostasis. The brain's motivation center integrates both satiety signals from the caudal brainstem and adiposity signals from the hypothalamic homeostatic system to regulate feeding motivation. Both peripheral peptide hormones and neuropeptides produced by neurons located in the hypothalamus and caudal brainstem play crucial roles in the control of both homeostatic food intake and hedonic eating. Neuropeptides modulate the activity of neurons located in feeding-related neural circuits by opening or closing the coupled ion channels following the activation of neuropeptide receptors, the G-protein coupled receptors (GPCRs) expressed on the neuronal membrane. This chapter summarizes central neuropeptide signaling in the control of food intake, and ion channels in mediating neuropeptide modulation of feeding-related neurons in the brain.

8.1   CENTRAL NEUROPEPTIDE-PRODUCING NEURONS IN THE CONTROL OF FOOD

In the brain, a large number of neuropeptide-expressing neurons participate in the control of food intake through sensing peripheral satiety and energy signals from the gut and adipose tissue. Some of these neuropeptide neurons belong to the first-order neurons that are directly modulated by peripherally released hormones such as ghrelin, leptin, and insulin (Balthasar et al. 2004). The typical representatives are neuropeptide-expressing neurons in the arcuate nucleus of the hypothalamus including agouti-related protein (AgRP) and proopiomelanocortin (POMC) neurons. However, many other neurons may receive information from the first-order neurons in the arcuate nucleus rather than a direct control by hormones secreted by gut and adipose tissue. The second-order neurons typically include oxytocin neurons in the paraventricular nucleus of the hypothalamus and melanin-concentrating hormone (MCH) neurons in the lateral hypothalamus. In addition to the hypothalamus, the caudal brainstem, especially the nucleus of the solitary tract, is also a region with neurons that produce neuropeptides including glucagon-like peptide-1 (GLP-1) and cholecystokinin (CCK) for satiety control. Although these neuropeptide-expressing neurons are located in separated brain areas, they work together via long-distance neural circuit connections to integrate satiety or hunger signals to the brain's motivation center for the control of food intake.

8.1.1   NEURONS THAT SYNTHESIZE AGRP AND POMC IN THE ARCUATENUCLEUS OF THE HYPOTHALAMUS

The arcuate nucleus, one of the most popular areas for homeostatic regulation of food intake, is located at the bottom of the hypothalamus around the third ventricle. There are two major populations of neurons in the arcuate nucleus that respond to energy metabolism and act to regulate food intake. The anorexigenic POMC neurons inhibit food intake when they are activated, while the orexigenic AgRP neurons stimulate food consumption when their activity is increased. Optogenetic and chemogenetic activation of AgRP neurons rapidly induces food intake, while stimulation of POMC neurons takes hours to decrease food consumption (Aponte, Atasoy, and Sternson 2011). To sense the energy states of the body, AgRP and POMC neurons in the arcuate nucleus differently respond to the metabolic signals from adipose tissue and gut. Leptin is the satiety signal that arises from adipose tissue and the circulating levels of leptin correlate with adipose mass. The leptin receptors (ObRs) are expressed in many brain regions including the arcuate nucleus for the control of energy and glucose homeostasis (Xu et al. 2018; Varela and Horvath 2012). When leptin level goes up, it stimulates POMC neurons but inhibits AgPR neurons to reduce food intake through the activation of leptin receptors (Varela and Horvath 2012). Ghrelin is another hormone that is released by the stomach In the hungry state. Both AgRP and POMC neurons express ghrelin receptors, which contribute to sense the energy deficit. Increased ghrelin level activates AgRP neurons but inhibits POMC neurons to increase appetite (Cowley et al. 2003).

AgRP neurons in the arcuate nucleus express not only AgPR but also neuropeptide Y (NPY) and GABA as neurotransmitters. To orchestrate food intake, AgRP neurons project to multiple downstream areas such as the paraventricular nucleus of the hypothalamus (PVN), parabrachial nucleus (PBN), and paraventricular thalamus (PVT). In addition to the targets outside of the arcuate nucleus, AgRP neurons also innervate POMC neurons locally through inhibitory projections. To respond to the released AgRP, NPY, and GABA from presynaptic AgRP neurons, the postsynaptic neurons express melanocortin receptors (MC3R and MC4R), NPY receptors (Y1, Y2, Y4, and Y5), and GABA receptors. Activation of AgPR neurons in the arcuate nucleus hyperpolarizes and inhibits the postsynaptic neurons by releasing the above three neurotransmitters. Particularly, AgRP is a high-affinity endogenous antagonist of MC3/4R receptors that competitively inhibit melanocortin receptors (MC3R and MC4R) to promote food intake (Corander et al. 2011; Ollmann et al. 1997). MC4R deficiency in both mice and humans leads to increased food intake and obesity (Butler et al. 2001; Yeo et al. 2003).

POMC is the precursor of melanocyte-stimulating hormone (MSH), adrenocorticotrophin (ACTH), and beta-endorphin. Activation of POMC neurons mainly excites postsynaptic MC4R-expressing neurons in the PVN to produce inhibitory regulation on food intake. Due to the high-density MC4R expression, PVN becomes a major target for both AgRP and POMC neurons to maintain energy homeostasis. In addition to PVN, POMC neurons in the arcuate nucleus also innervate other brain areas such as the lateral hypothalamus and ventromedial hypothalamus for the functional regulation of food intake (Sohn 2015).

8.1.2 Oxytocin Neurons in the Paraventricular Nucleus of the Hypothalamus

Oxytocin is a nine-amino-acid neuropeptide that is produced by neurons in the PVN and supraoptic nucleus of the hypothalamus (SON). Magnocellular oxytocin neurons in both PVN and SON project to the posterior pituitary where oxytocin is secreted for peripheral functions, while parvocellular neurons in the PVN project widely to many brain regions for the central control of stress response, social motivation, and food intake. Oxytocin plays an important role in the inhibition of food intake, and oxytocin neurons are activated following food consumption (Johnstone, Fong, and Leng 2006; Hume, Sabatier, and Menzies 2017). Oxytocin neurons in the PVN express MC4R, which is the target of AgRP and POMC neurons. Thus, oxytocin neurons of the PVN are considered to be the second-order anorexigenic neurons. Optogenetic activation of PVN oxytocin neurons abolishes the potentiation on food intake evoked by the activation of projections from AgRP neurons to PVN (Atasoy et al. 2012), suggesting the importance of second-order oxytocin neurons in controlling homeostatic food intake. Hypothalamic oxytocin mRNA expression is reduced by fasting and recovers after refeeding or the administration of leptin (Kublaoui et al. 2008; Tung et al. 2008), suggesting the involvement of oxytocin in regulating energy homeostasis. Oxytocin neurons in the PVN are also critical for the maintenance of normal body weight on a high-fat diet (Wu et al. 2012; Camerino 2009).

Postsynaptic brain areas targeted by PVN oxytocin neurons express oxytocin receptors, which are important for the functional regulation of oxytocin in food intake. The downstream brain areas that mediate oxytocin regulation of food intake including the arcuate nucleus, the nucleus of the solitary tract (NTS), and the brain's reward and motivation system. The ablation of oxytocin receptor-expressing neurons or knockdown of oxytocin receptors in the NTS induces hyperphagia (Baskin et al. 2010; Ong, Bongiorno, et al. 2017), while activation of oxytocin receptors in arcuate nucleus causes satiety (Fenselau et al. 2017). Intra-VTA injection of oxytocin decreases sucrose intake in rats and administration of oxytocin to nucleus accumbens (NAc) reduces food and palatable sucrose intake (Mullis, Kay, and Williams 2013; Olszewski et al. 2010; Herisson et al. 2016).

It is noteworthy that oxytocin does not always decrease food intake and its effect\ on appetite is affected by specific emotional and social contexts (Olszewski, Klockars, and Levine 2016). Oxytocin antagonists increased sugar intake in dominant mice but not subordinate animals (Olszewski, Allen, and Levine 2015). Intracerebroventricular injection of oxytocin did not decrease food intake in pregnant rats within the first-hour post-injection. However, food intake in pregnant dams12 hours after oxytocin administration was even higher than in saline controls (Douglas, Johnstone, and Leng 2007). Surprisingly, intraperitoneal injection of selective oxytocin agonists increased food intake in a novel environment (Olszewski et al. 2014). In addition to animal researches, clinical studies also reported oxytocin only reduces food intake in the fasted state in obese but not in normal-weight men (Thienel et al. 2016). Oxytocin dysfunction is also involved in anorexia nervosa (AN), a common eating disorder characterized by severe hypophagia and high anxiety (Oldershaw et al. 2011). In women with AN, oxytocin levels in both cerebrospinal fluid and blood are lower than control subjects (Demitrack et al. 1990; Lawson et al. 2011). The dysfunction of the brain oxytocin system might underline elevated anxiety and decreased motivation for food in AN (Fetissov et al. 2005). Oxytocin has been reported to potentially treat AN by clinical studies and some clinical trials are still under investigation (Maguire et al. 2013; Russell et al. 2018; Kim, Kim, Cardi, et al. 2014; Kim, Kim, Park, et al. 2014). Our latest studies indicated that oxytocin activation of PVT oxytocin receptors promotes feeding motivation to attenuate stress-induced hypophagia, suggesting a potential mechanism that oxytocin increases food intake during stress (Barrett et al. 2021). However, future studies are necessary to further reveal the central mechanism by which oxytocin increases the motivation to eat in specific conditions.

8.1.3 NEURONS THAT SYNTHESIZE MELANIN-CONCENTRATING HORMONE AND OREXIN IN LATERAL HYPOTHALAMUS

The lateral hypothalamus is a region with second-order neurons that synthesize neuropeptides such as melanin-concentrating hormone (MCH) and orexin for the control of food intake. Both AgRP/NPY and POMC neurons in the ARC project densely to the lateral hypothalamus with immunoreactive fibers in close apposition to MCH neurons (Elias et al. 1998). Electrophysiological studies in slices showed that NPY inhibits MCH neurons through activating NPY receptors in both post-synaptic MCH neurons and presynaptic glutamate and GABA neurons (van den Pol et al. 2004). Melanin-concentrating hormone (MCH) is an orexigenic hormone that exerts its modulation on MCH-innervated neurons in different brain areas through the activation of MCH receptors. There are two MCH receptor subtypes, MCHR1 and MCHR2, which were detected primarily in the brain, eye, and skeletal muscle, and low levels in the tongue and pituitary. However, only MCHR1 was found in rodents and predominantly expressed in the brain for mediating the central functions in energy homeostasis and sleep. Through activating MCHR1, MCH increases food intake and reduces energy expenditure to promote overall weight gain in rodents. The important role of the MCH system in controlling energy balance is supported by findings that rodent lacking MCH or MCHR1 were lean (Marsh et al. 2002; Shimada et al. 1998), whereas MCH overexpression and central administration of MCH led to increased food intake and body weight gain (Della-Zuana et al. 2002; Qu et al. 1996). Although MCH neurons are highly restricted in the lateral hypothalamus and zona incerta, MCH fibers are found widespread throughout the brain such as the arcuate nucleus and PVN of the hypothalamus, the nucleus accumbens (NAc), the septal nuclei, and the brainstem. Among those MCH-innervated areas, feeding neurocircuits in the arcuate nucleus, PVN, and NAc were found to contribute to MCH regulation of feeding (Abbott et al. 2003; Georgescu et al. 2005; Sears et al. 2010). In addition to direct site-to-site projections, MCH neurons also regulate feeding behavior by cerebral ventricular volume transmission (Noble et al. 2018).

Orexin/hypocretin is another orexigenic neuropeptide that is produced and released by neurons located in the lateral hypothalamus. Despite the restricted expression of orexin cell body within the hypothalamus, orexin projections are found in a wide range of brain areas involved in feeding and arousal. The lateral hypothalamus synthesizes and releases orexin-A (or hypocretin-1) and orexin-B (or hypocretin-2), two structurally analogous neuropeptides, to exert functional modulation of postsynaptic neurons expressed with orexin receptors (OX1R and OX2R). In addition to orexin, orexin neurons also release dynorphin from the same synaptic vesicles which lead to a complex effect on the postsynaptic neurons (Muschamp et al. 2014; Chou et al. 2001).

Orexin was found to regulate both homeostatic food intake and the motivation for palatable food containing high fat and sugar. The target for orexin-A to increase homeostatic food intake is OX1R-expressing neurons in the lateral hypothalamus but not in other feeding-related areas including the ventromedial hypothalamus, amygdala, or the nucleus of the solitary tract (Dube, Kalra, and Kalra 1999; Sweet et al. 1999; Kotz et al. 2002). However, orexin-B has little effect on feeding. These findings suggest orexin neurons form local neural circuits to innervate feeding-promoting neurons in the lateral hypothalamus including themselves to regulate homeostatic food intake. Similar to its role in controlling the consumption of regular diets, orexin also increases palatable food consumption and promotes the motivation to obtain food rewards through activating OX1R in the extra-hypothalamic brain areas including NAc, VTA, and paraventricular thalamus (PVT) (Meffre et al. 2019).

8.1.4 NEUROPEPTIDE-PRODUCING NEURONS IN THE NUCLEUS OF THE SOLITARY TRACT

The hypothalamus is not the only region that produces neuropeptides for the control of food intake. Neurons in the caudal nucleus of the solitary tract (NTS) synthesizing pre-proglucagon (PPG), the precursor of glucagon-lie peptide-1 (GLP-1), are the main source for the GLP-1 to regulate food intake, reward, and stress response. To control food intake, PPG neurons project widely to the hypothalamus and extrahypothalamic areas.

Intracerebroventricular GLP-1 powerfully inhibits feeding in fasted rats and the inhibitory effect is blocked by specific GLP-1 receptor antagonist exendin 9–39 (Turton et al. 1996). Following intracerebroventricular GLP-1 injection, the c-fos expression is significantly increased in PVN and central amygdala (CeA), suggesting neurons in these two regions are activated by GLP-1 for the control of food intake. Direct GLP-1 injection in PVN causes anorexia which confirms the contribution of PVN in the control of food intake by central GLP-1 signaling. The inhibitory effect of peripheral administration of GLP-1 receptor agonist exendin-4 is also attenuated by lesions of the CeA (Qiao et al. 2019). In addition to PVN, other hypothalamic areas are also involved in the GLP-1 regulation of food intake. Intracerebral injections of GLP-1 significantly suppress food intake in the lateral hypothalamus, dorsomedial hypothalamus, and ventromedial hypothalamus (Schick et al. 2003).

GLP-1-expressing neurons also suppress the motivation for food reward by projecting to extrahypothalamic areas such as the bed nucleus of the stria terminalis (BNST), lateral septum, and PVT. Intra-BNST GLP-1 injection potently reduces both regular and high-fat food intake, whereas injection of GLP-1 receptor antagonist in BNST has no effect on regular food consumption but suppresses hypophagia induced by acute stress (Williams et al. 2018). In addition to BNST, both PVT and lateral septum are innervated by GLP-1 neurons in NTS. Injections of GLP-1 receptor agonists in PVT and lateral septum both decrease the motivation for food reward in rodents (Ong, Liu, et al. 2017; Terrill et al. 2016; Terrill, Maske, and Williams 2018). These findings together indicate that activation of endogenous GLP-1 signaling suppresses the motivation for food which contributes to stress-induced hypophagia.

Cholecystokinin (CCK) is a well-studied peptide known for its role in inhibiting food intake both peripherally and centrally (Della Fera and Baile 1979; Schick et al. 1990; Schwartz and Moran 1994). CCK-expressing neurons within the NTS can be activated by satiety signals following a meal. Both parabrachial nucleus (PBN) and PVN receive dense innervation from CCK-expressing NTS neurons. Activation of CCK-expressing NTS neurons or their projections to PBN and PVN inhibit food intake in mice even in energy-depleted states (D'Agostino et al. 2016; Roman, Derkach, and Palmiter 2016).

In addition to GLP-1 and CCK-expressing neurons, POMC neurons in the NTS are also involved in the control of food intake. Different from POMC neurons in the arcuate nucleus, activation of POMC neurons in the NTS acutely suppresses food intake (Zhan et al. 2013). POMC neurons in the NTS express 5-HT_{2C} receptors that are responsible for serotonin inhibition of food intake (D'Agostino et al. 2018).

8.2   ION CHANNELS IN THE NEUROPEPTIDE MODULATION OF NEURONAL ACTIVITY AND NEUROTRANSMITTER RELEASE

The main difference between classic small-molecule neurotransmitters and neuropeptides is that neuropeptides are synthesized and mainly packaged into large dense-core vesicles in the cell body while neurotransmitters are packaged in the synapse. Due to their location, synaptic vesicles release with smaller changes in intracellular calcium concentrations while dense core vesicles require higher intracellular calcium concentrations to release (van den Pol 2012). Neurons are often named for the primary neuropeptide or neurotransmitter they release. The actions of a neuron depend on its synaptic connectivity. In the control of feeding behaviors and energy metabolism, hypothalamic neuropeptide transmission plays a critical role in integrating nonhypothalamic and peripheral signals by activation of neuropeptide receptors which are metabotropic G-protein coupled receptors (GPCRs). Intracellular second messengers and G proteins are activated following neuropeptide binding to the coupled GPCRs, which opens or closes ion channels in the neuronal membrane to change the activity of the neurons and modulate the release of fast neurotransmitters (Figure 8.1). Thus, both GPCRs and ion channels on the membrane work together to modulate the neuronal activity when neuropeptides are released from feeding-related neurons in the hypothalamus and brainstem.

8.2.1   G-PROTEIN COUPLED RECEPTORS

The most common type of membrane receptors, GPCRs characteristically have seven transmembrane α-helices that form the receptor with an internal coupling to a heterotrimeric G-protein complex (Fredriksson et al. 2003). GPCRs can be activated by a variety of stimuli including ions, photons, neuropeptides, and small neurotransmitters. These extracellular signaling molecules act on GPCRs to induce intracellular responses. The responding actions to ligand binding are determined by the heterotrimeric G-protein binding to the GPCR. When activated, the GPCR assists in the exchange of GDP for GTP on the Gα protein. GTP-Gα dissociates from the dimeric Gβ/γ. Separately, free Gα and Gβ/γ act locally to trigger intracellular cascades before rejoining at the receptor (Mahoney and Sunahara 2016). There are four main families of Gα proteins: the stimulatory $G\alpha_s$, the inhibitory $G\alpha_i$,

(a)   (b)

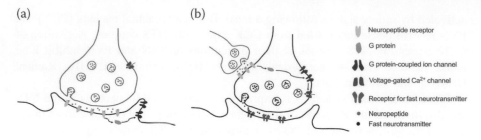

FIGURE 8.1   **Presynaptic and postsynaptic modulation of neuropeptide on neuronal activity.** **A**, A neuropeptide neuron innervates a soma or dendrite of another neuron for postsynaptic modulation of the membrane activity. Released neuropeptide binds to postsynaptic neuropeptide receptors and then affects the function of ion channels through G protein and intracellular second messengers. The open or close of G protein-coupled ion channels finally modulates the activity of the postsynaptic neurons. **B**, A neuropeptide neuron innervates the presynaptic terminals to modulate fast neurotransmission. Released neuropeptide binds to its receptors on the membrane of the axonal terminal which leads to activation of G protein to modulate the function of the coupled ion channels. The membrane potential of the axonal terminal is changed by the activity of G protein-coupled ion channels, which affects Ca^{2+} influx via voltage-gated Ca^{2+} channels to modulate vesicular neurotransmitter release.

$G\alpha_q$, and $G\alpha_{12}$. The largest and most diverse family is $G\alpha_i$. The β/γ subunits are less diverse with 5 $G\beta$ and 12 $G\gamma$ genes (Syrovatkina et al. 2016). The associated G proteins determine neuropeptide regulation of the neuronal activity through coupling the neuropeptide receptors to following ion channels on the membrane.

8.2.2   INWARDLY RECTIFYING POTASSIUM CHANNELS

Inwardly rectifying potassium (Kir) channels include seven subfamilies named Kir1-Kir7. Activation of Kir channels causes hyperpolarization following potassium channel opening (Glaaser and Slesinger 2015). Kir subfamily Kir3 is better known as G protein-gated inwardly rectifying potassium channels (GIRKs). When bound by agonists, GPCR activation leads to inhibitory G protein-dependent opening of GIRK channels for the hyperpolarizing response on neurons. Agonist binding induces an exchange of guanidine diphosphate (GDP) for the activated guanidine triphosphate (GTP) by the $G\alpha_{i/o}$ subunit. Free GTP bound $G\alpha_{i/o}$ disassociates from the $G\beta/\gamma$ subunit complex. The $G\beta/\gamma$ binds to the GIRK channel and, with PIP_2 as a cofactor, the channel opening allows potassium ions to cross the membrane of the cell. GDP-bound $G\alpha$ re-associates with the $G\beta/\gamma$ and the channel closes (Glaaser and Slesinger 2015). When GIRK channels are activated, efflux of positively charged potassium ions results in hyperpolarization. In the brain, a variety of intercellular signaling molecules such as neuropeptides can activate GIRK channels so long as the GPCR contains the $G_{i/o}$ (Glaaser and Slesinger 2015). GPCR kinases inactivate GIRKs through the competitive binding of $G\beta/\gamma$ (van den Pol 2012).

A large number of feeding-related neuropeptides modulate the neuronal activity through the opening or closing of the GIRK channels. NPY is a typical inhibitory neuropeptide that decreases neuronal activity when it binds to NPY receptors widely expressed in the brain. The hyperpolarizing effects of NPY are mainly mediated by the increased activity of GIRK channels following the activation of NPY receptors in many areas including the arcuate nucleus, lateral hypothalamus, ventromedial hypothalamus, hippocampus, lateral amygdala, and thalamus (Acuna-Goycolea et al. 2005; Sun, Huguenard, and Prince 2001; Roseberry et al. 2004; Fu, Acuna-Goycolea, and van den Pol 2004). Endogenous opioid peptides are also involved in the control of food intake. Both dynorphin and met-enkephalin are released in the hypothalamus for regulating the activity of feeding-related neural circuits. In the arcuate nucleus, dynorphin inhibits both POMC and dopamine neurons through the opening of GIRK channels after the activation of *kappa* opioid receptors (Zhang and van den Pol 2013, 2015). In addition to dynorphin, met-enkephalin potentiates GIRK currents in POMC neurons through activating *mu* opioid receptors (Zhang and van den Pol 2015).

The inhibition of GIRK channels is also responsible for the excitation of neurons by some neuropeptides. Orexin is an excitatory neuropeptide that modulates multiple ion channels following the activation of OX1R and OX2R. Orexin excites neurons mainly through modulating non-selective cation channels which will be introduced in the following paragraph. However, orexin also blocks the activation of GIRK channels. In cultured neurons from the locus coeruleus and tuberomammillary nucleus, orexin activation of both OX1R and OX2R inhibits GIRK currents evoked by other neuropeptides such as somatostatin and nociceptin (Hoang et al. 2003). Pertussis toxin (PTX)-insensitive G-protein (such as $G_{i/o}$) contributes to orexin-evoked inhibition of GIRK channels (Hoang et al. 2003, 2004).

Interestingly, both α-MSH and AgRP modulate Kir1.7 in MC4R-positive PVN neurons through a G protein-independent coupling (Ghamari-Langroudi et al. 2015). α-MSH depolarizes MC4R-positive neurons in the PVN and the current-voltage relationship demonstrates that the conductance involved in α-MSH-induced depolarization is mediated by Kir7.1. However, the depolarizing effect is not affected by blocking intracellular G protein signaling. In HEK293 cells with MC4R and Kir7.1 transfection, α-MSH inhibits but AgRP increases the currents mediated by Kir7.1, suggesting both α-MSH and AgRP couple MC4R to Kir7.1 channel without G protein for the regulation of the neuronal activity.

8.2.3 ATP-SENSITIVE POTASSIUM CHANNELS

The ability for neurons to rapidly respond to changes in glucose levels is critical for energy homeostasis. When glucose enters the cell, it undergoes glycolysis, the tricarboxylic acid (TCA) cycle, and the electron transport chain, to activate the synthesis of adenosine triphosphate (ATP). ATP is the "energy-carrying molecule" that acts as fuel to stimulate or propagate an action (Fernie, Carrari, and Sweetlove 2004). The more glucose consumed, the more energy is readily available in the cell for use or storage in adipocytes. The sensing of ATP allows neurons to modulate activity based on intracellular energy levels. ATP-sensitive potassium channels (K_{ATP} channels) are expressed ubiquitously in the brain and peripheral cells. In the central nervous system, K_{ATP} channels are found in many feeding-related neurons including ventromedial hypothalamic neurons, POMC and AgRP neurons of the arcuate nucleus, and MCH and orexin neurons of the lateral hypothalamus. In POMC neurons, the activity of the K_{ATP} channel was found to be altered in mice with high-fat-diet-induced obesity (Parton et al. 2007). Another study also reported that an age-dependent increase in mammalian target of rapamycin (mTOR) signaling silences hypothalamic POMC neurons by elevating K_{ATP} channel activity, suggesting increased K_{ATP} channel activity contributes to age-dependent obesity (Yang et al. 2012).

In both POMC and AgRP neurons, K_{ATP} channels are modulated by ATP as well as leptin and insulin. Insulin activates K_{ATP} channels on AgRP neurons by stimulating the phosphoinositide 3-kinase (PI3K) signaling pathway (Qiu et al. 2014). Leptin also utilizes the PI3K pathway to hyperpolarize cells via the activation of K_{ATP} channels (Belgardt, Okamura, and Bruning 2009). In orexin neurons, leptin electrically silences the cell via indirect activation of K_{ATP} channels (Goforth et al. 2014). In addition to receptors for peripherally secreted peptides, some neuropeptide receptors are also coupled to K_{ATP} channels for the modulation of neuronal activity in the nervous system. In the arcuate nucleus of the hypothalamus, GLP-1 receptor activation by the selective agonist liraglutide inhibits NPY neurons through activation of protein kinase A (PKA)-dependent K_{ATP} channels (He et al. 2019). In the dorsal motor nucleus of the vagus, activation of MC4R by agonist MTII and THIQ hyperpolarizes cholinergic neurons through activating K_{ATP} channels (Sohn et al. 2013). Also, K_{ATP} channels are required for GLP-1 modulation of peripheral cells (Light et al. 2002; Aizawa et al. 1998).

8.2.4 Non-Selective Cation Channels

Named for their low exclusivity, non-selective cation channels (NSCs) are permeable to multiple cations including potassium, calcium, and sodium ions through a voltage-independent way. NSCs are diverse with distinct activation requirements and ion permeability (Siemen 1993). Two of the four subgroups, transient receptor potential (TRP) channels and calcium-activated non-selective channels, can be blocked by flufenamic acid while the other two, acid-sensitive cation channels and hyperpolarization-activated cation channels, do not respond to flufenamic acid (Palus-Chramiec et al. 2019). This makes the identification of specific subgroups challenging. NSCs act to maintain the spontaneous firing of neurons following neuropeptide receptor activation. Orexin in particular has been associated with the opening of NSCs. In the subthalamic nucleus, the neuronal response to orexin is mediated by the coactivation of NSCs (Li et al. 2019). In the intergeniculate leaflet of the thalamus, NSC is critical for orexin-evoked depolarization (Palus-Chramiec et al. 2019). CCK depolarizes orexin neurons by activating NSCs linked to the CCK_A receptor while oxytocin and arginine vasopressin excite orexin neurons through oxytocin and V1a receptor, respectively (Tsujino et al. 2005; Tsunematsu et al. 2008). In the lateral hypothalamus, both oxytocin and vasopressin excite MCH neurons by activating both NSCs and sodium/calcium exchangers (Yao et al. 2012). In the arcuate nucleus, orexigenic NPY neurons are depolarized by bombesin-related peptides including gastrin-releasing peptide and neuromedin B that act on NSCs as well as the sodium/calcium exchanger. Bombesin-related peptides also excite the anorexigenic POMC neurons of the arcuate nucleus partially through activating NSCs, suggesting a complex mechanism of bombesin-related peptides in the regulation of food intake (van den Pol et al. 2009). In the lateral hypothalamus, thyrotropin-releasing hormone (TRH) inhibits MCH neurons by increasing synaptic GABA release from local GABA neurons. Selective TRP-canonical (TRPC) channel blockers attenuate the TRH effect, suggesting that NSCs contribute to TRH modulation of lateral hypothalamic GABA and MCH neurons (Zhang and van den Pol 2012).

8.2.5 Voltage-Gated Calcium Channel and Neurotransmitter Release

Neuropeptide receptors are expressed not only on the postsynaptic area for modulating the activity of the postsynaptic membrane but also on presynaptic terminals for affecting both glutamate and GABA release (Figure 8.1) (Miller 1998; Willis 2006). According to the properties of presynaptic neuropeptide receptors and their modulation on the coupled ion channels, some neuropeptide such as NPY and dynorphin typically reduce transmitter release, whereas others including orexin and GLP-1 tend to increase the release (Colmers, Lukowiak, and Pittman 1988; van den Pol et al. 1998; Acuna-Goycolea and van den Pol 2004). In the lateral hypothalamus, TRH inhibits MCH neurons through increasing presynaptic GABA release without a direct effect on the membrane properties of MCH neurons (Zhang and van den Pol 2012). In the arcuate nucleus, dynorphin not only directly hyperpolarizes both POMC and NPY neurons but also decreases neurotransmitter release (Zhang and van den Pol 2013). It decreases GABA release onto both POMC and NPY neurons, whereas it only reduces glutamate release onto POMC neurons. In addition to POMC and NPY neurons, endogenous opioid peptides dynorphin and met-enkephalin also reduce GABA release onto arcuate DA neurons (Zhang and van den Pol 2015). Together, the complex effect of dynorphin on feeding-related neural networks in the arcuate nucleus may contribute to the regulation of dynorphin on food intake.

Neuropeptides may regulate neurotransmitter release through presynaptic facilitation or inhibition, which is determined by the effect on calcium influx into the axonal terminals (Colmers, Lukowiak, and Pittman 1988). Activation of presynaptic neuropeptide receptors either increases or decreases intracellular calcium levels in presynaptic axonal terminals, which affect the calcium-dependent transmitter release process including vesicle movement and fusion. Voltage-gated calcium channels play an important role in the modulation of cytoplasmic calcium by neuropeptide receptor activation. For instance, neuropeptide nociceptin reduces glutamate release from the retinohypothalamic tract to the suprachiasmatic nucleus (SCN) neurons by indirectly attenuating N-type calcium currents and to a lesser degree P/Q-type calcium currents (Gompf et al. 2005). In this study, the retinal ganglion cell bodies were cut and dissociated by the slice preparation. Thus, nociceptin treatment in slices reduces glutamate release by inhibiting voltage-gated calcium channels in the axonal terminals but not the cell body of retinal ganglion cells. Similarly, nociceptin also inhibits presynaptic GABA release in the central amygdala (Roberto and Siggins 2006). In lateral hypothalamic neurons, MCH depresses both presynaptic glutamate and GABA release (Gao and van den Pol 2001). The possible mechanism is that MCH reduces calcium influx through voltage-gated L, N, and P/Q calcium channels (Gao and van den Pol 2002). In addition to MCH, dynorphin depresses voltage-gated calcium currents and reduces presynaptic glutamate release in lateral hypothalamic orexin neurons (Li and van den Pol 2006).

8.3 NEUROPEPTIDE RECEPTORS AND ION CHANNELS IN DEVELOPING PHARMACOTHERAPY IN OBESITY

Obesity has become a global health threat due to the increased risk of obesity-related complications such as coronary heart disease, diabetes, and end-stage renal disease (GBD 2015 Obesity Collaborators et al. 2017; Ogden et al. 2015). In the United States, the age-adjusted prevalence of obesity in adults has increased to over 40% in 2018. The destroyed balance between food intake and energy expenditure is the main reason forobesity. However, the high prevalence of obesity is probably caused by the dysfunction of feeding-related neural circuits that leads to overeating of easily available calorie-dense food. Especially, dysfunction in neuropeptide modulation and neuropeptide receptor signaling has been widely reported to be involved in obesity-related conditions. Neuropeptide receptors also become one of the most promising biological targets for developing effective pharmacotherapies in obesity.

8.3.1 DYSFUNCTIONAL NEUROPEPTIDE SIGNALING IN OBESITY

In the hypothalamus, the melanocortin system plays an important role in controlling food intake and energy balance. Activation of melanocortin receptors (MC3R and MC4R) in the PVN decreases food intake, while inhibition of those receptors by AgRP induces feeding behavior. Mice with genetic knockouts of MC3R or MC4R revealed that the absence of MC3Rs results in minor weight gain while MC4R knockouts are hyperphagic and develop obesity (Chen et al. 2000). MC4Rs are activated by the POMC derivative peptide α-MSH and are antagonized by AgRP. POMC derivatives are anorexigenic and mice with knockouts of the POMC gene develop obesity whereas ablation of AgRP neurons in the arcuate nucleus leads to starvation (Challis et al. 2004; Gropp et al. 2005; Luquet et al. 2005). Interestingly, double knockout of POMC and AgRP results in a phenotype indistinguishable from POMC knockout (Corander et al. 2011). The early-onset synaptic changes probably counteract the lethality of AgRP dysfunction. Studies also indicated that NPY/AgRP ablation in neonates, but not adults, leads to the development of compensatory mechanisms in feeding regulation (Luquet et al. 2005). The fatality associated with AgRP neuron dysfunction is indicative of the importance of these neurons in initiating food intake. In humans, frameshift mutations in the MC4R gene are known to cause early-onset obesity presented with increased fat and lean mass (Anderson et al. 2016). The MC4R mutations in obese individuals were first reported by two research groups in 1998 (Yeo et al. 1998; Vaisse et al. 1998). Compared to noncarriers, individuals that carry pathogenic variants of MC4R have a 4.5-fold increase in the risk of developing obesity (MacKenzie 2006). To date, more than 200 MC4R variants have been identified, primarily heterozygous dominant acting missense variants (Kuhnen, Krude, and Biebermann 2019). Heterozygous variants are also the most common genetic form of obesity in the pediatric age group and are found in 2%–5% of subjects with extreme pediatric obesity (Vaisse et al. 1998; Farooqi et al. 2000). Homozygous MC4R variants have been identified in the offspring of consanguineous families (Farooqi et al. 2000, 2003).

Dysfunction in other neuropeptide systems is also involved in eating disorders and obesity. Transgenic mice with orexin knockouts have reduced food intake but late-onset obesity due to the dramatically reduced energy expenditure from decreased motor activity and metabolic rate. Mice lacking pro-MCH genes are hypophagic and lean, whereas mice with MCH overexpression in the lateral hypothalamus are hyperphagic and obese (Shimada et al. 1998; Ludwig et al. 2001). Two infrequent missense variations (Y181H and R248Q) were identified in a mutation screen of the MCHR1 gene in individuals with severe and early-onset obesity although in vitro studies showed no evidence for functional implications (Gibson et al. 2004). The human genetic disorder, Prader-Willi, results in the dysfunction of oxytocin signaling in the brain. Key symptoms of the mutation include mental retardation, hyperphagia, and severe obesity (Kabasakalian, Ferretti, and Hollander 2018; Martin et al. 1998; Swaab, Purba, and Hofman 1995). Interestingly, ablation of hypothalamic oxytocin neurons has no significant effect on food intake though the decreased expression of oxytocin receptors results in mild obesity in male mice (Onaka and Takayanagi 2019).

8.3.2 Dysfunctional Neuropeptide Modulation of Ion Channels in Obesity

Ion channels tightly control the neuronal activity through transmembrane ion flux for depolarization or hyperpolarization. Changes in the protein structure of ion channels can drastically change how the neuron responds to extracellular signals such as neuropeptides. For instance, dysfunctional K_{ATP} channels, possibly from high-fat diets, would be unable to close under elevated glucose levels, resulting in constitutively active inhibition of the anorexigenic POMC neurons, leading to the development of obesity (Parton et al. 2007). Mutations in K_{ATP} channels are also associated with congenital diabetes and hyperinsulinism (Tinker et al. 2018). Tonically elevated PIP_3, a signaling molecule naturally activated by insulin, acts as a sexually dimorphic inhibitor of POMC neurons via stimulation of K_{ATP} channels. Female mice have a larger weight gain than males when PIP_3 is perpetually elevated (Plum et al. 2006). Kir6.2 is a key pore-forming subunit of K_{ATP} channels (Miki et al. 2001). The defective Kir6.2 prevents ATP blockade of K_{ATP} channels, which results in increased food intake and obesity due to loss of glucose sensitivity of Kir6.2-expressing hypothalamic neurons (Sohn 2013; Miki et al. 2001). Kir6.2 knockout mice also showed a blunted hypothalamic response to glucose loading and elevated hypothalamic

NPY expression accompanied by hyperphagia, while they are resistant to obesity (Park et al. 2011).

8.3.3 DRUGS TARGET NEUROPEPTIDE RECEPTORS AND ION CHANNELS FOR THE TREATMENT OF OBESITY

Animal research discovers new targets for clinical scientists to develop therapies and treatments in an attempt to better human health. Obesity is a complex disease that develops for reasons specific to each person. To treat such an individualistic disease, the leading risk factor must be identified. One of the current strategies in obesity treatment is to develop drugs targeting ion channels and neuropeptide receptors in neurons of feeding neural circuits.

Dysfunctional melanocortin signaling caused by MC4R mutation is linked to obesity. Setmelanotide is an MC4R agonist that has been tested clinically for decreasing food intake and body weight control. In patients with leptin receptor deficiency, setmelanotide treatment significantly reduced food intake and body weight (Kuhnen et al. 2016). Unlike other MC4R agonists, no increase in blood pressure or severe adverse effect related to setmelanotide was observed in the study. However, in obese MC4R mutation carriers, setmelanotide treatment produces no significant difference in body weight change compared to control (Collet et al. 2017). These findings suggest that MC4R is required for setmelanotide to produce an inhibitory effect on food intake and body weight. In POMC-deficient patients, treatment with setmelanotide restores the MC4R signaling pathway, which leads to severe weight loss (Kuhnen et al. 2016; Clement et al. 2018). According to the mechanism for the MC4R-activated signaling in controlling neuronal activity, both Kir7.1 and K_{ATP} channels may contribute to the decrease in food intake by setmelanotide.

GLP-1R is another important target for anti-obesity treatment. Liraglutide is a GLP-1R agonist that has been approved by the FDA for obesity treatment (Mehta, Marso, and Neeland 2017). In adults with obesity or overweight who had either hypertension or dyslipidemia, treatment with liraglutide resulted in body weight loss and improved metabolic control (Pi-Sunyer et al. 2015). In adolescents with obesity, the use of liraglutide plus lifestyle therapy led to a significantly greater reduction in body-mass index (BMI) than placebo plus lifestyle therapy (Kelly et al. 2020). Both peripheral and central GLP-1Rs contribute to the liraglutide-mediated reduction in food intake and body weight (Adams et al. 2018; Burmeister et al. 2017; Beiroa et al. 2014; Sisley et al. 2014; He et al. 2019). In the arcuate nucleus of the hypothalamus, both TRPC5 and K_{ATP} channels contribute to liraglutide activation of POMC neurons and inhibition of NPY/AgRP neurons (He et al. 2019), suggesting these two channels play a role in liraglutide reduction of food intake and body weight of obese patients.

Topiramate is a sulfamate modified fructose diacetonide that affects multiple ion channels and receptors including voltage-gated sodium and calcium channels, $GABA_A$ receptor, and AMPA/kainite glutamate receptors. Topiramate has been approved by the FDA for weight loss management. Topiramate as monotherapy causes 6%–7% more body weight loss compared to placebo, whereas it has

9%–10% more weight loss compared to placebo after 1 year of treatment when combined with phentermine (Kopelman et al. 2010; Wong, Sullivan, and Heap 2012). Although the mechanism remains unclear, topiramate may inhibit food intake through direct modulation of ion channels in feeding-related neurons and indirect increase in the expression of feeding-related neuropeptides such as POMC and TRH, as well as potentiation of insulin and leptin signaling in the hypothalamus (Caricilli et al. 2012).

REFERENCES

Abbott, C. R., A. R. Kennedy, A. M. Wren, M. Rossi, K. G. Murphy, L. J. Seal, J. F. Todd, M. A. Ghatei, C. J. Small, and S. R. Bloom. 2003. "Identification of hypothalamic nuclei involved in the orexigenic effect of melanin-concentrating hormone." *Endocrinology* 144 (9):3943–3949. doi: 10.1210/en.2003-0149.

Acuna-Goycolea, C., N. Tamamaki, Y. Yanagawa, K. Obata, and A. N. van den Pol. 2005. "Mechanisms of neuropeptide Y, peptide YY, and pancreatic polypeptide inhibition of identified green fluorescent protein-expressing GABA neurons in the hypothalamic neuroendocrine arcuate nucleus." *J Neurosci* 25 (32):7406–7419. doi: 10.1523/JNEUROSCI.1008-05.2005.

Acuna-Goycolea, C., and A. van den Pol. 2004. "Glucagon-like peptide 1 excites hypocretin/orexin neurons by direct and indirect mechanisms: implications for viscera-mediated arousal." *J Neurosci* 24 (37):8141–8152. doi: 10.1523/JNEUROSCI.1607-04.2004.

Adams, J. M., H. Pei, D. A. Sandoval, R. J. Seeley, R. B. Chang, S. D. Liberles, and D. P. Olson. 2018. "Liraglutide modulates appetite and body weight through glucagon-like peptide 1 receptor-expressing glutamatergic neurons." *Diabetes* 67 (8):1538–1548. doi: 10.2337/db17-1385.

Aizawa, T., M. Komatsu, N. Asanuma, Y. Sato, and G. W. Sharp. 1998. "Glucose action 'beyond ionic events' in the pancreatic beta cell." *Trends Pharmacol Sci* 19 (12):496–499. doi: 10.1016/s0165-6147(98)01273-5.

Anderson, E. J., I. Cakir, S. J. Carrington, R. D. Cone, M. Ghamari-Langroudi, T. Gillyard, L. E. Gimenez, and M. J. Litt. 2016. "60 YEARS OF POMC: regulation of feeding and energy homeostasis by alpha-MSH." *J Mol Endocrinol* 56 (4):T157–T174. doi: 10.1530/JME-16-0014.

Aponte, Y., D. Atasoy, and S. M. Sternson. 2011. "AGRP neurons are sufficient to orchestrate feeding behavior rapidly and without training." *Nat Neurosci* 14 (3):351–355. doi: 10.1038/nn.2739.

Atasoy, D., J. N. Betley, H. H. Su, and S. M. Sternson. 2012. "Deconstruction of a neural circuit for hunger." *Nature* 488 (7410):172–177. doi: 10.1038/nature11270.

Balthasar, N., R. Coppari, J. McMinn, S. M. Liu, C. E. Lee, V. Tang, C. D. Kenny, R. A. McGovern, S. C. Chua, Jr., J. K. Elmquist, and B. B. Lowell. 2004. "Leptin receptor signaling in POMC neurons is required for normal body weight homeostasis." *Neuron* 42 (6):983–991. doi: 10.1016/j.neuron.2004.06.004.

Barrett, L. R., J. Nunez, and X. Zhang. 2021. Oxytocin activation of paraventricular thalamic neurons promotes feeding motivation to attenuate stress-induced hypophagia. *Neuropsychopharmacology* 46 (5):1045-1056. doi: 10.1038/s41386-021-00961-3.

Baskin, D. G., F. Kim, R. W. Gelling, B. J. Russell, M. W. Schwartz, G. J. Morton, H. N. Simhan, D. H. Moralejo, and J. E. Blevins. 2010. "A new oxytocin-saporin cytotoxin for lesioning oxytocin-receptive neurons in the rat hindbrain." *Endocrinology* 151 (9):4207–4213. doi: 10.1210/en.2010-0295.

Beiroa, D., M. Imbernon, R. Gallego, A. Senra, D. Herranz, F. Villarroya, M. Serrano, J. Ferno, J. Salvador, J. Escalada, C. Dieguez, M. Lopez, G. Fruhbeck, and R. Nogueiras.

2014. "GLP-1 agonism stimulates brown adipose tissue thermogenesis and browning through hypothalamic AMPK." *Diabetes* 63 (10):3346–3358. doi: 10.2337/db14-0302.

Belgardt, B. F., T. Okamura, and J. C. Bruning. 2009. "Hormone and glucose signalling in POMC and AgRP neurons." *J Physiol* 587 (Pt 22):5305–5314. doi: 10.1113/jphysiol.2009.179192.

Burmeister, M. A., J. E. Ayala, H. Smouse, A. Landivar-Rocha, J. D. Brown, D. J. Drucker, D. A. Stoffers, D. A. Sandoval, R. J. Seeley, and J. E. Ayala. 2017. "The hypothalamic glucagon-like peptide 1 receptor is sufficient but not necessary for the regulation of energy balance and glucose homeostasis in mice." *Diabetes* 66 (2):372–384. doi: 10.2337/db16-1102.

Butler, A. A., D. L. Marks, W. Fan, C. M. Kuhn, M. Bartolome, and R. D. Cone. 2001. "Melanocortin-4 receptor is required for acute homeostatic responses to increased dietary fat." *Nat Neurosci* 4 (6):605–611. doi: 10.1038/88423.

Camerino, C. 2009. "Low sympathetic tone and obese phenotype in oxytocin-deficient mice." *Obesity (Silver Spring)* 17 (5):980–984. doi: 10.1038/oby.2009.12.

Caricilli, A. M., E. Penteado, L. L. de Abreu, P. G. Quaresma, A. C. Santos, D. Guadagnini, D. Razolli, F. C. Mittestainer, J. B. Carvalheira, L. A. Velloso, M. J. Saad, and P. O. Prada. 2012. "Topiramate treatment improves hypothalamic insulin and leptin signaling and action and reduces obesity in mice." *Endocrinology* 153 (9):4401–4411. doi: 10.1210/en.2012-1272.

Challis, B. G., A. P. Coll, G. S. Yeo, S. B. Pinnock, S. L. Dickson, R. R. Thresher, J. Dixon, D. Zahn, J. J. Rochford, A. White, R. L. Oliver, G. Millington, S. A. Aparicio, W. H. Colledge, A. P. Russ, M. B. Carlton, and S. O'Rahilly. 2004. "Mice lacking pro-opiomelanocortin are sensitive to high-fat feeding but respond normally to the acute anorectic effects of peptide-YY(3-36)." *Proc Natl Acad Sci USA* 101 (13):4695–4700. doi: 10.1073/pnas.0306931101.

Chen, A. S., D. J. Marsh, M. E. Trumbauer, E. G. Frazier, X. M. Guan, H. Yu, C. I. Rosenblum, A. Vongs, Y. Feng, L. Cao, J. M. Metzger, A. M. Strack, R. E. Camacho, T. N. Mellin, C. N. Nunes, W. Min, J. Fisher, S. Gopal-Truter, D. E. MacIntyre, H. Y. Chen, and L. H. Van der Ploeg. 2000. "Inactivation of the mouse melanocortin-3 receptor results in increased fat mass and reduced lean body mass." *Nat Genet* 26 (1):97–102. doi: 10.1038/79254.

Chou, T. C., C. E. Lee, J. Lu, J. K. Elmquist, J. Hara, J. T. Willie, C. T. Beuckmann, R. M. Chemelli, T. Sakurai, M. Yanagisawa, C. B. Saper, and T. E. Scammell. 2001. "Orexin (hypocretin) neurons contain dynorphin." *J Neurosci* 21 (19):RC168.

Clement, K., H. Biebermann, I. S. Farooqi, L. Van der Ploeg, B. Wolters, C. Poitou, L. Puder, F. Fiedorek, K. Gottesdiener, G. Kleinau, N. Heyder, P. Scheerer, U. Blume-Peytavi, I. Jahnke, S. Sharma, J. Mokrosinski, S. Wiegand, A. Muller, K. Weiss, K. Mai, J. Spranger, A. Gruters, O. Blankenstein, H. Krude, and P. Kuhnen. 2018. "MC4R agonism promotes durable weight loss in patients with leptin receptor deficiency." *Nat Med* 24 (5):551–555. doi: 10.1038/s41591-018-0015-9.

Collet, T. H., B. Dubern, J. Mokrosinski, H. Connors, J. M. Keogh, E. Mendes de Oliveira, E. Henning, C. Poitou-Bernert, J. M. Oppert, P. Tounian, F. Marchelli, R. Alili, J. Le Beyec, D. Pepin, J. M. Lacorte, A. Gottesdiener, R. Bounds, S. Sharma, C. Folster, B. Henderson, S. O'Rahilly, E. Stoner, K. Gottesdiener, B. L. Panaro, R. D. Cone, K. Clement, I. S. Farooqi, and L. H. T. Van der Ploeg. 2017. "Evaluation of a melanocortin-4 receptor (MC4R) agonist (setmelanotide) in MC4R deficiency." *Mol Metab* 6 (10):1321–1329. doi: 10.1016/j.molmet.2017.06.015.

Colmers, W. F., K. Lukowiak, and Q. J. Pittman. 1988. "Neuropeptide Y action in the rat hippocampal slice: site and mechanism of presynaptic inhibition." *J Neurosci* 8 (10):3827–3837.

Corander, M. P., D. Rimmington, B. G. Challis, S. O'Rahilly, and A. P. Coll. 2011. "Loss of agouti-related peptide does not significantly impact the phenotype of murine POMC deficiency." *Endocrinology* 152 (5):1819–1828. doi: 10.1210/en.2010-1450.

Cowley, M. A., R. G. Smith, S. Diano, M. Tschop, N. Pronchuk, K. L. Grove, C. J. Strasburger, M. Bidlingmaier, M. Esterman, M. L. Heiman, L. M. Garcia-Segura, E. A. Nillni, P. Mendez, M. J. Low, P. Sotonyi, J. M. Friedman, H. Liu, S. Pinto, W. F. Colmers, R. D. Cone, and T. L. Horvath. 2003. "The distribution and mechanism of action of ghrelin in the CNS demonstrates a novel hypothalamic circuit regulating energy homeostasis." *Neuron* 37 (4):649–661. doi: 10.1016/s0896-6273(03)00063-1.

D'Agostino, G., D. Lyons, C. Cristiano, M. Lettieri, C. Olarte-Sanchez, L. K. Burke, M. Greenwald-Yarnell, C. Cansell, B. Doslikova, T. Georgescu, P. B. Martinez de Morentin, M. G. Myers, Jr., J. J. Rochford, and L. K. Heisler. 2018. "Nucleus of the solitary tract serotonin 5-HT2C receptors modulate food intake." *Cell Metab* 28 (4):619–630.e5. doi: 10.1016/j.cmet.2018.07.017.

D'Agostino, G., D. J. Lyons, C. Cristiano, L. K. Burke, J. C. Madara, J. N. Campbell, A. P. Garcia, B. B. Land, B. B. Lowell, R. J. Dileone, and L. K. Heisler. 2016. "Appetite controlled by a cholecystokinin nucleus of the solitary tract to hypothalamus neurocircuit." *Elife* 5. doi: 10.7554/eLife.12225.

Della Fera, M. A., and C. A. Baile. 1979. "CCK-octapeptide injected in CSF causes satiety in sheep." *Ann Rech Vet* 10 (2-3):234–236.

Della-Zuana, O., F. Presse, C. Ortola, J. Duhault, J. L. Nahon, and N. Levens. 2002. "Acute and chronic administration of melanin-concentrating hormone enhances food intake and body weight in Wistar and Sprague-Dawley rats." *Int J Obes Relat Metab Disord* 26 (10):1289–1295. doi: 10.1038/sj.ijo.0802079.

Demitrack, M. A., M. D. Lesem, S. J. Listwak, H. A. Brandt, D. C. Jimerson, and P. W. Gold. 1990. "CSF oxytocin in anorexia nervosa and bulimia nervosa: clinical and pathophysiologic considerations." *Am J Psychiatry* 147 (7):882–886. doi: 10.1176/ajp.147.7.882.

Douglas, A. J., L. E. Johnstone, and G. Leng. 2007. "Neuroendocrine mechanisms of change in food intake during pregnancy: a potential role for brain oxytocin." *Physiol Behav* 91 (4):352–365. doi: 10.1016/j.physbeh.2007.04.012.

Dube, M. G., S. P. Kalra, and P. S. Kalra. 1999. "Food intake elicited by central administration of orexins/hypocretins: identification of hypothalamic sites of action." *Brain Res* 842 (2):473–477. doi: 10.1016/s0006-8993(99)01824-7.

Elias, C. F., C. B. Saper, E. Maratos-Flier, N. A. Tritos, C. Lee, J. Kelly, J. B. Tatro, G. E. Hoffman, M. M. Ollmann, G. S. Barsh, T. Sakurai, M. Yanagisawa, and J. K. Elmquist. 1998. "Chemically defined projections linking the mediobasal hypothalamus and the lateral hypothalamic area." *J Comp Neurol* 402 (4):442–459.

Farooqi, I. S., J. M. Keogh, G. S. Yeo, E. J. Lank, T. Cheetham, and S. O'Rahilly. 2003. "Clinical spectrum of obesity and mutations in the melanocortin 4 receptor gene." *N Engl J Med* 348 (12):1085–1095. doi: 10.1056/NEJMoa022050.

Farooqi, I. S., G. S. Yeo, J. M. Keogh, S. Aminian, S. A. Jebb, G. Butler, T. Cheetham, and S. O'Rahilly. 2000. "Dominant and recessive inheritance of morbid obesity associated with melanocortin 4 receptor deficiency." *J Clin Invest* 106 (2):271–279. doi: 10.1172/JCI9397.

Fenselau, H., J. N. Campbell, A. M. Verstegen, J. C. Madara, J. Xu, B. P. Shah, J. M. Resch, Z. Yang, Y. Mandelblat-Cerf, Y. Livneh, and B. B. Lowell. 2017. "A rapidly acting glutamatergic ARC→PVH satiety circuit postsynaptically regulated by alpha-MSH." *Nat Neurosci* 20 (1):42–51. doi: 10.1038/nn.4442.

Fernie, A. R., F. Carrari, and L. J. Sweetlove. 2004. "Respiratory metabolism: glycolysis, the TCA cycle and mitochondrial electron transport." *Curr Opin Plant Biol* 7 (3):254–261. doi: 10.1016/j.pbi.2004.03.007.

Fetissov, S. O., J. Harro, M. Jaanisk, A. Jarv, I. Podar, J. Allik, I. Nilsson, P. Sakthivel, A. K. Lefvert, and T. Hokfelt. 2005. "Autoantibodies against neuropeptides are associated with psychological traits in eating disorders." *Proc Natl Acad Sci USA* 102 (41):14865–14870. doi: 10.1073/pnas.0507204102.

Fredriksson, R., M. C. Lagerstrom, L. G. Lundin, and H. B. Schioth. 2003. "The G-protein-coupled receptors in the human genome form five main families. Phylogenetic analysis, paralogon groups, and fingerprints." *Mol Pharmacol* 63 (6):1256–1272. doi: 10.1124/mol.63.6.1256.

Fu, L. Y., C. Acuna-Goycolea, and A. N. van den Pol. 2004. "Neuropeptide Y inhibits hypocretin/orexin neurons by multiple presynaptic and postsynaptic mechanisms: tonic depression of the hypothalamic arousal system." *J Neurosci* 24 (40):8741–8751. doi: 10.1523/JNEUROSCI.2268-04.2004.

Gao, X. B., and A. N. van den Pol. 2001. "Melanin concentrating hormone depresses synaptic activity of glutamate and GABA neurons from rat lateral hypothalamus." *J Physiol* 533 (Pt 1):237–252. doi: 10.1111/j.1469-7793.2001.0237b.x.

Gao, X. B., and A. N. van den Pol. 2002. "Melanin-concentrating hormone depresses L-, N-, and P/Q-type voltage-dependent calcium channels in rat lateral hypothalamic neurons." *J Physiol* 542 (Pt 1):273–286. doi: 10.1113/jphysiol.2002.019372.

GBD 2015 Obesity Collaborators A. Afshin, M. H. Forouzanfar, M. B. Reitsma, P. Sur, K. Estep, A. Lee, L. Marczak, A. H. Mokdad, M. Moradi-Lakeh, M. Naghavi, J. S. Salama, T. Vos, K. H. Abate, C. Abbafati, M. B. Ahmed, Z. Al-Aly, A. Alkerwi, R. Al-Raddadi, A. T. Amare, A. Amberbir, A. K. Amegah, E. Amini, S. M. Amrock, R. M. Anjana, J. Arnlov, H. Asayesh, A. Banerjee, A. Barac, E. Baye, D. A. Bennett, A. S. Beyene, S. Biadgilign, S. Biryukov, E. Bjertness, D. J. Boneya, I. Campos-Nonato, J. J. Carrero, P. Cecilio, K. Cercy, L. G. Ciobanu, L. Cornaby, S. A. Damtew, L. Dandona, R. Dandona, S. D. Dharmaratne, B. B. Duncan, B. Eshrati, A. Esteghamati, V. L. Feigin, J. C. Fernandes, T. Furst, T. T. Gebrehiwot, A. Gold, P. N. Gona, A. Goto, T. D. Habtewold, K. T. Hadush, N. Hafezi-Nejad, S. I. Hay, M. Horino, F. Islami, R. Kamal, A. Kasaeian, S. V. Katikireddi, A. P. Kengne, C. N. Kesavachandran, Y. S. Khader, Y. H. Khang, J. Khubchandani, D. Kim, Y. J. Kim, Y. Kinfu, S. Kosen, T. Ku, B. K. Defo, G. A. Kumar, H. J. Larson, M. Leinsalu, X. Liang, S. S. Lim, P. Liu, A. D. Lopez, R. Lozano, A. Majeed, R. Malekzadeh, D. C. Malta, M. Mazidi, C. McAlinden, S. T. McGarvey, D. T. Mengistu, G. A. Mensah, G. B. M. Mensink, H. B. Mezgebe, E. M. Mirrakhimov, U. O. Mueller, J. J. Noubiap, C. M. Obermeyer, F. A. Ogbo, M. O. Owolabi, G. C. Patton, F. Pourmalek, M. Qorbani, A. Rafay, R. K. Rai, C. L. Ranabhat, C. Reinig, S. Safiri, J. A. Salomon, J. R. Sanabria, I. S. Santos, B. Sartorius, M. Sawhney, J. Schmidhuber, A. E. Schutte, M. I. Schmidt, S. G. Sepanlou, M. Shamsizadeh, S. Sheikhbahaei, M. J. Shin, R. Shiri, I. Shiue, H. S. Roba, D. A. S. Silva, J. I. Silverberg, J. A. Singh, S. Stranges, S. Swaminathan, R. Tabares-Seisdedos, F. Tadese, B. A. Tedla, B. S. Tegegne, A. S. Terkawi, J. S. Thakur, M. Tonelli, R. Topor-Madry, S. Tyrovolas, K. N. Ukwaja, O. A. Uthman, M. Vaezghasemi, T. Vasankari, V. V. Vlassov, S. E. Vollset, E. Weiderpass, A. Werdecker, J. Wesana, R. Westerman, Y. Yano, N. Yonemoto, G. Yonga, Z. Zaidi, Z. M. Zenebe, B. Zipkin, and C. J. L. Murray. 2017. "Health effects of overweight and obesity in 195 countries over 25 years." *N Engl J Med* 377 (1):13–27. doi: 10.1056/NEJMoa1614362.

Georgescu, D., R. M. Sears, J. D. Hommel, M. Barrot, C. A. Bolanos, D. J. Marsh, M. A. Bednarek, J. A. Bibb, E. Maratos-Flier, E. J. Nestler, and R. J. DiLeone. 2005. "The hypothalamic neuropeptide melanin-concentrating hormone acts in the nucleus accumbens to modulate feeding behavior and forced-swim performance." *J Neurosci* 25 (11):2933–2940. doi: 10.1523/JNEUROSCI.1714-04.2005.

Ghamari-Langroudi, M., G. J. Digby, J. A. Sebag, G. L. Millhauser, R. Palomino, R. Matthews, T. Gillyard, B. L. Panaro, I. R. Tough, H. M. Cox, J. S. Denton, and R. D. Cone. 2015. "G-protein-independent coupling of MC4R to Kir7.1 in hypothalamic neurons." *Nature* 520 (7545):94–98. doi: 10.1038/nature14051.

Gibson, W. T., P. Pissios, D. J. Trombly, J. Luan, J. Keogh, N. J. Wareham, E. Maratos-Flier, S. O'Rahilly, and I. S. Farooqi. 2004. "Melanin-concentrating hormone receptor mutations and human obesity: functional analysis." *Obes Res* 12 (5):743–749. doi: 10.103 8/oby.2004.89.

Glaaser, I. W., and P. A. Slesinger. 2015. "Structural insights into GIRK channel function." *Int Rev Neurobiol* 123:117–160. doi: 10.1016/bs.irn.2015.05.014.

Goforth, P. B., G. M. Leininger, C. M. Patterson, L. S. Satin, and M. G. Myers, Jr. 2014. "Leptin acts via lateral hypothalamic area neurotensin neurons to inhibit orexin neurons by multiple GABA-independent mechanisms." *J Neurosci* 34 (34):11405–11415. doi: 10.1523/JNEUROSCI.5167-13.2014.

Gompf, H. S., M. G. Moldavan, R. P. Irwin, and C. N. Allen. 2005. "Nociceptin/orphanin FQ (N/OFQ) inhibits excitatory and inhibitory synaptic signaling in the suprachiasmatic nucleus (SCN)." *Neuroscience* 132 (4):955–965. doi: 10.1016/j.neuroscience.2004.11. 057.

Gropp, E., M. Shanabrough, E. Borok, A. W. Xu, R. Janoschek, T. Buch, L. Plum, N. Balthasar, B. Hampel, A. Waisman, G. S. Barsh, T. L. Horvath, and J. C. Bruning. 2005. "Agouti-related peptide-expressing neurons are mandatory for feeding." *Nat Neurosci* 8 (10):1289–1291. doi: 10.1038/nn1548.

He, Z., Y. Gao, L. Lieu, S. Afrin, J. Cao, N. J. Michael, Y. Dong, J. Sun, H. Guo, and K. W. Williams. 2019. "Direct and indirect effects of liraglutide on hypothalamic POMC and NPY/AgRP neurons – implications for energy balance and glucose control." *Mol Metab* 28:120–134. doi: 10.1016/j.molmet.2019.07.008.

Herisson, F. M., J. R. Waas, R. Fredriksson, H. B. Schioth, A. S. Levine, and P. K. Olszewski. 2016. "Oxytocin acting in the nucleus accumbens core decreases food intake." *J Neuroendocrinol* 28 (4). doi: 10.1111/jne.12381.

Hoang, Q. V., D. Bajic, M. Yanagisawa, S. Nakajima, and Y. Nakajima. 2003. "Effects of orexin (hypocretin) on GIRK channels." *J Neurophysiol* 90 (2):693–702. doi: 10.1152/ jn.00001.2003.

Hoang, Q. V., P. Zhao, S. Nakajima, and Y. Nakajima. 2004. "Orexin (hypocretin) effects on constitutively active inward rectifier K+ channels in cultured nucleus basalis neurons." *J Neurophysiol* 92 (6):3183–3191. doi: 10.1152/jn.01222.2003.

Hume, C., N. Sabatier, and J. Menzies. 2017. "High-sugar, but not high-fat, food activates supraoptic nucleus neurons in the male rat." *Endocrinology* 158 (7):2200–2211. doi: 10.1210/en.2016-1640.

Johnstone, L. E., T. M. Fong, and G. Leng. 2006. "Neuronal activation in the hypothalamus and brainstem during feeding in rats." *Cell Metab* 4 (4):313–321. doi: 10.1016/j.cmet.2 006.08.003.

Kabasakalian, A., C. J. Ferretti, and E. Hollander. 2018. "Oxytocin and Prader-Willi syndrome." *Curr Top Behav Neurosci* 35:529–557. doi: 10.1007/7854_2017_28.

Kelly, A. S., P. Auerbach, M. Barrientos-Perez, I. Gies, P. M. Hale, C. Marcus, L. D. Mastrandrea, N. Prabhu, S. Arslanian, and N. N. Trial Investigators. 2020. "A randomized, controlled trial of liraglutide for adolescents with obesity." *N Engl J Med* 382 (22):2117–2128. doi: 10.1056/NEJMoa1916038.

Kim, Y. R., C. H. Kim, V. Cardi, J. S. Eom, Y. Seong, and J. Treasure. 2014. "Intranasal oxytocin attenuates attentional bias for eating and fat shape stimuli in patients with anorexia nervosa." *Psychoneuroendocrinology* 44:133–142. doi: 10.1016/j.psyneuen.2 014.02.019.

Kim, Y. R., C. H. Kim, J. H. Park, J. Pyo, and J. Treasure. 2014. "The impact of intranasal oxytocin on attention to social emotional stimuli in patients with anorexia nervosa: a double blind within-subject cross-over experiment." *PLoS One* 9 (6):e90721. doi: 10.1371/journal.pone.0090721.

Kopelman, P., H. Groot Gde, A. Rissanen, S. Rossner, S. Toubro, R. Palmer, R. Hallam, A. Bryson, and R. I. Hickling. 2010. "Weight loss, HbA1c reduction, and tolerability of cetilistat in a randomized, placebo-controlled phase 2 trial in obese diabetics: comparison with orlistat (Xenical)." *Obesity (Silver Spring)* 18 (1):108–115. doi: 10.1038/oby.2009.155.

Kotz, C. M., J. A. Teske, J. A. Levine, and C. Wang. 2002. "Feeding and activity induced by orexin A in the lateral hypothalamus in rats." *Regul Pept* 104 (1-3):27–32. doi: 10.1016/s0167-0115(01)00346-9.

Kublaoui, B. M., T. Gemelli, K. P. Tolson, Y. Wang, and A. R. Zinn. 2008. "Oxytocin deficiency mediates hyperphagic obesity of Sim1 haploinsufficient mice." *Mol Endocrinol* 22 (7):1723–1734. doi: 10.1210/me.2008-0067.

Kuhnen, P., K. Clement, S. Wiegand, O. Blankenstein, K. Gottesdiener, L. L. Martini, K. Mai, U. Blume-Peytavi, A. Gruters, and H. Krude. 2016. "Proopiomelanocortin deficiency treated with a melanocortin-4 receptor agonist." *N Engl J Med* 375 (3):240–246. doi: 10.1056/NEJMoa1512693.

Kuhnen, P., H. Krude, and H. Biebermann. 2019. "Melanocortin-4 receptor signalling: importance for weight regulation and obesity treatment." *Trends Mol Med* 25 (2):136–148. doi: 10.1016/j.molmed.2018.12.002.

Lawson, E. A., D. A. Donoho, J. I. Blum, E. M. Meenaghan, M. Misra, D. B. Herzog, P. M. Sluss, K. K. Miller, and A. Klibanski. 2011. "Decreased nocturnal oxytocin levels in anorexia nervosa are associated with low bone mineral density and fat mass." *J Clin Psychiatry* 72 (11):1546–1551. doi: 10.4088/JCP.10m06617.

Li, G. Y., Q. X. Zhuang, X. Y. Zhang, J. J. Wang, and J. N. Zhu. 2019. "Ionic mechanisms underlying the excitatory effect of orexin on rat subthalamic nucleus neurons." *Front Cell Neurosci* 13:153. doi: 10.3389/fncel.2019.00153.

Li, Y., and A. N. van den Pol. 2006. "Differential target-dependent actions of coexpressed inhibitory dynorphin and excitatory hypocretin/orexin neuropeptides." *J Neurosci* 26 (50):13037–13047. doi: 10.1523/JNEUROSCI.3380-06.2006.

Light, P. E., J. E. Manning Fox, M. J. Riedel, and M. B. Wheeler. 2002. "Glucagon-like peptide-1 inhibits pancreatic ATP-sensitive potassium channels via a protein kinase A- and ADP-dependent mechanism." *Mol Endocrinol* 16 (9):2135–2144. doi: 10.1210/me.2002-0084.

Ludwig, D. S., N. A. Tritos, J. W. Mastaitis, R. Kulkarni, E. Kokkotou, J. Elmquist, B. Lowell, J. S. Flier, and E. Maratos-Flier. 2001. "Melanin-concentrating hormone overexpression in transgenic mice leads to obesity and insulin resistance." *J Clin Invest* 107 (3):379–386. doi: 10.1172/JCI10660.

Luquet, S., F. A. Perez, T. S. Hnasko, and R. D. Palmiter. 2005. "NPY/AgRP neurons are essential for feeding in adult mice but can be ablated in neonates." *Science* 310 (5748):683–685. doi: 10.1126/science.1115524.

MacKenzie, R. G. 2006. "Obesity-associated mutations in the human melanocortin-4 receptor gene." *Peptides* 27 (2):395–403. doi: 10.1016/j.peptides.2005.03.064.

Maguire, S., A. O'Dell, L. Touyz, and J. Russell. 2013. "Oxytocin and anorexia nervosa: a review of the emerging literature." *Eur Eat Disord Rev* 21 (6):475–478. doi: 10.1002/erv.2252.

Mahoney, J. P., and R. K. Sunahara. 2016. "Mechanistic insights into GPCR-G protein interactions." *Curr Opin Struct Biol* 41:247–254. doi: 10.1016/j.sbi.2016.11.005.

Marsh, D. J., D. T. Weingarth, D. E. Novi, H. Y. Chen, M. E. Trumbauer, A. S. Chen, X. M. Guan, M. M. Jiang, Y. Feng, R. E. Camacho, Z. Shen, E. G. Frazier, H. Yu, J. M.

Metzger, S. J. Kuca, L. P. Shearman, S. Gopal-Truter, D. J. MacNeil, A. M. Strack, D. E. MacIntyre, L. H. Van der Ploeg, and S. Qian. 2002. "Melanin-concentrating hormone 1 receptor-deficient mice are lean, hyperactive, and hyperphagic and have altered metabolism." *Proc Natl Acad Sci USA* 99 (5):3240–3245. doi: 10.1073/pnas.052706899.

Martin, A., M. State, G. M. Anderson, W. M. Kaye, J. M. Hanchett, C. W. McConaha, W. G. North, and J. F. Leckman. 1998. "Cerebrospinal fluid levels of oxytocin in Prader-Willi syndrome: a preliminary report." *Biol Psychiatry* 44 (12):1349–1352. doi: 10.1016/s0006-3223(98)00190-5.

Meffre, J., M. Sicre, M. Diarra, F. Marchessaux, D. Paleressompoulle, and F. Ambroggi. 2019. "Orexin in the posterior paraventricular thalamus mediates hunger-related signals in the nucleus accumbens core." *Curr Biol* 29 (19):3298–3306.e4. doi: 10.1016/j.cub.2019.07.069.

Mehta, A., S. P. Marso, and I. J. Neeland. 2017. "Liraglutide for weight management: a critical review of the evidence." *Obes Sci Pract* 3 (1):3–14. doi: 10.1002/osp4.84.

Miki, T., B. Liss, K. Minami, T. Shiuchi, A. Saraya, Y. Kashima, M. Horiuchi, F. Ashcroft, Y. Minokoshi, J. Roeper, and S. Seino. 2001. "ATP-sensitive K+ channels in the hypothalamus are essential for the maintenance of glucose homeostasis." *Nat Neurosci* 4 (5):507–512. doi: 10.1038/87455.

Miller, R. J. 1998. "Presynaptic receptors." *Annu Rev Pharmacol Toxicol* 38:201–227. doi: 10.1146/annurev.pharmtox.38.1.201.

Mullis, K., K. Kay, and D. L. Williams. 2013. "Oxytocin action in the ventral tegmental area affects sucrose intake." *Brain Res* 1513:85–91. doi: 10.1016/j.brainres.2013.03.026.

Muschamp, J. W., J. A. Hollander, J. L. Thompson, G. Voren, L. C. Hassinger, S. Onvani, T. M. Kamenecka, S. L. Borgland, P. J. Kenny, and W. A. Carlezon, Jr. 2014. "Hypocretin (orexin) facilitates reward by attenuating the antireward effects of its cotransmitter dynorphin in ventral tegmental area." *Proc Natl Acad Sci USA* 111 (16):E1648–E1655. doi: 10.1073/pnas.1315542111.

Noble, E. E., J. D. Hahn, V. R. Konanur, T. M. Hsu, S. J. Page, A. M. Cortella, C. M. Liu, M. Y. Song, A. N. Suarez, C. C. Szujewski, D. Rider, J. E. Clarke, M. Darvas, S. M. Appleyard, and S. E. Kanoski. 2018. "Control of feeding behavior by cerebral ventricular volume transmission of melanin-concentrating hormone." *Cell Metab* 28 (1):55–68.e7. doi: 10.1016/j.cmet.2018.05.001.

Ogden, C. L., M. D. Carroll, C. D. Fryar, and K. M. Flegal. 2015. "Prevalence of obesity among adults and youth: United States, 2011–2014." *NCHS Data Brief* (219):1–8.

Oldershaw, A., J. Treasure, D. Hambrook, K. Tchanturia, and U. Schmidt. 2011. "Is anorexia nervosa a version of autism spectrum disorders?" *Eur Eat Disord Rev* 19 (6):462–474. doi: 10.1002/erv.1069.

Ollmann, M. M., B. D. Wilson, Y. K. Yang, J. A. Kerns, Y. Chen, I. Gantz, and G. S. Barsh. 1997. "Antagonism of central melanocortin receptors in vitro and in vivo by agouti-related protein." *Science* 278 (5335):135–138. doi: 10.1126/science.278.5335.135.

Olszewski, P. K., K. Allen, and A. S. Levine. 2015. "Effect of oxytocin receptor blockade on appetite for sugar is modified by social context." *Appetite* 86:81–87. doi: 10.1016/j.appet.2014.10.007.

Olszewski, P. K., A. Klockars, and A. S. Levine. 2016. "Oxytocin: a conditional anorexigen whose effects on appetite depend on the physiological, behavioural and social contexts." *J Neuroendocrinol* 28 (4). doi: 10.1111/jne.12376.

Olszewski, P. K., A. Klockars, H. B. Schioth, and A. S. Levine. 2010. "Oxytocin as feeding inhibitor: maintaining homeostasis in consummatory behavior." *Pharmacol Biochem Behav* 97 (1):47–54. doi: 10.1016/j.pbb.2010.05.026.

Olszewski, P. K., C. Ulrich, N. Ling, K. Allen, and A. S. Levine. 2014. "A non-peptide oxytocin receptor agonist, WAY-267,464, alleviates novelty-induced hypophagia in mice: insights into changes in c-Fos immunoreactivity." *Pharmacol Biochem Behav* 124:367–372. doi: 10.1016/j.pbb.2014.07.007.

Onaka, T., and Y. Takayanagi. 2019. "Role of oxytocin in the control of stress and food intake." *J Neuroendocrinol* 31 (3):e12700. doi: 10.1111/jne.12700.

Ong, Z. Y., D. M. Bongiorno, M. A. Hernando, and H. J. Grill. 2017. "Effects of endogenous oxytocin receptor signaling in nucleus tractus solitarius on satiation-mediated feeding and thermogenic control in male rats." *Endocrinology* 158 (9):2826–2836. doi: 10.121 0/en.2017-00200.

Ong, Z. Y., J. J. Liu, Z. P. Pang, and H. J. Grill. 2017. "Paraventricular thalamic control of food intake and reward: role of glucagon-like peptide-1 receptor signaling." *Neuropsychopharmacology* 42 (12):2387–2397. doi: 10.1038/npp.2017.150.

Palus-Chramiec, K., L. Chrobok, M. Kepczynski, and M. H. Lewandowski. 2019. "Orexin A depolarises rat intergeniculate leaflet neurons through non-selective cation channels." *Eur J Neurosci* 50 (4):2683–2693. doi: 10.1111/ejn.14394.

Park, Y. B., Y. J. Choi, S. Y. Park, J. Y. Kim, S. H. Kim, D. K. Song, K. C. Won, and Y. W. Kim. 2011. "ATP-sensitive potassium channel-deficient mice show hyperphagia but are resistant to obesity." *Diabetes Metab J* 35 (3):219–225. doi: 10.4093/dmj.2011 .35.3.219.

Parton, L. E., C. P. Ye, R. Coppari, P. J. Enriori, B. Choi, C. Y. Zhang, C. Xu, C. R. Vianna, N. Balthasar, C. E. Lee, J. K. Elmquist, M. A. Cowley, and B. B. Lowell. 2007. "Glucose sensing by POMC neurons regulates glucose homeostasis and is impaired in obesity." *Nature* 449 (7159):228–232. doi: 10.1038/nature06098.

Pi-Sunyer, X., A. Astrup, K. Fujioka, F. Greenway, A. Halpern, M. Krempf, D. C. Lau, C. W. le Roux, R. Violante Ortiz, C. B. Jensen, J. P. WildingSCALE Obesity and Prediabetes NN8022-1839 Study Group. 2015. "A randomized, controlled trial of 3.0 mg of liraglutide in weight management." *N Engl J Med* 373 (1):11–22. doi: 10.1056/ NEJMoa1411892.

Plum, L., X. Ma, B. Hampel, N. Balthasar, R. Coppari, H. Munzberg, M. Shanabrough, D. Burdakov, E. Rother, R. Janoschek, J. Alber, B. F. Belgardt, L. Koch, J. Seibler, F. Schwenk, C. Fekete, A. Suzuki, T. W. Mak, W. Krone, T. L. Horvath, F. M. Ashcroft, and J. C. Bruning. 2006. "Enhanced PIP3 signaling in POMC neurons causes KATP channel activation and leads to diet-sensitive obesity." *J Clin Invest* 116 (7):1886–1901. doi: 10.1172/JCI27123.

Qiao, H., W. N. Ren, H. Z. Li, and Y. X. Hou. 2019. "Inhibitory effects of peripheral administration of exendin-4 on food intake are attenuated by lesions of the central nucleus of amygdala." *Brain Res Bull* 148:131–135. doi: 10.1016/j.brainresbull.201 9.03.002.

Qiu, J., C. Zhang, A. Borgquist, C. C. Nestor, A. W. Smith, M. A. Bosch, S. Ku, E. J. Wagner, O. K. Ronnekleiv, and M. J. Kelly. 2014. "Insulin excites anorexigenic proopiomelanocortin neurons via activation of canonical transient receptor potential channels." *Cell Metab* 19 (4):682–693. doi: 10.1016/j.cmet.2014.03.004.

Qu, D., D. S. Ludwig, S. Gammeltoft, M. Piper, M. A. Pelleymounter, M. J. Cullen, W. F. Mathes, R. Przypek, R. Kanarek, and E. Maratos-Flier. 1996. "A role for melanin-concentrating hormone in the central regulation of feeding behaviour." *Nature* 380 (6571):243–247. doi: 10.1038/380243a0.

Roberto, M., and G. R. Siggins. 2006. "Nociceptin/orphanin FQ presynaptically decreases GABAergic transmission and blocks the ethanol-induced increase of GABA release in central amygdala." *Proc Natl Acad Sci USA* 103 (25):9715–9720. doi: 10.1073/ pnas.0601899103.

Roman, C. W., V. A. Derkach, and R. D. Palmiter. 2016. "Genetically and functionally defined NTS to PBN brain circuits mediating anorexia." *Nat Commun* 7:11905. doi: 10.1038/ncomms11905.

Roseberry, A. G., H. Liu, A. C. Jackson, X. Cai, and J. M. Friedman. 2004. "Neuropeptide Y-mediated inhibition of proopiomelanocortin neurons in the arcuate nucleus shows enhanced desensitization in ob/ob mice." *Neuron* 41 (5):711–722. doi: 10.1016/s0896-6273(04)00074-1.

Russell, J., S. Maguire, G. E. Hunt, A. Kesby, A. Suraev, J. Stuart, J. Booth, and I. S. McGregor. 2018. "Intranasal oxytocin in the treatment of anorexia nervosa: randomized controlled trial during re-feeding." *Psychoneuroendocrinology* 87:83–92. doi: 10.1016/j.psyneuen.2017.10.014.

Schick, R. R., G. J. Harty, T. L. Yaksh, and V. L. Go. 1990. "Sites in the brain at which cholecystokinin octapeptide (CCK-8) acts to suppress feeding in rats: a mapping study." *Neuropharmacology* 29 (2):109–118. doi: 10.1016/0028-3908(90)90050-2.

Schick, R. R., J. P. Zimmermann, T. vorm Walde, and V. Schusdziarra. 2003. "Peptides that regulate food intake: glucagon-like peptide 1-(7-36) amide acts at lateral and medial hypothalamic sites to suppress feeding in rats." *Am J Physiol Regul Integr Comp Physiol* 284 (6):R1427–R1435. doi: 10.1152/ajpregu.00479.2002.

Schwartz, G. J., and T. H. Moran. 1994. "CCK elicits and modulates vagal afferent activity arising from gastric and duodenal sites." *Ann N Y Acad Sci* 713:121–128. doi: 10.1111/j.1749-6632.1994.tb44058.x.

Sears, R. M., R. J. Liu, N. S. Narayanan, R. Sharf, M. F. Yeckel, M. Laubach, G. K. Aghajanian, and R. J. DiLeone. 2010. "Regulation of nucleus accumbens activity by the hypothalamic neuropeptide melanin-concentrating hormone." *J Neurosci* 30 (24):8263–8273. doi: 10.1523/JNEUROSCI.5858-09.2010.

Shimada, M., N. A. Tritos, B. B. Lowell, J. S. Flier, and E. Maratos-Flier. 1998. "Mice lacking melanin-concentrating hormone are hypophagic and lean." *Nature* 396 (6712): 670–674. doi: 10.1038/25341.

Siemen, D. 1993. "Nonselective cation channels." *EXS* 66:3–25. doi: 10.1007/978-3-0348-7327-7_1.

Sisley, S., R. Gutierrez-Aguilar, M. Scott, D. A. D'Alessio, D. A. Sandoval, and R. J. Seeley. 2014. "Neuronal GLP1R mediates liraglutide's anorectic but not glucose lowering effect." *J Clin Invest* 124 (6):2456–2463. doi: 10.1172/JCI72434.

Sohn, J. W. 2013. "Ion channels in the central regulation of energy and glucose home-ostasis." *Front Neurosci* 7:85. doi: 10.3389/fnins.2013.00085.

Sohn, J. W. 2015. "Network of hypothalamic neurons that control appetite." *BMB Rep* 48 (4):229–233. doi: 10.5483/bmbrep.2015.48.4.272.

Sohn, J. W., L. E. Harris, E. D. Berglund, T. Liu, L. Vong, B. B. Lowell, N. Balthasar, K. W. Williams, and J. K. Elmquist. 2013. "Melanocortin 4 receptors reciprocally regulate sympathetic and parasympathetic preganglionic neurons." *Cell* 152 (3):612–619. doi: 10.1016/j.cell.2012.12.022.

Sun, Q. Q., J. R. Huguenard, and D. A. Prince. 2001. "Neuropeptide Y receptors differentially modulate G-protein-activated inwardly rectifying K^+ channels and high-voltage-activated Ca^{2+} channels in rat thalamic neurons." *J Physiol* 531 (Pt 1):67–79. doi: 10.1111/j.1469-7793.2001.0067j.x.

Swaab, D. F., J. S. Purba, and M. A. Hofman. 1995. "Alterations in the hypothalamic paraventricular nucleus and its oxytocin neurons (putative satiety cells) in Prader-Willi syndrome: a study of five cases." *J Clin Endocrinol Metab* 80 (2):573–579. doi: 10.1210/jcem.80.2.7852523.

Sweet, D. C., A. S. Levine, C. J. Billington, and C. M. Kotz. 1999. "Feeding response to central orexins." *Brain Res* 821 (2):535–538. doi: 10.1016/s0006-8993(99)01136-1.

Syrovatkina, V., K. O. Alegre, R. Dey, and X. Y. Huang. 2016. "Regulation, signaling, and physiological functions of G-proteins." *J Mol Biol* 428 (19):3850–3868. doi: 10.1016/j.jmb.2016.08.002.

Terrill, S. J., C. M. Jackson, H. E. Greene, N. Lilly, C. B. Maske, S. Vallejo, and D. L. Williams. 2016. "Role of lateral septum glucagon-like peptide 1 receptors in food intake." *Am J Physiol Regul Integr Comp Physiol* 311 (1):R124–R132. doi: 10.1152/ajpregu.00460.2015.

Terrill, S. J., C. B. Maske, and D. L. Williams. 2018. "Endogenous GLP-1 in lateral septum contributes to stress-induced hypophagia." *Physiol Behav* 192:17–22. doi: 10.1016/j.physbeh.2018.03.001.

Thienel, M., A. Fritsche, M. Heinrichs, A. Peter, M. Ewers, H. Lehnert, J. Born, and M. Hallschmid. 2016. "Oxytocin's inhibitory effect on food intake is stronger in obese than normal-weight men." *Int J Obes (Lond)* 40 (11):1707–1714. doi: 10.1038/ijo.2016.149.

Tinker, A., Q. Aziz, Y. Li, and M. Specterman. 2018. "ATP-sensitive potassium channels and their physiological and pathophysiological roles." *Compr Physiol* 8 (4):1463–1511. doi: 10.1002/cphy.c170048.

Tsujino, N., A. Yamanaka, K. Ichiki, Y. Muraki, T. S. Kilduff, K. Yagami, S. Takahashi, K. Goto, and T. Sakurai. 2005. "Cholecystokinin activates orexin/hypocretin neurons through the cholecystokinin A receptor." *J Neurosci* 25 (32):7459–7469. doi: 10.1523/JNEUROSCI.1193-05.2005.

Tsunematsu, T., L. Y. Fu, A. Yamanaka, K. Ichiki, A. Tanoue, T. Sakurai, and A. N. van den Pol. 2008. "Vasopressin increases locomotion through a V1a receptor in orexin/hypocretin neurons: implications for water homeostasis." *J Neurosci* 28 (1):228–238. doi: 10.1523/JNEUROSCI.3490-07.2008.

Tung, Y. C., M. Ma, S. Piper, A. Coll, S. O'Rahilly, and G. S. Yeo. 2008. "Novel leptin-regulated genes revealed by transcriptional profiling of the hypothalamic paraventricular nucleus." *J Neurosci* 28 (47):12419–12426. doi: 10.1523/JNEUROSCI.3412-08.2008.

Turton, M. D., D. O'Shea, I. Gunn, S. A. Beak, C. M. Edwards, K. Meeran, S. J. Choi, G. M. Taylor, M. M. Heath, P. D. Lambert, J. P. Wilding, D. M. Smith, M. A. Ghatei, J. Herbert, and S. R. Bloom. 1996. "A role for glucagon-like peptide-1 in the central regulation of feeding." *Nature* 379 (6560):69–72. doi: 10.1038/379069a0.

Vaisse, C., K. Clement, B. Guy-Grand, and P. Froguel. 1998. "A frameshift mutation in human MC4R is associated with a dominant form of obesity." *Nat Genet* 20 (2):113–114. doi: 10.1038/2407.

van den Pol, A. N. 2012. "Neuropeptide transmission in brain circuits." *Neuron* 76 (1):98–115. doi: 10.1016/j.neuron.2012.09.014.

van den Pol, A. N., C. Acuna-Goycolea, K. R. Clark, and P. K. Ghosh. 2004. "Physiological properties of hypothalamic MCH neurons identified with selective expression of reporter gene after recombinant virus infection." *Neuron* 42 (4):635–652. doi: 10.1016/s0896-6273(04)00251-x.

van den Pol, A. N., X. B. Gao, K. Obrietan, T. S. Kilduff, and A. B. Belousov. 1998. "Presynaptic and postsynaptic actions and modulation of neuroendocrine neurons by a new hypothalamic peptide, hypocretin/orexin." *J Neurosci* 18 (19):7962–7971.

van den Pol, A. N., Y. Yao, L. Y. Fu, K. Foo, H. Huang, R. Coppari, B. B. Lowell, and C. Broberger. 2009. "Neuromedin B and gastrin-releasing peptide excite arcuate nucleus neuropeptide Y neurons in a novel transgenic mouse expressing strong Renilla green fluorescent protein in NPY neurons." *J Neurosci* 29 (14):4622–4639. doi: 10.1523/JNEUROSCI.3249-08.2009.

Varela, L., and T. L. Horvath. 2012. "Leptin and insulin pathways in POMC and AgRP neurons that modulate energy balance and glucose homeostasis." *EMBO Rep* 13 (12):1079–1086. doi: 10.1038/embor.2012.174.

Williams, D. L., N. A. Lilly, I. J. Edwards, P. Yao, J. E. Richards, and S. Trapp. 2018. "GLP-1 action in the mouse bed nucleus of the stria terminalis." *Neuropharmacology* 131:83–95. doi: 10.1016/j.neuropharm.2017.12.007.

Willis, W. D. 2006. "John Eccles' studies of spinal cord presynaptic inhibition." *Prog Neurobiol* 78 (3-5):189–214. doi: 10.1016/j.pneurobio.2006.02.007.

Wong, D., K. Sullivan, and G. Heap. 2012. "The pharmaceutical market for obesity therapies." *Nat Rev Drug Discov* 11 (9):669–670. doi: 10.1038/nrd3830.

Wu, Z., Y. Xu, Y. Zhu, A. K. Sutton, R. Zhao, B. B. Lowell, D. P. Olson, and Q. Tong. 2012. "An obligate role of oxytocin neurons in diet induced energy expenditure." *PLoS One* 7 (9):e45167. doi: 10.1371/journal.pone.0045167.

Xu, J., C. L. Bartolome, C. S. Low, X. Yi, C. H. Chien, P. Wang, and D. Kong. 2018. "Genetic identification of leptin neural circuits in energy and glucose homeostases." *Nature* 556 (7702):505–509. doi: 10.1038/s41586-018-0049-7.

Yang, S. B., A. C. Tien, G. Boddupalli, A. W. Xu, Y. N. Jan, and L. Y. Jan. 2012. "Rapamycin ameliorates age-dependent obesity associated with increased mTOR signaling in hypothalamic POMC neurons." *Neuron* 75 (3):425–436. doi: 10.1016/j.neuron.2012.03.043.

Yao, Y., L. Y. Fu, X. Zhang, and A. N. van den Pol. 2012. "Vasopressin and oxytocin excite MCH neurons, but not other lateral hypothalamic GABA neurons." *Am J Physiol Regul Integr Comp Physiol* 302 (7):R815–R824. doi: 10.1152/ajpregu.00452.2011.

Yeo, G. S., I. S. Farooqi, S. Aminian, D. J. Halsall, R. G. Stanhope, and S. O'Rahilly. 1998. "A frameshift mutation in MC4R associated with dominantly inherited human obesity." *Nat Genet* 20 (2):111–112. doi: 10.1038/2404.

Yeo, G. S., E. J. Lank, I. S. Farooqi, J. Keogh, B. G. Challis, and S. O'Rahilly. 2003. "Mutations in the human melanocortin-4 receptor gene associated with severe familial obesity disrupts receptor function through multiple molecular mechanisms." *Hum Mol Genet* 12 (5):561–574. doi: 10.1093/hmg/ddg057.

Zhan, C., J. Zhou, Q. Feng, J. E. Zhang, S. Lin, J. Bao, P. Wu, and M. Luo. 2013. "Acute and long-term suppression of feeding behavior by POMC neurons in the brainstem and hypothalamus, respectively." *J Neurosci* 33 (8):3624–3632. doi: 10.1523/JNEUROSCI.2742-12.2013.

Zhang, X., and A. N. van den Pol. 2012. "Thyrotropin-releasing hormone (TRH) inhibits melanin concentrating hormone neurons: implications for TRH-mediated anorexic and arousal actions." *J Neurosci* 32 (9):3032–3043. doi: 10.1523/JNEUROSCI.5966-11.2012.

Zhang, X., and A. N. van den Pol. 2013. "Direct inhibition of arcuate proopiomelanocortin neurons: a potential mechanism for the orexigenic actions of dynorphin." *J Physiol* 591 (7):1731–1747. doi: 10.1113/jphysiol.2012.248385.

Zhang, X., and A. N. van den Pol. 2015. "Dopamine/tyrosine hydroxylase neurons of the hypothalamic arcuate nucleus release GABA, communicate with dopaminergic and other arcuate neurons, and respond to dynorphin, met-enkephalin, and oxytocin." *J Neurosci* 35 (45):14966–14982. doi: 10.1523/JNEUROSCI.0293-15.2015.

9 Prefrontal Inhibitory Signaling in the Control of Social Behaviors

Qian Yang, Jun Wang, and Han Xu

CONTENTS

9.1   INTRODUCTION

Humans are social animals. Social interaction accounts for a large amount of our daily life and is fundamental to our physical and mental health (Hawkley and Cacioppo 2010; Kahneman et al. 2004). On the other hand, social impairments are commonly observed in various neurodegenerative, neuropsychiatric and neurodevelopmental disorders (Bicks et al. 2015; Hari and Kujala 2009; Hari et al. 2015). Deciphering the neural mechanisms underlying social interaction behavior is therefore of great importance in terms of both basic science and clinical medicine.

To support sophisticated social behavior, a number of brain structures have been discovered to be involved in social perception, social cognition, and social interaction in the past decades (Adolphs 2001; Anderson 2016; Chen and Hong 2018; Frith and Frith 2012; Walum and Young 2018). Among the specific brain areas identified, the medial prefrontal cortex (mPFC) plays a pivotal role for top-down control of animal social behavior (Schilbach et al. 2006; Sliwa and Freiwald 2017). Particularly, the inhibitory signaling mediated by GABAergic interneurons (INs) is crucial to maintain prefrontal balance between excitation and inhibition, and to ensure normal social behavioral manifestation (Yizhar et al. 2011). Furthermore, the most recent studies reveal that distinct subtypes of inhibitory INs have differential roles in cortical computation and behavioral modulation. This review focuses on the function of the mPFC in the regulation of social interaction behavior, with a special emphasis on inhibitory signaling mediated by subtypes of GABAergic INs.

9.2   THE FUNCTION OF THE MPFC IN SOCIAL INTERACTION BEHAVIOR

Social interaction is a complex behavior that involves multiple neural processes, including sensory perception, motivation, learning and memory, reward-seeking, and motor generation. Not surprisingly, a variety of brain structures have been identified to participate in social behavior, such as the olfactory bulb (Sanchez-Andrade and Kendrick 2009; Montagrin et al. 2018), the amygdala (Adolphs 2010), the hippocampus (Montagrin et al. 2018), the paraventricular nucleus of hypothalamus (PVN) (Resendez et al. 2020; Tang et al. 2020; Hung et al. 2017), the cerebellum (Carta et al. 2019), the ventral tegmental area (VTA) (Gunaydin et al. 2014), the nucleus accumbens (NAc) (Dolen et al. 2013), and the mPFC (Ko 2017). Of these, the mPFC is a higher-order cortical area that constantly receives and processes incoming information from numerous upstream structures and transmits integrated output to down-stream brain structures for top-down behavioral control. Therefore, the mPFC represents an ideal hub to combine external social cues with animal's internal states to produce proper output and thus generate appropriate social behavior.

9.2.1   CORRELATION BETWEEN THE MPFC AND SOCIAL INTERACTION

Indeed, findings from both human and animal studies support the notion that the mPFC serves as a key brain region in the control of social interaction behavior. Functional magnetic resonance imaging (fMRI) studies observed enhanced neural activities of the mPFC in human adults when engaging in social interaction with virtual others or viewing scenes involving social interactions between characters (Schilbach et al. 2006; Wagner et al. 2016). Interestingly, by using near-infrared spectroscopy (NIRS), it was found that this activation profile is also present during social plays in human infants (Urakawa et al. 2015). This observation suggests that the mPFC is already tuned for computation of social information in early life in humans (Grossmann 2013, 2015). Furthermore, whole-brain fMRI in macaque monkeys revealed that the mPFC is exclusively engaged in social interaction analysis but not physical interaction with objects, suggesting its functional specialization for interaction processing in the social domain (Sliwa and Freiwald 2017; Shepherd and Freiwald 2018).

In support of human and non-human primate studies, socially driven neuronal activation is also reported in the mPFC of rodents. For instance, the expression of the immediate early gene, c-fos, in mouse mPFC is significantly increased following social interaction with a conspecific (Avale et al. 2011; Kim et al. 2016). Consistent with the biochemical responses, in vivo electrophysiological recordings revealed elevated discharge rates in a subset of mPFC excitatory neurons during real-time social interaction in rats and mice (Lee et al. 2016; Jodo et al. 2010; Liu et al. 2020). Recently, calcium imaging via endoscope was successfully incorporated to optically record mPFC neural population activity in freely social interacting animals. This technique advancement not only allows simultaneous recordings from a large number of neurons but also enables the test of stability of observed neuronal activity across multiple days. With this technique, Liang et al. monitored calcium activities of principal neurons in the mPFC when mice freely explored social targets and found that distinct ON and OFF neuronal ensembles encode social exploration despite epoch-to-epoch variability in activities of individual neurons (Liang et al. 2018). In addition, combined with retrograde labeling, Murugan et al. further recorded from distinct mPFC neuronal subpopulations that target different downstream regions (Murugan et al. 2017). They found that NAc-projecting mPFC neurons but not amygdala or VTA-projecting subpopulations convey combined social and spatial information (Murugan et al. 2017). These electrophysiological or optical recording studies in rodents corroborated findings from previous imaging studies in humans and non-human primates, and elucidated the complex activity dynamics of prefrontal neurons during social interaction behavior at single-cellular resolution.

9.2.2 CAUSAL ROLE OF THE mPFC IN SOCIAL INTERACTION

The above studies provide convincing evidence that mPFC neuronal activities are good correlates of social interaction behavior. Echoing these observations, complex human social interaction is compromised in patients whose frontal lobes are damaged as a consequence of diseases (Bechara et al. 2000; Saver and Damasio 1991). Among subdivisions of the frontal lobe, selective lesion of the anterior cingulate gyrus in male macaque monkeys disrupts the normal patterns of social interest in other male or female macaque individuals, indicating that the anterior cingulate cortex (ACC, a subregion of the mPFC) is particularly important for social evaluation (Rudebeck et al. 2006). In contrast, focal lesions of the medial orbitofrontal cortex (mOFC) of macaque monkeys, another subregion of the mPFC, induce mild impairments in decision making but do not induce alterations in evaluation of social information (Noonan et al. 2010). These lesion studies suggest that mPFC subregions do not uniformly contribute to social behavior regulation but there exists a division of labor. In agreement with lesion studies in monkeys, pharmacological inactivation of the prefrontal cortical region with GABA receptor agonist reduces frequency and duration of social play in rats (van Kerkhof et al. 2013), and decreases the time of social interaction in mice assessed with a three-chamber social preference test (Xu et al. 2019). Together, on top of correlative observations, these findings establish a causal relationship between the mPFC and social interaction behavior.

9.3 PREFRONTAL INHIBITION IS CRUCIAL FOR SOCIAL INTERACTION

Cortical neurons can be classified into two broad categories: glutamatergic neurons and GABAergic neurons (Peters and Jones 1984). Glutamatergic neurons are projection neurons that release the excitatory neurotransmitter glutamate and their dendrites are spiny. In comparison, GABAergic neurons are typically local-circuit interneurons (INs) that release the inhibitory neurotransmitter GABA and their dendrites are aspiny or sparsely spiny. Although GABAergic INs represent only a minority of all cortical neurons, the synaptic inhibition they generate is essential to balance cortical excitation, and hence to ensure normal cortical computations and also animal behaviors. Consistently, finely tuned inhibition in the mPFC is necessary in generating coordinated cortical network activities required for appropriate social interaction. Conversely, aberrant inhibition in the mPFC disrupts the balance between excitation and inhibition and results in abnormality in cortical functioning and thus leads to social behavioral impairments.

9.3.1 EVIDENCE FROM HUMAN CLINICAL STUDIES

Social cognition and corresponding behavior is a sensitive domain that is vulnerable to environmental insults or pathological degradations in the central nervous system. It is therefore not surprising that social dysfunctions are commonly observed in a number of neuropsychiatric disorders, notably autism spectrum disorder (ASD), schizophrenia, and depression (Green et al. 2015; Fernandez et al. 2018; Kupferberg et al. 2016).

In patients of those brain disorders with shared social deficits, alterations in mPFC inhibitory signaling are a consistent finding. With direct in vivo measurement using proton magnetic resonance spectroscopy, GABA has been found to be lower in the frontal cortex of children with ASD (Harada et al. 2011). Consistently, post-mortem examinations find that the number of parvalbumin-positive (PV) GABAergic INs is decreased in the mPFC of autism individuals (Hashemi et al. 2017). Likewise, growing studies demonstrate that schizophrenia is associated with both presynaptic and postsynaptic alterations in prefrontal cortical GABAergic INs, particularly PV INs (Lewis et al. 2012, 2005). Moreover, combining transcranial magnetic stimulation with electroencephalography to measure GABAergic receptor-mediated inhibitory neurotransmission, it is found that prefrontal long-interval cortical inhibition was significantly reduced in patients with schizophrenia compared to healthy subjects (Radhu et al. 2015). Similarly, reductions in GABA and GAD67 levels, decreases in expression of GABAergic IN markers, and alterations in GABAergic receptor levels have been reported in patients with major depressive disorder (Fogaca and Duman 2019). Together, abnormalities in inhibitory neurotransmitters, decreases in the number of inhibitory neurons, and also reductions in functional synaptic inhibition are consistently observed in the mPFC of major neuropsychiatric disorders. The weakening of inhibitory signaling and the resultant disruption in excitation and inhibition balance likely play an essential role in the pathophysiology of those disorders with shared social deficits.

9.3.2 EVIDENCE FROM ANIMAL MODELS OF DISEASES

Animal models offer valuable tools for understanding the biological mechanisms underlying human diseases and also searching for more effective therapeutic targets. Over the past decades, a large range of animal models, particularly mouse models, have been developed to recapitulate the prominent symptoms of different neuropsychiatric disorders.

The above discussed findings from human clinical studies suggest that impaired cortical inhibition is a key regulator in the pathogenesis of psychiatric disorders with shared social deficits. In line with those clinical findings, impaired inhibition in the cerebral cortex, the mPFC in particular, is routinely found in mouse models carrying genetic modifications known to cause autism in humans. For instance, mice with heterozygous loss-of-function mutations in the *SCN1A* gene ($Scn1a^{+/-}$) have reduced Na^+ currents and impaired action potential firing in GABAergic neurons and decreased inhibitory transmission in hippocampal CA1 and prefrontal cortex (Han et al. 2012). Similarly, the frequency of spontaneous GABAergic neurotransmission is significantly reduced in the hippocampus of BTBR mice, a model of idiopathic autism (Han et al. 2014). Also, in vitro patch-clamp recordings revealed decreased neuronal excitability in mPFC PV INs of neuroligin 3 R451C knockin mice (Cao et al. 2018), and major reductions in prefrontal synaptic inhibition in neuroligin-2 conditional knockout mice (Liang et al. 2015). In addition, compared to wild-type mice, the neural activity of mPFC PV INs was directly measured using fiber photometry and was found diminished in contactin-associated protein-like 2 (*CNTNAP2*) knockout mice during social interaction (Selimbeyoglu et al. 2017).

Similar to findings from mouse models of ASD, both structural and functional impairments in mPFC inhibition were also observed in schizophrenia and depression animal models. In methylazoxymethanol acetate (MAM)-treated rats, a verified animal model of schizophrenia, the density of PV INs is decreased throughout the mPFC (Lodge et al. 2009). Functionally, the synaptic inhibition onto mPFC layer 2/3 excitatory pyramidal cells is reduced in *disrupted-in-schizophrenia-1 (Disc1)* locus impairment mice (Delevich et al. 2020). In mice expressing truncated Disc1, a genetic mouse model of depression, prelimibic PV INs are reduced in number and they receive fewer presynaptic excitatory inputs and form fewer release sites on postsynaptic targets (Sauer et al. 2015). Apart from the genetic mouse model of depression, mouse models of chronic stress also show defects in GABAergic transmission of layer V mPFC pyramidal neurons, concomitant with depressive-like behaviors (Ghosal et al. 2020).

9.3.3 RESTORATION OF PREFRONTAL CORTICAL INHIBITION RESCUES SOCIAL IMPAIRMENTS

The findings from both human and animal studies indicate that impairment in prefrontal inhibition is tightly linked to social deficits and therefore suggest a sound correlation between prefrontal inhibition and social interaction behavior. Besides, these observations point out that targeting prefrontal inhibition could serve as a potential therapeutic option to rescue social dysfunctions under disease conditions. Consistent with this idea, increasing studies demonstrate that restoration in cortical GABAergic inhibitory tone, particularly in the mPFC, is sufficient to ameliorate social defects in animal models of diseases. For example, treatment with low doses of clonazepam, a positive allosteric modulator of the GABAa receptor, fully rescued the abnormal social behaviors in Scn1a$^{+/-}$ mice, a mouse model with marked autistic-like behavior (Han et al. 2012). Likewise, treatment with low-dose benzodiazepines increases inhibitory neurotransmission in hippocampus and significantly improves social impairments in BTBR mice, and this behavioral improvement is GABAa receptor α2,3-subunit specific (Han et al. 2014). In the above two studies, the enhancement of inhibitory neurotransmission was both achieved through positive allosteric modulation of GABAa receptors with pharmacological chemicals. In addition, optogenetically increasing the excitability of PV INs in the mPFC is also efficient to rescue deficits in social behavior and hyperactivity in adult mice lacking *CNTNAP2* (Selimbeyoglu et al. 2017). Similarly, chemogenetic activation of mPFC PV INs in the adult mice mitigates social deficits induced by social isolation during the juvenile critical window (Bicks et al. 2020).

The aforementioned pharmacological or optogenetic/chemogenetic manipulation studies further support a causal relationship between prefrontal inhibition and social interaction behavior. Indeed, aside from animal models of diseases, artificial increase in neuronal activity of excitatory neurons across the mPFC elevates excitation inhibition balance and decreases time of social interaction in wild-type mice (Yizhar et al. 2011). Interestingly, compensatory elevation of inhibitory cell excitability with advanced optogenetic tools partially rescued social impairments caused by disruption of the balance between excitation and inhibition (Yizhar et al. 2011). Together, these observations provide direct evidence that maintaining proper inhibition within the mPFC neuronal network is required to maintain functional E/I balance and to ensure normal behavior of social interaction.

9.4 IN SUBTYPE-DIFFERENTIAL MODULATION ON SOCIAL BEHAVIOR

Cortical synaptic inhibition is mediated by GABAergic INs, which come with different forms in anatomy, intrinsic membrane properties, synaptic connectivity, and the expression of specific chemical markers (Ascoli et al. 2008; Fishell and Rudy 2011; Rudy et al. 2011; Somogyi and Klausberger 2005). The striking diversity in cellular and synaptic features strongly suggests the existence of functional differentiation among subtypes of cortical INs. However, previous methodological limitations hampered the dissection of functional specialization of subpopulations of GABAergic INs in animal behaviors including social interaction. With the advent of transgenetic mice with subtypes of INs labeled and the development of multiple experimental techniques such as optogenetics, pharmacogenetics, and electrophysiological recordings, neuroscientists are now able to parse out the contribution of specific IN subtypes in cortical computations.

9.4.1 A COMPARISON BETWEEN PV INs AND SST INs

To accomplish the complex functions of the cerebral cortex, the mammalian brain has evolved a large diversity of GABAergic INs. In rodent neocortex, PV INs and SST INs are the two largest subpopulations and account for about 40% and 30% of all GABAergic neurons, respectively (Lee et al. 2010; Rudy et al. 2011; Tremblay et al. 2016). Despite sharing the same developmental origin (Hu et al. 2017; Wonders and Anderson 2005), these two IN subtypes exhibit remarkable differences in morphological and physiological properties as described in the following text (Ascoli et al. 2008; Rudy et al. 2011; Somogyi and Klausberger 2005).

PV INs are readily distinguishable from other IN subtypes due to their characteristic capability in firing action potentials at high frequencies (up to hundreds of Hertz) and hence the most well studied IN subtype. For this very reason, PV INs are also well known as fast spiking (FS) INs. In terms of synaptic contacts, on one hand, PV INs receive strong excitatory inputs from both cortical and subcortical regions. On the other hand, PV INs preferentially innervate the somatic and perisomatic compartments of pyramidal cells. As a consequence of their membrane and synaptic properties, PV INs are able to mediate robust feedforward inhibition onto pyramidal cells and to powerfully control the spike timing of the output neurons. Besides, interestingly, PV INs frequently form synapses with each other both chemically and electrically. Such unique synaptic features are important in synchronizing the activities among PV INs and therefore generating cortical network gamma oscillations, which are believed to be essential for many cognitive functions.

In contrast to PV INs, SST INs typically do not fire at high frequencies. As for synaptic inputs, SST INs receive facilitating glutamatergic excitatory inputs from local pyramidal cells. Output wise, SST INs usually send their axon fibers to the upper layers and target the distal apical dendrites of pyramidal cells. Due to the input and output synaptic configurations, SST INs are likely recruited by recurrent network activities and are well positioned to gate the excitatory inputs targeting the distal dendrites of pyramidal cells. Moreover, as another distinction from PV INs, SST INs rarely form chemical synapses with each other although they form electrical synapses with each other as PV INs do.

9.4.2 PV INs IN THE mPFC ARE NECESSARY FOR THE CONTROL OF SOCIABILITY

The striking differences in neuronal morphologies, synaptic physiologies, and also membrane properties between PV INs and SST INs strongly suggest their functional differentiation in cortical network operations and hence behavioral control. In fact, growing studies identify functional specification of PV INs and SST INs in distinct brain functions such as sensory perception (Lee et al. 2012), motor integration (Lee et al. 2013), space coding (Miao et al. 2017), working memory (Kim et al. 2016; Kamigaki and Dan 2017), as well as reward processing (Kvitsiani et al. 2013) and fear expression (Xu et al. 2019; Cummings and Clem 2020). Most recently, studies start to reveal cell type-differential modulation of prefrontal cortical PV INs and SST INs on social interaction behavior.

With optogenetic tagging techniques, our group successfully identified PV INs and SST INs in the mPFC of adult male mice by using PV-Cre and SST-Cre mouse lines, respectively. And then we directly measured the activities of these two IN subtypes during real-time social interaction with chronic extracellular electrophysiological recordings. Interestingly, we observed distinct firing patterns of PV INs and SST INs in response to the same social stimuli (Liu et al. 2020). A majority of PV INs recorded displayed an elevated firing frequency when test mice were interacting with stimulus mouse, and a subset of PV INs increased their firing rates even when the test animals were approaching the social target (Liu et al. 2020; Xu et al. 2019). Our study is the first demonstration of enhanced activities of mPFC PV INs at single unit level, and is consistent with a previous study measuring bulk calcium signals employing fiber photometry (Selimbeyoglu et al. 2017). In striking contrast, SST INs maintained their firing rates either when the test mice were approaching the social target or interacting with the stimulus mouse (Liu et al. 2020), corroborating an earlier study from our group employing fiber photometry (Xu et al. 2019). In agreement with the differential correlates between the activities of PV INs and SST INs with social interaction behavior, pharmacogenetic inhibition of PV INs but not SST INs significantly reduced the time that the experimental mice spent with the stimulus mouse. Therefore, the particular subpopulation of mPFC GABAergic INs, namely PV INs, is the key player in the top-down control of sociability.

9.4.3 SST INs in the mPFC Are Required for Emotion Discrimination

The ability to accurately detect affective states in others is crucial to build social relationships and avoid dangers. The mPFC plays an important role in emotional discrimination via top-down control of the limbic system (Dal Monte et al. 2013; Hiser and Koenigs 2018). To further understand the underlying neuronal substrates, Scheggia et al. recorded neural activities in the mPFC of observer mice during a behavioral task for emotional discrimination (Scheggia et al. 2020). They found that the narrow-spiking neurons, that is, putative inhibitory INs, are the most responsive during interactions with either a stressed or a relieved demonstrator mouse. In line with findings from our pharmacogenetic inhibition experiments (Liu et al. 2020), Scheggia et al. found that optogenetic inhibition of PV INs but not SST INs compromised the observer's general sociability. However, interestingly, photoinhibition of SST INs abolished affective state discrimination, yet inhibition of PV INs had no effect. Consistently, when Ca^{2+} signals of individual SST INs were examined with in vivo single-cell microendoscopy, increased synchronous activity of SST INs was noticed when the observer was interacting with an emotionally altered demonstrator. Thus, contrary to not being a determinant in regulating sociability, mPFC SST INs are required for the prefrontal gating of emotion discrimination.

Together, the above recent studies demonstrated distinct activity patterns and manipulation effects of two major subtypes of mPFC INs, i.e. PV INs and SST INs, in general sociability and emotion discrimination, respectively (please see Figure 9.1). Findings from these new studies strengthen the notion that cortical inhibition is

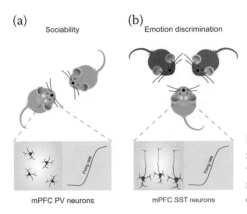

(a) Sociability

(b) Emotion discrimination

mPFC PV neurons

mPFC SST neurons

FIGURE 9.1 PV INs and SST INs in the mPFC differentially modulate social behaviors. (a) PV INs play a determinant role in regulating sociability, and (b) SST INs are essential for emotion discrimination.

essential to balance the excitation and hence to support normal cortical functions. What's more, these findings draw new pictures regarding the functional specialization among IN subtypes of the mPFC in social behavioral regulation. In other words, different subtypes of GABAergic INs do not act uniformly but contribute differentially to cortical social information processing depending on the specific social task and the exact social context. These new findings are of great importance in bringing us closer to the neural mechanism behind complex social behaviors. From a translational point of view, they also spark the hope for precise intervention in the treatment of diverse social deficits in patients with neuropsychiatric disorders.

9.5   PERSPECTIVES

It is well accepted that understanding the functional diversity of cortical GABAergic INs is essential to understand cortical functions. In addition to PV INs and SST INs, those expressing the ionotropic serotonin receptor 5HT3a (5HT3aR) represent another subpopulation of GABAergic INs (Lee et al. 2013). 5HT3aR INs are not homogeneous and can be further divided into several subtypes. 5HT3aR INs form synapses with both excitatory neurons and inhibitory neurons as PV INs and SST INs do, yet their precise functions in social behavior are to be determined in the future.

The present review piece focuses on the role of prefrontal inhibitory signaling mediated by GABAergic INs via GABAa receptors, a class of ligand-gated ion channels. In addition to classical ion channels such as voltage-gated ion channels and ligand-gated ion channels, nonclassical ion channels are also highly expressed in brain regions crucial for social behavior including the mPFC. These nonclassical channels have been drawing extensive attention and have been found to be involved in a range of physiological functions, and dysfunctions of these channels have been implicated in many pathological conditions. For example, it has recently been shown that the acid-sensing ion channel 1a (ASIC1a) in the ventral hippocampus acts as a crucial molecular regulator in fear extinction through long-term modification of the specific projection from the ventral hippocampus to the mPFC (Wang et al. 2018). Despite their abundance and potential significance, the precise roles of nonclassical ion channels in the regulation of cortical network functions and hence social behavior manifestations are interesting topics worthy of exploration.

ACKNOWLEDGMENTS

This work is supported by grants from the National Key R&D Program of China (2016YFA0501000), the National Natural Science Foundation of China (31471025, 91432110), the Zhejiang Provincial Natural Science Foundation of China (LR17H090002), and the Fundamental Research Funds for the Central Universities (2019QNA5001) to Han Xu.

REFERENCES

Adolphs, R. 2001. "The neurobiology of social cognition." *Curr Opin Neurobiol* no. 11 (2):231–239. doi: 10.1016/s0959-4388(00)00202-6.

Adolphs, R. 2010. "What does the amygdala contribute to social cognition?" *Ann N Y Acad Sci* no. 1191:42–61. doi: 10.1111/j.1749-6632.2010.05445.x.

Anderson, D. J. 2016. "Circuit modules linking internal states and social behaviour in flies and mice." *Nat Rev Neurosci* no. 17 (11):692–704. doi: 10.1038/nrn.2016.125.

Ascoli, G. A., L. Alonso-Nanclares, S. A. Anderson, G. Barrionuevo, R. Benavides-Piccione, A. Burkhalter, G. Buzsaki, B. Cauli, J. Defelipe, A. Fairen, D. Feldmeyer, G. Fishell, Y. Fregnac, T. F. Freund, D. Gardner, E. P. Gardner, J. H. Goldberg, M. Helmstaedter, S. Hestrin, F. Karube, Z. F. Kisvarday, B. Lambolez, D. A. Lewis, O. Marin, H. Markram, A. Munoz, A. Packer, C. C. Petersen, K. S. Rockland, J. Rossier, B. Rudy, P. Somogyi, J. F. Staiger, G. Tamas, A. M. Thomson, M. Toledo-Rodriguez, Y. Wang, D. C. West, and R. Yuste. 2008. "Petilla terminology: nomenclature of features of GABAergic interneurons of the cerebral cortex." *Nat Rev Neurosci* no. 9 (7):557–568. doi: 10.1038/nrn2402.

Avale, M. E., J. Chabout, S. Pons, P. Serreau, F. De Chaumont, J. C. Olivo-Marin, J. P. Bourgeois, U. Maskos, J. P. Changeux, and S. Granon. 2011. "Prefrontal nicotinic receptors control novel social interaction between mice." *FASEB J* no. 25 (7):2145–2155. doi: 10.1096/fj.10-178558.

Bechara, A., H. Damasio, and A. R. Damasio. 2000. "Emotion, decision making and the orbitofrontal cortex." *Cereb Cortex* no. 10 (3):295–307. doi: 10.1093/cercor/10.3.295.

Bicks, L. K., H. Koike, S. Akbarian, and H. Morishita. 2015. "Prefrontal cortex and social cognition in mouse and man." *Front Psychol* no. 6:1805. doi: 10.3389/fpsyg.2015.01 805.

Bicks, L. K., K. Yamamuro, M. E. Flanigan, J. M. Kim, D. Kato, E. K. Lucas, H. Koike, M. S. Peng, D. M. Brady, S. Chandrasekaran, K. J. Norman, M. R. Smith, R. L. Clem, S. J. Russo, S. Akbarian, and H. Morishita. 2020. "Prefrontal parvalbumin interneurons require juvenile social experience to establish adult social behavior." *Nat Commun* no. 11 (1):1003. doi: 10.1038/s41467-020-14740-z.

Cao, W., S. Lin, Q. Q. Xia, Y. L. Du, Q. Yang, M. Y. Zhang, Y. Q. Lu, J. Xu, S. M. Duan, J. Xia, G. Feng, J. Xu, and J. H. Luo. 2018. "Gamma oscillation dysfunction in mPFC leads to social deficits in neuroligin 3 R451C knockin mice." *Neuron* no. 97 (6):1253–1260.e7. doi: 10.1016/j.neuron.2018.02.001.

Carta, I., C. H. Chen, A. L. Schott, S. Dorizan, and K. Khodakhah. 2019. "Cerebellar modulation of the reward circuitry and social behavior." *Science* no. 363 (6424). doi: 10.1126/science.aav0581.

Chen, P., and W. Hong. 2018. "Neural circuit mechanisms of social behavior." *Neuron* no. 98 (1):16–30. doi: 10.1016/j.neuron.2018.02.026.

Cummings, K. A., and R. L. Clem. 2020. "Prefrontal somatostatin interneurons encode fear memory." *Nat Neurosci* no. 23 (1):61–74. doi: 10.1038/s41593-019-0552-7.

Dal Monte, O., F. Krueger, J. M. Solomon, S. Schintu, K. M. Knutson, M. Strenziok, M. Pardini, A. Leopold, V. Raymont, and J. Grafman. 2013. "A voxel-based lesion study on facial emotion recognition after penetrating brain injury." *Soc Cogn Affect Neurosci* no. 8 (6):632–639. doi: 10.1093/scan/nss041.

Delevich, K., H. Jaaro-Peled, M. Penzo, A. Sawa, and B. Li. 2020. "Parvalbumin interneuron dysfunction in a thalamo-prefrontal cortical circuit in Disc1 locus impairment mice." *eNeuro* no. 7 (2). doi: 10.1523/ENEURO.0496-19.2020.

Dolen, G., A. Darvishzadeh, K. W. Huang, and R. C. Malenka. 2013. "Social reward requires coordinated activity of nucleus accumbens oxytocin and serotonin." *Nature* no. 501 (7466):179–184. doi: 10.1038/nature12518.

Fernandez, M., I. Mollinedo-Gajate, and O. Penagarikano. 2018. "Neural circuits for social cognition: implications for autism." *Neuroscience* no. 370:148–162. doi: 10.1016/j.neuroscience.2017.07.013.

Fishell, G., and B. Rudy. 2011. "Mechanisms of inhibition within the telencephalon: 'where the wild things are'." *Annu Rev Neurosci* no. 34:535–567. doi: 10.1146/annurev-neuro-061010-113717.

Fogaca, M. V., and R. S. Duman. 2019. "Cortical GABAergic dysfunction in stress and depression: new insights for therapeutic interventions." *Front Cell Neurosci* no. 13:87. doi: 10.3389/fncel.2019.00087.

Frith, C. D., and U. Frith. 2012. "Mechanisms of social cognition." *Annu Rev Psychol* no. 63:287–313. doi: 10.1146/annurev-psych-120710-100449.

Ghosal, S., C. H. Duman, R. J. Liu, M. Wu, R. Terwilliger, M. J. Girgenti, E. Wohleb, M. V. Fogaca, E. M. Teichman, B. Hare, and R. S. Duman. 2020. "Ketamine rapidly reverses stress-induced impairments in GABAergic transmission in the prefrontal cortex in male rodents." *Neurobiol Dis* no. 134:104669. doi: 10.1016/j.nbd.2019.104669.

Green, M. F., W. P. Horan, and J. Lee. 2015. "Social cognition in schizophrenia." *Nat Rev Neurosci* no. 16 (10):620–631. doi: 10.1038/nrn4005.

Grossmann, T. 2013. "The role of medial prefrontal cortex in early social cognition." *Front Hum Neurosci* no. 7:340. doi: 10.3389/fnhum.2013.00340.

Grossmann, T. 2015. "The development of social brain functions in infancy." *Psychol Bull* no. 141 (6):1266–1287. doi: 10.1037/bul0000002.

Gunaydin, L. A., L. Grosenick, J. C. Finkelstein, I. V. Kauvar, L. E. Fenno, A. Adhikari, S. Lammel, J. J. Mirzabekov, R. D. Airan, K. A. Zalocusky, K. M. Tye, P. Anikeeva, R. C. Malenka, and K. Deisseroth. 2014. "Natural neural projection dynamics underlying social behavior." *Cell* no. 157 (7):1535–1551. doi: 10.1016/j.cell.2014.05.017.

Han, S., C. Tai, C. J. Jones, T. Scheuer, and W. A. Catterall. 2014. "Enhancement of inhibitory neurotransmission by GABAA receptors having alpha2,3-subunits ameliorates behavioral deficits in a mouse model of autism." *Neuron* no. 81 (6):1282–1289. doi: 10.1016/j.neuron.2014.01.016.

Han, S., C. Tai, R. E. Westenbroek, F. H. Yu, C. S. Cheah, G. B. Potter, J. L. Rubenstein, T. Scheuer, H. O. de la Iglesia, and W. A. Catterall. 2012. "Autistic-like behaviour in Scn1a$^{+/-}$ mice and rescue by enhanced GABA-mediated neurotransmission." *Nature* no. 489 (7416):385–390. doi: 10.1038/nature11356.

Harada, M., M. M. Taki, A. Nose, H. Kubo, K. Mori, H. Nishitani, and T. Matsuda. 2011. "Non-invasive evaluation of the GABAergic/glutamatergic system in autistic patients observed by MEGA-editing proton MR spectroscopy using a clinical 3 tesla instrument." *J Autism Dev Disord* no. 41 (4):447–454. doi: 10.1007/s10803-010-1065-0.

Hari, R., L. Henriksson, S. Malinen, and L. Parkkonen. 2015. "Centrality of social interaction in human brain function." *Neuron* no. 88 (1):181–193. doi: 10.1016/j.neuron.2015 .09.022.

Hari, R., and M. V. Kujala. 2009. "Brain basis of human social interaction: from concepts to brain imaging." *Physiol Rev* no. 89 (2):453–479. doi: 10.1152/physrev.00041.2007.

Hashemi, E., J. Ariza, H. Rogers, S. C. Noctor, and V. Martinez-Cerdeno. 2017. "The number of parvalbumin-expressing interneurons is decreased in the prefrontal cortex in autism." *Cereb Cortex* no. 27 (3):1931–1943. doi: 10.1093/cercor/bhw021.

Hawkley, L. C., and J. T. Cacioppo. 2010. "Loneliness matters: a theoretical and empirical review of consequences and mechanisms." *Ann Behav Med* no. 40 (2):218–227. doi: 10.1007/s12160-010-9210-8.

Hiser, J., and M. Koenigs. 2018. "The multifaceted role of the ventromedial prefrontal cortex in emotion, decision making, social cognition, and psychopathology." *Biol Psychiatry* no. 83 (8):638–647. doi: 10.1016/j.biopsych.2017.10.030.

Hu, J. S., D. Vogt, M. Sandberg, and J. L. Rubenstein. 2017. "Cortical interneuron development: a tale of time and space." *Development* no. 144 (21):3867–3878. doi: 10.1242/ dev.132852.

Hung, L. W., S. Neuner, J. S. Polepalli, K. T. Beier, M. Wright, J. J. Walsh, E. M. Lewis, L. Luo, K. Deisseroth, G. Dolen, and R. C. Malenka. 2017. "Gating of social reward by oxytocin in the ventral tegmental area." *Science* no. 357 (6358):1406–1411. doi: 10.1126/science.aan4994.

Jodo, E., T. Katayama, M. Okamoto, Y. Suzuki, K. Hoshino, and Y. Kayama. 2010. "Differences in responsiveness of mediodorsal thalamic and medial prefrontal cortical neurons to social interaction and systemically administered phencyclidine in rats." *Neuroscience* no. 170 (4):1153–1164. doi: 10.1016/j.neuroscience.2010.08.017.

Kahneman, D., A. B. Krueger, D. A. Schkade, N. Schwarz, and A. A. Stone. 2004. "A survey method for characterizing daily life experience: the day reconstruction method." *Science* no. 306 (5702):1776–1780. doi: 10.1126/science.1103572.

Kamigaki, T., and Y. Dan. 2017. "Delay activity of specific prefrontal interneuron subtypes modulates memory-guided behavior." *Nat Neurosci* no. 20 (6):854–863. doi: 10.1038/ nn.4554.

Kim, D., H. Jeong, J. Lee, J. W. Ghim, E. S. Her, S. H. Lee, and M. W. Jung. 2016. "Distinct roles of parvalbumin- and somatostatin-expressing interneurons in working memory." *Neuron* no. 92 (4):902–915. doi: 10.1016/j.neuron.2016.09.023.

Ko, J. 2017. "Neuroanatomical substrates of rodent social behavior: the medial prefrontal cortex and its projection patterns." *Front Neural Circuits* no. 11:41. doi: 10.3389/ fncir.2017.00041.

Kupferberg, A., L. Bicks, and G. Hasler. 2016. "Social functioning in major depressive disorder." *Neurosci Biobehav Rev* no. 69:313–332. doi: 10.1016/j.neubiorev.201 6.07.002.

Kvitsiani, D., S. Ranade, B. Hangya, H. Taniguchi, J. Z. Huang, and A. Kepecs. 2013. "Distinct behavioural and network correlates of two interneuron types in prefrontal cortex." *Nature* no. 498 (7454):363–366. doi: 10.1038/nature12176.

Lee, E., I. Rhim, J. W. Lee, J. W. Ghim, S. Lee, E. Kim, and M. W. Jung. 2016. "Enhanced neuronal activity in the medial prefrontal cortex during social approach behavior." *J Neurosci* no. 36 (26):6926–6936. doi: 10.1523/JNEUROSCI.0307-16.2016.

Lee, S. H., A. C. Kwan, S. Zhang, V. Phoumthipphavong, J. G. Flannery, S. C. Masmanidis, H. Taniguchi, Z. J. Huang, F. Zhang, E. S. Boyden, K. Deisseroth, and Y. Dan. 2012. "Activation of specific interneurons improves V1 feature selectivity and visual perception." *Nature* no. 488 (7411):379–383. doi: 10.1038/nature11312.

Lee, S., J. Hjerling-Leffler, E. Zagha, G. Fishell, and B. Rudy. 2010. "The largest group of superficial neocortical GABAergic interneurons expresses ionotropic serotonin receptors." *J Neurosci* no. 30 (50):16796–16808. doi: 10.1523/JNEUROSCI.1869-10.2 010.

Lee, S., I. Kruglikov, Z. J. Huang, G. Fishell, and B. Rudy. 2013. "A disinhibitory circuit mediates motor integration in the somatosensory cortex." *Nat Neurosci* no. 16 (11): 1662–1670. doi: 10.1038/nn.3544.

Lewis, D. A., A. A. Curley, J. R. Glausier, and D. W. Volk. 2012. "Cortical parvalbumin interneurons and cognitive dysfunction in schizophrenia." *Trends Neurosci* no. 35 (1):57–67. doi: 10.1016/j.tins.2011.10.004.

Lewis, D. A., T. Hashimoto, and D. W. Volk. 2005. "Cortical inhibitory neurons and schizophrenia." *Nat Rev Neurosci* no. 6 (4):312–324. doi: 10.1038/nrn1648.

Liang, B., L. Zhang, G. Barbera, W. Fang, J. Zhang, X. Chen, R. Chen, Y. Li, and D. T. Lin. 2018. "Distinct and dynamic ON and OFF neural ensembles in the prefrontal cortex code social exploration." *Neuron* no. 100 (3):700–714.e9. doi: 10.1016/j.neuron.201 8.08.043.

Liang, J., W. Xu, Y. T. Hsu, A. X. Yee, L. Chen, and T. C. Sudhof. 2015. "Conditional neuroligin-2 knockout in adult medial prefrontal cortex links chronic changes in synaptic inhibition to cognitive impairments." *Mol Psychiatry* no. 20 (7):850–859. doi: 10.1038/mp.2015.31.

Liu, L., H. Xu, J. Wang, J. Li, Y. Tian, J. Zheng, M. He, T. L. Xu, Z. Y. Wu, X. M. Li, S. M. Duan, and H. Xu. 2020. "Cell type-differential modulation of prefrontal cortical GABAergic interneurons on low gamma rhythm and social interaction." *Sci Adv* no. 6 (30):eaay4073. doi: 10.1126/sciadv.aay4073.

Lodge, D. J., M. M. Behrens, and A. A. Grace. 2009. "A loss of parvalbumin-containing interneurons is associated with diminished oscillatory activity in an animal model of schizophrenia." *J Neurosci* no. 29 (8):2344–2354. doi: 10.1523/JNEUROSCI.5419-08.2009.

Miao, C., Q. Cao, M. B. Moser, and E. I. Moser. 2017. "Parvalbumin and somatostatin interneurons control different space-coding networks in the medial entorhinal cortex." *Cell* no. 171 (3):507–521.e17. doi: 10.1016/j.cell.2017.08.050.

Montagrin, A., C. Saiote, and D. Schiller. 2018. "The social hippocampus." *Hippocampus* no. 28 (9):672–679. doi: 10.1002/hipo.22797.

Murugan, M., H. J. Jang, M. Park, E. M. Miller, J. Cox, J. P. Taliaferro, N. F. Parker, V. Bhave, H. Hur, Y. Liang, A. R. Nectow, J. W. Pillow, and I. B. Witten. 2017. "Combined social and spatial coding in a descending projection from the prefrontal cortex." *Cell* no. 171 (7):1663–1677.e16. doi: 10.1016/j.cell.2017.11.002.

Noonan, M. P., J. Sallet, P. H. Rudebeck, M. J. Buckley, and M. F. Rushworth. 2010. "Does the medial orbitofrontal cortex have a role in social valuation?" *Eur J Neurosci* no. 31 (12):2341–2351. doi: 10.1111/j.1460-9568.2010.07271.x.

Peters, A., and E. G. Jones. 1984. "Classification of cortical neurons." In *Cerebral Cortex*, edited by E. G. Jones and A. Peters, 107–120. Plenum Press, New York, NY.

Radhu, N., L. Garcia Dominguez, F. Farzan, M. A. Richter, M. O. Semeralul, R. Chen, P. B. Fitzgerald, and Z. J. Daskalakis. 2015. "Evidence for inhibitory deficits in the prefrontal cortex in schizophrenia." *Brain* no. 138 (Pt 2):483–497. doi: 10.1093/brain/awu360.

Resendez, S. L., V. M. K. Namboodiri, J. M. Otis, L. E. H. Eckman, J. Rodriguez-Romaguera, R. L. Ung, M. L. Basiri, O. Kosyk, M. A. Rossi, G. S. Dichter, and G. D. Stuber. 2020. "Social stimuli induce activation of oxytocin neurons within the paraventricular nucleus of the hypothalamus to promote social behavior in male mice." *J Neurosci* no. 40 (11):2282–2295. doi: 10.1523/JNEUROSCI.1515-18.2020.

Rudebeck, P. H., M. J. Buckley, M. E. Walton, and M. F. Rushworth. 2006. "A role for the macaque anterior cingulate gyrus in social valuation." *Science* no. 313 (5791): 1310–1312. doi: 10.1126/science.1128197.

Rudy, B., G. Fishell, S. Lee, and J. Hjerling-Leffler. 2011. "Three groups of interneurons account for nearly 100% of neocortical GABAergic neurons." *Dev Neurobiol* no. 71 (1):45–61. doi: 10.1002/dneu.20853.

Sanchez-Andrade, G., and K. M. Kendrick. 2009. "The main olfactory system and social learning in mammals." *Behav Brain Res* no. 200 (2):323–335. doi: 10.1016/j.bbr.2 008.12.021.

Sauer, J. F., M. Struber, and M. Bartos. 2015. "Impaired fast-spiking interneuron function in a genetic mouse model of depression." *Elife* no. 4. doi: 10.7554/eLife.04979.

Saver, J. L., and A. R. Damasio. 1991. "Preserved access and processing of social knowledge in a patient with acquired sociopathy due to ventromedial frontal damage." *Neuropsychologia* no. 29 (12):1241–1249. doi: 10.1016/0028-3932(91)90037-9.

Scheggia, D., F. Manago, F. Maltese, S. Bruni, M. Nigro, D. Dautan, P. Latuske, G. Contarini, M. Gomez-Gonzalo, L. M. Requie, V. Ferretti, G. Castellani, D. Mauro, A. Bonavia, G. Carmignoto, O. Yizhar, and F. Papaleo. 2020. "Somatostatin interneurons in the prefrontal cortex control affective state discrimination in mice." *Nat Neurosci* no. 23 (1):47–60. doi: 10.1038/s41593-019-0551-8.

Schilbach, L., A. M. Wohlschlaeger, N. C. Kraemer, A. Newen, N. J. Shah, G. R. Fink, and K. Vogeley. 2006. "Being with virtual others: neural correlates of social interaction." *Neuropsychologia* no. 44 (5):718–730. doi: 10.1016/j.neuropsychologia.2005.07.017.

Selimbeyoglu, A., C. K. Kim, M. Inoue, S. Y. Lee, A. S. O. Hong, I. Kauvar, C. Ramakrishnan, L. E. Fenno, T. J. Davidson, M. Wright, and K. Deisseroth. 2017. "Modulation of prefrontal cortex excitation/inhibition balance rescues social behavior in CNTNAP2-deficient mice." *Sci Transl Med* no. 9 (401). doi: 10.1126/ scitranslmed.aah6733.

Shepherd, S. V., and W. A. Freiwald. 2018. "Functional networks for social communication in the Macaque monkey." *Neuron* no. 99 (2):413–420.e3. doi: 10.1016/j.neuron.201 8.06.027.

Sliwa, J., and W. A. Freiwald. 2017. "A dedicated network for social interaction processing in the primate brain." *Science* no. 356 (6339):745–749. doi: 10.1126/science.aam6383.

Somogyi, P., and T. Klausberger. 2005. "Defined types of cortical interneurone structure space and spike timing in the hippocampus." *J Physiol* no. 562 (Pt 1):9–26. doi: 10.1113/jphysiol.2004.078915.

Tang, Y., D. Benusiglio, A. Lefevre, L. Hilfiger, F. Althammer, A. Bludau, D. Hagiwara, A. Baudon, P. Darbon, J. Schimmer, M. K. Kirchner, R. K. Roy, S. Wang, M. Eliava, S. Wagner, M. Oberhuber, K. K. Conzelmann, M. Schwarz, J. E. Stern, G. Leng, I. D. Neumann, A. Charlet, and V. Grinevich. 2020. "Social touch promotes interfemale communication via activation of parvocellular oxytocin neurons." *Nat Neurosci* no. 23 (9):1125–1137. doi: 10.1038/s41593-020-0674-y.

Tremblay, R., S. Lee, and B. Rudy. 2016. "GABAergic interneurons in the neocortex: from cellular properties to circuits." *Neuron* no. 91 (2):260–292. doi: 10.1016/j.neuron.201 6.06.033.

Urakawa, S., K. Takamoto, A. Ishikawa, T. Ono, and H. Nishijo. 2015. "Selective Medial prefrontal cortex responses during live mutual gaze interactions in human infants: an fNIRS study." *Brain Topogr* no. 28 (5):691–701. doi: 10.1007/s10548-014-0414-2.

van Kerkhof, L. W., R. Damsteegt, V. Trezza, P. Voorn, and L. J. Vanderschuren. 2013. "Social play behavior in adolescent rats is mediated by functional activity in medial prefrontal cortex and striatum." *Neuropsychopharmacology* no. 38 (10):1899–1909. doi: 10.1038/npp.2013.83.

Wagner, D. D., W. M. Kelley, J. V. Haxby, and T. F. Heatherton. 2016. "The dorsal medial prefrontal cortex responds preferentially to social interactions during natural viewing." *J Neurosci* no. 36 (26):6917–6925. doi: 10.1523/JNEUROSCI.4220-15.2016.

Walum, H., and L. J. Young. 2018. "The neural mechanisms and circuitry of the pair bond." *Nat Rev Neurosci* no. 19 (11):643–654. doi: 10.1038/s41583-018-0072-6.

Wang, Q., Q. Wang, X. L. Song, Q. Jiang, Y. J. Wu, Y. Li, T. F. Yuan, S. Zhang, N. J. Xu, M. X. Zhu, W. G. Li, and T. L. Xu. 2018. "Fear extinction requires ASIC1a-dependent regulation of hippocampal-prefrontal correlates." *Sci Adv* no. 4 (10):eaau3075. doi: 10.1126/sciadv.aau3075.

Wonders, C., and S. A. Anderson. 2005. "Cortical interneurons and their origins." *Neuroscientist* no. 11 (3):199–205. doi: 10.1177/1073858404270968.

Xu, H., L. Liu, Y. Tian, J. Wang, J. Li, J. Zheng, H. Zhao, M. He, T. L. Xu, S. Duan, and H. Xu. 2019. "A disinhibitory microcircuit mediates conditioned social fear in the prefrontal cortex." *Neuron* no. 102 (3):668–682.e5. doi: 10.1016/j.neuron.2019.02.026.

Yizhar, O., L. E. Fenno, M. Prigge, F. Schneider, T. J. Davidson, D. J. O'Shea, V. S. Sohal, I. Goshen, J. Finkelstein, J. T. Paz, K. Stehfest, R. Fudim, C. Ramakrishnan, J. R. Huguenard, P. Hegemann, and K. Deisseroth. 2011. "Neocortical excitation/inhibition balance in information processing and social dysfunction." *Nature* no. 477 (7363):171–178. doi: 10.1038/nature10360.

10 Studying Brain Function Using Non-human Primate Models

Neng Gong

CONTENTS

Rodents have been the leading model in neuroscience research because of advantages such as low cost, short reproductive cycle, well-developed techniques, multiple transgenic lines, etc. However, they also have many shortcomings, especially in high-level cognitive function studies, for examples, their simplistic social behavior and relatively poor cognitive competence. Non-human primates (NHPs) have attracted more and more attention from neuroscientists, because they are phylogenetically closer to humans and share similar neuroanatomical structures and neural function. Compared to rodents, NHPs are more likely to possess some high-level cognitive abilities, for examples, self-awareness, empathy, and theory of mind (1). In this chapter, we will introduce the application of NHP in studying visual and auditory function, prosocial behavior and self-awareness, and the development of the NHP transgenic model.

10.1 STUDYING HIGH-LEVEL VISUAL PROCESSING RESEARCH USING NON-HUMAN PRIMATE MODELS

Like humans, NHPs have a highly developed and sophisticated visual system, which help them to perceive visual information about objects, search for food, and interact with conspecifics. Unsurprisingly, the visual system of NHPs has been a popular research topic in neuroscience for decades.

In primates, the visual stream begins at the retina for low-level feature processing, such as light intensity and color, then the information is sent to the thalamic nuclei relay (lateral geniculate nucleus and pulvinar), and finally to the cerebral cortex for further processing (2). There are two distinct visual pathways: the ventral and the dorsal visual pathway, which are also called the "What" pathway and "Where" pathway. The former encodes object's identity and the latter encodes object's location and is involved in visual-motor control (2–4). The two pathways are interconnected to communicate visual information, and both contribute to visual perception (5,6).

10.1.1 OBJECT AND FACE RECOGNITION IN NHPS

Object recognition has been extensively studied in primates' ventral pathway, in which the information stream from the primary visual cortex (V1) hierarchically extends to the V2, V4, temporal-occipital junction (TEO), and inferior temporal cortex (IT). In the 1880s, Sanger Brown found that after inducing a lesion to a rhesus monkey's temporal lobe, the monkey could not recognize objects. The finding was later confirmed and further explored by Heinrich Kluver and Paul Bucy (7,8). A series of questions arose: "How can neurons code the object's global information?", "Are there any neurons coding specifically to object identity?", etc. In the ventral pathway, the IT cortex receives diverse information from preceding relays and its neurons have a very large receptive field (9). In an experiment that recorded visual properties of a macaque's IT cortex, Charles Gross found that a neuron unit had no response to any light stimuli, whereas it showed vigorous response when the experimenter waved his hand at the screen (10). Thereafter, subgroups of neurons encoding complex features, such as shape of hand, were found in the IT cortex. Even more interestingly, subpopulations of neurons in the temporal cortex (superior temporal sulcus and IT) show selective response to a wide variety of face stimuli, instead of simple basic features such as light and color (11–13). Further studies using functional magnetic resonance imaging (fMRI) and electrophysiological single-unit recording revealed that macaques (Macaca mulatta) had face-selective patches in the temporal cortex extending from V4 to rostral TE, which is similar in relative size and number to those of humans. The electrophysiology data indicated that 97% of visually responsive neurons in the middle face patch were face-selective (14,15). How do the neuron population in the face patch encode the facial property and even identity? The image of a face can be viewed as a combination of physical features in a multidimensional space. The neurons in the middle fundus (MF) patch and middle lateral (ML) patch detect distinct face parts and are also tuned to geometrical features of the face, resulting in population coding of coarse facial features (16). In contrast, although neurons in the anterior medial patch-the top of face recognition processing hierarchy- also use feature axis to encode information of a face, they are more sensitive to finer facial features and can represent an individual face using a small cell ensemble (~200 cells) (17). What happens to face-selective neurons during the process of attuning to a new face? How can neurons interact with each other or even with neurons in other patches to encode facial information? The common marmoset (*Callithrix jacchus*) might be a suitable model for exploring these questions, since marmoset has similar face patches in its IT cortex, and more importantly, the smooth surface of the cortex with well-developed calcium indicator system enables long-term large-scale calcium imaging of neural activity in the IT cortex in marmosets (18–21).

10.1.2   Visual-Guide Motion in NHPs

Visuomotor integration is one of the most important function of the dorsal pathways, which provides the motor system with information to guide movement (22–24). Humans and NHPs have much finer hand motions than other mammals, which is a critical skill to develop the ability for tool use. In both human and macaque, the dorsal stream from the visual cortex extends to the posterior parietal cortex (PPC) and to the frontal cortex. Muscimol injection to inhibit neural activity in the anterior intraparietal area (AIP) of the PPC or area F5 (PMv, ventral premotor area) of the frontal cortex makes a macaque unable to preshape the hand when grasping an object (25,26). The results indicate that AIP and F5 might play an important role in visuomotor integration for guiding grasping (given that AIP has been demonstrated demonstrated to have robustly reciprocal connection with the F5 area by retrograde tracing) (27,28). By using fMRI and electrophysiology approaches, AIP neurons are found not only to exhibit highly selective response to specific 3-D shapes and visual depth disparities in 2-D images of objects, but also are active during grasping (29,30). The anterior F5 area (F5a) also encodes information about 3-D shape, in which neurons have highly selective response to convex and concave surfaces (31). In addition, these neurons are also activated when subjects grasp an object under a visual guide, but weaken in the dark environment. In a delayed grasping task, where the macaque performed power or precision grips on a handle in different orientations, neurons in the AIP robustly responded to the visual signal of handles, while neurons in F5 had higher response during movement preparation and execution (32,33). Additional research is needed to elucidate how the AIP and F5 interact with each other to perform grasping, and to better understand the cross-talk between the dorsal and ventral pathways. This will expand our knowledge on the neural mechanism of the visual system and promote the development of brain-computer interface and robotics.

10.2 STUDYING VOCAL COMMUNICATION AND AUDITORY FUNCTION USING THE MARMOSET MONKEY

10.2.1 MARMOSETS' VOCAL COMMUNICATION

Vocal communication is one of the most important cognitive functions of human beings. It is related to intrinsic vocal production, acquired learning, environment, and also evolution (34). The past few decades have witnessed a rise of interest in the common marmoset as a promising vocal communication model for neuroscience research (35–37). They are native to north-eastern Brazil and have adapted a large vocal repertoire to communicate through the dense forest (38). Marmosets are a vocally rich species and use diverse vocal calls to communicate within their relatively stable family structure and with other conspecifics (38,39). Their vocal communication occurs nearly constantly during daytime even in the captive environment, which is in contrast to many African and Asian primate species (39). As technology advances, scientists are able to classify most of the marmoset repertoire and have found that vocal signals depend on the particular social and ecological context, which shows the importance of vocal communication in the social behavior of marmosets (35,40,41). For example, reproductively suppressed marmosets, due to their inferior status in the group, produced significantly more territorial Phee call after pairing with another opposite-sex conspecifics. The nonreproductive marmosets may suppress territorial call to convey their subordinate status to the breeding pairs of the groups and show less altruistic behavior towards the infant marmosets (41,42). Phee call is the most well studied vocalization behavior in the common marmoset. The acoustic structure of this species-typical calling is one or more long slow-frequency modulated "whistle"-like pulse(s). This species-typical calling bears long-distance communication function and conveys a variety of social information (41,43,44). Antiphonal calling is another important featured vocal behavior of the marmoset (35,44). During antiphonal communication, marmosets produce Phee calls alternately, without interrupting others' calling. It is worth noting that this phenomenon is shown only in matured marmosets, suggesting a potential auditory-vocal learning ability during marmosets' ontogeny (45,46). Captive marmosets' vocalization shows greater frequency and temporal variability when they were recorded in isolation than in home cage. The increased variability in vocalization may help conspecifics to localize the isolated caller (41,43).

10.2.2 MARMOSETS' AUDITORY SYSTEM

As NHPs, the marmoset shares a similar auditory cortex architecture with other primates, including humans (47–50). As yet, the specific behaviors and cognitive ways of this species and the neural mechanisms behind these particular vocal behaviors have not yet been fully studied. The auditory center of the marmoset includes the auditory nerve, cochlear nucleus, trapezoid body, periolivary nuclei, superior olives, lateral lemniscus, inferior and superior colliculus, medial geniculate body, and auditory cortices (51). The auditory cortex can be divided into core area, belt area, and para-belt area according to the hierarchical organization of anatomy and function (47,48). The core area can be divided into three detailed areas, named RT (rostro-temporal), R (rostral), and A1 (primary cortex) from ventral to dorsal. These three areas are interconnected and receive projections mainly from the medial geniculate body, while A1 also has projections to R and RT areas. The output from the core area goes through the belt area and para-belt area, and finally projects to higher brain regions, including vlPFC and dlPFC to participate in the formation of auditory perception (52).

One of the most well-studied features of auditory neurons is frequency tuning along the auditory center (53–55), which had long been a controversial topic until Wang et al. demonstrated the frequency tuning ability of marmoset during both awake and anesthetized states (56). Compared to relatively transient responses to acoustic stimulation in the auditory cortex of anesthetized animals, neurons in the auditory cortex of awake marmoset monkeys were able to response in a higher and more sustained manner for both short and long duration sounds throughout the duration of the stimulus, especially when the neurons were driven by their preferred stimuli. In contrast, responses became more transient when auditory cortex neurons responded to non-preferred stimuli (56).

In the superficial layers of A1 area of awake marmosets, neurons exhibit high stimulus selectivity to particular features of acoustic stimuli and were commonly found to be more selective to species-specific marmoset vocalizations and environmental sounds such as natural trill-twitter call (57). Interestingly, these nonlinear combination-sensitive neurons were once seen as unresponsive neurons (58). Such neurons may provide a bridge between tone-tuned thalamus-recipient A1 neurons and neurons in the secondary auditory cortex that exhibit complex response properties.

As for representation of sound pressure level, in A1 of awake marmosets, most neurons (~60%) had O-shaped FRAs (frequency response areas), as opposed to V-shaped FRAs (59). However, O-shaped units were not reported in unanesthetized conditions because of their narrow tuning and low maximum driven rate (60). O-shaped FRAs represent the nonmonotonic rate-level function of a neuron when they respond to pure tones of varying frequency and sound pressure level. Given that there is a large proportion of nonmonotonic neurons found in the auditory thalamus (53), this might be the result of a potential link between the thalamus and auditory cortex.

The marmoset also serves as a great model for pitch processing and harmonic perception. Feng et al. have found a unique population of harmonic template neurons in the core region of the auditory cortex in marmosets (61). These neurons show nonlinear facilitation to harmonic complex sounds over inharmonic sounds and selectivity for particular harmonic structures. It shows that marmoset may have the potential to determine whether sound is harmonic or not, which is important for auditory perception. Marmosets can behaviorally perceive pitch of harmonic complex sounds and extract pitch by using temporal envelope cues for lower-pitch sounds composed of higher-order harmonics and spectral cues for higher-pitch sounds composed of lower-order harmonics (62,63). An important structure in which neurons with pitch-selective response is the pitch center, clustered near the anterolateral border of A1 (64–66). However, where the inputs of pitch-selective neurons come from, and whether the pitch-selective neurons also play a role in pitch perception, still remain unknown.

To conclude, marmosets show great value in neuroscience as a model of neural basis behind vocal behaviors. The marmoset model offers a promising approach to study language-related or musical aspects of vocal communication that may be unique to primates. Though not without shortcomings, marmosets still show advantages in the research of acoustic fields. As technology advances, many new optical and molecular methods are being developed to improve the level of observation. Two-photon Ca^{2+} imaging in marmosets is a trending way to acquire higher spatial resolution of the marmoset cortex (21,68). Adeno-associated viruses (AAV) express robustly in marmosets (21,67,69), providing opportunities to apply modern optogenetic and pharmacogenetic techniques to the study of primate brain circuitry. Additionally, in combination with freely moving single neuron recording, science will continue discovering more natural ways to explore the neural processes underlying natural vocal behaviors (46,70).

10.3 STUDYING PROSOCIAL BEHAVIOR USING THE MARMOSET MONKEY

10.3.1 PROSOCIAL BEHAVIORS IN MARMOSET

Prosocial behaviors have been reported in marmosets in recent decades. Maternal care is widely observed in mammals; however, paternal and alloparental care is reported in fewer species, including marmoset. As a cooperative breeding species, marmosets live in family groups in the wild, which consist of several adults and their offspring (71). In the family, both parents and siblings are engaged in infant caring (72). In the first eight weeks after birth of an infant, parents are the major participants in infant carrying. The mother contributes most in the first two weeks, and its relative contribution (compared with other group members) peaks at the third week; the father carries the infant most often on the first day and then his contribution declines after reaching a peak at the fourth week. Siblings take part in the carrying actively in the first week and gradually reduce their involvement and stop. The contribution of the siblings also depends on their experience: if an individual is experienced, it will contribute more to carrying (39).

Marmoset infants rely on food from other group members during weaning. When an infant asks for food, parents and siblings will transfer familiar foods more frequently than novel ones, to help the infant to learn about ordinary diet (73). The food transfer behavior can be reactive or proactive. Reactive food transfer refers to that an adult tolerates the immature taking its food away. In a proactive food transfer, the food possessor releases a food call and waits until the food is taken (74). Interestingly, when no audience is present, adults share more food to the immatures. This reverse audience effect is parallel to humans' diffusion-of-responsibility effect- people offer less help when bystanders are present but feel more responsible for the person who needs help when alone (75).

To investigate prosocial behaviors in marmosets, Judith M. Bukart et al. set up a food provision paradigm, in which a donor marmoset could provide food to its

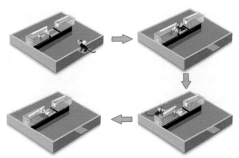

FIGURE 10.1   The paradigm for evaluating marmosets' prosocial behavior.

partner by pulling the tray within its reach without any rewards for itself. The result showed that a marmoset adult would help its unrelated partner getting the food, even if the receiver was stranger (76). Recently, in a novel rescue paradigm, marmoset adults and older siblings, in most cases, would jump to rescue a young infant trapped on an island surrounded by water (Figure 10.1). However, parents hardly rescued each other, but the presence of infant calls could induce the parents' mutual rescue. Moreover, having young infants in the family group promoted both infant- and mate-rescue behavior. fMRI study on awake marmosets demonstrates the specific cortical areas, of marmosets with young infants, activated by infant calls (42).

10.3.2   THE ROLE OF HORMONES IN PROSOCIAL BEHAVIOR

Little is known about the neural mechanism of prosocial behavior. What is known is that prosocial behaviors are related to hormonal changes. Oxytocin, a nonapeptide hormone, plays an important role in marmoset's maternal and alloparental care (77,78). The urinary oxytocin levels of all breeders and helpers increased after an infant's birth in the first day and decreased as the infant grew up. The oxytocin levels of the mother reached a peak then remained at high level until infants began to eat solid food. The urinary oxytocin levels of breeders and helpers increase again when they shared food proactively with the infants (78). It is worth mentioning that the oxytocin (Pro^8-OT) in marmosets is a variant; the leucine at the eighth amino-acid position is a proline instead of the conserved Leu^8-OT; this results in structural distinction in marmoset oxytocin.

Treated with Pro^8-OT, males responded faster to the infant stimulus, but their sustained interest in infant stimulus was not affected by the treatment (77). By administering oxytocin into the central nervous system, researchers found that the times males refused their offsprings' request of food transfer decreased (79). However, Pro^8-OT may reduce a marmoset's prosociality, as shown in the task of pulling a tray to help a stranger to get the food (80). Oxytocin also helps maintaining the social bond between well-established marmoset pairs. Treated with Pro^8-OT, females and males all reduced the time spent closely with an opposite-sex stranger. Males reduced the time spent in close proximity with all opposite-sex individuals. Females preferred more social interaction with their partners (81).

Therefore, it seems that the role of oxytocin in social behavior is not monolithic in enhancing prosociality but varies depending on social context, and more research is needed to explore the mechanism of oxytocin's regulation in prosocial behavior.

In addition to oxytocin, arginine-vasopressin and prolactin may be involved in prosocial behavior. Intranasal arginine-vasopressin facilitates response to infant stimuli in female but not male marmosets (77). For fathers and older offspring, prolactin levels were elevated after infant carrying (82). For marmosets who were parentally inexperienced, their prolactin levels increase after infant retrievals (72).

To some extent, marmoset society is similar to human society, especially in terms of family structure and cooperative breeding. It was presumed that prosocial behavior may have been involved in the optimization of the cooperative breeding system (76). Prosocial behavior may improve the living condition of the group members and thus enhance the survival ability of the whole group. For example, in case of food shortage, food sharing between group members can help the group get through the difficulties until new food sources are found.

In humans, only behavior that is beneficial to others and is done intentionally can be classified as prosocial behavior in psychology. Likewise, the criteria for intentionality in marmoset prosociality has been established, including flexibility, audience effects, and goal directedness. Flexibility refers to the subject's adjustment of its behavior to specific conditions and the usage of multiple means to achieve a goal. Audience effects refer to 1) behavior that occurs only when potential recipients are present, and 2) behavior is influenced by whether other helpers are present. Goal directedness refers to the persistence of the behavior until the goal is achieved and the distinct reaction shown by the helper when the expected goal is not achieved (74). But is it appropriate to mechanically apply what "intentionality" is in humans to marmosets? Maybe we should establish more sophisticated behavioral paradigms to investigate marmosets' prosocial behavior in a more naturalistic way instead of creating descriptive and subjective criteria.

As mentioned above, hormones partake in modulating prosocial behavior and act as an affected molecule. However, the synergy and antagonism effects of hormones in generating and regulating prosocial behavior remain unclear. Many more question should be explored in the future.

10.4   STUDYING SELF-AWARENESS USING THE NON-HUMAN PRIMATE MODEL

Self-awareness is a critical function in humans, and the potential for self-awareness in NHP has been tested extensively. The classical mirror self-recognition (MSR) test, is generally considered as a reflection of self-awareness. MSR is detected by using the mark test (83), in which an odorless and non-irritant dye is placed on the subject's face and can only be seen in the mirror. Subjects pass the test if they touch the dye mark after seeing themselves in the mirror, but not if in the absence of the mirror. This shows that the subjects can recognize themselves and have mirror self-recognition ability.

10.4.1   THE HISTORY OF MSR IN NON-HUMAN PRIMATES

Since the establishment of the mark test four decades ago, MSR research revealed that most great apes (humans, chimpanzees, orangutans, and bonobos but not the gorilla) passed the mark test and convincingly showed the capacity to recognize themselves by mirrors (84–87). The putative discontinuity in phylogeny of the ability suggests the existence of a so-called cognitive gap between great apes and the rest of the animal kingdom (88). However, the mark test has some methodological flaws (89). Failing a mark test is likely to result in a false negative result. It remains controversial whether failing the mark test is a result of the lack of self-recognition ability or the inadequacy of the mark test. Firstly, the negative results may be explained by a lack of motivation. More than one-half of the chimpanzees failed to use the mirror for spontaneous inspection of directly visible body parts (90). Secondly, some species are naturally averse to eye contact, i.e., gorillas (91) and macaques (92), these animals cannot be fully exposed to the reflected face. Thirdly, failing the test may simply resulted from poor eyesight (88).

Therefore, researchers began to improve their training method, hoping for positive results in many species who did not pass the mirror test before. They trained subjects to touch visible marks by increasing motivation (93), replacing common dye with colored food that subjects like, such as chocolate (94), changing the shape and size of the mirror to avoid direct gaze with mirror image (92), and training subjects to use mirrors as a tool to get food (93–95).However, all these efforts failed. As a result, scientists concluded that monkeys cannot recognize themselves in the mirror. Interestingly, rhesus monkeys with a head implant did not pass the mark test, although they showed self-directed genital-related behaviors in front of the mirror (96). The head near the implanted area could produce a persistent irritant sensation. Perhaps it is the sensation of wounds that facilitates and triggers the MSR ability of macaques.

Gong and his team designed an ingenious visual-somatosensory training method in 2015 (97) (Figure 10.2). Rhesus monkeys passed the mark test after being exposed to mild stimulation of the body; the monkey trained eventually passed the mirror test. However, the experiment has been largely questioned. One query is

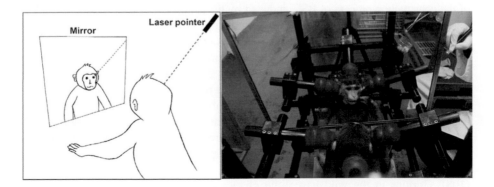

FIGURE 10.2   Visual-somatosensory training in macaque by laser pointer.

that the monkey touching a mark on its face can be considered to be an "engineered behavior" rather than spontaneous MSR (87). Therefore, Gong and his team re-designed the experiment (98).Rhesus monkeys were only taught how to use mirrors as tools to find light spots around them, without any training involving the mon-keys' face and body. In another words, monkeys were trained to learn precise visual-proprioceptive association for mirror images. This study suggests that rhesus monkeys can spontaneously pass the mark test after mirror-usage training. Strangely, in other studies, in which monkeys have been trained to obtain food only through mirrors (93–95), and appeared to use mirror for behavior guiding (99), none of them passed the mark test.

10.4.2 THE SECRET OF MSR: MULTISENSORY INTEGRATION

Why were the training methods designed by Chang and colleagues effective? Firstly, monkeys are chair-restrained and forced to be exposed to their mirror images for a longer time. They have to face the mirror directly to complete the training task, which made them to overcome their aversion to eye contact, rather than avoiding eye contact by changing the shape and size of the mirror (92), Secondly, the training process in-volved multisensory integration of proprioceptive, vestibular, and visual bodily inputs within peripersonal space (PPS) and satisfied the four major requirements of generation of body self-consciousness (BSC)-proprioceptive, body-related visual information, PPS, and embodiment conditions (100). Meanwhile, some studies show that rubber hand illusion, a phenomenon related to hand-centered aspects of BSC, can be induced in macaques (101–104). Nevertheless, in the mirror training process, the subjects could simultaneously see their own real bodies and mirror bodies, in contrast to the rubber hand and virtual body illusion experiment. The prolonged synchronous integration of multiple senses induces the mirror image illusion like the hand illusion, while the co-existence of the monkey's real body and mirror body confused it. It is only by under-standing the magic of mirrors that the brain can understand and explain this confusing phenomenon. The third reason is that monkeys were trained to use the mirror as a tool more intensively and accurately than the previous training (93–95); they were provided the opportunities to further understand the relationship between the positions inside and outside the mirrors. Therefore, we hypothesized that MSR results from the combination of the construction process of BSC, induced by the simultaneous integration of multiple senses for a mirror image, and the understanding of the mirror principle.

MSR ability might not be an innate ability, and more intelligent animals are more likely to acquire this ability gradually. On the one hand, this may explain why most children recognize themselves in the mirror by 2 years of age (105), while children from rural non-modern societies only "pass" the mark test after 6 years (106). On the other hand, this is likely to account for the observation that not all chimpanzees passed the mark test spontaneously (107). In contrast to chimpanzees and orangutans, the MSR testing results of gorillas are much more controversial. Wild gorillas usually do not drink water directly from a reflective pond but absorb water by drinking morning dew and eating succulent vegetation, half of which is made up of water (108). Gorillas in the wild have little chance to understand mirrors. Another key reason is that gorillas, like macaques, are averse to eye contact. It is reported that a gorilla without aversion to direct eye contact could recognized himself in a mirror (91). Similarly, some gorillas with extensive contact with humans from an early age passed the mark test (109). These results indicate that gorillas may have the ability to pass MSR. Rhesus monkeys are considered to be intelligently inferior to the great apes, so they need extensive training to acquire the MSR ability (97,98). This also suggests that a small number of other animals with high intelligence, such as elephants, dolphins, and magpies, could possibly pass the mark test (110–112). This hypothesis remains to be tested and research on the neural basis and mechanism of the macaque MSR model can shed light on this question.

In conclusion, self-awareness, a high-level cognitive ability, may exist in more species, and more appropriate experimental methods are needed to reveal them and explored related neural mechanism.

10.5 TRANSGENIC RESEARCH IN THE NON-HUMAN PRIMATE MODEL

10.5.1 TRANSGENIC NHPs HISTORY

As mentioned above, rodent is one of the most successful animal models in modern biology because of an array of well-developed genetic tools to directly and precisely modify and manipulate their genomes. However, due to the distant evolutionary relations, human and mice are very different in anatomy, physiology, cognitive ability, and behaviors. These differences limit the transfer of knowledge derived from mice study to humans (113). The development of NHP model by genetic tools can help scientists to better understand human biology. Similar to the transgenic strategies in mice, the first exogenous DNA transfer in monkeys was carried out by intracytoplasmic sperm injection (ICSI) with plasmid-bound spermatozoa (114). It is shown that transgene can be expressed successfully after injecting sperm-bound plasmid DNA, which encoded the GFP gene under the control of a CMV promotor, into mature rhesus macaques' oocytes. However, fluorescence is not observed in the three newborns with transgenic genes in this study, including a healthy one and two stillborns; and the GFP gene is negative according to the PCR analysis of tissues of the newborns. Theses results show that injecting exogenous DNA into monkeys' embyros directly is an inefficient method. However, more importantly, the research demonstrates that it is possible to produce transgenic monkey. This group of researchers then injected pseudotyped replication-defective retroviral vector into monkeys' embryo to facilitate DNA transfer, since retrovirus had been shown to assist integration of exogenous DNA with the genome (115). 224 mature monkeys' oocytes were injected with retrovirus, which overexpressed the GFP gene under the control of the CMV promoter. Yet, among the three offspring produced, only one is qualified to be transgenic. Although germline transmission and the presence of transgenic sperm were not attested (115), this study demonstrated that it is feasible to produce transgenic NHPs with retroviral vector injection. Germline transmission was achieved with common marmoset (Callithrix jacchus) by Sasaki et al. in 2009 (116). Transgenic marmosets with expressed EGFP gene were produced by self-inactivating lentiviral vectors, which were constructed based on human immunodeficiency virus type 1 (HIV-1) and contained promotor of CAG and CMV. The feasibility of constructed transgenic NHP model was confirmed by the production of stable transgenic offspring. However, the production efficiency of transgenic NHP remained low (114); it is therefore essential to optimize monkey's transgenesis productivity. Niu's group looked for the optimized virus type and injection timing (117). researchers used simian immunodeficiency virus-based vector and showed that using 1-cell stage or 8-cell stage transgenic embryos yielded no significant differences for virus injection timing. Although the multi-cells stage embryo was more resistant in virus injection, injection to multi-stage embryos is more likely to generate chimeirc newborns than fully transgenic primates. In another study, researchers improved the methodology by injecting lentivirus before fertilization (118). Under the control of CAG promoter, GFP gene is expressed in the entire embryo, including amnion and fibroblast.

Hitherto, although the development of transgenic NHP model is limited by primate pregnancy, sophisticated micro-injecting, and transplant surgery, the advancement of transgenic NHP engineering techniques can facilitate the development of NHP model and contribute to further understanding of human biology.

10.5.2 THE APPLICATION OF TRANSGENIC NHPS

The most direct goal of developing transgenic NHP is to produce disease models that can be applied to humans as closely as possible, particularly those of neural diseases i.e., Huntington's disease (HD), Parkinson's disease (PD), Alzheimer's disease (AD), and Autism Spectrum Disorder (ASD). The first NHP disease model involved Huntington's disease (HD) (119). Researchers constructed a lentivirus vector that express exon1 of the human HTT gene, which contained a common HD patients' mutant HTT genome- the 84 CAG repeats. The two live transgenic newborns showed typical clinical features of HD, including dystonia, chorea, and motor deficiency. A high level of mutant HTT aggregation in the striatum, cerebral cortex, hippocampus, and cerebellum was found in both HD monkeys and HD rodent model (120). HD monkeys' germline transmission was accomplished in 2015 (121). These results suggest promising prospect for the replacement of transgenic rodents by NHP, whose cognition and behaviour are more like those of humans. With the usage of similar strategy, the PD NHP model (122) and ASD NHP model (123) were successfully generated in 2015 and 2016 respectively. Methyl-CpG binding protein 2 (MeCP2) transgenic cynomolgus monkeys were produced by using lentiviral vector controlled by human Synapsin promotor; the MeCP2 was confirmed in monkeys' brain tissue and these monkeys showed autism-like characteristics, including high frequency of repetitive circular locomotion, anxiety, low social interactions, and mild impairment of cognition. Overexpressed MeCP2 germline transmission was also reported in this study. The NHPs transgenic techniques would continue to progress and may become as mature as transgenic engineering techniques in rodents.

Besides the construction of the disease model, the transgenic engineering technique can also serve as an exploitation tool for neurobiology. GCaMP, a genetically encoded calcium indicator fused with GFP, can monitor the neuronal activity in the brain of a living animal via the optical technique. Park's group utilized a high tier

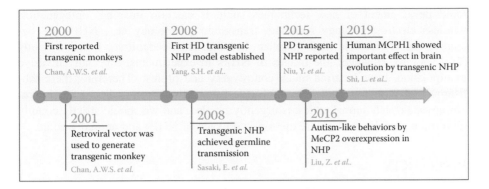

FIGURE 10.3   The development and application history of transgenic NHPs.

lentiviral vector to generate the GCaMP transgenic marmosets under the control of the CMV promotor and Synapsin promotor (124). In additions, Tomioka's group created a tetracyclin-inducible transgene expression (tet-on) system and achieved target exogenous gene conditional control (125). The ever-expanding development of optogenetics and chemogenetics techniques enbales more precise and deeper understanding of the human nerve system by transgenic NHP model.

More recently, Chinese researchers demonstrated that rhesus monkeys carrying the human MCPH1 gene showed a delayed neural maturation similar to human brain development trajectory (126). No difference is observed between transgenic and wild type monkeys in self-injury, stereotype, feeding, locomotion, resting, wake, and sleep; the only difference lies at short-term memory test, with the transgenic monkeys showing higher scores. This result suggest that a single human-specific molecular genetic mutation could play an important role in human evolution (127).

NHP genome editing techniques are still developing. Recently, precise exogenous gene knock-in NHPs via CPRISR/Cas9 128,129 has been tested in recent study (128,129). This technique can overcome the limitation of lentiviral vector-mediated gene transferring at random insertion and uncontrollable expression levels (130). More conditional expression lines, i.e., Cre lines, can be generated in the near future with the establishment of Tet-on system. These versatile NHP models demonstrate huge potential for providing researchers with further insight into the mechanisms of human development and human diseases125 (Figure 10.3).

10.6 CONCLUSION

Suitable animal model is important for the understanding of human brain function and diseases. Given their phylogenetic proximity to humans, NHPs play a critical role in high-level cognitive function research. Some national science projects, such as the Japanese National Brain Project and China Brain Project, adopt NHPs as their major experimental animal to explore the neural mechanism of brain functions and to pave the way for neurological disease therapies (1,131). Nowadays, with the rapid development of new techniques such as calcium imaging, optogenetics, wireless electrophysiology recoding, transgenic technology etc., NHPs are becoming more feasible and popular in high-level cognitive function study (21,132–136). However, the usages of NHPs in experiments, especially invasive manipulation, have incurred much controversy and protest. Therefore, the establishment of common guidelines of NHPs in neuroscience, in line with the requirements of lab animal ethics, is urgently needed, and we should strictly comply with these guidelines to ensure appropriate usage of NHPs and animal welfare.

REFERENCES

1. M. M. Poo *et al.*, China Brain Project: basic neuroscience, brain diseases, and brain-inspired computing. *Neuron* **92**, 591–596 (2016).
2. J. J. Nassi, E. M. Callaway, Parallel processing strategies of the primate visual system. *Nature Reviews Neuroscience* **10**, 360–372 (2009).
3. L. G. Ungerleider, Two cortical visual systems. In *Analysis of Visual Behavior*, Cambridge, MA: MIT Press, 549–586 (1982).
4. M. A. Goodale, A. D. Milner, Separate visual pathways for perception and action. *Trends in Neurosciences* **15**, 20–25 (1992).
5. E. Freud, D. C. Plaut, M. Behrmann, 'What' is happening in the dorsal visual pathway. *Trends in Cognitive Sciences* **20**, 773–784 (2016).
6. I. C. Van Dromme, E. Premereur, B. E. Verhoef, W. Vanduffel, P. Janssen, Posterior parietal cortex drives inferotemporal activations during three-dimensional object vision. *PLoS Biology* **14**, e1002445 (2016).
7. H. Klüver, P. C. Bucy, Preliminary analysis of functions of the temporal lobes in monkeys. *Archives of Neurology & Psychiatry* **42**, 979–1000 (1939).
8. S. Brown, E. A. Sharpey-Schafer, XI. An investigation into the functions of the occipital and temporal lobes of the monkey's brain. *Philosophical Transactions of the Royal Society of London B*, **179**, 303–327 (1888).
9. C. G. Gross, D. B. Bender, C. D. Rocha-Miranda, Visual receptive fields of neurons in inferotemporal cortex of the monkey. *Science* **166**, 1303–1306 (1969).
10. C. G. Gross, C. D. Rocha-Miranda, D. Bender, Visual properties of neurons in inferotemporal cortex of the macaque. *Journal of Neurophysiology* **35**, 96–111 (1972).
11. D. I. Perrett, E. T. Rolls, W. Caan, Visual neurones responsive to faces in the monkey temporal cortex. *Experimental Brain Research* **47**, 329–342 (1982).
12. R. Desimone, T. D. Albright, C. G. Gross, C. Bruce, Stimulus-selective properties of inferior temporal neurons in the macaque. *Journal of Neuroscience* **4**, 2051–2062 (1984).
13. W. A. Freiwald, The neural mechanisms of face processing: cells, areas, networks, and models. *Current Opinion in Neurobiology* **60**, 184–191 (2020).
14. D. Y. Tsao, W. A. Freiwald, R. B. Tootell, M. S. Livingstone, A cortical region consisting entirely of face-selective cells. *Science* **311**, 670–674 (2006).
15. D. Y. Tsao, S. Moeller, W. A. Freiwald, Comparing face patch systems in macaques and humans. *Proceedings of the National Academy of Sciences of the United States of America* **105**, 19514–19519 (2008).
16. W. A. Freiwald, D. Y. Tsao, M. S. Livingstone, A face feature space in the macaque temporal lobe. *Nature Neuroscience* **12**, 1187–1196 (2009).
17. L. Chang, D. Y. Tsao, The code for facial identity in the primate brain. *Cell* **169**, 1013–1028.e1014 (2017).
18. C.-C. Hung *et al.*, Functional mapping of face-selective regions in the extrastriate visual cortex of the marmoset. *Journal of Neuroscience* **35**, 1160–1172 (2015).

19. J. F. Mitchell, D. A. Leopold, The marmoset monkey as a model for visual neuroscience. *Neuroscience Research* **93**, 20–46 (2015).
20. J. E. Park *et al.*, Generation of transgenic marmosets expressing genetically encoded calcium indicators. *Scientific Reports* **6**, 34931 (2016).
21. O. Sadakane *et al.*, Long-term two-photon calcium imaging of neuronal populations with subcellular resolution in adult non-human primates. *Cell Reports* **13**, 1989–1999 (2015).
22. P. Janssen, H. Scherberger, Visual guidance in control of grasping. *Annual Review of Neuroscience* **38**, 69–86 (2015).
23. A. Ledberg, S. L. Bressler, M. Ding, R. Coppola, R. Nakamura, Large-scale visuomotor integration in the cerebral cortex. *Cerebral Cortex* **17**, 44–62 (2007).
24. P. R. Roelfsema, A. K. Engel, P. König, W. Singer, Visuomotor integration is associated with zero time-lag synchronization among cortical areas. *Nature* **385**, 157–161 (1997).
25. L. Fogassi *et al.*, Cortical mechanism for the visual guidance of hand grasping movements in the monkey: a reversible inactivation study. *Brain* **124**, 571–586 (2001).
26. V. Gallese, A. Murata, M. Kaseda, N. Niki, H. Sakata, Deficit of hand preshaping after muscimol injection in monkey parietal cortex. *Neuroreport: An International Journal for the Rapid Communication of Research in Neuroscience* **5**, 1525–1529 (1994).
27. E. Borra *et al.*, Cortical connections of the macaque anterior intraparietal (AIP) area. *Cerebral Cortex* **18**, 1094–1111 (2008).
28. M. Gerbella, A. Belmalih, E. Borra, S. Rozzi, G. Luppino, Cortical connections of the anterior (F5a) subdivision of the macaque ventral premotor area F5. *Brain Structure and Function* **216**, 43–65 (2011).
29. J.-B. Durand *et al.*, Anterior regions of monkey parietal cortex process visual 3D shape. *Neuron* **55**, 493–505 (2007).
30. S. Srivastava, G. A. Orban, P. A. De Mazière, P. Janssen, A distinct representation of three-dimensional shape in macaque anterior intraparietal area: fast, metric, and coarse. *Journal of Neuroscience* **29**, 10613–10626 (2009).
31. T. Theys, P. Pani, J. van Loon, J. Goffin, P. Janssen, Selectivity for three-dimensional shape and grasping-related activity in the macaque ventral premotor cortex. *Journal of Neuroscience* **32**, 12038–12050 (2012).
32. M. A. Baumann, M.-C. Fluet, H. Scherberger, Context-specific grasp movement representation in the macaque anterior intraparietal area. *Journal of Neuroscience* **29**, 6436–6448 (2009).
33. J. A. Michaels, H. Scherberger, Population coding of grasp and laterality-related information in the macaque fronto-parietal network. *Scientific Reports* **8**, 1–15 (2018).
34. S. E. R. Egnor, M. D. Hauser, A paradox in the evolution of primate vocal learning. *Trends in Neurosciences* **27**, 649–654 (2004).
35. H.-C. Chen, G. Kaplan, L. J. Rogers, Contact calls of common marmosets (*Callithrix jacchus*): influence of age of caller on antiphonal calling and other vocal responses. *American Journal of Primatology* **71**, 165–170 (2009).
36. S. J. Eliades, C. T. Miller, Marmoset vocal communication: behavior and neurobiology: Marmoset vocal communication. *Developmental Neurobiology* **77**, 286–299 (2017).
37. A. Saito, The marmoset as a model for the study of primate parental behavior. *Neuroscience Research* **93**, 99–109 (2015).
38. G. Epple, Comparative studies on vocalization in marmoset monkeys (Hapalidae). *Folia Primatologica*, **8**, 1–40 (1968).
39. D. A. Mills, C. P. Windle, H. F. Baker, R. M. Ridley, Analysis of infant carrying in large, well-established family groups of captive marmosets (*Callithrix jacchus*). *Primates* **45**, 259–265 (2004).

40. J. A. Agamaite, C.-J. Chang, M. S. Osmanski, X. Wang, A quantitative acoustic analysis of the vocal repertoire of the common marmoset (*Callithrix jacchus*). *The Journal of the Acoustical Society of America* **138**, 2906–2928 (2015).
41. J. L. Norcross, J. D. Newman, Context and gender-specific differences in the acoustic structure of common marmoset (*Callithrix jacchus*) phee calls. *American Journal of Primatology* **30**, 37–54 (1993).
42. J. Huang *et al.*, Having infants in the family group promotes altruistic behavior of marmoset monkeys. *Current Biology* **30**, 4047–4055.e4043 (2020).
43. J. L. Norcross, J. D. Newman, Social context affects phee call production by non-reproductive common marmosets (*Callithrix jacchus*). *American Journal of Primatology*, **43**, 135–146 (1997).
44. C. T. Miller, K. Mandel, X. Wang, The communicative content of the common marmoset phee call during antiphonal calling. *American Journal of Primatology* **72**, 974–980 (2010).
45. C. P. Chow, J. F. Mitchell, C. T. Miller, Vocal turn-taking in a non-human primate is learned during ontogeny. *Proceedings of the Royal Society B: Biological Sciences* **282**, 20150069 (2015).
46. S. J. Eliades, X. Wang, Neural substrates of vocalization feedback monitoring in primate auditory cortex. *Nature* **453**, 1102–1106 (2008).
47. L. A. de la Mothe, S. Blumell, Y. Kajikawa, T. A. Hackett, Cortical connections of auditory cortex in marmoset monkeys: lateral belt and parabelt regions. *The Anatomical Record: Advances in Integrative Anatomy and Evolutionary Biology* **295**, 800–821 (2012).
48. J. H. Kaas, T. A. Hackett, Subdivisions of auditory cortex and processing streams in primates. *Proceedings of the National Academy of Sciences of the United States of America* **97**, 11793–11799 (2000).
49. J. H. Kaas, T. A. Hackett, Subdivisions of auditory cortex and levels of processing in primates. *Audiology and Neurotology* **3**, 73–85 (1998).
50. T. A. Chaplin, H.-H. Yu, J. G. M. Soares, R. Gattass, M. G. P. Rosa, A conserved pattern of differential expansion of cortical areas in simian primates. *Journal of Neuroscience* **33**, 15120–15125 (2013).
51. B. E. Butler, S. G. Lomber, Functional and structural changes throughout the auditory system following congenital and early-onset deafness: implications for hearing restoration. *Frontiers in Systems Neuroscience* **7**, (2013).
52. F. Aboitiz, A brain for speech. Evolutionary continuity in primate and human auditory-vocal processing. *Frontiers in Neuroscience* **12**, 174 (2018).
53. E. L. Bartlett, S. Sadagopan, X. Wang, Fine frequency tuning in monkey auditory cortex and thalamus. *Journal of Neurophysiology* **106**, 849–859 (2011).
54. L. Bouffaut, S. Madhusudhana, V. Labat, A.-O. Boudraa, H. Klinck, A performance comparison of tonal detectors for low-frequency vocalizations of Antarctic blue whales. *The Journal of the Acoustical Society of America* **147**, 260–266 (2020).
55. T. Tani *et al.*, Sound frequency representation in the auditory cortex of the common marmoset visualized using optical intrinsic signal imaging. *eneuro* **5**, ENEURO.0078-0018.2018 (2018).
56. X. Wang, T. Lu, R. K. Snider, L. Liang, Sustained firing in auditory cortex evoked by preferred stimuli. *Nature* **435**, 341–346 (2005).
57. S. Sadagopan, X. Wang, Nonlinear spectrotemporal interactions underlying selectivity for complex sounds in auditory cortex. *Journal of Neuroscience* **29**, 11192–11202 (2009).
58. B. Philibert *et al.*, Functional organization and hemispheric comparison of primary auditory cortex in the common marmoset (*Callithrix jacchus*). *The Journal of Comparative Neurology* **487**, 391–406 (2005).

59. S. Sadagopan, X. Wang, Level invariant representation of sounds by populations of neurons in primary auditory cortex. *Journal of Neuroscience* **28**, 3415–3426 (2008).
60. G. H. Recanzone, D. C. Guard, M. L. Phan, Frequency and intensity response properties of single neurons in the auditory cortex of the behaving macaque monkey. *Journal of Neurophysiology* **83**, 2315–2331 (2000).
61. L. Feng, X. Wang, Harmonic template neurons in primate auditory cortex underlying complex sound processing. *Proceedings of the National Academy of Sciences of the United States of America* **114**, E840–E848 (2017).
62. M. S. Osmanski, X. Song, X. Wang, The role of harmonic resolvability in pitch perception in a vocal nonhuman primate, the common marmoset (*Callithrix jacchus*). *Journal of Neuroscience* **33**, 9161–9168 (2013).
63. M. S. Osmanski, X. Wang, Measurement of absolute auditory thresholds in the common marmoset (*Callithrix jacchus*). *Hearing Research* **277**, 127–133 (2011).
64. D. Bendor, X. Wang, The neuronal representation of pitch in primate auditory cortex. *Nature* **436**, 1161–1165 (2005).
65. M. Belyk, P. Q. Pfordresher, M. Liotti, S. Brown, The neural basis of vocal pitch imitation in humans. *Journal of Cognitive Neuroscience* **28**, 621–635 (2016).
66. M. J. Tramo, Neurophysiology and neuroanatomy of pitch perception: auditory cortex. *Annals of the New York Academy of Sciences* **1060**, 148–174 (2005).
67. M. Li, F. Liu, H. Jiang, T. S. Lee, S. Tang, Long-term two-photon imaging in awake Macaque monkey. *Neuron* **93**, 1049–1057.e1043 (2017).
68. H.-H. Zeng *et al.*, Local homogeneity of tonotopic organization in the primary auditory cortex of marmosets. *Proceedings of the National Academy of Sciences of the United States of America* **116**, 3239–3244 (2019).
69. A. Watakabe *et al.*, Application of viral vectors to the study of neural connectivities and neural circuits in the marmoset brain: application of virus vectors to marmosets. *Developmental Neurobiology* **77**, 354–372 (2017).
70. S. Iwano *et al.*, Single-cell bioluminescence imaging of deep tissue in freely moving animals. *Science* **359**, 935–939 (2018).
71. Digby L. J., Barreto C. E., Social organization in a wild population of *Callithrix jacchus*. *Folia Primatologica* **61**, 123–134 (1993).
72. M. T. da Silva Mota, C. R. Franci, M. B. C. de Sousa, Hormonal changes related to paternal and alloparental care in common marmosets (*Callithrix jacchus*). *Hormones and Behavior* **49**, 293–302 (2006).
73. G. R. Brown *et al.*, Adult–infant food transfer in common marmosets: an experimental study. *American Journal of Primatology*, **65**, 301–312 (2005).
74. J. M. Burkart, C. P. V. Schaik, Marmoset prosociality is intentional. *Animal Cognition* **23**, (2020).
75. R. K. Brügger, T. Kappeler-Schmalzriedt, J. M. Burkart, Reverse audience effects on helping in cooperatively breeding marmoset monkeys. *Biology Letters* **14**, 20180030 (2018).
76. J. M. Burkart, E. Fehr, C. Efferson, C. P. V. Schaik, Other-regarding preferences in a nonhuman primate: common marmosets provision food altruistically. *Proceedings of the National Academy of Sciences of the United States of America* **104**, 1962–19766 (2007).
77. J. H. Taylor, J. A. French, Oxytocin and vasopressin enhance responsiveness to infant stimuli in adult marmosets. *Hormones and Behavior* **75**, 154–159 (2015).
78. C. Finkenwirth, E. Martins, T. Deschner, J. M. Burkart, Oxytocin is associated with infant-care behavior and motivation in cooperatively breeding marmoset monkeys. *Hormones and Behavior* **80**, 10–18 (2016).
79. A. Saito, K. Nakamura, Oxytocin changes primate paternal tolerance to offspring in food transfer. *Journal of Comparative Physiology A* **197**, 329–337 (2011).

80. A. C. Mustoe, J. Cavanaugh, A. M. Harnisch, B. E. Thompson, J. A. French, Do marmosets care to share? Oxytocin treatment reduces prosocial behavior toward strangers. *Hormones and Behavior* **71**, 83–90 (2015).

81. J. Cavanaugh, A. C. Mustoe, J. H. Taylor, J. A. French, Oxytocin facilitates fidelity in well-established marmoset pairs by reducing sociosexual behavior toward opposite-sex strangers. *Psychoneuroendocrinology* **49**, 1–10 (2014).

82. R. L. Roberts *et al.*, Prolactin levels are elevated after infant carrying in parentally inexperienced common marmosets. *Physiology & Behavior* **72**, 713–720 (2001).

83. G. G. Gallup, Chimpanzees: self-recognition. *Science* **167**, 86–87 (1970).

84. S. D. Suarez, G. Gallupjr, Self-recognition in chimpanzees and orangutans, but not gorillas. *Journal of Human Evolution* **10**, 175–188 (1981).

85. V. Walraven, L. V. Elsacker, R. Verheyen, Reactions of a group of pygmy chimpanzees (*Pan paniscus*) to their mirror-images: evidence of self-recognition. *Primates* **36**, 145–150 (1995).

86. D. J. Povinelli, J. Gorodon, G. Gallup, T. J. Eddy, Chimpanzees recognize themselves in mirrors. *Animal Behaviour* **53**, 1083–1088 (1997).

87. J. R. Anderson, G. G. Gallup, Mirror self-recognition: a review and critique of attempts to promote and engineer self-recognition in primates. *Primates; Journal of Primatology* **56**, 317–326 (2015).

88. T. S. S. Schilhab, What mirror self-recognition in nonhumans can tell us about aspects of self. *Biology & Philosophy* **19**, 111–126 (2004).

89. C. M. Heyes, Theory of mind in nonhuman primates. *Behavioral & Brain Sciences* **21**, 101–114 (1998).

90. K. B. Swartz, D. Sarauw, S. Evans, Comparative aspects of mirror self-recognition in great apes. In *Mentalities of Gorillas & Orangutans:* Comparative perspectives, Cambridge University Press (1999).

91. S. Posada, M. Colell, Another gorilla (*Gorilla gorilla*) recognizes himself in a mirror. *American Journal of Primatology* **69**, 576–583 (2010).

92. S. Macellini, P. F. Ferrari, L. Bonini, L. Fogassi, A. Paukner, A modified mark test for own-body recognition in pig-tailed macaques (*Macaca nemestrina*). *Animal cognition* **13**, 631–639 (2010).

93. P. G. Roma *et al.*, Mark Tests for mirror self-recognition in capuchin monkeys (*Cebus apella*) trained to touch marks. *American Journal of Primatology*, **69**, 989–1000 (2007).

94. A. Heschl, J. Burkart, A new mark test for mirror self-recognition in non-human primates. *Primates* **47**, 187–198 (2006).

95. J. R. Anderson, Mirror-mediated finding of hidden food by monkeys (*Macaca tonkeana* and *M. fascicularis*). *Journal of Comparative Psychology* **100**, 237–242 (1986).

96. A. Z. Rajala, K. R. Reininger, K. M. Lancaster, L. C. Populin, L. Jan, Rhesus Monkeys (*Macaca mulatta*) do recognize themselves in the mirror: implications for the evolution of self-recognition. *PLoS One* **5**, (2010).

97. L. Chang, Q. Fang, S. Zhang, M. M. Poo, N. Gong, Mirror-induced self-directed behaviors in rhesus monkeys after visual-somatosensory training. *Current Biology* **25**, 212–217 (2015).

98. L. Chang, S. Zhang, M. M. Poo, N. Gong, Spontaneous expression of mirror self-recognition in monkeys after learning precise visual-proprioceptive association for mirror images. *Proceedings of the National Academy of Sciences of the United States of America* **114**, 3258–3263 (2017).

99. S. Itakura, Mirror guided behavior in Japanese monkeys (*Macaca fuscata fuscata*). *Primates* **28**, 149–161 (1987).

100. O. Blanke, M. Slater, A. Serino, Behavioral, neural, and computational principles of bodily self-consciousness. *Neuron* **88**, 145–166 (2015).

101. M. S. A. Graziano, Where is my arm? The relative role of vision and proprioception in the neuronal representation of limb position. *Proceedings of the National Academy of Sciences of the United States of America*, **96**, 10418–10421 (1999).
102. M. S. Graziano *et al.*, Coding the location of the arm by sight. *Science* **290**, 1782–1786 (2000).
103. M. S. A. Graziano, D. F. Cooke, Parieto-frontal interactions, personal space, and defensive behavior. *Neuropsychologia* **44**, 2621–2635 (2006).
104. W. Fang *et al.*, From the cover: statistical inference of body representation in the macaque brain. *Proceedings of the National Academy of Sciences of the United States of America* **116**, 20151–20157 (2019)
105. B. Amsterdam, Mirror self-image reactions before age two. *Developmental Psychobiology* **5**, 297–305 (2010).
106. T. Broesch, T. Callaghan, J. Henrich, C. Murphy, P. Rochat, Cultural variations in children's mirror self-recognition. *Journal of Cross-Cultural Psychology* **42**, 1018–1029 (2011).
107. G. G. Gallup, M. K. Mcclure, S. D. Hill, R. A. Bundy, Capacity for self-recognition in differentially reared chimpanzees. *Psychological Record* **21**, 69–74 (1971).
108. M. E. Rogers *et al.*, Gorilla diet in the lope reserve, gabon. *Oecologia* **84**, 326–339 (1990).
109. F. G. Patterson, R. H. Cohn, *Self-recognition and self-awareness in lowland gorillas.* (1994), Cambridge University Press.
110. D. Reiss, L. Marino, Mirror self-recognition in the bottlenose dolphin: a case of cognitive convergence. *Proceedings of the National Academy of Sciences of the United States of America* **98**, 5937–5942 (2001).
111. J. Plotnik, F. De Waal, D. Reiss, Self-recognition in an Asian elephant. *Proceedings of the National Academy of Sciences of the United States of America* **103**, 17053–17057 (2006).
112. H. Prior, A. Schwarz, O. Güntürkün, Mirror-induced behavior in the magpie (*Pica pica*): evidence of self-recognition. *PLoS Biology* **6**, e202 (2008).
113. J. C. Izpisua Belmonte *et al.*, Brains, genes, and primates, *Neuron* **86**, 617–631 (2015).
114. A. W. S. Chan *et al.*, Foreign DNA transmission by ICSI: injection of spermatozoa bound with exogenous DNA results in embryonic GFP expression and live Rhesus monkey births. *Molecular Human Reproduction* **6**, 26–33 (2000).
115. A. W. S. Chan, K. Y. Chong, C. Martinovich, C. Simerly, G. Schatten, Transgenic monkeys produced by retroviral gene transfer into mature oocytes. *Science* **291**, 309–312 (2001).
116. E. Sasaki *et al.*, Generation of transgenic non-human primates with germline transmission. *Nature* **459**, 523–527 (2009).
117. Y. Niu *et al.*, Transgenic rhesus monkeys produced by gene transfer into early-cleavage-stage embryos using a simian immunodeficiency virus-based vector. *Proceedings of the National Academy of Sciences of the United States of America* **107**, 17663–17667 (2010).
118. Y. Seita *et al.*, Generation of transgenic cynomolgus monkeys that express green fluorescent protein throughout the whole body. *Scientific Reports* **6**, 24868 (2016).
119. S. H. Yang *et al.*, Towards a transgenic model of Huntington's disease in a non-human primate. *Nature* **453**, 921–924 (2008).
120. B. R. Snyder, A. W. S. Chan, Progress in developing transgenic monkey model for Huntington's disease. *Journal of Neural Transmission (Vienna)* **125**, 401–417 (2018).
121. S. Moran *et al.*, Germline transmission in transgenic Huntington's disease monkeys. *Theriogenology* **84**, 277–285 (2015).
122. Y. Niu *et al.*, Early Parkinson's disease symptoms in alpha-synuclein transgenic monkeys. *Human Molecular Genetics* **24**, 2308–2317 (2015).
123. Z. Liu *et al.*, Autism-like behaviours and germline transmission in transgenic monkeys overexpressing MeCP2. *Nature* **530**, 98–102 (2016).

124. J. E. Park *et al.*, Generation of transgenic marmosets expressing genetically encoded calcium indicators. *Scientific Reports* **6**, 34931 (2016).
125. I. Tomioka *et al.*, Generation of transgenic marmosets using a tetracyclin-inducible transgene expression system as a neurodegenerative disease model. *Biology of Reproduction* **97**, 772–780 (2017).
126. L. Shi *et al.*, Transgenic rhesus monkeys carrying the human MCPH1 gene copies show human-like neoteny of brain development. *National Science Review* **6**, 480–493 (2019).
127. L. Shi, B. Su, A transgenic monkey model for the study of human brain evolution. *Zoological Research* **40**, 236–238 (2019).
128. Y. Cui *et al.*, Generation of a precise Oct4-hrGFP knockin cynomolgus monkey model via CRISPR/Cas9-assisted homologous recombination. *Cell Research* **28**, 383–386 (2018).
129. X. Yao *et al.*, Generation of knock-in cynomolgus monkey via CRISPR/Cas9 editing. *Cell Research* **28**, 379–382 (2018).
130. H. Okano, N. Kishi, Investigation of brain science and neurological/psychiatric disorders using genetically modified non-human primates. *Current Opinion in Neurobiology* **50**, 1–6 (2018).
131. H. Okano *et al.*, Brain/MINDS: a Japanese National Brain Project for Marmoset Neuroscience. *Neuron* **92**, 582–590, (2016).
132. M. Berger, N. S. Agha, A. Gail, Wireless recording from unrestrained monkeys reveals motor goal encoding beyond immediate reach in frontoparietal cortex. *Elife* **9**, e51322 (2020).
133. M. M. Chernov, R. M. Friedman, G. Chen, G. R. Stoner, A. W. Roe, Functionally specific optogenetic modulation in primate visual cortex. *Proceedings of the National Academy of Sciences of the United States of America* **115**, 10505–10510 (2018).
134. D. A. Schwarz *et al.*, Chronic, wireless recordings of large-scale brain activity in freely moving rhesus monkeys. *Nature Methods* **11**, 670–676 (2014).
135. H. Watanabe *et al.*, Forelimb movements evoked by optogenetic stimulation of the macaque motor cortex. *Nature Communications* **11**, 1–9 (2020).
136. L. Xin, L. Min, S. Bing, Application of the genome editing tool CRISPR/Cas9 in non-human primates. *Zoological Research* **37**, 241 (2016).

11 Application of In Vivo Ca^{2+} Imaging in the Pathological Study of Autism Spectrum Disorders

Pan Xu, Yuanlei Yue, and Hui Lu

CONTENTS

11.1 THE BIOLOGY OF CA^{2+}

Ca^{2+} is an important second messenger for neurotransmitter release, membrane depolarization, and homeostasis of all cells in the brain. In neurons, action potentials lead to calcium transients in the cytosol through voltage-gated calcium channels, and this rise is reversed as calcium is buffered, extruded, and pumped back into internal stores (Koester and Sakmann, 2000). While neurons are at rest, the concentration of the free calcium ion is low (about 50–100 nM) in the cytoplasm but much higher (about 1000 nM) in extracellular fluid. Organelles, such as mitochondria and endoplasmic reticulum, also contain a large number of Ca^{2+} ions. When neurons are active, voltage-gated calcium ion channels are activated, allowing calcium ions to flux into the cell. The intracellular Ca^{2+} ions stored in organelles are also released to the cytoplasm. Thus, the concentration of the free calcium ions in the cytoplasm increases about tenfold during action compared to resting (Figure 11.1).

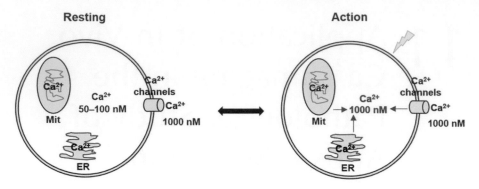

FIGURE 11.1   The concentration of Ca^{2+} ions in the cytoplasm sharply rises during action potential firing.

11.2   GENETICALLY ENCODED CA^{2+} INDICATORS (GECIS)

Since neuronal activities involve the fluctuation of intracellular Ca^{2+} concentration, such fluctuation can carry information on neuronal activities that is closely correlated with signal processing in the brain. To date, multiple types of fluorescent calcium probes have been developed (Tsien, 1988). These fluorescent probes can either change their fluorescence intensity or wavelength of their emission light in response to the change of the Ca^{2+} concentration. Thus, neuronal activity can be recorded by detecting the change of the fluorescence with fluorescence microscopies, such as two-photon or one-photon microscopy (Stosiek et al., 2003).

In 2001, a genetically encoded calcium indicator (GECI) GCaMP, was first developed by Junichi Nakai, based on a green fluorescent protein (Nakai et al., 2001). It is composed of green fluorescent protain (GFP,) calmodulin (CaM) and M13 domains. Calcium binding with GCaMP results in a structural shift and brighter fluorescenceMiyawaki 1997. Now, many different versions of GECIs have been developed. Among them, the green and red Ca^{2+} indicators (GCaMP and RCaMP, respectively) are most commonly used. Without a calcium ion, GCaMP cannot be excited to emit light. However, when the calcium ion appears, the structure of the GCaMP can be changed because of the combining of Ca^{2+}. Then, it can be excited to emit green light (Figure 11.2). The repertoire and properties of GECI are constantly improving to further expand the palette of GECIs and to more precisely reflect complex neural dynamics. For example, a quadripolar GECI suite (XCaMP) and a near-infrared Ca^{2+} indicator were developed recently (Inoue et al., 2019; Qian et al., 2019).

Through transgenic or viral labeling, GECIs can be expressed in neurons, and are now widely used in the research of brain functions with several advantages. First, the activity of dozens, hundreds, or even thousands of neurons can be simultaneously imaged with sub-cellular resolution. Second, it could be combined with cell-type-specific genetic manipulations, fulfilling to record from specific cell populations. Third, it allows chronic imaging in freely moving animals over a period of weeks or months using photon based-microscope. Fourth, it makes neural activity imaging less invasive than traditional methods (e.g., tetrodes, silicon probes).

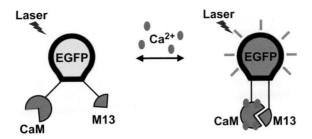

FIGURE 11.2 Green fluorescent protein can be excited to emit green fluorescence when combining with Ca^{2+} to change its structure.

Fifth, it could be used to manipulate network activity, by combining with other approaches, such as optogenetics, chemogenetics (Packer et al., 2013).

11.3 TWO-PHOTON CA^{2+} IMAGING

Along with the development of GECI, two-photon calcium imaging technology is widely used in characterizing neural dynamics within the intact, living brain. It can record the neuronal activity in the brain of head-fixed awake animals. Visual, ol-factory, and sensory stimulation can be applied while simultaneous imaging is performed. Motor behavior, such as paw, whisker, and tongue movement, can be recorded at the same time. Together, the correlation between sensory input, animal behavior, and neuronal activity of neural ensembles can be analyzed. With viral labeling technology in transgenic mouse lines, GECI can be labeled in specific cell types, providing the opportunity to characterize the neuronal activity of those specific cell populations. For instance, chronic two-photon imaging is used to monitor the activity of sparsely labeled substantia nigra-originated dopaminergic axons in the dorsal striatum, which reveals distinct temporal and spatial encoding of reward and motor signals (Howe and Dombeck, 2016).

Two-photon microscopy is a useful technique to study the circuit level me-chanism of neurological disorders, such as autism spectrum disorders (ASDs). For example, GCaMP6s was expressed in the primary visual cortical excitatory pyr-amidal cells and inhibitory parvalbumin (PV)-expressing interneurons, respectively, in a mouse model of Fragile X syndrome, one of the ASD mouse models (Nakai et al., 2001). These neurons' response to orientation visual stimulation is char-acterized in detail via in vivo two-photon Ca^{2+} imaging. This study showed that the percentage of orientation-selective pyramidal cells decreases in Fragile X syndrome mice and the visually evoked activity is reduced in their PV neurons, leading to orientation-tuning deficits. Another group recorded whisker touch-related activity of excitatory and inhibitory neuronal populations in layer 2/3 of the somatosensory cortex by two-photon calcium imaging (Michaelson et al., 2018). They found re-duced sensory responses in both excitatory and inhibitory neurons in mice with SYNGAP1 mutation, another ASD mouse model. Their findings suggest that SYNGAP1 mutation disrupts sensory processing within the somatosensory cortical circuits. Two-photon imaging is also used in ex vivo study with hippocampal slices of ASD mouse models with mutations in methyl-CpG-binding protein 2 (MECP2) (Lu et al., 2016). This study covers three mouse models of MECP2 disorders: constitutive Mecp2 null, mosaic Mecp2+/−, and MECP2 duplication mice. It discovers that the hippocampal circuits in these three lines of mice share three novel signs of hippocampal circuit dysfunction. First, the hippocampal CA1 pyramidal neurons tended strongly to fire in synchrony. Second, the circuit was not able to restore homeostasis after perturbations of excitatory-inhibitory balance. Third, the inhibitory neurons showed abnormally low excitatory synaptic responses. Thus, even in the context of very different transcriptional effects, the effect at the circuit level can be shared—perhaps not just between Rett and MECP2 duplication syndrome, but among other intellectual disability disorders.

Two-photon microscopy is not only used to record the activity of neurons but also the Ca^{2+} fluctuation in glial cells of ASD models. Calcium activities in astrocytes were measured in a mouse model of Rett syndrome, an autism-associated neurodevelopmental disorder (Dong et al., 2018). The researchers found that the frequency of spontaneous intracellular calcium activity in the astrocytes of mutant mice is higher compared with that of wild type mice. Both percentages of cells with calcium oscillation and the oscillation frequency increase in mutant astrocytes. Those findings reveal abnormal calcium homeostasis in the glial cells of the ASD model.

11.4   MINIATURE SINGLE-PHOTON FLUORESCENCE MICROSCOPE

To investigate how neurons process information and represent various behaviors, neuroscientists seek to characterize the neuronal activity in the brain of freely behaving animals, which enable them to correlate the neuronal activities with specific behaviors, sensory inputs, and cognitive processes. Although extracellular electrophysiological recordings have been used for such studies in the last several decades, which now could record data from large neuronal populations using multi-tetrode arrays or silicon probes in freely behaving rodents (Dong et al., 2018; Jun et al., 2017), the limitation is the lack of cell-type specificity.

With the utilization of GECIs, two-photon imaging allows the structural and functional imaging of large neuronal networks at cellular resolution (Denk and Svoboda, 1997; Dombeck et al., 2007, 2010). However, this imaging technique requires the animals to be head-fixed, thus limits the behavioral repertoire and the use of a large battery of sophisticated behavioral tests. Recently, owing to its relatively straightforward optical setup, a miniature single-photon fluorescence microscope (miniscope) has become the most widely used approach for imaging from freely behaving animals.

The early developed designs were optical fiber-based (Ferezou et al., 2006), which relay illumination and emission light from a regular fluorescence system. In a landmark work in 2011, Ghosh et al. reported a fully integrated miniature single-photon wide-field fluorescence microscope, with fluorescence excitation and detection combined onboard the microscope housing (Ghosh et al., 2011). Its components made use of relatively accessible technologies such as complementary metal-oxide-semiconductor (CMOS) imaging sensors, LEDs as light source, off-the-shelf optical components, and gradient refractive index (or GRIN) lenses, overcoming cost limitations of table-top lasers and expensive detectors. A typical design for such a miniaturized microscope is shown in Figure 11.3 (Aharoni and Hoogland, 2019). Its optical path is similar to that of a conventional, wide-field fluorescence microscope, except for the difference of using the GRIN lenses. GRIN lenses provide an optical interface to the brain, with its advantage of short working distance, flat bottom, and multiple ranges of lengths and diameters. For superficial neuron imaging, a single objective GRIN lens is placed directly on the brain surface. For deeper areas, an objective GRIN lens is combined with a small-diameter relay GRIN lens, which is implanted above the region of interest. Signals from individual neurons could be effectively extracted from the images recorded by the miniature single-photon microscope, using analytical techniques, including a combination of principal and independent component analysis (PCA/ICA) and more recently non-negative matrix factorization (CNMF) with an added term to model a time-varying fluorescence background signal (Lu et al., 2018; Zhou et al., 2018). Therefore, the miniaturized head-mounted fluorescence microscope enables cellular-resolution imaging in awake behaving animals, while leveraging the use of cheap components.

The innovation of the miniscope has been accelerated dramatically with open-source sharing of miniscope projects (Aharoni and Hoogland, 2019), such as UCLA miniscope, FinchScope of Boston University, and CHEndoscope of University of Toronto Princeton. Some functional improvements have been made recently to guide the new frontiers. Senarathna et al. reported an improved miniaturized

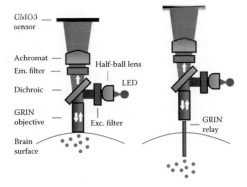

FIGURE 11.3 Miniaturized single-photon excitation microscope design, which is used in combination with gradient refractive index (GRIN) lenses. It is comprised of an excitation LED light source, half-ball lens light collimator, excitation filter (Exc. filter) and dichroic mirror for reflecting excitation light down toward the specimen and transmitting emitted fluorescence up to the imaging detector. Emitted light is focused onto a CMOS imaging sensor using an achromatic lens after passing through an emission filter (Em. filter). The use of GRIN lenses permits imaging from both superficial (left) or deep-lying (right) brain structures, when it is placed directly on the brain surface, or combined with a thinner relay GRIN lens that is implanted into the brain, respectively. Cells transduced with a fluorescent activity reporter (green dots) could be imaged. Adjustment of the focal plane in the specimen is achieved by moving the image sensor toward or away from the achromatic lens.

microscope, which could be reduced to 3 g with a suspension mechanism (Senarathna et al., 2019). The three-dimensional imaging of neuronal activity was achieved with light-field microscopy, by which the spatial intensity and the direction of light of different imaging depths could be extracted together from each frame with a computational de-mixing strategy (Prevedel et al., 2014). Besides, Barbera et al. reported a wireless model of the miniature microscope that contained a micro SD card for storage, and a battery backpack (Barbera et al., 2019). The wireless design not only resolves the problem of wire entanglement but also removes many other constraints on the design of behavioral paradigms. But its battery life limits the time of continuous recording. Notably, the miniscope has been integrated with optogenetic techniques to fulfill manipulating cellular activity while imaging (Stamatakis et al., 2018). Krabbe, et al. utilized this method to verify that that US-induced VIP activity promotes PN depolarization through disinhibition and thus ultimately facilitates associative learning (Krabbe et al., 2019).

11.5 THE APPLICATION OF MINIATURE SINGLE-PHOTON FLUORESCENCE MICROSCOPES IN NEUROSCIENCE RESEARCH

Over the past several years, single-photon microscope imaging of live animals has been widely adopted in neuroscience research. A large number of brain areas in freely behaving animals have been imaged to explore the brain functions including spatial navigation (Ghosh et al., 2011), complex motor tasks (Wang et al., 2017), memory formation and linking (Cai et al., 2016), and social behaviors (Murugan et al., 2017). For example, through recording calcium activities of the mPFC via miniscope while mice freely explored social targets, Bo Liang et al. identified distinct and dynamic ON and OFF neural ensembles that displayed opposing activities to code real-time social exploration, which carries information of salience and novelty for social targets (Liang et al., 2018). Kinsky et al. employed single-photon calcium imaging in freely moving mice to observe the trajectory-dependent activity through learning and mastery of a continuous spatial alternation task. They found that splitter neurons emerge rapidly, remain active, and retain consistent spatial information across multiple days, all of which support memory-guided behavior (Kinsky et al., 2020).

Recently, the miniature fluorescence microscope has also been used to uncover dysfunctional neural circuits underlying behavioral deficits of neurological disorders. Neurodegenerative diseases usually are characterized by the accumulation of distinct protein-based macroscopic deposits that cause neuronal loss. Meanwhile, synaptic dysfunction, altered calcium homeostasis, abnormal neural activity, and large-scale neural circuit impairment are also well-observed features of many ASDs (Werner et al., 2019). Given that growing evidence suggests that the neurological deficits in ASDs may reflect functional impairment in neural circuits rather than neuronal loss, the neural circuits research deserves substantial consideration (Seeley et al., 2009). Miniature microscope recording allows longitudinal studies on individual neurons in large-scale populations from subcortical brain regions in freely moving animals. ASDs have multiple behavioral phenotypes, including stereotyped movements, perseverant interests, impaired attention and executive function, and deficits of social interactions (Nakai et al., 2018). The circuit-centered research that can survey functional connectivity patterns in the neural circuit is more suitable for the study of ASDs. In vivo calcium imaging with miniscope is used to investigate the circuit alterations underlying atypical sensory processing in a mouse model of Fragile X syndrome, and proved that the neural circuitry abnormalities contribute to the tactile defensiveness and impaired visual discrimination in Fmr1 KO mice (Cantu, 2019). We employ this technique to Rett syndrome mouse models and demonstrate that MeCP2 mutation-caused hypoactivity leads to the impaired neural coding of mouse anxiety state and social stimuli discrimination by the mouse medial prefrontal cortical circuit. Optogenetic stimulation to restore this hypoactivity locally can rescue both the anxiety phenotype and social deficit of the Rett mouse model (work in Lu lab).

Overall, in vivo Ca^{2+} imaging with one- or two-photon microscopy is a fascinating technology to decipher the neuronal coding of behaviors in the intact brain. Its application in the study of ASD pathology provides unpreceded insights on the neural circuit-level mechanism underlying behavioral abnormalities. More and more evidence shows that many ASDs with different genetic mutations share common circuit-level deficits. Given that the presence of so many autism-associated genes brings dramatic challenges to ASD treatment with gene therapy, the neural circuit is of promising potential to be a common target for improving the behavioral deficits shared by many with ASD.

REFERENCES

Aharoni, D., and Hoogland, T. M. (2019). Circuit investigations with open-source miniaturized microscopes: past, present and future. Frontiers in Cellular Neuroscience *13*, 141.

Barbera, G., Liang, B., Zhang, L., Li, Y., and Lin, D.-T. (2019). A wireless miniScope for deep brain imaging in freely moving mice. Journal of Neuroscience Methods *323*, 56–60.

Cai, D. J., Aharoni, D., Shuman, T., Shobe, J., Biane, J., Song, W., Wei, B., Veshkini, M., La-Vu, M., and Lou, J. (2016). A shared neural ensemble links distinct contextual memories encoded close in time. Nature *534*, 115–118.

Cantu, D. A. (2019). Analysis of fluorescent calcium signals in the detection of neural circuitry abnormalities in a mouse model of autism. UCLA Electronic Theses and Dissertations.

Denk, W., and Svoboda, K. (1997). Photon upmanship: why multiphoton imaging is more than a gimmick. Neuron *18*, 351–357.

Dombeck, D. A., Harvey, C. D., Tian, L., Looger, L. L., and Tank, D. W. (2010). Functional imaging of hippocampal place cells at cellular resolution during virtual navigation. Nature Neuroscience *13*, 1433–1440.

Dombeck, D. A., Khabbaz, A. N., Collman, F., Adelman, T. L., and Tank, D. W. (2007). Imaging large-scale neural activity with cellular resolution in awake, mobile mice. Neuron *56*, 43–57.

Dong, Q., Liu, Q., Li, R., Wang, A., Bu, Q., Wang, K. H., and Chang, Q. (2018). Mechanism and consequence of abnormal calcium homeostasis in Rett syndrome astrocytes. eLife *7*, e33417.

Ferezou, I., Bolea, S., and Petersen, C. C. (2006). Visualizing the cortical representation of whisker touch: voltage-sensitive dye imaging in freely moving mice. Neuron *50*, 617–629.

Ghosh, K. K., Burns, L. D., Cocker, E. D., Nimmerjahn, A., Ziv, Y., Gamal, A. E., and Schnitzer, M. J. (2011). Miniaturized integration of a fluorescence microscope. Nature Methods *8*, 871–878.

Howe, M. W., and Dombeck, D. A. (2016). Rapid signalling in distinct dopaminergic axons during locomotion and reward. Nature *535*, 505–510.

Inoue, M., Takeuchi, A., Manita, S., Horigane, S.-i., Sakamoto, M., Kawakami, R., Yamaguchi, K., Otomo, K., Yokoyama, H., and Kim, R. (2019). Rational engineering of XCaMPs, a multicolor GECI suite for in vivo imaging of complex brain circuit dynamics. Cell *177*, 1346–1360.e1324.

Jun, J. J., Steinmetz, N. A., Siegle, J. H., Denman, D. J., Bauza, M., Barbarits, B., Lee, A. K., Anastassiou, C. A., Andrei, A., and Aydın, Ç. (2017). Fully integrated silicon probes for high-density recording of neural activity. Nature *551*, 232–236.

Kinsky, N. R., Mau, W., Sullivan, D. W., Levy, S. J., Ruesch, E. A., and Hasselmo, M. E. (2020). Trajectory-modulated hippocampal neurons persist throughout memory-guided navigation. Nature Communications *11*, 2443.

Koester, H. J., and Sakmann, B. (2000). Calcium dynamics associated with action potentials in single nerve terminals of pyramidal cells in layer 2/3 of the young rat neocortex. The Journal of Physiology *529*, 625–646.

Krabbe, S., Paradiso, E., d'Aquin, S., Bitterman, Y., Courtin, J., Xu, C., Yonehara, K., Markovic, M., Muller, C., Eichlisberger, T., et al. (2019). Adaptive disinhibitory gating by VIP interneurons permits associative learning. Nature Neuroscience *22*, 1834–1843.

Liang, B., Zhang, L., Barbera, G., Fang, W., Zhang, J., Chen, X., Chen, R., Li, Y., and Lin, D. T. (2018). Distinct and dynamic ON and OFF neural ensembles in the prefrontal cortex code social exploration. Neuron *100*, 700–714.

Lu, H., Ash, R. T., He, L., Kee, S. E., Wang, W., Yu, D., Hao, S., Meng, X., Ure, K., Ito-Ishida, A., et al. (2016). Loss and gain of MeCP2 cause similar hippocampal circuit dysfunction that is rescued by deep brain stimulation in a Rett Syndrome Mouse Model. Neuron *91*, 739–747.

Lu, J., Li, C., Singh-Alvarado, J., Zhou, Z. C., Fröhlich, F., Mooney, R., and Wang, F. (2018). MIN1PIPE: a miniscope 1-photon-based calcium imaging signal extraction pipeline. Cell Reports *23*, 3673–3684.

Michaelson, S. D., Ozkan, E. D., Aceti, M., Maity, S., Llamosas, N., Weldon, M., Mizrachi, E., Vaissiere, T., Gaffield, M. A., Christie, J. M., et al. (2018). SYNGAP1 heterozygosity disrupts sensory processing by reducing touch-related activity within somatosensory cortex circuits. Nature Neuroscience *21*, 1–13.

Miyawaki, A., Llopis, J., Heim, R., McCaffery, J. M., Adams, J. A., Ikura, M., and Tsien, R. Y. (1997). Fluorescent indicators for Ca^{2+} based on green fluorescent proteins and calmodulin. Nature *388*, 882–887.

Murugan, M., Jang, H. J., Park, M., Miller, E. M., Cox, J., Taliaferro, J. P., Parker, N. F., Bhave, V., Hur, H., and Liang, Y. (2017). Combined social and spatial coding in a descending projection from the prefrontal cortex. Cell *171*, 1663–1677.e1616.

Nakai, J., Ohkura, M., and Imoto, K. (2001). A high signal-to-noise Ca^{2+} probe composed of a single green fluorescent protein. Nature Biotechnology *19*, 137–141.

Nakai, N., Takumi, T., Nakai, J., and Sato, M. (2018). Common defects of spine dynamics and circuit function in neurodevelopmental disorders: a systematic review of findings from in vivo optical imaging of mouse models. Frontiers in Neuroscience *12*, 412.

Packer, A. M., Roska, B., and Hausser, M. (2013). Targeting neurons and photons for optogenetics. Nature Neuroscience *16*, 805–815.

Prevedel, R., Yoon, Y.-G., Hoffmann, M., Pak, N., Wetzstein, G., Kato, S., Schrödel, T., Raskar, R., Zimmer, M., Boyden, E. S., et al. (2014). Simultaneous whole-animal 3D imaging of neuronal activity using light-field microscopy. Nature Methods *11*, 727–730.

Qian, Y., Piatkevich, K. D., Mc Larney, B., Abdelfattah, A. S., Mehta, S., Murdock, M. H., Gottschalk, S., Molina, R. S., Zhang, W., and Chen, Y. (2019). A genetically encoded near-infrared fluorescent calcium ion indicator. Nature Methods *16*, 171–174.

Seeley, W. W., Crawford, R. K., Zhou, J., Miller, B. L., and Greicius, M. D. (2009). Neurodegenerative diseases target large-scale human brain networks. Neuron *62*, 42–52.

Senarathna, J., Yu, H., Deng, C., Zou, A. L., Issa, J. B., Hadjiabadi, D. H., Gil, S., Wang, Q., Tyler, B. M., Thakor, N. V., et al. (2019). A miniature multi-contrast microscope for functional imaging in freely behaving animals. Nature Communications *10*, 99.

Stamatakis, A. M., Schachter, M. J., Gulati, S., Zitelli, K. T., Malanowski, S., Tajik, A., Fritz, C., Trulson, M., and Otte, S. L. (2018). Simultaneous optogenetics and cellular resolution calcium imaging during active behavior using a miniaturized microscope. Frontiers in Neuroscience *12*, 496.

Stosiek, C., Garaschuk, O., Holthoff, K., and Konnerth, A. (2003). *In vivo* two-photon calcium imaging of neuronal networks. Proceedings of the National Academy of Sciences of the United States of America *100*, 7319.

Tsien, R. Y. (1988). Fluorescence measurement and photochemical manipulation of cytosolic free calcium. Trends in Neurosciences *11*, 419–424.

Wang, X., Liu, Y., Li, X., Zhang, Z., Yang, H., Zhang, Y., Williams, P. R., Alwahab, N. S., Kapur, K., and Yu, B. (2017). Deconstruction of corticospinal circuits for goal-directed motor skills. Cell *171*, 440–455.c414.

Werner, C. T., Williams, C. J., Fermelia, M. R., Lin, D. T., and Li, Y. (2019). Circuit mechanisms of neurodegenerative diseases: a new frontier with miniature fluorescence microscopy. Frontiers in Neuroscience *13*, 1174.

Zhou, P., Resendez, S. L., Rodriguez-Romaguera, J., Jimenez, J. C., Neufeld, S. Q., Giovannucci, A., Friedrich, J., Pnevmatikakis, E. A., Stuber, G. D., Hen, R., et al. (2018). Efficient and accurate extraction of in vivo calcium signals from microendoscopic video data. eLife *7*, e28728.

12 Nonclassical Ion Channels in Depression

Zahra Farzinpour and Zhi Zhang

CONTENTS

12.1 INTRODUCTION

There are diverse ion channels in the mammalian genome that conduct ions across the cell membrane, e.g. Na^+, K^+, Ca^{2+}, and Cl^-, in order to execute a wide range of biological functions such as neuronal activity, initiation of muscle contraction and relaxation, and generation of action potential and synaptic transmission. The human genome has been reported to encode at least 55 ligand-gated ion channels, 145 voltage-gated ion channels, and 27 so-called other channels, such as Piezo and store-operated channels[1]. The 2003 Nobel Prize in Chemistry was awarded to Peter Agre and Roderick MacKinnon for "the discovery of water channels" and "structural and mechanistic studies of ion channels," respectively, thereby illustrating the wide acknowledgment among biologists that study of ion channels is a topic of utmost importance.

In recent years, an accumulating body of evidence has shown that nonclassical ion channels, in particular, play a critical role in the regulation of the development of several emotional disorders, including depression and anxiety. In order to bring awareness to these specialized proteins in the quest to develop effective future treatment strategies, this chapter provides a brief overview of some nonclassical ion channels in the central nervous system (CNS), including hyperpolarization-activated cyclic nucleotide-gated (HCN) channels, acid-sensing ion (ASIC) channels, ligand-gated purinergic ion (P2XR) channels, transient receptor potential (TRP) channels, and finally the TWIK-related potassium channel-1 (TREK1), with special emphasis on their contribution to depression-associated behaviors.

12.2 NEUROPLASTICITY FOR DEPRESSION

The evidence of neuroplasticity in the CNS includes alterations in dendritic function, synapticremodeling, axonal sprouting, neurite extension, synaptogenesis, and neurogenesis. Intracellular signaling pathways coupled to ion channels adjust acute and long-term neuroplastic events that are involved in cognitive processes and complex psychological disorders. In particular, phenomena such as the loss of dendritic spines, dendritic atrophy, and loss of synapses in specific brain areas are all reportedly associated with depression[2]. Ketamine and lithium are two well-known mood stabilizers used for the treatment of mood disorders such as bipolar disorder and depression. Understanding the mechanism of action of these medications provides a chance to clarify the different manifestations of synaptic plasticity that can be targeted to alleviate symptoms of depression. Acute and chronic changes in the environment can induce adaptive (short-term synaptic plasticity) and homeostatic forms (long-term potentiation and long-term depression) of plasticity in neuronal networks. Genetic manipulation in animal models or pharmacological small molecule targeting of ion channels (i.e. blockers and activators) provide means to study homeostatic plasticity and its function in chronic mental disorders like depression. While ion channels have been established as functional components in synaptic plasticity, their dysfunction similarly entails the development or maintenance of depression.

12.3 HCN CHANNELS IN DEPRESSION

The HCN channel is primarily expressed in the brain, heart, and retina[3], and modulates neuronal excitability and activity through a hyperpolarization-activated current (I_h) consisting of Na^+ and K^+ cations[4,5]. HCN channels are voltage-gated and modulated by the endogenous ligand cyclic adenosine 3',5'-monophosphate (cAMP), the opening of which (typically at potentials below −50 mV) causes membrane depolarization and decreases membrane resistance. Neuronal network activity can be affected by HCN expression levels, its localization to subcellular compartments, and the composition of its subunits[6]. Given the role of I_h and HCN channels in controlling synaptic transmission and rhythmic oscillatory activity in the brain[7], disruption of HCN channel function can be involved in the development of depression or targeted as a therapeutic strategy.

Evidence has shown that abnormality in I_h is associated with depression-associated behaviors[8,9–10]. Experiments in HCN1, HCN2, and TRIP8b-null mice showed that reduced I_h accompanies depression-related behaviors, indicated by declines in immobility behavior in the forced swimming test (FST) and latency to immobility in the tail suspension test (TST)[11]. In genetic studies of HCN channels in patients suffering from depression, no correlation was observed between single-nucleotide polymorphisms in HCN channel genes and depression or stress sensitivity[12]. However, in rats, HCN knockdown using shRNA specifically in the CA1 region of the dorsal hippocampus alleviated depressive-like behaviors[13], whereas rats that overexpressed HCN channels in the CA1 displayed depressive-like behavior[14]. Hence, HCN channels represent viable therapeutic targets for the healing of affective disorders, since the inhibition of HCN1 activity in cortical regions and enhancement of HCN2 activity in subcortical structures may be advantageous for patients suffering from anxiety and depression[15].

HCN channels are also expressed in reward-involved brain regions, including the prefrontal cortex (PFC), hippocampus, nucleus accumbens (NAc), and ventral tegmental area (VTA)[16,17]. In cholinergic interneurons in the NAc of a chronic stress mouse model, HCN2 channels were found to be reduced in number, leading to a commensurate reduction in the neuronal tonic firing rate, accompanied by depressive-like behavior[10]. HCN2 overexpression in these neurons is sufficient to decrease the animals' depressive-related behaviors. In the same model, I_h was reduced in NAc–projecting dopaminergic neurons in the VTA, thus contributing to decreased neuronal firing, while in contrast, elevated HCN2 channel function prevented an increase in depressive-like behavior[9]. Interestingly, I_h up-regulation in NAc-projecting neurons of the VTA might be a major element in the development of natural resilience to stress[8]. In susceptible mice, HCN2 overexpression decreased the risk for stress-induced depressive-like behaviors. Although there is not enough evidence supporting the requirement of HCN channels in the development of depression, manipulation of either HCN expression or function is potentially an effective strategy for alleviating depressive symptoms.

Disruption of the I_h current in HCN1[13], HCN2, or TRIP8b[18] -transgenic mice was also shown to affect anxiety-like behaviors. Moreover, researchers identified a manifest anxiogenic effect in mice via hairpin RNA silencing of hcn4, indicating that HCN4 plays an essential role in anxiety-like behaviors[19]. The chemogenetic inhibition of neurons in dorsal bed nucleus of the stria terminalis (BNST) imitates aspects of α-2A adrenergic receptor signaling. These receptors generate HCN-dependent excitatory actions and induce anxiety-like behaviors[20], determined by the elevated zero maze (EZM) and open field (OF) tests. Two well-known I_h current potentiators, Gabapentin and lamotrigine (LTG), are used to treat different neurological disorders. Gabapentin is an effective treatment for anxiety, nightmares, and insomnia in individuals who have post-traumatic stress disorder (PTSD)[21,22]. Gabapentin is also used as an adjunctive treatment for the management of anxiety disorders, especially for social phobia and panic disorder[23,24–25]. Lamotrigine efficacy is limited to bipolar disorder treatment[26].

The pharmacological regulation of I_h currents or HCN channels (Table 12.1) represents a promising therapeutic approach for treatment of depression. Using ZD7288 as an HCN blocker inhibits neuronal firing activity in the PFC, while improving working memory[27]. Additionally, HCN1 in the hippocampus plays a vital role in the antidepressant properties of ketamine[28]. Interaction between ketamine and homodimers or heterodimers of HCN1 specifically inhibits I_h currents. This I_h inhibition could be involved in the pharmacological effects of ketamine for treatment of depression, since this antidepressant directly blocks HCN1 channels[29]. Unfortunately, the current lack of three-dimensional structure of the transmembrane core has hindered the design of high selectivity HCN channel blockers. In addition, the

TABLE 12.1

HCN Channels as Drug Targets

Activity	Medications	Mechanism of Action
Inhibitors	ZD7288	A selective HCN channel/I_h blocker through putative glutamate-receptor mediated signaling
	Propofol	HCN1 channel inhibitor by shifting the voltage-dependence to more
	Ketamine	hyperpolarized voltages and declining the peak current amplitude
	Cilobradine (DK-AH269)	An open channel blocker of neuronal I_h; a heart rate-reducing agent with off-target effects potentially used as a starting point for the development of novel pain-relieving medications via putative glutamate receptor-mediated signaling
	Zatebradine (UL-FS49)	A potent inhibitor of HCN channels; a heart rate-reducing agent
	Halothane	Through hyperpolarizing shift in the activation curve and a decrease
	Isoflurane	in maximal current amplitude
	Lidocaine	HCN channel inhibitor local anesthetic
	Loperamide	I_h inhibitor in dorsal root ganglion (DRG) neurons
Potentiator	Lamotrigine	Potentiating I_h currents regulation of voltage-gated sodium and
	Gabapentin	calcium channels to upregulates the activity of HCN channels
Indirect modulator	Guanfacine	Inhibition of cAMP–HCN signaling via activation of α_{2A}-adrenoreceptors

development of selective HCN inhibitors with the ability to cross the blood-brain barrier is necessary for the treatment of distinct neurological and neuropsychiatric conditions such as depression.

12.4   ASICS IN DEPRESSION

Acid-sensing ion channels (ASICs) are proton-gated trimeric channel proteins that belong to the degenerin/epithelial sodium channel (DEG/ENaC) family[30,31]. Central neurons express six ASIC subunits[32]. ASIC1a and ASIC2 are expressed in the CNS, while almost all other subtypes (excluding ASIC4) are found in sensory neurons of the peripheral nervous system[33]. ASIC gating depends on both extracellular and intracellular pH. By activation of these channels, an inward current depolarizes the cell membrane, assisting in the initiation of dendritic spikes and action potentials, as well as influencing intracellular second-messenger pathways. In recent years, great progress has been made in clarifying the relationship between protons, ASICs, and neurological disease.

ASICs are distributed on the neuron cell body, dendrites, and spines to regulate synaptic plasticity[34], which has been implicated in a variety of physiological and pathological conditions, including depression and anxiety[35]. Significant, detectable expression of ASICs in brain structures relevant to emotional behavior, such as the amygdala, BNST, NAc, habenula, and periaqueductal grey (PAG), indicates that ASICs are well-positioned to induce emotional symptoms[36–38–39]. Thus, modulation of ASICs may also serve as a viable, novel therapeutic strategy for treatment of neurological and psychiatric pathologies.

ASIC1a was found to be involved in depression-related behaviors in animal models. Genetically or pharmacologically interrupting ASIC1a reduced depression-related behaviors in FST and TST assays. ASIC1a blockade and activation in the amygdala suppressed or increased anxiety-associated behavior, respectively. Exposure to ammonium increased the spontaneous inhibitory postsynaptic currents recorded from principal cells notably more than spontaneous excitatory postsynaptic currents, by dropping the extracellular pH, which elicits an inward current in these neurons[40]. The effects of ASIC1a disruption on synaptic plasticity and neurophysiology support a likely role for ASICs in anxiety and depression behavior.

ASIC1a gene deletion or inhibition has been demonstrated to induce an anti-depressive effect[41,42], while restoring ASIC1a expression in the amygdala reduced the anti-depressant-like behavior in ASIC1a$^{-/-}$ mice[41], thus suggesting that ASIC1a antagonists may be a promising avenue for alleviating depressive symptoms. Pharmacologically, treatment with A-317567 and psalmotoxin (PcTx1), both potent ASIC1a inhibitors, also results in the induction of anti-depressant effects[41]. Amiloride is a blocker of epithelial sodium channels. Studies in a rat model showed that amiloride decreased symptoms of chronic stress-induced anhedonia with elevation of serum adrenocorticotropic hormone and corticosterone. These findings suggest that inhibition of ASICs can affect the hypothalamic-pituitary-adrenal axis to relieve depressive-like behaviors[43].

In addition to ASIC1a, ASIC3 knockout moderated anxiety- and aggression-like behaviors. Although ASIC3 knockout did not affect synaptic plasticity of the hippocampus or depression-like behaviors[44], ASIC3 performed a fundamental role in pain signaling associated with tissue acidosis, as it is expressed predominantly in pain-sensing neurons (nociceptors)[45] and in proprioceptors, which are low-threshold mechanoreceptors[46].

ASIC4 knockout increased fear and symptoms associated with anxiety[33]. ASIC4 may also regulates the membrane expression and activity of ASIC1a and ASIC3 in a heterologous expression system[47]. However, there is currently no evidence of any effect related to ASIC2 knockout in these neurological disorders. These studies imply that the functions of ASICs in depression are likely dependent on specific ASIC subtype and their localization in certain brain regions.

Although ASIC inhibitors could be used as pharmacological tools to study the function of these channels, there is still a lack of ASIC-specific antidepressants, and our current understanding of the mechanistic role of these channels in depression remains incomplete. The role of ASIC1a in neuronal plasticity, its distribution in the limbic circuit, and its effects on anxiety-[35] and depression-related behavior[41] suggests that ASICs may be a relatively strong target for further investigation and potential treatment of depression. For example, an association has been established (i.e. 463 cases) between a haplotype of the ASIC1a-containing locus and major depression disorder (MDD)[41]. In order to find novel antidepressants, further exploration of the relationship between ASIC1a and depression-associated behavior are needed. Additionally, ASIC1a is the most widely studied ASIC in depression and other emotion-related behaviors, although other subtypes may also perform key functions in depression and its therapy, therefore deserve closer scrutiny. Furthermore, it remains unresolved which ASIC subtype is dominant in the development of depression, and which ASIC can be manipulated to greatest effect in the alleviation of depressive symptoms is also a topic of considerable interest.

12.5 P2XRS IN DEPRESSION

Purinergic P2XRs are ligand-gated ion channels that are responsive to adenosine 5'-triphosphate (ATP) and other related nucleotides. Seven subtypes of P2XRs (P2X1Rs–P2X7Rs) have been identified, which form homomultimeric or heteromultimeric pores. P2X7Rs are expressed in the bone, lung, kidney, pancreas, muscle, liver, spleen, and other organs, predominantly expressed in immune system cells including monocytes and macrophages[48]. Extracellular ATP, a fast neurotransmitter, activates surface ionotropic P2XRs and G protein-coupled P2YRs, another type of purinergic receptors[49], to participate in communication between neurons and glia in the CNS. In addition, during ATP synaptic transmission by P2XRs, Ca^{2+} entry affects the frequency dependence for the induction of synaptic plasticity[50]. These receptors also regulate cell growth and proliferation, causing apoptosis and cell death[51,52]. Furthermore, they are responsible for initiation and release of inflammatory factors including tumor necrosis factor-α and interleukin-1β (IL-1β) from microglia and macrophages, which are involved in the progression of neurodegenerative disorders[53–56]. P2XRs therefore play a functional role in synaptic plasticity via multiple pathways.

Among P2XRs, the P2X7R subtype has received the most research attention in studies of depression. In humans, the gene encoding the P2X7Rs subunit is positioned nearby the tip of the long arm of chromosome 12. Earlier genetic findings identified a depression susceptibility locus in the region of chromosome 12q23–24 that was significantly linked with mood disorders such as MDD[57–59]. Indeed, activation of P2X7Rs was found in animal models to be involved in the progress of depression[60,61], and P2X7R knockout mice displayed antidepressant effects. P2X7Rs also participated in antidepressant action and the development of mood disorders through their role in neuro-protection and their influence on neurotransmission in the hippocampus[62]. For example, P2X7R blocker improved scape-oriented behavior in FST. In addition, P2X7Rs in the hippocampus were shown to be involved in the comorbidity of diabetic neuropathic pain and depression[63,64]. However, it remains unclear whether P2X7Rs are necessary for the development of depression and how they may underlie depressive symptoms.

In contrast with the profound involvement of P2X7Rs in a wide variety of depression-associated behaviors, P2X2R and P2X3R knockout mice displayed no deficits in normal motor function, learning and memory behaviors, exploratory behavior, anxiety-like behaviors, and passive coping response to behavioral trials[65]. However, P2X2Rs were found to contribute fundamental roles in functions of the peripheral sensory system, such as pain[66]. Moreover, a reversible and selective P2X3R and P2X2/3R antagonist (MK-7264, gefapixant) is highly effective against hyperalgesia[67]. Despite the significant advances in understanding the role of the P2X7R subtype in depression, there is still a lack of compelling evidence for contributions by other subtypes, although future studies are likely to resolve clear functions for other subtypes in neurological and psychological functions and disorders.

While substantial research has gone into understanding how these P2X ion channels contribute to depression, there is also substantial overlap in neurological pathways related to development of anxiety disorders, and hence P2X channel function in anxiety. Compared to wild type, P2X7R knockout mice did not exhibit any obvious anxiety-like behaviors[68]. Additionally, the administration of the P2X7R antagonist A-804598 did not result in significant impact on anxiety-like behaviors in the OFT and elevated plus maze (EPM)[69]. However, very high relative expression of P2X4Rs was observed in the PAG, an essential region for fear and anxiety[70]. Interestingly, P2X7R has been described as a critical communication link between the nervous and immune systems[71]. P2X7R activity may be related to the pathophysiology of psychiatric disorders such as depressive and anxiety-like behaviors through regulation of pro-inflammatory cytokine release and cell death[72]. Brain glial cells obtained from P2X7R knockout mice treated with LPS produced less IL-1β than wild type[73], while other research revealed that IL-1β-deficient mice exhibited reduced anxiety compared to control animals[74].

The pharmacological effects of the PX2R antagonist have been observed in several animal models for depression. For example, P2X7R antagonists such as AZ10606120 and A804598 showed significant anti-depressant profiles[28,68,75]. Activation of the extracellular-signal-regulated kinase (ERK) pathway is involved in P2XR-mediated depression[75,76], suggesting that the P2XR-ERK signaling pathway could serve as a viable target for therapeutic drug development. An increasing body of evidence, summarized in Table 12.2, supports the development of novel P2X7R antagonists for interference in P2X7R function in

TABLE 12.2
P2XR Channels as Drug Targets

Subtypes	Antagonists	Activity and Function
P2X7	JNJ-54175446	A selective, high-affinity, novel and brain penetrant antagonist of the ATP-gated P2X7 channel.
	JNJ-55308942	A selective, potent, brain penetrant P2X7 functional antagonist.
	AZ 10606120 dihydrochloride	Binds in a positive cooperative manner to sites distinct from, but coupled to, the ATP binding site and functions as a negative allosteric modulator of human P2X7.Prevents tumor growth and exhibits antagonistic effects in animal models.
	JNJ 47965567	A potent, brain penetrant P2X7 antagonist.
	A 804598	A potent, competitive P2X7 receptor antagonist.
	GW791343	A negative allosteric modulator of human P2X7; a positive allosteric modulator of rat P2X7.
P2X4	BX430	Noncompetitive, allosteric antagonist of human P2X4Rs; BX430 has no effect on mouse and rat P2X4Rs.

mood disorders. Although no evidence yet exists of a role for P2X4Rs in the development of depression, several antidepressants have been found to inhibit P2X4Rs[77], while RNA interference targeting P2X2Rs in the medial PFC may also mediate antidepressant effects[78]. In addition, carbamazepine-based compounds may also antagonize P2X4Rs, indicating their possible utility in the treatment of neuropathic pain[79]. The identification of previously unknown P2XR functions in depression and their cognate antidepressant interaction partners, particularly for the less characterized P2XR subtypes, therefore poses a highly relevant research challenge that can ultimately resolve the mechanisms underlying depression.

12.6   TRP CHANNELS IN DEPRESSION

Transient receptor potential (TRP) channels—the second largest class of ion channels in humans—act as sensors in a variety of physiological processes. The TRP channels are subdivided into seven main subfamilies based on amino acid homology: TRPC (canonical), TRPM (melastatin), TRPA (ankyrin), TRPV (vanilloid), TRPN (nomp), TRPML (mucolipin), and TRPP (polycystin)[80]. Several TRP channels are regulated by phosphatidylinositol 4,5-bisphosphate, while most TRP channels serve as Ca^{2+} import pathways. Some of these channels are constitutively open, while some are typically gated by sensing the Ca^{2+} concentration of intracellular Ca^{2+} stores[81].

The pathophysiology of depression and anxiety is regulated by a variety of TRP channels. Individuals suffering from bipolar-I have shown a reduction in TRPC7 mRNA levels and an increase in basal Ca^{2+} levels in a patient-derived lymphoblast cell line[82]. The TRPC3 subtypes 3, 6, and, 7 are deficient in B lymphocytes of individuals suffering from bipolar disorder[83]. Moreover, single nucleotide polymorphisms (SNPs) in the *TRPM2* promoter region were notably found to be associated with bipolar-II, strongly suggesting that TRPM2 polymorphisms may increase the risk of this neurological disorder[84,85]. Genome-wide association studies in patients with MDD have shown that TRPV4 mutations are a risk factor for depression[86]. In an animal model of anxiety, genetic deletion of TRPC4 or TRPC5 decreases anxious behaviors, while null mice demonstrate increased exploratory behaviors. TRPC5 knockout mice displayed deficits in their innate fear and anxiety behavior[87,88]. Consequently, TRPC4 and TRPC5 have been proposed as potentially useful drug targets for psychiatric disorders such as MDD, obsessive-compulsive disorder, and generalized anxiety disorder (GAD)[89]. TRPV4-deficient mice showed a substantial reduction in the social and depressive-associated behaviors observed in wild type[90], signifying that the constitutive activation of TRPV4 by physiological brain temperature is essential for normal brain function. Moreover, disruption of TRPV4 function could lead to psychiatric symptoms.

Pharmacologically acute inhibition of TRPC4 and TRPC5 through HC-070 administration was found to decrease anxiety- and depressive-related behavior[89]. Leptin, which interacts with TRPC1 (and possibly TRPC3) via orexin1, exerts antidepressant effects, and impaired leptin production in a rat model thus contributes to depression-associated behaviors[91]. Similarly, mice carrying a constitutive knockout allele of

TRPC4 were found to exhibit reduced anxiety levels[87]. Inhibitors of TRPV1 are primarily categorized as either capsaicin or proton blockers; for example SB-705498, which antagonizes capsaicin signaling, was developed as an intranasal medicine for management of neuropathic and inflammatory pain management. Currently, clinical trials are underway to investigate the compound's efficacy in treating rhinitis, alleviating dental pain, and relieving acute migraine headaches. Research on TRP channels is presently intensifying as further studies emerge on their potential value as drug targets. As more TRP channel subtypes are examined in animal models and clinical trials of psychiatric and neurological diseases, more advanced engineering approaches, such as structural modifications and computer-assisted rational drug design, will also increase the potency and effectiveness of drugs that block these channels.

It is apparent that multiple subtypes of TRP channels are involved in depression and anxiety, and in particular, the clarification of which specific subtypes of TRP channels are responsible for certain symptoms of depression and anxiety. It also remains unclear which subtype of TRP channels contribute the predominant roles in the development and maintenance of depression and anxiety. It is also unknown whether the manipulation of a single subtype is enough to alleviate symptoms of depression and anxiety, or if it is necessary to combine multiple channel subtypes to achieve an effective treatment strategy. Another possibility is that the TRP channels controlling depressive symptoms are possibly specific to regions of the brain. These questions represent the metaphorical tip of the iceberg in terms of the many possible research directions needed to fully resolve the signaling, regulation, and mechanistic function of TRP channels, with the ultimate goal of alleviating or curing symptoms of depression, anxiety, and other emotional disorders.

12.7 TREK1 IN DEPRESSION

In addition to the channels discussed in the above sections, depression is also regulated through mechano-gated and arachidonic acid-activated TWIK-related potassium 1 (TREK1) channels, which are expressed in the PFC, hippocampus, hypothalamus, dorsal raphe nucleus (DRN), and sensory neurons of the dorsal root ganglia[92]. TREK1 activity is initiated via intracellular acidosis, warm temperature (optimally between $32°C$ and $37°C$), and mechanical stimulation (at approximately -50 mm Hg)[93]. This channel is reversibly opened by polyunsaturated fatty acids, such as arachidonic acid[94]. TREK1 is also regulated via several chemical stimuli, including cellular lipids (polylysine or spermine), and inner leaflet phospholipids like phosphatidylinositol, phosphatidylethanolamine, or phosphatidylserine[95].

Studies of genetic inactivation of TREK1 have revealed the involvement of this two-pore-domain potassium (K_{2P}) channel in pain, ischemia, epilepsy, and depression[92]. TREK2 and TRAAK are two other neuronal K_{2P} channels with similar biophysical properties to TREK1; however, a depression-resistant phenotype has only been induced in *Trek*1-deficient mice[92,96]. This finding has shown that TREK1 knockout or treatment with TREK1 inhibitors can produce antidepressant effects in mice. In addition, TREK1 is resistant to all of the traditional blockers of one-pore-domain potassium channels[94]. The antidepressant effects of TREK1 are characterized by an escalation in the serotonin concentration in the brain[97].

Inhibition of the TREK1 channel can increase the activity of the neurons in which it is expressed, e.g. serotonin secretion is increased by suppression of TREK1 in serotonin neurons, which is an essential target for depression.

Moreover, the firing rate of serotonergic neurons in $Trek1^{-/-}$ mice is double that of wild-type mice[92]. The connection between serotonin levels and antidepressant effects is controversial[98]. On the other hand, K_{2P} mechanosensitive channels are primarily expressed in DRN serotonin neurons and neurons of the PFC and the hippocampus, which mediate the cognitive aspects of depressive symptoms, such as memory impairment, feeling of worthlessness, guilt, and suicidality[99,100]. The application of fluoxetine to block TREK1 generates an antidepressant effect and may contribute to a boost in presynaptic excitability. TREK1 inhibition is accompanied by inhibition of serotonin reuptake, which thus increases serotonin neurotransmission[92,101,102]. Consequently, TREK1 antagonists could provide advantageous effects in the treatment of depression, and this channel is considered an important pharmacological target for future studies.

Serotonin (5-HT) plays a significant role in the neurobiology of depression as part of the "monoaminergic theory of depression."[103] The use of selective serotonin reuptake inhibitors for targeting TREK1 activity has enabled major advances in understanding the role of K_{2P} channels in the pathophysiology of depression[92]. Administration of the antidepressants fluoxetine and paroxetine both reduced the immobility time of wild-type mice in the FST, but did not affect $Trek1^{-/-}$ mice[94]. In addition, $Trek1^{-/-}$ mice were found to exhibit a decrease in negative serotonin feedback in presynaptic neurons, followed by an increase in serotonin neurotransmission and antidepressant behaviors. Due to the critical function of TREK1 in the 5-HT system, and especially in the pathophysiology of depression, TREK1 and its associated signaling pathways remain promising targets for the development of antidepressant drugs.

In both animal and human studies, elevated transcription of $Trek2$ mRNA has been observed in the main anatomic loci of anxiety-related behavior, i.e. the hippocampus and amygdala[104–106]. The TREK1 inhibitor spadin produces a satisfactory antidepressant effect via enhanced phosphorylation of cAMP response element-binding protein and neurogenesis in the hippocampus. Furthermore, $Trek1$-deficient mice treated with spadin had a depression-resistant phenotype, although an acute treatment of spadin did not alleviate anxiety-related behaviors[107]. These findings agreed with the former results that showed $Trek1$-deficient mice do not exhibit an anxiety-resistant phenotype[92]. Given the relationship between TREK1 and sortilin, examination of $Sort1$-deficient mice revealed a corticosterone-independent anxiety-like behavior in various anxiety-related behavioral screens, such as the marble burying and the light-dark tests[108]. Additionally, the genetic ablation of $Trek$ led to the discovery of its gender-dependent influence on anxious behaviors, as female $Trek2$-deficient mice spent more time than males and $Trek1/2/Traak^{-/-}$ mice spent significantly more time than wild type in the center of the OF, and also displayed fewer anxiety behaviors. However, $Trek2$-deficient mice and $Trek1/2/Traak^{-/-}$ mice behaved like the wild-type controls in the EPM and light/dark box tests, suggesting that neither genotype nor gender were contributing factors to performance in these tests[96]. In contrast with the results of experiments conducted with animals, riluzole, a putative activator of TREK1 channels[109], may represent a promising direction for future treatment of mood and anxiety disorders in humans[110]. However, there is currently insufficient information regarding TREK1 to be considered a reliable target of treatments for anxiety.

12.8 IMPLICATIONS

Since 1982 when the first ion channels were cloned, the importance of these structures in the body has been made increasingly clear. Several ion channels have been classified that contribute to the onset of numerous chronic neurological disorders, including depression. Despite long-term efforts to resolve the regulatory mechanisms of classical ion channels in the control of synaptic plasticity- related to depression, as well as the discovery and development of antidepressants that target these channels, the overall progress toward understanding how their dysfunction contributes to psychiatric disorders remains remarkably limited. Current and future research efforts expanding our knowledge of the biological bases for nonclassical ion channels in development and maintenance of depression will provide deeper insights into the mechanisms underlying depression. In this system, nonclassical ion channels and their related intracellular signaling pathways could be beneficial and promising targets for depression therapy and antidepressant development.

REFERENCES

1. Brown, B. M., Nguyen, H. M. & Wulff, H. Recent advances in our understanding of the structure and function of more unusual cation channels. *F1000Research* **8** (2019), 123.
2. Serafini, G. Neuroplasticity and major depression, the role of modern antidepressant drugs. *World Journal of Psychiatry* **2**, 49 (2012).
3. Santoro, B. & Shah, M. M. Hyperpolarization-activated cyclic nucleotide-gated channels as drug targets for neurological disorders. *Annual Review of Pharmacology and Toxicology* **60**, 109–131 (2020).
4. Lee, S.-Y. *et al.* PRMT7 deficiency causes dysregulation of the HCN channels in the CA1 pyramidal cells and impairment of social behaviors. *Experimental & Molecular Medicine*, **52**:604–614 (2020).
5. Piskorowski, R., Santoro, B. & Siegelbaum, S. A. TRIP8b splice forms act in concert to regulate the localization and expression of HCN1 channels in CA1 pyramidal neurons. *Neuron* **70**, 495–509 (2011).
6. Postea, O. & Biel, M. Exploring HCN channels as novel drug targets. *Nature Reviews Drug Discovery* **10**, 903–914 (2011).
7. Benarroch, E. E. HCN channels: function and clinical implications. *Neurology* **80**, 304–310 (2013).
8. Friedman, A. K. *et al.* Enhancing depression mechanisms in midbrain dopamine neurons achieves homeostatic resilience. *Science* **344**, 313–319 (2014).
9. Zhong, P. *et al.* HCN2 channels in the ventral tegmental area regulate behavioral responses to chronic stress. *Elife* **7**, e32420 (2017).
10. Cheng, J., Umschweif, G., Leung, J., Sagi, Y. & Greengard, P. HCN2 channels in cholinergic interneurons of nucleus accumbens shell regulate depressive behaviors. *Neuron* **101**, 662–672.e665 (2019).

11. Lewis, A. S. *et al.* Deletion of the hyperpolarization-activated cyclic nucleotide-gated channel auxiliary subunit TRIP8b impairs hippocampal Ih localization and function and promotes antidepressant behavior in mice. *Journal of Neuroscience* **31**, 7424–7440 (2011).
12. Mcintosh, A. M. Genetic variation in hyperpolarization-activated cyclic nucleotide-gated channels and its relationship with neuroticism, cognition and risk of depression. *Frontiers in Genetics* **3**, 116 (2012).
13. Kim, C. S., Chang, P. Y. & Johnston, D. Enhancement of dorsal hippocampal activity by knockdown of HCN1 channels leads to anxiolytic-and antidepressant-like behaviors. *Neuron* **75**, 503–516 (2012).
14. Kim, C. S., Brager, D. H. & Johnston, D. Perisomatic changes in h-channels regulate depressive behaviors following chronic unpredictable stress. *Molecular Psychiatry* **23**, 892–903 (2018).
15. Ku, S. M. & Han, M.-H. HCN channel targets for novel antidepressant treatment. *Neurotherapeutics* **14**, 698–715 (2017).
16. Notomi, T. & Shigemoto, R. Immunohistochemical localization of Ih channel subunits, HCN1–4, in the rat brain. *Journal of Comparative Neurology* **471**, 241–276 (2004).
17. Moosmang, S. *et al.* Cellular expression and functional characterization of four hyperpolarization-activated pacemaker channels in cardiac and neuronal tissues. *European Journal of Biochemistry* **268**, 1646–1652 (2001).
18. Lewis, A. S. *et al.* Deletion of the HCN channel auxiliary subunit TRIP8b impairs hippocampal Ih localization and function and promotes antidepressant behavior in mice. *Journal of Neuroscience* **31**, 7424 (2011).
19. Günther, A., Luczak, V., Gruteser, N., Abel, T. & Baumann, A. HCN4 knockdown in dorsal hippocampus promotes anxiety-like behavior in mice. *Genes, Brain and Behavior* **18**, e12550 (2019).
20. Harris, N. A. *et al.* Dorsal BNST α2A-adrenergic receptors produce HCN-dependent excitatory actions that initiate anxiogenic behaviors. *Journal of Neuroscience* **38**, 8922–8942 (2018).
21. Brannon, N., Labbate, L. & Huber, M. Gabapentin treatment for posttraumatic stress disorder. *Canadian Journal of Psychiatry. Revue Canadienne de Psychiatrie* **45**, 84 (2000).
22. Hamner, M. B., Brodrick, P. S. & Labbate, L. A. Gabapentin in PTSD: a retrospective, clinical series of adjunctive therapy. *Annals of Clinical Psychiatry* **13**, 141–146 (2001).
23. Pollack, M. H., Matthews, J. & Scott, E. L. Gabapentin as a potential treatment for anxiety disorders. *American Journal of Psychiatry* **155**, 992–993 (1998).
24. Pande, A. C. *et al.* Treatment of social phobia with gabapentin: a placebo-controlled study. *Journal of Clinical Psychopharmacology* **19**, 341–348 (1999).
25. Pande, A. C. *et al.* Placebo-controlled study of gabapentin treatment of panic disorder. *Journal of Clinical Psychopharmacology* **20**, 467–471 (2000).
26. Calabrese, J. R. *et al.* Spectrum of activity of lamotrigine in treatment-refractory bipolar disorder. *American Journal of Psychiatry* **156**, 1019–1023 (1999).
27. Wang, M. *et al.* α2A-adrenoceptors strengthen working memory networks by in-hibiting cAMP-HCN channel signaling in prefrontal cortex. *Cell* **129**, 397–410 (2007).
28. Zhang, K. *et al.* Essential roles of AMPA receptor GluA1 phosphorylation and pre-synaptic HCN channels in fast-acting antidepressant responses of ketamine. *Science Signalling* **9**, ra123-ra123 (2016).
29. Chen, X., Shu, S. & Bayliss, D. A. HCN1 channel subunits are a molecular substrate for hypnotic actions of ketamine. *Journal of Neuroscience* **29**, 600–609 (2009).
30. Krishtal, O. The ASICs: signaling molecules? modulators? *Trends in Neurosciences* **26**, 477–483 (2003).

31. Kellenberger, S. & Schild, L. Epithelial sodium channel/degenerin family of ion channels: a variety of functions for a shared structure. *Physiological Reviews* **82**, 735–767 (2002).
32. Wemmie, J. A., Price, M. P. & Welsh, M. J. Acid-sensing ion channels: advances, questions and therapeutic opportunities. *Trends in Neurosciences* **29**, 578–586 (2006).
33. Lin, S. H. *et al.* Genetic mapping of ASIC 4 and contrasting phenotype to ASIC 1a in modulating innate fear and anxiety. *European Journal of Neuroscience* **41**, 1553–1568 (2015).
34. de la Rosa, D. A. *et al.* Distribution, subcellular localization and ontogeny of ASIC1 in the mammalian central nervous system. *The Journal of Physiology* **546**, 77–87 (2003).
35. Ziemann, A. E. *et al.* The amygdala is a chemosensor that detects carbon dioxide and acidosis to elicit fear behavior. *Cell* **139**, 1012–1021 (2009).
36. Coryell, M. W. *et al.* Targeting ASIC1a reduces innate fear and alters neuronal activity in the fear circuit. *Biological Psychiatry* **62**, 1140–1148 (2007).
37. Deval, E. *et al.* Acid-sensing ion channels (ASICs): pharmacology and implication in pain. *Pharmacology & Therapeutics* **128**, 549–558 (2010).
38. Wemmie, J. A. *et al.* Acid-sensing ion channel 1 is localized in brain regions with high synaptic density and contributes to fear conditioning. *Journal of Neuroscience* **23**, 5496–5502 (2003).
39. Zha, X.-m., Wemmie, J. A., Green, S. H. & Welsh, M. J. Acid-sensing ion channel 1a is a postsynaptic proton receptor that affects the density of dendritic spines. *Proceedings of the National Academy of Sciences of the United States of America* **103**, 16556–16561 (2006).
40. Pidoplichko, V. I. *et al.* ASIC1a activation enhances inhibition in the basolateral amygdala and reduces anxiety. *Journal of Neuroscience* **34**, 3130–3141 (2014).
41. Coryell, M. W. *et al.* Acid-sensing ion channel-1a in the amygdala, a novel therapeutic target in depression-related behavior. *Journal of Neuroscience* **29**, 5381–5388 (2009).
42. Coryell, M. W. *et al.* Restoring Acid-sensing ion channel-1a in the amygdala of knock-out mice rescues fear memory but not unconditioned fear responses. *Journal of Neuroscience* **28**, 13738–13741 (2008).
43. Zhou, W., Ye, S., Luo, R., Wu, L.-M. & Wang, W. Inhibition of acid-sensing ion channels reduces the hypothalamus–pituitary–adrenal axis activity and ameliorates depression-like behavior in rats. *RSC Advances* **9**, 8707–8713 (2019).
44. Wu, W. L., Lin, Y. W., Min, M. Y. & Chen, C. C. Mice lacking Asic3 show reduced anxiety-like behavior on the elevated plus maze and reduced aggression. *Genes, Brain and Behavior* **9**, 603–614 (2010).
45. Deval, E. *et al.* ASIC3, a sensor of acidic and primary inflammatory pain. *The EMBO Journal* **27**, 3047–3055 (2008).
46. Cheng, Y.-R., Jiang, B.-Y. & Chen, C.-C. Acid-sensing ion channels: dual function proteins for chemo-sensing and mechano-sensing. *Journal of Biomedical Science* **25**, 46 (2018).
47. Donier, E., Rugiero, F., Jacob, C. & Wood, J. N. Regulation of ASIC activity by ASIC4–new insights into ASIC channel function revealed by a yeast two-hybrid assay. *European Journal of Neuroscience* **28**, 74–86 (2008).
48. Hillman, K., Burnstock, G. & Unwin, R. The P2X7 ATP receptor in the kidney: a matter of life or death? *Nephron Experimental Nephrology* **101**, e24–e30 (2005).
49. Khakh, B. S. & North, R. A. Neuromodulation by extracellular ATP and P2X receptors in the CNS. *Neuron* **76**, 51–69 (2012).
50. Khakh, B. S. & North, R. A. P2X receptors as cell-surface ATP sensors in health and disease. *Nature* **442**, 527–532 (2006).
51. Abbracchio, M. P. & Burnstock, G. Purinergic signalling: pathophysiological roles. *The Japanese Journal of Pharmacology* **78**, 113–145 (1998).

52. Greig, A. V., Linge, C., Cambrey, A. & Burnstock, G. Purinergic receptors are part of a signaling system for keratinocyte proliferation, differentiation, and apoptosis in human fetal epidermis. *Journal of Investigative Dermatology* **121**, 1145–1149 (2003).

53. Adinolfi, E. *et al.* Expression of P2X7 receptor increases in vivo tumor growth. *Cancer Research* **72**, 2957–2969 (2012).

54. Chessell, I. P. *et al.* Disruption of the P2X7 purinoceptor gene abolishes chronic inflammatory and neuropathic pain. *Pain* **114**, 386–396 (2005).

55. Ferrari D. P. C., Adinolfi E., Lemoli R. M., Curti A., Idzko M., et al. The P2X7 receptor: a key player in IL-1 processing and release. *Journal of Immunology* **176**: 3877–3883 (2006).

56. Hide, I. *et al.* Extracellular ATP triggers tumor necrosis factor-α release from rat microglia. *Journal of Neurochemistry* **75**, 965–972 (2000).

57. Curtis, D. *et al.* Genome scan of pedigrees multiply affected with bipolar disorder provides further support for the presence of a susceptibility locus on chromosome 12q23-q24, and suggests the presence of additional loci on 1p and 1q. *Psychiatric Genetics* **13**, 77–84 (2003).

58. Abkevich, V. *et al.* Predisposition locus for major depression at chromosome 12q22-12q23. 2. *The American Journal of Human Genetics* **73**, 1271–1281 (2003).

59. Lucae, S. *et al.* P2RX7, a gene coding for a purinergic ligand-gated ion channel, is associated with major depressive disorder. *Human Molecular Genetics* **15**, 2438–2445 (2006).

60. Zhang, K. *et al.* P2X7 as a new target for chrysophanol to treat lipopolysaccharide-induced depression in mice. *Neuroscience Letters* **613**, 60–65 (2016).

61. Otrokocsi, L., Kittel, Á. & Sperlágh, B. P2X7 receptors drive spine synapse plasticity in the learned helplessness model of depression. *International Journal of Neuropsychopharmacology* **20**, 813–822 (2017).

62. Suzuki, T. *et al.* Production and release of neuroprotective tumor necrosis factor by P2X7 receptor-activated microglia. *The Journal of Neuroscience* **24**, 1–7, (2004).

63. Shen, Y. *et al.* Effects of palmatine on rats with comorbidity of diabetic neuropathic pain and depression. *Brain Research Bulletin* **139**, 56–66 (2018).

64. Armstrong, J. N., Brust, T. B., Lewis, R. G. & MacVicar, B. A. Activation of presynaptic P2X7-like receptors depresses mossy fiber–CA3 synaptic transmission through p38 mitogen-activated protein kinase. *Journal of Neuroscience* **22**, 5938–5945 (2002).

65. Kong, Y. *et al.* Involvement of P2X2 receptor in the medial prefrontal cortex in ATP modulation of the passive coping response to behavioral challenge. *Genes, Brain and Behavior*, 2020, **19**, e12691.

66. Zheng, X.-b. *et al.* Effects of 1, 8-cineole on neuropathic pain mediated by P2X2 receptor in the spinal cord dorsal horn. *Scientific Reports* **9**, 1–8 (2019).

67. Richards, D., Gever, J. R., Ford, A. P. & Fountain, S. J. Action of MK-7264 (gefapixant) at human P2X3 and P2X2/3 receptors and in vivo efficacy in models of sensitisation. *British Journal of Pharmacology* **176**, 2279–2291 (2019).

68. Basso, A. M. *et al.* Behavioral profile of P2X7 receptor knockout mice in animal models of depression and anxiety: relevance for neuropsychiatric disorders. *Behavioural Brain Research* **198**, 83–90 (2009).

69. Iwata, M. *et al.* Psychological stress activates the inflammasome via release of adenosine triphosphate and stimulation of the purinergic type 2X7 receptor. *Biological Psychiatry* **80**, 12–22 (2016).

70. Behbehani, M. M. Functional characteristics of the midbrain periaqueductal gray. *Progress in Neurobiology* **46**, 575–605 (1995).

71. Volonte, C., Apolloni, S., D Skaper, S. & Burnstock, G. P2X7 receptors: channels, pores and more. *CNS & Neurological Disorders - Drug Targets* **11**, 705–721 (2012).

72. Xie, B. *et al.* The expression of P2X7 receptors on peripheral blood mononuclear cells in patients with primary Sjögren's syndrome and its correlation with anxiety and depression. *Clinical and Experimental Rheumatology* **32**, 354–360 (2014).

73. Boufidou, F. *et al.* CSF and plasma cytokines at delivery and postpartum mood disturbances. *Journal of Affective Disorders* **115**, 287–292 (2009).

74. Koo, J. W. & Duman, R. S. Interleukin-1 receptor null mutant mice show decreased anxiety-like behavior and enhanced fear memory. *Neuroscience Letters* **456**, 39–43 (2009).

75. Ito-Ishida, A., Kakegawa, W. & Yuzaki, M. ERK1/2 but not p38 MAP kinase is essential for the long-term depression in mouse cerebellar slices. *European Journal of Neuroscience* **24**, 1617–1622 (2006).

76. Jo, Y.-H. & Schlichter, R. Synaptic corelease of ATP and GABA in cultured spinal neurons. *Nature Neuroscience* **2**, 241–245 (1999).

77. Nagata, K. *et al.* Antidepressants inhibit P2X4 receptor function: a possible involvement in neuropathic pain relief. *Molecular Pain* **5**, 20 (2009).

78. Cao, X. *et al.* Astrocyte-derived ATP modulates depressive-like behaviors. *Nature Medicine* **19**, 773 (2013).

79. Tian, M. *et al.* Arbamazepine derivatives with P2X4 receptor-blocking activity. *Bioorganic & Medicinal Chemistry* **22**, 1077–1088 (2014).

80. Nilius, B. & Owsianik, G. The transient receptor potential family of ion channels. *Genome Biology* **12**, 218 (2011).

81. Clapham, D. E. TRP channels as cellular sensors. *Nature* **426**, 517–524 (2003).

82. Yoon, I.-S. *et al.* Altered TRPC7 gene expression in bipolar-I disorder. *Biological Psychiatry* **50**, 620–626 (2001).

83. Roedding, A. S., Li, P. P. & Warsh, J. J. Characterization of the transient receptor potential channels mediating lysophosphatidic acid-stimulated calcium mobilization in B lymphoblasts. *Life Sciences* **80**, 89–97 (2006).

84. Xu, C. *et al.* Association of the putative susceptibility gene, transient receptor potential protein melastatin type 2, with bipolar disorder. *American Journal of Medical Genetics Part B: Neuropsychiatric Genetics* **141**, 36–43 (2006).

85. Xu, C. *et al.* Association of the transient receptor potential TRPM2 gene with bipolar II disorder. *Journal of Medical Genetics. Part B, Neuropsychiatric Genetics* **141**,36–43 (2006).

86. Wong, M.-L. *et al.* The PHF21B gene is associated with major depression and modulates the stress response. *Molecular Psychiatry* **22**, 1015–1025 (2017).

87. Riccio, A. *et al.* Decreased anxiety-like behavior and Gαq/11-dependent responses in the amygdala of mice lacking TRPC4 channels. *Journal of Neuroscience* **34**, 3653–3667 (2014).

88. Riccio, A. *et al.* Essential role for TRPC5 in amygdala function and fear-related behavior. *Cell* **137**, 761–772 (2009).

89. Just, S. *et al.* Treatment with HC-070, a potent inhibitor of TRPC4 and TRPC5, leads to anxiolytic and antidepressant effects in mice. *PLoS One* **13** (2018).

90. Shibasaki, K. *et al.* TRPV4 activation at the physiological temperature is a critical determinant of neuronal excitability and behavior. *Pflügers Archiv - European Journal of Physiology* **467**, 2495–2507 (2015).

91. Lu, X.-Y., Kim, C. S., Frazer, A. & Zhang, W. Leptin: a potential novel antidepressant. *Proceedings of the National Academy of Sciences of the United States of America* **103**, 1593–1598 (2006).

92. Heurteaux, C. *et al.* Deletion of the background potassium channel TREK-1 results in a depression-resistant phenotype. *Nature Neuroscience* **9**, 1134–1141 (2006).

93. Maingret, F. *et al.* TREK-1 is a heat-activated background K+ channel. *The EMBO Journal* **19**, 2483–2491 (2000).

94. Patel, A. J. *et al.* A mammalian two pore domain mechano-gated S-like K+ channel. *The EMBO Journal* **17**, 4283–4290 (1998).
95. Kim, D. Fatty acid-sensitive two-pore domain K+ channels. *Trends in Pharmacological Sciences* **24**, 648–654 (2003).
96. Mirkovic, K., Palmersheim, J., Lesage, F. & Wickman, K. Behavioral characterization of mice lacking Trek channels. *Frontiers in Behavioral Neuroscience* **6**, 60 (2012).
97. Ye, D. *et al.* TREK1 channel blockade induces an antidepressant-like response synergizing with 5-HT1A receptor signaling. *European Neuropsychopharmacology* **25**, 2426–2436 (2015).
98. Cowen, P. J. & Browning, M. What has serotonin to do with depression? *World Psychiatry* **14**, 158 (2015).
99. Nestler, E. J. *et al.* Neurobiology of depression. *Neuron* **34**, 13–25 (2002).
100. Otte, C. *et al.* Major depressive disorder. *Nature Reviews Disease Primers* **2**, 1–20 (2016).
101. Kennard, L. E. *et al.* Inhibition of the human two-pore domain potassium channel, TREK-1, by fluoxetine and its metabolite norfluoxetine. *British Journal of Pharmacology* **144**, 821–829 (2005).
102. Gordon, J. A. & Hen, R. TREKing toward new antidepressants. *Nature Neuroscience* **9**, 1081–1083 (2006).
103. Duman, R. S., Heninger, G. R. & Nestler, E. J. A molecular and cellular theory of depression. *Archives of General Psychiatry* **54**, 597–606 (1997).
104. Medhurst, A. D. *et al.* Distribution analysis of human two pore domain potassium channels in tissues of the central nervous system and periphery. *Molecular Brain Research* **86**, 101–114 (2001).
105. Lein, E. S. *et al.* Genome-wide atlas of gene expression in the adult mouse brain. *Nature* **445**, 168–176 (2007).
106. Mirkovic, K. & Wickman, K. Identification and characterization of alternative splice variants of the mouse Trek2/Kcnk10 gene. *Neuroscience* **194**, 11–18 (2011).
107. Mazella, J. *et al.* Spadin, a sortilin-derived peptide, targeting rodent TREK-1 channels: a new concept in the antidepressant drug design. *PLoS Biology* **8**, e1000355 (2010).
108. Moreno, S. *et al.* Altered Trek-1 function in sortilin deficient mice results in decreased depressive-like behavior. *Frontiers in Pharmacology* **9**, 863 (2018).
109. Duprat, F. *et al.* The neuroprotective agent riluzole activates the two P domain K+ channels TREK-1 and TRAAK. *Molecular Pharmacology* **57**, 906–912 (2000).
110. Pittenger, C. *et al.* Riluzole in the treatment of mood and anxiety disorders. *CNS Drugs* **22**, 761–786 (2008).

13 Ion Channels of Reward Pathway in Drug Abuse

Wen Zhang

CONTENTS

Drug abuse is a worldwide affliction. It is often characterized as addicts who cannot stop the abuse of drugs even with clear knowledge of the serious negative consequences of abuse. Addicts often show a cluster of cognitive, behavior, and physiological symptoms, which suggests that drug abuse is a brain disease. Studies have shown that compromised functions of a group of brain regions are associated with drug abuse. These brain regions are interconnected, exerting important functions of reward-related activities, and the connections make a large-scale brain circuit called the "reward pathway." The reward pathway includes the ventral tegmental area (VTA), substantia nigra, nucleus accumbens (NAc), prefrontal cortex (PFC), orbitofrontal cortex, amygdala, hippocampus, and associated structures. Among these brain regions, VTA is important for reward and motivation, NAc is important for habitation and locomotion, the amygdala is critical for fear and negative emotions, and the cortical regions are important for memory processing and emotions. Beside projections to the NAc, VTA dopaminergic neurons also project to several other brain regions in the reward pathway, such as the PFC, central and basolateral amygdala, and the hippocampus (Robison and Nestler 2011). These brain regions in the reward pathway are also interconnected besides via dopaminergic innervation from VTA. For example, NAc receives extensive glutamatergic inputs from the PFC, ventral hippocampus, and amygdala. And the PFC, hippocampus, and amygdala form reciprocal excitatory glutamatergic connections with one another. Besides the two neurotransmitters mentioned above, brain regions in the reward pathway are also modulated by cholinergic interneurons in those regions, the serotonergic inputs from dorsal raphe, and noradrenergic inputs from the locus coeruleus. Furthermore, studies have shown that the VTA dopaminergic neurons could also corelease glutamate or GABA with dopamine at terminals (Tritsch, Ding, and Sabatini

235

2012; Hnasko et al. 2012). Of the ion channels, most ion channel families are expressed in these brain regions. However, compared with other brain functions and diseases, non-ion channels still are under-investigated for their roles in drug abuse. This chapter discusses the functions of several nonclassical ion channels in drug abuse, such as transient receptor potential channels, acid-sensing ion channels, hyperpolarization-activated cyclic nucleotide-gated channels, and P2X ion channels. This chapter also discusses the role of γ-Aminobutyric acid type A receptors in drug abuse, especially their role in benzodiazepines abuse.

13.1   TRANSIENT RECEPTOR POTENTIAL CHANNELS

Transient receptor potential (TRP) channels are a large family of non-selective cation channels. Besides traditional ligand-gated opening, mechanical force, chemical stress, and temperature also can lead to channel opening. Most TRP channels are permeable to Ca^{2+} with the exceptions of TRPM4 and TRPM5, which are only permeable to monovalent cations. The TRP superfamily contains six groups: TRPA (ankyrin), TRPC (canonical), TRPM (melastatin), TRPML (mucolipin), TRPP (polycystin), and TRPV (vanilloid). TRP channels were first identified in *Drosophila melanogaster*, and studies have found around 30 channels in mammals belonging to this superfamily. The diverse activation mechanisms and expression of these channels make them involved in a wide range of activities of the central and peripheral nervous systems. However, the roles of these channels in drug abuse have only recently begun to be explored, with one of the most thoroughly studied families of TRP channels in drug abuse being the TRPV family. The TRPV family contains six mammalian members: TRPV1-6. TRPV1-4 are all heat-activated channels, which exhibit cation non-selectivity and modest Ca^{2+} permeability. TRPV5 and TRPV6 are the only highly Ca^{2+}-selective channels in the TRP family, and both are regulated by $[Ca^{2+}]_i$. In contrast with other TRPVs, the temperature sensitivity of TRPV5 and TRPV6 is relatively low. TRPV1-4 are also sensitive to a broad array of endogenous and synthetic ligands. For example, TRPV1 is activated by capsaicin, heat ($\geq 43°C$), and many other chemicals, including an endocannabinoid, anandamide; the topical analgesic, camphor; piperine in black pepper; and allicin in garlic (Caterina et al. 1997; Zygmunt et al. 1999; Xu, Blair, and Clapham 2005; McNamara, Randall, and Gunthorpe 2005; Macpherson et al. 2005). Activation of TRPV1 leads to membrane depolarization, and TRPV1-mediated current can be facilitated by extracellular acidification (Xu, Blair, and Clapham 2005). A pH change to <6, ethanol, and nicotine also have similar effects on TRPV1 activity (Trevisani et al. 2002; Liu et al. 2004).

A study has shown that knockout of *Trpv1* led to higher alcohol addiction susceptibility in mice (Blednov and Harris 2009), which suggests a possible role of TRPV1 in drug addiction. Besides alcohol, another study on methamphetamine addiction found that the expression of *Trpv1* mRNA increased specifically in the frontal cortex but not in the striatum or the hippocampus of mice up to a week after three consecutive methamphetamine injections (Tian et al. 2010). Interestingly, mice showing methamphetamine-induced conditioned place preference exhibited *Trpv1* mRNA and protein expression increase in NAc (Tian et al. 2018), which might indicate different patterns of TRPV1 activation in the reward pathway. Furthermore, inhibition of TRPV1 with capsazepine resulted in an inhibition of addiction-related behaviors, and similar results were also reported on cocaine- and morphine-induced conditioned place preference or self-administration (Hong et al. 2017; You et al. 2019; Ma et al. 2018).

Recent studies have begun to explore the underlying mechanisms of TRPV1 in addiction. In the reward pathway, the dopamine system is essential for reward-related physiological activities and drug abuse. It is reported that activation of D1-like subtypes of dopamine receptors led to opening of TRPV1 and a calcium influx, while the specific inhibitor of phospholipase C (PLC) blocked this effect (Lee et al. 2015; Chakraborty et al. 2016). These results indicate possible crosstalk between D1- or D5-containing dopamine receptors with TRPV1. The D1 subtype of dopamine receptors is directly related to reward-related signals in the brain. In naïve animals, stimulation of D1 receptors enhanced acquisition and maintenance of drug self-administration behavior (Self 2010). Blockade of D1 receptors reduced the rewarding effects of psychostimulant drugs such as cocaine and amphetamine, and D1 dopamine receptor knockout mice fail to acquire cocaine self-administration (Self 2014). Dopamine receptors are GPCRs, and activation of D1-like dopamine receptors facilitates excitatory input to D1-expressing neurons in NAc. A recent optogenetic study supports this notion that naïve mice with direct optogenetic excitation of D1-containing striatal neurons can induce drug-like self-administration behavior (Kravitz, Tye, and Kreitzer 2012). Considering the aforementioned role of D1-like dopamine receptors in drug addiction, the crosstalk between TRPV1 and D1R could further enhance neuron excitation as TRPV1 is a Ca^{2+} permeable non-selective cation channel. However, further studies are needed to confirm the existence of this crosstalk in the central nervous system, especially in the reward pathway, and the prominence of this crosstalk on neuron activity and drug abuse.

13.2   ACID-SENSING ION CHANNELS

Acid-sensing ion channels (ASICs) are cation channels with a proton as the ligand. They are widely expressed in the central and peripheral nervous systems, and numerous studies have shown their important roles in brain diseases, such as ischemia, chronic pain, and post-traumatic stress disorder (Gao et al. 2005; Duan et al. 2007; Li et al. 2016). The ASIC family has six subunits in rodents (ASIC1a/b, ASIC2a/b, ASIC3, and ASIC4) with ASIC1a/b and ASIC2a/b being splice variants that differ in the first third of the protein. Of these subunits, ASIC1a, ASIC2, ASIC3, and ASIC4 are widely expressed in the brain. ASICs are Na^+-selective channels, and activation of these channels leads to depolarization of the cell, which can result in firing of action potential in excitatory neurons (Vukicevic and Kellenberger 2004).

Several studies identified that drug abuse led to a subtype-specific change of ASICs. Consecutive cocaine administration led to an increase in protein expression of ASIC1 but not of ASIC2, and such change was limited to the striatum but not the prefrontal cortex or the hippocampus (Zhang et al. 2009). A study on ASIC expression in the reward pathway of amphetamine-treated rats reported similar findings, and it also showed that surface expression of ASIC2 in the prefrontal cortex decreased while that of ASIC1 did not change (Suman et al. 2010). A study utilizing ASIC1 and ASIC2 knockout mice showed that deletion of ASIC1 significantly compromised behavior sensitization to cocaine, while ASIC2 knockout only affected that behavior at a high dose of cocaine (30 mg/kg) (Jiang et al. 2013). Interestingly, another study reported that *Asic1a* knockout mice showed increased cocaine-conditioned place preference, which could be restored following viral overexpression of *Asic1a* in NAc (Kreple et al. 2014). This study showed an interesting role of ASIC1a on the excitatory transmission onto medium spiny neurons (MSNs) in the NAc: the deletion of ASIC1a increased dendritic spine density, excitatory transmission, and Ca^{2+}-permeable AMPA receptors of the medium spiny neurons in the NAc. In another study from the same group, they reported that deletion of *Asic1a* with *synapsin I* promoter reduced ASIC1a expression in most brain regions while in the NAc ASIC1a expression was relatively normal. They found that cocaine-conditioned place preference (CPP) was normal for these *syn1-Asic1a* deletion mice (Taugher et al. 2017). They also showed that overexpression of *ASIc1a* driven by *CMV* promoter in the NAc core of rats enhanced drug addiction-related behaviors with self-administration (Gutman et al. 2020). The differences from these studies offer several intriguing questions for the function of ASIC1 in drug addiction to be explored. First, does ASIC1 or ASIC1a expression in the NAc have a dosage effect on the postsynaptic excitatory activity and addiction-related behaviors of rodents? Second, are similar behavior changes with either deletion or over-expression of *Asic1a* due to different roles of the NAc core and shell in drug addiction-related behaviors? Third, does activity of ASIC1a have a similar effect on two types of MSNs in the NAc, namely *Drd1*$^+$- and *Drd2*$^+$-MSNs?

13.3 HYPERPOLARIZATION-ACTIVATED CYCLIC NUCLEOTIDE–GATED (HCN) CHANNELS

HCNs are voltage-gated cation channels that are activated with hyperpolarization. They open at membrane potentials more negative than −50 mV, thus they are active at the resting membrane potential of neurons. These channels are permeable to both Na^+ and K^+, resulting in a depolarizing current at rest, which inevitably also influences the membrane resistance. The current mediated by HCNs is also called I_h current; I_h has significant influences on neuron activity. For example, I_h plays an important role in the tonic firing of Purkinje cells in the cerebellum (Williams et al. 2002). As I_h decreases membrane resistance, its activity limits low threshold Ca^{2+} channel activity and both the amplitude and kinetics of EPSP in dendrites. These effects lead to the decrease of EPSP summation and dendritic excitability in neurons (Tsay, Dudman, and Siegelbaum 2007; Shah et al. 2004; Magee 1998). The HCN family contains four subunits, HCN1–4. The HCN1 subunit is expressed in the cortex, hippocampus, cerebellum, and brain stem. The HCN2 subunit is expressed in high abundance in the thalamus and brain stem. HCN3 is expressed relatively low in the CNS, and HCN4 is expressed specifically to the olfactory bulb (Moosmang et al. 1999).

HCN2 expression increased in VTA, PFC, NAc, and the hippocampus of the mice with cocaine-induced behavioral sensitization (Santos-Vera et al. 2013), and further examination revealed the HCN2 surface/intracellular (S/I) ratio in the VTA decreased whereas in the PFC the ratio increased. In addition, HCN4 total expression in the VTA decreased after cocaine sensitization, although the S/I ratio did not alter (Santos-Vera et al. 2019). Chronic morphine exposure led to a decrease in HCN1, and an increase in HCN2 on cell membrane of hippocampal CA1 area of rats (Zhou et al. 2015). Blockage of HCN channels reduced extracellular dopamine concentration in the NAc and drug-seeking behaviors of rats with methamphetamine-induced self-administration (Cao et al. 2016). Re-exposure to methamphetamine after 5 days withdrawal from methamphetamine injection led to a decrease in the protein expression of HCN1 channels in both the hippocampus and the PFC (Zhou et al. 2019). Another addictive substance, nicotine, directly affects the function of HCN channels in the lateral septum (Kodirov, Wehrmeister, and Colom 2014), suggesting a possible role of HCN1 in nicotine addiction. Considering that HCN channels show a selective expression pattern in neurons and brain regions (Calejo et al. 2014), more studies are needed to fully understand the role of HCN channels in drug abuse.

13.4   P2X RECEPTORS

P2X receptors are cation-permeable ionotropic receptors. There are seven subunits of P2X, namely P2X1-7. These subunits are widely distributed in the brain, being present in both neurons and glial cells. P2X receptors bind with extracellular ATP, which is also co-released from the presynaptic site with other neurotransmitters. Studies have shown that ATP co-releases with acetylcholine (Zhang et al. 2000), noradrenaline (Sperlagh et al. 1998), γ-aminobutyric acid (Jo and Schlichter 1999), dopamine (Krugel et al. 2001), and glutamate (Thyssen et al. 2010). On the other hand, it is suggested that ATP is primarily stored and released from a distinct pool of vesicles and that the release of ATP is not synchronized with the co-transmitters, such as GABA or glutamate (Pankratov et al. 2006). All P2X receptors are permeable to small monovalent cations, but the monovalent cation permeability differs between these subunits. For example, P2X1 shows little selectivity for Na^+ over K^+, but P2X2 shows higher permeability for K^+ than for Na^+ (For review, see North 2002). Some of these subunits have significant calcium permeability, such as P2X1 and P2X4. P2X2, P2X4, P2X6, and P2X7 receptors are widely expressed in the CNS. Recent studies are beginning to show the roles of P2X receptors in substance abuse. It is shown that P2X4 knockout mice exhibited higher alcohol intake, and spatial-specific deletion of P2X4 in the NAc core also significantly increased alcohol intake and preference (Khoja et al. 2018). On the other hand, deletion of P2X7 in the midbrain periaqueductal gray with antisense oligodeoxynucleotide reduced the development of chronic morphine tolerance in rats (Xiao, Li, and Sun 2015).

13.5   γ-AMINOBUTYRIC ACID TYPE A (GABA$_A$) RECEPTORS

GABA$_A$ receptors (GABAARs) are GABA-gated Cl⁻-permeable anion channels. They are the major inhibitory receptors on the neuron, and they act with glutamate receptors to control the excitation/inhibition balance of the central nervous system. Due to its important function and ubiquitous expression in the brain, many studies have found changes in GABAergic transmission in the brain associated with neurological disorders, such as schizophrenia, depression, epilepsy, chronic pain, and other diseases (Cohen et al. 2002; Luscher, Shen, and Sahir 2011; Moore et al. 2002; Zhang et al. 2014, 2016, 2017).

Compared with other ion channels mentioned previously, a unique feature regarding GABAARs in drug abuse is that benzodiazepines (BDZs), a class of drugs commonly used to treat anxiety and insomnia, are often abused and they act strongly as positive allosteric modulators of GABAARs. Of the commonly prescribed BDZs, diazepam and zolpidem increase the ligand-binding affinity of GABAARs. Flunitrazepam also decreased receptor deactivation, which increases GABAAR-mediated Cl^- current. Such actions of BDZs are different from other drugs commonly abused. Furthermore, addiction of BDZs is less frequent than dependence on them, which manifests as a withdrawal syndrome and is often characterized by sleep disturbances, irritability, increased tension and anxiety, panic attacks, and sweating (Petursson 1994). Furthermore, a serious side-effect of BDZ withdrawal is suicide (Colvin 1995).

GABAARs are heteropentamers consisting of eight distinct subunit families, namely α, β, γ, δ, ε, π, θ, and ρ. There are at least 19 subunits identified: α1–6, β1–3, γ1–3, δ, ε, π, θ, and ρ1–3, and it is shown that most $GABA_A$ receptors in the brain are comprised of two α-, two β-, and one γ-subunits. BDZs bind to a pocket formed by α and γ subunits that is distinct from the binding site of the agonist, GABA which locates between α and β subunits. A histidine residue (H101) of α1 subunit and the homologous residues in α2, 3, 5 subunits are crucial for BDZ action (Benson et al. 1998). GABAAR α4 and α6 subunits lack such a residue, and they are insensitive to BDZs (Smith et al. 1998). Genetically engineered mice further help us to understand the roles of individual subunits in the symptoms associated with BDZ use. For example, it is shown that the α1 subunit mediates the sedative amnesic, and partially mediates the anticonvulsive effects of diazepam and also the addictive properties of BDZs (Heikkinen, Moykkynen, and Korpi 2009; Tan et al. 2010; Rudolph et al. 1999; McKernan et al. 2000). The α2 subunit mediates anxiolytic actions and myorelaxant effects, the α3 subunit contributes to myorelaxant actions, and the α5 subunit modulates myorelaxant actions and effects of BDZs on memory (Low et al. 2000; Crestani et al. 2001, 2002; Collinson et al. 2002; Cheng et al. 2006).

Besides the actions of BDZs on the $GABA_A$ receptor-mediated current, chronic BDZ use also regulates GABAAR expression in the central nervous system. It is shown that chronic BDZ use increased GABAAR internalization from the membrane, and decreased α2 subunit surface expression (Jacob et al. 2012). It also decreased α1 and γ2 subunit mRNAs in neurons (Kang and Miller 1991), and upon withdrawal α1 and γ2 mRNAs decreased while α4, a BDZ-insensitive subunit mRNA, increased (Follesa et al. 2001). Furthermore, chronic treatment with BDZs reduced receptor sensitivity to GABA (Hernandez et al. 1989; Roca et al. 1990), which indicates development of tolerance to BDZs.

Of substances often abused, alcohol use also produces similar symptoms as BDZs, suggesting it might also act on GABAARs. Indeed, mouse models showed the roles of specific subunits in physiological and behavioral responses to alcohol consumption. α1 knockout led to a decrease in alcohol consumption and an increased aversion to alcohol (Blednov et al. 2003). α1 and β2 double-knockout mice exhibited shorter periods of loss of righting reflex following alcohol consumption and increased locomotor stimulant effects from alcohol exposure (Blednov et al. 2003; June et al. 2007). Knockdown of α2 reduced binge-like drinking activity (Liu et al. 2011), similar to the effect of viral-mediated deletion of α4 subunit expression in the NAc shell (Rewal et al. 2009). Mice lacking the δ subunit exhibited less preference for alcohol and reduced hyperexcitability from withdrawal after chronic alcohol exposure (Mihalek et al. 2001).

While many other abused drugs do not directly bind to GABAARs, changes in GABAAR subunit expression are associated with the abuse of nicotine (Agrawal et al. 2008; Balan et al. 2018), cocaine (Dixon et al. 2010), and amphetamines (Wearne et al. 2016). In rodents and addicts, cocaine abuse led to change of γ2 in the hippocampus (Enoch et al. 2012) and deletion of *Gabra2* gene abolished cocaine-induced conditioned reinforcement (Dixon et al. 2010). Cocaine use also decreased α2 subunit expression in the NAc shell but not in the NAc core (Chen et al. 2007). Interestingly, sensitization to methamphetamine in rats led to decreased α2 expression in both the NAc core and NAc shell (Zhang et al. 2006).

Besides these GABA$_A$ receptor subunit-specific roles in drug abuse, recent studies also are beginning to show that different neurons in the reward pathway have cell type-specific expression of GABAAR subunits. Dopaminergic neurons in VTA express mRNAs for α2, α3, α4, β1, β3, and γ2 subunit isoforms (Okada et al. 2004), whereas GABAergic interneurons in VTA express α1, β2, and γ2 subunit isoforms (Gao et al. 1993; Gao and Fritschy 1994). A recent study showed that such a cell type-specific expression pattern of GABAAR subunits is important for the addictive effect of BDZs (Tan et al. 2010). BDZs target the α1-containing GABAergic interneurons, which project to the dopaminergic neurons in VTA. Thus, BDZs disinhibit dopaminergic neurons in VTA, and increase dopamine release. Mutation of the α1 subunit (H101R) abolished BDZ-induced synaptic plasticity in VTA, disinhibition of dopaminergic neurons, and self-administration of BDZ. These results suggest an important role of the α1 subunit in reinforcement. However, whether this subunit- and cell type-specific effect is unique to VTA or exists in other brain regions of the reward pathway is still not clear.

ACKNOWLEDGMENT

This work was partially supported by National Key R&D Program of China (2019YFA0706201), and National Natural Science Foundation of China Grants 31741060 and 91732109.

REFERENCES

Agrawal, A., M. L. Pergadia, S. F. Saccone, A. L. Hinrichs, C. N. Lessov-Schlaggar, N. L. Saccone, R. J. Neuman, N. Breslau, E. Johnson, D. Hatsukami, G. W. Montgomery, A. C. Heath, N. G. Martin, A. M. Goate, J. P. Rice, L. J. Bierut, and P. A. Madden. 2008.

Gamma-aminobutyric acid receptor genes and nicotine dependence: evidence for association from a case-control study. *Addiction* 103 (6):1027–1038.

Balan, I., K. T. Warnock, A. Puche, M. C. Gondre-Lewis, H. June, and L. Aurelian. 2018. The GABA$_A$ receptor alpha2 subunit activates a neuronal TLR4 signal in the ventral tegmental area that regulates alcohol and nicotine abuse. *Brain Sci* 8 (4):72.

Benson, J. A., K. Low, R. Keist, H. Mohler, and U. Rudolph. 1998. Pharmacology of recombinant gamma-aminobutyric acidA receptors rendered diazepam-insensitive by point-mutated alpha-subunits. *FEBS Lett* 431 (3):400–404.

Blednov, Y. A., and R. A. Harris. 2009. Deletion of vanilloid receptor (TRPV1) in mice alters behavioral effects of ethanol. *Neuropharmacology* 56 (4):814–820.

Blednov, Y. A., D. Walker, H. Alva, K. Creech, G. Findlay, and R. A. Harris. 2003. GABA$_A$ receptor alpha 1 and beta 2 subunit null mutant mice: behavioral responses to ethanol. *J Pharmacol Exp Ther* 305 (3):854–863.

Calejo, A. I., M. Reverendo, V. S. Silva, P. M. Pereira, M. A. Santos, R. Zorec, and P. P. Goncalves. 2014. Differences in the expression pattern of HCN isoforms among mammalian tissues: sources and implications. *Mol Biol Rep* 41 (1):297–307.

Cao, D. N., R. Song, S. Z. Zhang, N. Wu, and J. Li. 2016. Nucleus accumbens hyperpolarization-activated cyclic nucleotide-gated channels modulate methamphetamine self-administration in rats. *Psychopharmacology* 233 (15-16):3017–3029.

Caterina, M. J., M. A. Schumacher, M. Tominaga, T. A. Rosen, J. D. Levine, and D. Julius. 1997. The capsaicin receptor: a heat-activated ion channel in the pain pathway. *Nature* 389 (6653):816–824.

Chakraborty, S., M. Rebecchi, M. Kaczocha, and M. Puopolo. 2016. Dopamine modulation of transient receptor potential vanilloid type 1 (TRPV1) receptor in dorsal root ganglia neurons. *J Physiol* 594 (6):1627–1642.

Chen, Q., T. H. Lee, W. C. Wetsel, Q. A. Sun, Y. Liu, C. Davidson, X. Xiong, E. H. Ellinwood, and X. Zhang. 2007. Reversal of cocaine sensitization-induced behavioral sensitization normalizes GAD67 and GABA$_A$ receptor alpha2 subunit expression, and PKC zeta activity. *Biochem Biophys Res Commun* 356 (3):733–738.

Cheng, V. Y., L. J. Martin, E. M. Elliott, J. H. Kim, H. T. Mount, F. A. Taverna, J. C. Roder, J. F. Macdonald, A. Bhambri, N. Collinson, K. A. Wafford, and B. A. Orser. 2006. Alpha5GABA$_A$ receptors mediate the amnestic but not sedative-hypnotic effects of the general anesthetic etomidate. *J Neurosci* 26 (14):3713–3720.

Cohen, I., V. Navarro, S. Clemenceau, M. Baulac, and R. Miles. 2002. On the origin of interictal activity in human temporal lobe epilepsy in vitro. *Science* 298 (5597):1418–1421.

Collinson, N., F. M. Kuenzi, W. Jarolimek, K. A. Maubach, R. Cothliff, C. Sur, A. Smith, F. M. Otu, O. Howell, J. R. Atack, R. M. McKernan, G. R. Seabrook, G. R. Dawson, P. J. Whiting, and T. W. Rosahl. 2002. Enhanced learning and memory and altered GABAergic synaptic transmission in mice lacking the alpha 5 subunit of the GABA$_A$ receptor. *J Neurosci* 22 (13):5572–5580.

Colvin, Rod. 1995. *Prescription drug abuse: the hidden epidemic: a guide to coping and understanding*. Omaha, NE: Addicus Books.

Crestani, F., R. Keist, J. M. Fritschy, D. Benke, K. Vogt, L. Prut, H. Bluthmann, H. Mohler, and U. Rudolph. 2002. Trace fear conditioning involves hippocampal alpha5 GABA(A) receptors. *Proc Natl Acad Sci USA* 99 (13):8980–8985.

Crestani, F., K. Low, R. Keist, M. Mandelli, H. Mohler, and U. Rudolph. 2001. Molecular targets for the myorelaxant action of diazepam. *Mol Pharmacol* 59 (3):442–445.

Dixon, C. I., H. V. Morris, G. Breen, S. Desrivieres, S. Jugurnauth, R. C. Steiner, H. Vallada, C. Guindalini, R. Laranjeira, G. Messas, T. W. Rosahl, J. R. Atack, D. R. Peden, D. Belelli, J. J. Lambert, S. L. King, G. Schumann, and D. N. Stephens. 2010. Cocaine effects on mouse incentive-learning and human addiction are linked to alpha2 subunit-containing GABA$_A$ receptors. *Proc Natl Acad Sci USA* 107 (5):2289–2294.

Duan, B., L. J. Wu, Y. Q. Yu, Y. Ding, L. Jing, L. Xu, J. Chen, and T. L. Xu. 2007. Upregulation of acid-sensing ion channel ASIC1a in spinal dorsal horn neurons contributes to inflammatory pain hypersensitivity. *J Neurosci* 27 (41):11139–11148.

Enoch, M. A., Z. Zhou, M. Kimura, D. C. Mash, Q. Yuan, and D. Goldman. 2012. GABAergic gene expression in postmortem hippocampus from alcoholics and cocaine addicts; corresponding findings in alcohol-naive P and NP rats. *PLoS One* 7 (1):e29369.

Follesa, P., E. Cagetti, L. Mancuso, F. Biggio, A. Manca, E. Maciocco, F. Massa, M. S. Desole, M. Carta, F. Busonero, E. Sanna, and G. Biggio. 2001. Increase in expression of the GABA(A) receptor alpha(4) subunit gene induced by withdrawal of, but not by long-term treatment with, benzodiazepine full or partial agonists. *Brain Res Mol Brain Res* 92 (1-2):138–148.

Gao, B., and J. M. Fritschy. 1994. Selective allocation of GABA$_A$ receptors containing the alpha 1 subunit to neurochemically distinct subpopulations of rat hippocampal interneurons. *Eur J Neurosci* 6 (5):837–853.

Gao, B., J. M. Fritschy, D. Benke, and H. Mohler. 1993. Neuron-specific expression of GABA$_A$-receptor subtypes: differential association of the alpha 1- and alpha 3-subunits with serotonergic and GABAergic neurons. *Neuroscience* 54 (4):881–892.

Gao, J., B. Duan, D. G. Wang, X. H. Deng, G. Y. Zhang, L. Xu, and T. L. Xu. 2005. Coupling between NMDA receptor and acid-sensing ion channel contributes to ischemic neuronal death. *Neuron* 48 (4):635–646.

Gutman, A. L., C. V. Cosme, M. F. Noterman, W. R. Worth, J. A. Wemmie, and R. T. LaLumiere. 2020. Overexpression of ASIC1$_A$ in the nucleus accumbens of rats potentiates cocaine-seeking behavior. *Addict Biol* 25 (2):e12690.

Heikkinen, A. E., T. P. Moykkynen, and E. R. Korpi. 2009. Long-lasting modulation of glutamatergic transmission in VTA dopamine neurons after a single dose of benzodiazepine agonists. *Neuropsychopharmacology* 34 (2):290–298.

Hernandez, T. D., C. Heninger, M. A. Wilson, and D. W. Gallager. 1989. Relationship of agonist efficacy to changes in GABA sensitivity and anticonvulsant tolerance following chronic benzodiazepine ligand exposure. *Eur J Pharmacol* 170 (3):145–155.

Hnasko, T. S., G. O. Hjelmstad, H. L. Fields, and R. H. Edwards. 2012. Ventral tegmental area glutamate neurons: electrophysiological properties and projections. *J Neurosci* 32 (43):15076–15085.

Hong, S. I., T. L. Nguyen, S. X. Ma, H. C. Kim, S. Y. Lee, and C. G. Jang. 2017. TRPV1 modulates morphine-induced conditioned place preference via p38 MAPK in the nucleus accumbens. *Behav Brain Res* 334:26–33.

Jacob, T. C., G. Michels, L. Silayeva, J. Haydon, F. Succol, and S. J. Moss. 2012. Benzodiazepine treatment induces subtype-specific changes in GABA(A) receptor trafficking and decreases synaptic inhibition. *Proc Natl Acad Sci U S A* 109 (45):18595–18600.

Jiang, Q., C. M. Wang, E. E. Fibuch, J. Q. Wang, and X. P. Chu. 2013. Differential regulation of locomotor activity to acute and chronic cocaine administration by acid-sensing ion channel 1a and 2 in adult mice. *Neuroscience* 246:170–178.

Jo, Y. H., and R. Schlichter. 1999. Synaptic corelease of ATP and GABA in cultured spinal neurons. *Nat Neurosci* 2 (3):241–245.

June, H. L., Sr., K. L. Foster, W. J. Eiler, 2nd, J. Goergen, J. B. Cook, N. Johnson, B. Mensah-Zoe, J. O. Simmons, H. L. June, Jr., W. Yin, J. M. Cook, and G. E. Homanics. 2007. Dopamine and benzodiazepine-dependent mechanisms regulate the EtOH-enhanced locomotor stimulation in the GABA$_A$ alpha1 subunit null mutant mice. *Neuropsychopharmacology* 32 (1):137–152.

Kang, I., and L. G. Miller. 1991. Decreased GABA$_A$ receptor subunit mRNA concentrations following chronic lorazepam administration. *Br J Pharmacol* 103 (2):1285–1287.

Khoja, S., N. Huynh, L. Asatryan, M. W. Jakowec, and D. L. Davies. 2018. Reduced expression of purinergic P2X4 receptors increases voluntary ethanol intake in C57BL/6J mice. *Alcohol* 68:63–70.

Kodirov, S. A., M. Wehrmeister, and L. V. Colom. 2014. Modulation of HCN channels in lateral septum by nicotine. *Neuropharmacology* 81:274–282.

Kravitz, A. V., L. D. Tye, and A. C. Kreitzer. 2012. Distinct roles for direct and indirect pathway striatal neurons in reinforcement. *Nat Neurosci* 15 (6):816–818.

Kreple, C. J., Y. Lu, R. J. Taugher, A. L. Schwager-Gutman, J. Du, M. Stump, Y. Wang, A. Ghobbeh, R. Fan, C. V. Cosme, L. P. Sowers, M. J. Welsh, J. J. Radley, R. T. LaLumiere, and J. A. Wemmie. 2014. Acid-sensing ion channels contribute to synaptic transmission and inhibit cocaine-evoked plasticity. *Nat Neurosci* 17 (8):1083–1091.

Krugel, U., H. Kittner, H. Franke, and P. Illes. 2001. Stimulation of P2 receptors in the ventral tegmental area enhances dopaminergic mechanisms in vivo. *Neuropharmacology* 40 (8):1084–1093.

Lee, D. W., P. S. Cho, H. K. Lee, S. H. Lee, S. J. Jung, and S. B. Oh. 2015. Trans-activation of TRPV1 by D1R in mouse dorsal root ganglion neurons. *Biochem Biophys Res Commun* 465 (4):832–837.

Li, W. G., M. G. Liu, S. Deng, Y. M. Liu, L. Shang, J. Ding, T. T. Hsu, Q. Jiang, Y. Li, F. Li, M. X. Zhu, and T. L. Xu. 2016. ASIC1a regulates insular long-term depression and is required for the extinction of conditioned taste aversion. *Nat Commun* 7:13770.

Liu, J., A. R. Yang, T. Kelly, A. Puche, C. Esoga, H. L. June, Jr., A. Elnabawi, I. Merchenthaler, W. Sieghart, H. L. June, Sr., and L. Aurelian. 2011. Binge alcohol drinking is associated with GABA$_A$ alpha2-regulated Toll-like receptor 4 (TLR4) expression in the central amygdala. *Proc Natl Acad Sci USA* 108 (11):4465–4470.

Liu, L., W. Zhu, Z. S. Zhang, T. Yang, A. Grant, G. Oxford, and S. A. Simon. 2004. Nicotine inhibits voltage-dependent sodium channels and sensitizes vanilloid receptors. *J Neurophysiol* 91 (4):1482–1491.

Low, K., F. Crestani, R. Keist, D. Benke, I. Brunig, J. A. Benson, J. M. Fritschy, T. Rulicke, H. Bluethmann, H. Mohler, and U. Rudolph. 2000. Molecular and neuronal substrate for the selective attenuation of anxiety. *Science* 290 (5489):131–134.

Luscher, B., Q. Shen, and N. Sahir. 2011. The GABAergic deficit hypothesis of major depressive disorder. *Mol Psychiatry* 16 (4):383–406.

Ma, S. X., H. C. Kim, S. Y. Lee, and C. G. Jang. 2018. TRPV1 modulates morphine self-administration via activation of the CaMKII-CREB pathway in the nucleus accumbens. *Neurochem Int* 121:1–7.

Macpherson, L. J., B. H. Geierstanger, V. Viswanath, M. Bandell, S. R. Eid, S. Hwang, and A. Patapoutian. 2005. The pungency of garlic: activation of TRPA1 and TRPV1 in response to allicin. *Curr Biol* 15 (10):929–934.

Magee, J. C. 1998. Dendritic hyperpolarization-activated currents modify the integrative properties of hippocampal CA1 pyramidal neurons. *J Neurosci* 18 (19):7613–7624.

McKernan, R. M., T. W. Rosahl, D. S. Reynolds, C. Sur, K. A. Wafford, J. R. Atack, S. Farrar, J. Myers, G. Cook, P. Ferris, L. Garrett, L. Bristow, G. Marshall, A. Macaulay, N. Brown, O. Howell, K. W. Moore, R. W. Carling, L. J. Street, J. L. Castro, C. I. Ragan, G. R. Dawson, and P. J. Whiting. 2000. Sedative but not anxiolytic properties of benzodiazepines are mediated by the GABA(A) receptor alpha1 subtype. *Nat Neurosci* 3 (6):587–592.

McNamara, F. N., A. Randall, and M. J. Gunthorpe. 2005. Effects of piperine, the pungent component of black pepper, at the human vanilloid receptor (TRPV1). *Br J Pharmacol* 144 (6):781–790.

Mihalek, R. M., B. J. Bowers, J. M. Wehner, J. E. Kralic, M. J. VanDoren, A. L. Morrow, and G. E. Homanics. 2001. GABA(A)-receptor delta subunit knockout mice have multiple defects in behavioral responses to ethanol. *Alcohol Clin Exp Res* 25 (12):1708–1718.

Moore, K. A., T. Kohno, L. A. Karchewski, J. Scholz, H. Baba, and C. J. Woolf. 2002. Partial peripheral nerve injury promotes a selective loss of GABAergic inhibition in the superficial dorsal horn of the spinal cord. *J Neurosci* 22 (15):6724–6731.

Moosmang, S., M. Biel, F. Hofmann, and A. Ludwig. 1999. Differential distribution of four hyperpolarization-activated cation channels in mouse brain. *Biol Chem* 380 (7-8):975–980.

North, R. A. 2002. Molecular physiology of P2X receptors. *Physiol Rev* 82 (4):1013–1067.

Okada, H., N. Matsushita, K. Kobayashi, and K. Kobayashi. 2004. Identification of GABA$_A$ receptor subunit variants in midbrain dopaminergic neurons. *J Neurochem* 89 (1):7–14.

Pankratov, Y., U. Lalo, A. Verkhratsky, and R. A. North. 2006. Vesicular release of ATP at central synapses. *Pflugers Arch* 452 (5):589–597.

Petursson, H. 1994. The benzodiazepine withdrawal syndrome. *Addiction* 89 (11):1455–1459.

Rewal, M., R. Jurd, T. M. Gill, D. Y. He, D. Ron, and P. H. Janak. 2009. Alpha4-containing GABA$_A$ receptors in the nucleus accumbens mediate moderate intake of alcohol. *J Neurosci* 29 (2):543–549.

Robison, A. J., and E. J. Nestler. 2011. Transcriptional and epigenetic mechanisms of addiction. *Nat Rev Neurosci* 12 (11):623–637.

Roca, D. J., I. Rozenberg, M. Farrant, and D. H. Farb. 1990. Chronic agonist exposure induces down-regulation and allosteric uncoupling of the gamma-aminobutyric acid/benzodiazepine receptor complex. *Mol Pharmacol* 37 (1):37–43.

Rudolph, U., F. Crestani, D. Benke, I. Brunig, J. A. Benson, J. M. Fritschy, J. R. Martin, H. Bluethmann, and H. Mohler. 1999. Benzodiazepine actions mediated by specific gamma-aminobutyric acid(A) receptor subtypes. *Nature* 401 (6755):796–800.

Santos-Vera, B., A. D. C. Vaquer-Alicea, C. E. Maria-Rios, A. Montiel-Ramos, A. Ramos-Cardona, R. Vazquez-Torres, P. Sanabria, and C. A. Jimenez-Rivera. 2019. Protein and surface expression of HCN2 and HCN4 subunits in mesocorticolimbic areas after cocaine sensitization. *Neurochem Int* 125:91–98.

Santos-Vera, B., R. Vazquez-Torres, H. G. Marrero, J. M. Acevedo, F. Arencibia-Albite, M. E. Velez-Hernandez, J. D. Miranda, and C. A. Jimenez-Rivera. 2013. Cocaine sensitization increases I h current channel subunit 2 (HCN(2)) protein expression in structures of the mesocorticolimbic system. *J Mol Neurosci* 50 (1):234–245.

Self, D. W. 2010. "Dopamine Receptor Subtypes in Rewardand Relapse." In *The Dopamine Receptors*, edited byK. A. Neve, 479–523. Humana Press.

Self, D. W. 2014. Diminished role for dopamine D1 receptors in cocaine addiction? *Biol Psychiatry* 76 (1):2–3.

Shah, M. M., A. E. Anderson, V. Leung, X. Lin, and D. Johnston. 2004. Seizure-induced plasticity of h channels in entorhinal cortical layer III pyramidal neurons. *Neuron* 44 (3):495–508.

Smith, S. S., Q. H. Gong, F. C. Hsu, R. S. Markowitz, J. M. ffrench-Mullen, and X. Li. 1998. GABA(A) receptor alpha4 subunit suppression prevents withdrawal properties of an endogenous steroid. *Nature* 392 (6679):926–930.

Sperlagh, B., H. Sershen, A. Lajtha, and E. S. Vizi. 1998. Co-release of endogenous ATP and [3H]noradrenaline from rat hypothalamic slices: origin and modulation by alpha2-adrenoceptors. *Neuroscience* 82 (2):511–520.

Suman, A., B. Mehta, M. L. Guo, X. P. Chu, E. E. Fibuch, L. M. Mao, and J. Q. Wang. 2010. Alterations in subcellular expression of acid-sensing ion channels in the rat forebrain following chronic amphetamine administration. *Neurosci Res* 68 (1):1–8.

Tan, K. R., M. Brown, G. Labouebe, C. Yvon, C. Creton, J. M. Fritschy, U. Rudolph, and C. Luscher. 2010. Neural bases for addictive properties of benzodiazepines. *Nature* 463 (7282):769–774.

Taugher, R. J., Y. Lu, R. Fan, A. Ghobbeh, C. J. Kreple, F. M. Faraci, and J. A. Wemmie. 2017. ASIC1$_A$ in neurons is critical for fear-related behaviors. *Genes Brain Behav* 16 (8):745–755.

Thyssen, A., D. Hirnet, H. Wolburg, G. Schmalzing, J. W. Deitmer, and C. Lohr. 2010. Ectopic vesicular neurotransmitter release along sensory axons mediates neurovascular coupling via glial calcium signaling. *Proc Natl Acad Sci USA* 107 (34):15258–15263.

Tian, Y. H., S. Y. Lee, H. C. Kim, and C. G. Jang. 2010. Repeated methamphetamine treatment increases expression of TRPV1 mRNA in the frontal cortex but not in the striatum or hippocampus of mice. *Neurosci Lett* 472 (1):61–64.

Tian, Y. H., S. X. Ma, K. W. Lee, S. Wee, G. F. Koob, S. Y. Lee, and C. G. Jang. 2018. Blockade of TRPV1 inhibits methamphetamine-induced rewarding effects. *Sci Rep* 8 (1):882.

Trevisani, M., D. Smart, M. J. Gunthorpe, M. Tognetto, M. Barbieri, B. Campi, S. Amadesi, J. Gray, J. C. Jerman, S. J. Brough, D. Owen, G. D. Smith, A. D. Randall, S. Harrison, A. Bianchi, J. B. Davis, and P. Geppetti. 2002. Ethanol elicits and potentiates noci-ceptor responses via the vanilloid receptor-1. *Nat Neurosci* 5 (6):546–551.

Tritsch, N. X., J. B. Ding, and B. L. Sabatini. 2012. Dopaminergic neurons inhibit striatal output through non-canonical release of GABA. *Nature* 490 (7419):262–266.

Tsay, D., J. T. Dudman, and S. A. Siegelbaum. 2007. HCN1 channels constrain synaptically evoked Ca^{2+} spikes in distal dendrites of CA1 pyramidal neurons. *Neuron* 56 (6):1076–1089.

Vukicevic, M., and S. Kellenberger. 2004. Modulatory effects of acid-sensing ion channels on action potential generation in hippocampal neurons. *Am J Physiol Cell Physiol* 287 (3):C682–C690.

Wearne, T. A., L. M. Parker, J. L. Franklin, A. K. Goodchild, and J. L. Cornish. 2016. GABAergic mRNA expression is upregulated in the prefrontal cortex of rats sensitized to methamphetamine. *Behav Brain Res* 297:224–230.

Williams, S. R., S. R. Christensen, G. J. Stuart, and M. Hausser. 2002. Membrane potential bistability is controlled by the hyperpolarization-activated current I(H) in rat cerebellar Purkinje neurons in vitro. *J Physiol* 539 (Pt 2):469–483.

Xiao, Z., Y. Y. Li, and M. J. Sun. 2015. Activation of P2X7 receptors in the midbrain periaqueductal gray of rats facilitates morphine tolerance. *Pharmacol Biochem Behav* 135:145–153.

Xu, H., N. T. Blair, and D. E. Clapham. 2005. Camphor activates and strongly desensitizes the transient receptor potential vanilloid subtype 1 channel in a vanilloid-independent mechanism. *J Neurosci* 25 (39):8924–8937.

You, I. J., S. I. Hong, S. X. Ma, T. L. Nguyen, S. H. Kwon, S. Y. Lee, and C. G. Jang. 2019. Transient receptor potential vanilloid 1 mediates cocaine reinstatement via the D1 dopamine receptor in the nucleus accumbens. *J Psychopharmacol* 33 (12):1491–1500.

Zhang, G. C., L. M. Mao, J. Q. Wang, and X. P. Chu. 2009. Upregulation of acid-sensing ion channel 1 protein expression by chronic administration of cocaine in the mouse striatum in vivo. *Neurosci Lett* 459 (3):119–122.

Zhang, M., H. Zhong, C. Vollmer, and C. A. Nurse. 2000. Co-release of ATP and ACh mediates hypoxic signalling at rat carotid body chemoreceptors. *J Physiol* 525 Pt 1:143–158.

Zhang, W., K. M. Daly, B. Liang, L. Zhang, X. Li, Y. Li, and D. T. Lin. 2017. BDNF rescues prefrontal dysfunction elicited by pyramidal neuron-specific DTNBP1 deletion in vivo. *J Mol Cell Biol* 9 (2):117–131.

Zhang, W., M. Peterson, B. Beyer, W. N. Frankel, and Z. W. Zhang. 2014. Loss of MeCP2 from forebrain excitatory neurons leads to cortical hyperexcitation and seizures. *J Neurosci* 34 (7):2754–2763.

Zhang, W., L. Zhang, B. Liang, D. Schroeder, Z. W. Zhang, G. A. Cox, Y. Li, and D. T. Lin. 2016. Hyperactive somatostatin interneurons contribute to excitotoxicity in neurodegenerative disorders. *Nat Neurosci* 19 (4):557–559.

Zhang, X., T. H. Lee, X. Xiong, Q. Chen, C. Davidson, W. C. Wetsel, and E. H. Ellinwood. 2006. Methamphetamine induces long-term changes in GABA$_A$ receptor alpha2 subunit and GAD67 expression. *Biochem Biophys Res Commun* 351 (1):300–305.

Zhou, M., K. Lin, Y. Si, Q. Ru, L. Chen, H. Xiao, and C. Li. 2019. Downregulation of HCN1 channels in hippocampus and prefrontal cortex in methamphetamine re-exposed mice with enhanced working memory. *Physiol Res* 68 (1):107–117.

Zhou, M., P. Luo, Y. Lu, C. J. Li, D. S. Wang, Q. Lu, X. L. Xu, Z. He, and L. J. Guo. 2015. Imbalance of HCN1 and HCN2 expression in hippocampal CA1 area impairs spatial learning and memory in rats with chronic morphine exposure. *Prog Neuropsychopharmacol Biol Psychiatry* 56:207–214.

Zygmunt, P. M., J. Petersson, D. A. Andersson, H. Chuang, M. Sorgard, V. Di Marzo, D. Julius, and E. D. Hogestatt. 1999. Vanilloid receptors on sensory nerves mediate the vasodilator action of anandamide. *Nature* 400 (6743):452–457.

14 Ion Channel Conformational Coupling in Ischemic Neuronal Death

Yi-Zhi Wang and Michael X. Zhu

CONTENTS

14.1 INTRODUCTION

From neuroprotection to neural injury, ion channels play a variety of important roles in different stages of ischemic stroke (Kahle et al., 2009; Lai et al., 2014; Weilinger et al., 2013; Xiong et al., 2004). However, for a very long period of time, the majority of researchers have stereotyped ion channels as just tunnels for ion exchange. Nearly all hypotheses regarding mechanisms of ion channel-mediated neurological damage were rooted from a simple idea: how ionic balance is disturbed by a certain channel during ischemic stroke. For example, numerous publications have shown that N-methyl-D-aspartate receptors (NMDARs) (Ge et al., 2020; Hoyte et al., 2004) and α-amino-3-hydroxy-5-methyl-4-isoxazolepropionic acid receptors (AMPARs) (Akins and Atkinson, 2002) contribute to ischemic neuronal death by mediating Ca^{2+} influx that results in cytosolic Ca^{2+} overload (Table 14.1). There is no doubt that these are all important findings and they are very enlightening. The concern is that the dominating ion conducting-based hypotheses may narrow the scope of exploration. In living cells, proteins always form networks and function in groups (or modules). No protein can fulfill a physiological or pathological function solely on its own. Neither do ion channels. In fact, an increasing number of ion channel-binding proteins (ICBP) have been revealed by affinity purification–mass spectrometry (Dunham et al., 2012) and proximity labeling analysis (Branon et al., 2018). Some of the ICBPs serve as auxiliary subunits to support ion channel function, like protein interacting with C Kinase - 1 (PINK1) for AMPARs (Xia et al., 1999) and beta subunits of voltage-gated Na^+, Ca^{2+} and K^+ channels (Yu et al., 2005). Others are the substrates or downstream effectors, like receptor-interacting serine/threonine-protein kinase 1 (RIPK1) for acid-sensing ion channel 1a (ASIC1a) (Wang et al., 2015).

Activation of an ion channel leads to conformational changes (Tombola et al., 2006; Unwin, 1989). One obvious and direct consequence of conformational changes is to open the pore for ion flux. However, it is often neglected that conformational changes also alter the ability of the ion channel to interact with its binding partners, leading to reorganization of the protein-protein interaction (PPI) network in which the channel protein takes part (De Las Rivas and Fontanillo, 2010; Lee et al., 2014; Stoilova-McPhie et al., 2013). The reorganization of PPI usually alters the status of protein modifications, such as ubiquitination and phosphorylation, which impact not only the function and localization of ion channels but other proteins within the PPI network as well (Perkins et al., 2010). Therefore, an ion channel may have both ion conduction-dependent and ion conduction-independent functions. In recent years, ion conduction-independent functions have emerged for more and more ion channels (Kaczmarek, 2006). For

TABLE 14.1

Death Mechanisms of Ion Channel-Mediated Ischemic Brain Damage

Name	Ion Conducting-Dependent Death Mechanism	Ion Conducting-Independent Death Mechanism
AMPARs	Ca^{2+}, Zn^{2+} overloading	Unknown
ASIC1a	Ca^{2+} overloading	Exposure of CP-1 death motif and phosphorylation of RIPK1
KARs	Ca^{2+} overloading	Unknown
$K_V2.1$	K^+ efflux	Form ER-PM junction to promote new $K_V2.1$ insertion
NMDARs	Ca^{2+} overloading	Activation of NMDAR-Src-Panx1 signaling
TRPC3/4/6	Ca^{2+} overloading	Unknown
TRPM2/4/7	Ca^{2+}, Mg^{2+}, Zn^{2+} overloading	TRPM2/4, unknown; TRPM7, activation of Fas
TRPV1/4	Ca^{2+} overloading	Unknown

instance, the voltage-gated K^+ channel, Shab-related subfamily, member 1 (Kv2.1) not only shapes membrane excitability by conducting K^+ but also regulates lipid recycling and ion channel trafficking as a structural scaffold (Kirmiz et al., 2018). NMDARs and ASIC1a are also found to mediate ischemic neuronal death in ion conduction-independent manner, besides mediating ion conduction-dependent brain injuries (Wang et al., 2015; Weilinger et al., 2016). The relationship between the ion conduction-dependent and ion conduction-independent pathways in ischemic neuronal death is very complex. They can either be coupled, like Kv2.1, or independent, like ASIC1a. However, most ion channel-targeted therapeutic strategies are designed to block ion conductance, thereby only disrupting the ion conduction-dependent death (Carmichael, 2012; Lai et al., 2014). The revealing of ion conduction-independent neuronal death mechanisms of many ion channels has been broadening the list of neuroprotection targets and providing promising complementary approaches to the current treatment of ischemic stroke.

In this chapter, we will introduce several typical ion channels involved in ischemic brain damage and compare the differences between their ion conduction-dependent and -independent functions. The focus will be on ion channel conformational coupling in neurons under conditions of ischemic stroke and how this mechanism contributes to neuronal death.

14.2 ION CONDUCTION-INDEPENDENT FUNCTIONS OF ION CHANNELS

Intercellular communications are critical for almost all essential functions of an organism, such as breeding, metabolic activities, and learning/memory. Receptors in the plasma membrane (PM) sense extracellular cues and convert the information to various kinds of intracellular signals to regulate cellular response. Generally speaking, there are two major types of cell surface receptors, ionotropic receptors and metabotropic receptors (Purves et al., 2001). The former refers to transmembrane proteins with an ion conducting pore, i.e. ion channels. There are a variety of extracellular cues that can activate ion channels, including chemical ligands (e.g. neurotransmitters), temperature alterations, mechanical forces, voltage changes, and pH fluctuations. When activated, the conformation of the channel changes to open state to allow selective permeation of one or multiple types of ions, like Na^+, K^+, Ca^{2+} and Cl^-. Although ion channels respond to stimulation very quickly (in a few milliseconds), the open state usually only lasts for a short period of time (from tens of milliseconds to minutes) and then it switches either back to the closed state or an inactive conformation before returning to the closed state. Metabotropic receptors refer to transmembrane proteins that do not conduct ions. They function either through second messengers like cyclic AMP and calcium or via signaling cascades that involve protein-protein interactions and phosphorylation. It has long been believed that ionotropic receptors and metabotropic receptors are completely distinct protein types dedicated to only ion conduction-dependent and independent functions, respectively. However, accumulating evidence in recent years shows that both ionotropic and metabotropic functions can be carried out by the same transmembrane protein. Such a dual function has been demonstrated in nearly every major family of ion channels, including transient receptor potential (TRP) channels, K^+ channels, Na^+ channels, Ca^{2+} channels, NMDARs, and kainate receptors (KARs) (Gomez-Ospina et al., 2006; Kaczmarek, 2006; Kruger and Isom, 2016; Petrovic et al., 2017; Runnels et al., 2001; Weilinger et al., 2016).

TRP channels represent a large group of non-selective cation channels conducting not only monovalent cations such as Na^+ and K^+, but also divalent cations like Ca^{2+}, Zn^{2+}, or Mg^{2+} to different degrees (Nilius and Owsianik, 2011). The 28 mammalian TRP members are divided into six subfamilies and involved in diverse functions such as sensing temperature, osmotic pressure, or volume changes, and a variety of chemicals (Ramsey et al., 2006). TRP channels play important roles in many physical and pathological processes, such as taste, nociception, and ischemic injuries. Although most TRP channels fulfill their functions by regulating ionic balance across biomembranes, some of them, such as transient receptor potential-melastatin-like 2 (TRPM2), TRPM6, and TRPM7, also function in ion conduction-independent manners (Krapivinsky et al., 2014; Perraud et al., 2001; Riazanova et al., 2001; Runnels et al., 2001; Sano et al., 2001). For instance, TRPM6 and TRPM7 are chanzymes, which not only act as ion channels mediating cation fluxes including Ca^{2+}, Zn^{2+}, and Mg^{2+}, but also contain a serine/threonine kinase domain at the carboxyl terminus (CT) of each subunit. TRPM7 has been considered as a major regulator for cellular Mg^{2+} homeostasis in vertebrates. However, tissue-specific deletion of the *Trpm7* gene from mouse T cells impairs thymopoiesis without affecting either the acute Mg^{2+} uptake ability or the maintenance of cellular Mg^{2+} levels (Jin et al., 2008). Instead, syntheses of many essential growth factors are severely disrupted in *Trpm7*$^{-/-}$ thymocytes, which impairs the differentiation and maintenance of thymic epithelial cells. Therefore, TRPM7 regulates thymopoiesis through its CT kinase activity but not via Mg^{2+} conducting function. Interestingly, in some specific progenitors derived from mouse embryonic stem cells (mESCs), the TRPM7 kinase domain can be cleaved from the channel part in vivo in a cell type-specific fashion (Krapivinsky et al., 2014). The TRPM7 cleaved kinases (M7CKs) translocate to the nucleus and bind to the zinc-finger domains of multiple chromatin-remodeling complexes, which modify histone phosphorylation (e.g. H3S10, H3S28, and H3T3) to regulate gene expression pattern. This binding is Zn^{2+} dependent and largely dependent on TRPM7's Zn^{2+} conducting function, suggesting a synergistic action of the ionotropic and metabotropic functions of TRPM7. The cleavage-dependent nuclear action of TRPM7 resembles that of Cav1.2 L-type voltage-gated Ca^{2+} channel, which also regulates gene expression through proteolytic cleavage of its CT and translocation of the cleaved small fragment into nucleus to activate gene transcription in neurons (Gomez-Ospina et al., 2006).

In ischemic stroke, TRPM7 plays a key role in anoxic cell death and delayed neuronal death caused by transient brain hypoperfusion (Sun et al., 2009). Under oxygen and glucose deprivation (OGD) conditions, TRPM7 mediates a Ca^{2+}-permeable nonselective cation conductance (I_{OGD}). In primary cultured cortical neurons, suppression of TRPM7 expression significantly reduces reactive oxygen/ nitrogen species (ROS) generation, anoxic Ca^{2+} overload, and neuronal death. Furthermore, the knockdown of TRMP7 by short hairpin RNA (shRNA) in rat brain makes neurons resistant to global ischemic damages. However, it is hard to explain the TRPM7-mediated neuronal death only by its channel function, although to what extent the kinase function of TRMP7 plays a role in ischemic neuronal death remains largely unknown. Evidence exists for M7CKs to contribute to Fas-induced apoptosis, which occurs in peri-infarct areas of ischemic brain and is a major factor of ischemic damage (Desai et al., 2012). Therefore, it is possible that TRPM7 contributes to ischemic neuronal death by regulating the Fas pathway through M7CKs, instead of its ion conduction function.

Voltage-gated potassium (Kv) channels are transmembrane proteins regulating resting membrane potential and action potential firing patterns (Wulff et al., 2009; Yellen, 2002). Kv channels comprise α and β subunits. The α subunits are pore-forming subunits designated as Kv1.x, Kv2.x, etc., while the β subunits, such as MinK-related peptides and potassium channel-interacting protein, are auxiliary subunits that do not participate in pore-forming but are involved in regulating channel gating and/or ion conductance. Although accumulating evidence shows that both α and β subunits of Kv channels can have ion conduction-independent functions, some of the Kv α subunits are more like the chanzyme TRPM7 described above with converged ionotropic and metabotropic functions. Therefore, we will only focus on the α subunits here. Among the more than 100 Kv channel isoforms, only Kv10.1 (ether-à-go-go, EAG) and Kv2.1 have so far been clearly demonstrated to possess ion conduction-independent functions (Castro-Rodrigues et al., 2018; Cazares-Ordonez and Pardo, 2017; Feinshreiber et al., 2010; Hegle et al., 2006; Kaczmarek, 2006; O'Connell et al., 2010). Kv10.1 is highly expressed in brain neurons, myoblasts, and cancer tissues. Overexpression of Kv10.1 in NIH 3T3 fibroblasts or C2C12 myoblasts significantly increases p38 mitogen-activated protein kinase (MAPK) activity and leads to dramatic cell proliferation, whereas a knockdown of Kv10.1 by shRNA not only reduces p38 activity but also inhibits cell proliferation (Hegle et al., 2006). Interestingly, a mutation of the pore region to abolish its ion conduction does not suppress Kv10.1-induced proliferation, suggesting that the underlying mechanism is independent of the K^+ flux. However, a mutation that keeps the Kv10.1 channel in a constitutive open state almost eliminated the proliferation effect of Kv10.1. These findings suggest that Kv10.1 channels may serve as a voltage sensor to switch on p38 MAPK pathways. Moreover, in the nervous systems of *Drosophila*, Kv10.1 physically interacts with Ca^{2+}/calmodulin-dependent protein kinase II (CaMKII) (Castro-Rodrigues et al., 2018). Intriguingly, the CaMKII-binding domain of Kv10.1 is homologous to the CaMKII autoregulatory region. Thus, the binding of Kv10.1 results in constitutive activation of CaMKII without the presence of calmodulin or even when cytosolic Ca^{2+} levels are low. These findings suggest that Kv10.1 may also serve as a voltage sensor that relieves CaMKII from autoinhibition through an ion conduction-independent manner. In addition, there is evidence that Kv10.1 is expressed in the nucleus, where it may affect gene expression by interacting with heterochromatin or regulating the nuclear K^+ homeostasis (Chen et al., 2011). Unlike TRPM7, the ionotropic and metabotropic functions of Kv10.1 seem to be uncoupled.

Kv2.1 channels are major delayed rectifying channels widely expressed in brain neurons (Misonou et al., 2005). These channels are best known to shape neuronal excitability during high-frequency stimulation. Although a small amount of Kv2.1 channels are localized on the neuronal cell surface across the entire PM in a diffused pattern, a large population of the Kv2.1 proteins are found to interact with endoplasmic reticulum (ER)-localized vesicle-associated membrane protein-associated proteins, VAPA and VAPB, to form micron-sized clusters in the soma, proximal dendrites, and axon initial segments (Fox et al., 2015; Johnson et al., 2018; Kirmiz et al., 2018). The clustering is dependent on the phosphorylation of its carboxyl terminus, which is regulated by cyclin-dependent kinase 5 (Cdk5). Pharmacological inhibition of Cdk5 leads to the dispersal of Kv2.1 clusters. These clusters represent the sites of ER/PM junctions and serve as hubs for regulating lipid transfer/homeostasis and ion channel trafficking. For example, clustered Kv2.1 channels regulate the dynamic recruitment of cytoplasmic membrane-associated phosphatidylinositol transfer proteins to ER/PM junctions. They also promote the clustering and opening of L-type Ca^{2+} channels, which functionally associate with ryanodine receptors in the ER (Vierra et al., 2019). Compared to the diffusely expressed Kv2.1 channels, clustered Kv2.1 channels do not efficiently conduct K^+. Mutations bearing a loss of ion conducting activity do not disrupt the ability of Kv2.1 to facilitate the formation of functional ER/PM junctions, suggesting that the clustered Kv2.1 mainly serves as a structural scaffold for physically interacting with VAPs to form and maintain ER/PM junctions in an ion conduction-independent manner. Brain ischemia causes Kv2.1 dephosphorylation and rapid cluster dispersal, which are regulated by cytosolic free Zn^{2+} levels (Aras et al., 2009; Misonou et al., 2008; Shah et al., 2014; Yeh et al., 2017). Declustering shifts voltage dependency and promotes activation of Kv2.1 channels for K^+ efflux. At physiological concentrations, intracellular K^+ suppresses the catalytic activities of several apoptotic proteases and nucleases (Hughes and Cidlowski, 1999; Yu, 2003). Therefore, the efflux of neuronal K^+ disinhibits apoptotic machineries and initiates the cell death cascades. However, it seems that the declustered Kv2.1 channels do not contribute to the proapoptotic K^+ efflux. Some reports demonstrate that brain ischemia leads to rapid insertion of new Kv2.1 channels into the PM from a reserve pool, which is facilitated by the clustered Kv2.1 channels at ER/PM junctions (Yeh et al., 2017, 2019). Additionally, brain ischemia also rapidly increases the expression level of Kv2.1 channels (Misonou et al., 2008). The proapoptotic K^+ efflux is mostly mediated by those newly inserted, diffused Kv2.1 channels. A synthetic peptide, TAT-DP-2, derived from the carboxyl terminus of the cognate channel Kv2.2 and able to interrupt Kv2.1-VAPA interactions, decluster Kv2.1 channels, and disrupt ER-PM junctions, not only significantly prevented membrane insertion of new Kv2.1 channels but also demonstrated promising neuroprotection against ischemic damage in an in vivo mouse ischemia-reperfusion model (Schulien et al., 2020; Yeh et al., 2020). This provides a good example of using the non-conducting function of an ion channel as a therapeutic target for treatment of ischemic stroke.

14.3   CONFORMATIONAL CHANGES OF ASIC1a MEDIATE ISCHEMIC NEURONAL NECROPTOSIS

ASIC1a belongs to the H^+-gated subgroup of the degenerin/epithelial Na^+ channel (DEG/ENaC) family of non-selective cation channels (Wang and Xu, 2011; Wemmie et al., 2013). In mammals, ASIC1a is widely expressed in both central nervous system and peripheral nervous system. In neurons, ASIC1a is distributed across dendritic membranes of neurons, including postsynaptic membranes, where it is involved in regulating membrane excitability and neurotransmission (Kreple et al., 2014). In addition, ASIC1a is also found to exist in mitochondrial membrane of mouse cortical neurons and shown to regulate mitochondrial permeability transition (Wang et al., 2013). Since pH alteration is a common phenomenon in many physiological and pathological processes, ASIC1a broadly contributes to a variety of pathophysiological conditions such as nociception, seizure, and brain ischemic damage (Wang and Xu, 2011; Wemmie et al., 2013). Ischemic stroke usually leads to a persistent acidosis in the brain (Pignataro et al., 2007). The pH value of the ischemic brain rapidly drops from 7.4 to around 6.0 after the onset of ischemic stroke, and this lasts for hours. Treatment with sodium bicarbonate not only restores brain pH to the neutral value but also significantly reduces the extent of ischemic brain injury, suggesting that persistent acidosis is a leading cause of ischemic neuronal death (Pignataro et al., 2007). As the major proton sensor in the brain, pharmacological inhibition of ASIC1a by psalmotoxin (PcTX1) or amiloride showed a similar degree of neuroprotection as the sodium bicarbonate treatment (Xiong et al., 2004). On the contrary, enhancement of ASIC1a currents by phosphorylation of Ser478 and Ser479 at the carboxyl terminus (Gao et al., 2005), or sensitization of ASIC1a by extracellular spermine (Duan et al., 2011), exacerbated acidosis-induced ischemic neuronal death. Furthermore, mice without the *Asic1a* gene demonstrate significant resistance to ischemic injury in middle cerebral artery occlusion (MCAO) model (Xiong et al., 2004). All these findings strongly suggest that ASIC1a mediates acidosis-induced neuronal death in ischemic brain.

14.3.1 ASIC1a-Mediated Acidic Neuronal Death Is Independent from Ca^{2+} Signaling

For a very long time, it was believed that Ca^{2+} influx through the ASIC1a channel is the leading cause of acidosis-induced neuronal death (Xiong et al., 2004; Yermolaieva et al., 2004). However, there are many contradictions between the Ca^{2+} influx-based death theory and intrinsic properties of ASIC1a channels. On one hand, ASIC1a channels only open for a few seconds and then fall into a desensitized state (Duan et al., 2011). Desensitized channels are not able to respond to any further extracellular H$^+$ stimulation (even stronger) unless the previously bound protons in the H$^+$-sensing domain are fully removed. This phenomenon is termed steady-state desensitization of ASIC1a channels. In other words, ASIC1a channels only function continuously (or intermittently) under conditions of high frequency pH oscillations, such as those happening in the synaptic cleft (Kreple et al., 2014). In persistent acidified environment, like that in ischemic brain, ASIC1a channels are only briefly opened for several seconds at the onset of acidosis. Then, the channels shift to steady-state desensitization and remain quiescent to further acidic simulations in the ischemic brain. Therefore, the rapid steady-state desensitization of ASIC1a channels leads to a paradox that the tissue acidification level at the onset of ischemic stroke, rather than the stroke duration, should determine the degree of brain damage. Obviously, this is inconsistent with the observation that the longer the acidosis lasts, the worse the brain damage gets. On the other hand, homomeric ASIC1a channels are mainly permeable to Na^+, with a small amount of Ca^{2+} also being conducted (Samways et al., 2009). Compared to other major Ca^{2+}-permeable channels such as NMDARs and AMPARs, the Ca^{2+} component of ASIC1a currents, which last for only a few seconds, is very small. It is unlikely that this tiny amount of Ca^{2+} influx mediated by ASIC1a channels can account for all the damages to the brain caused by acidosis. Thus, other death mechanisms, instead of Ca^{2+} overload, may underlie the ASIC1a-mediated ischemic brain damage, which may or may not require the ion conduction function of the ASIC1a channels.

In a recent study, we provided several lines of evidence demonstrating that ion conduction is not necessary for ASIC1a-mediated neuronal death (Wang et al., 2015). Firstly, a replacement of Na^+, K^+, and Ca^{2+} with the nonpermeable organic cation N-methyl-D-glucamine in the extracellular solution completely abolished the acid (pH 6.0)-evoked ASIC1a current, but only showed very limited protection (~10%) against acidosis (pH 6.0, one-hour treatment)-induced neuronal death. Simply removing Ca^{2+} from the extracellular solution affected neither the ASIC1a current nor acidosis-induced neuronal death. Thus, the Ca^{2+}-mediated death cascade obviously does not take a critical part in ASIC1a-mediated acidotoxicity. Secondly, the steady-state desensitized ASIC1a channels are still able to mediate acidotoxicity. In primary cultured mouse cortical neurons, the longer the time of acidification, the more the neurons die. However, ASIC1a is still required for neuronal death induced by the persistent acidosis as deletion of the *Asic1a* gene rendered neurons resistant to acidotoxicity. Thirdly, a non-conducting ASIC1a mutant, HIF-ASIC1a, which has $H^{32}IF^{34}$ in the ion conducting pathway mutated to AAA and therefore a dysfunctional pore, mediated acidosis-induced neuronal death to the same level as the wild type (WT) ASIC1a. By contrast, a conducting mutant, RC-ASIC1a, which has the carboxyl terminus (CT) of ASIC1a replaced by a shortened scrambled amino acid sequence, KLRILQSTVPRARDDPDLDN, failed to mediate acidotoxicity despite conducting the acid-evoked current normally. This suggests that the CT of ASIC1a, while not required for the ion conducting function of the channel, plays an essential role in ASIC1a-mediated acidotoxicity.

14.3.2 Exposure of CP-1 Cytotoxic Motif Is Essential for ASIC1a-Mediated Neuronal Death

Surprisingly, we found that ASIC1a CT contains a cytotoxic motif, CP-1. Introduction of a synthetic membrane-permeable peptide representing the CP-1 fragment into primary cultures of mouse cortical neurons caused severe neuronal demise even under the physiological pH of 7.4. We further found that under resting conditions, the CP-1 motif is masked by the distal amino terminal (NT) region of ASIC1a and completely silenced (Wang et al., 2020). In the Rosetta model of closed state ASIC1a, the four negatively charged glutamate residues (E^6EEE^9) at the distal NT region of ASIC1a are juxtaposed to three positively charged lysine residues (K468, K471, and K474) contained within the CP-1 motif. Therefore, the electrostatic interactions between residues at NT and CT likely keep ASIC1a in an autoinhibited state in terms of executing cytotoxicity. Using Förster resonance energy transfer, we showed that acidosis (pH 6.0 treatment) disrupts the NT-CT binding in both WT-ASIC1a and the pore-dead mutant, HIF-ASIC1a (Figure 14.1a), suggesting that extracellular protons cause NT-CT dissociation of ASIC1a, and this does not require ionic flux through the channel. More importantly, the NT-CT dissociation occurred slowly but persisted for as

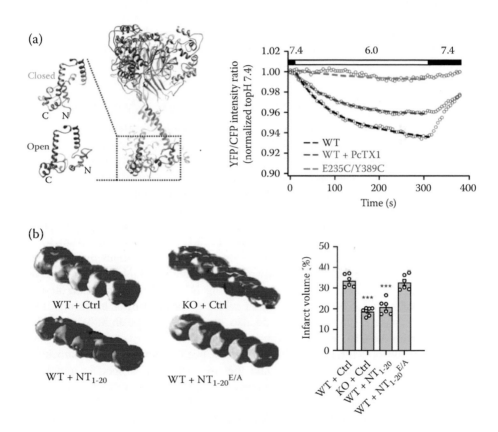

FIGURE 14.1 Inhibition of acidosis-induced NT-CT dissociation of ASIC1a suppressed ischemic brain damage in MCAO model (reprinted). a, Rosetta model of ASIC1a depicting an NT-CT interaction at the closed state and the dissociation between NT and CT of ASIC1a in the open state. b, FRET experiments showing that extracellular acidification induces ASIC1a-NT to dissociate from its CT. PcTX1 significantly slows down the NT-CT disassociation. Mutation of E235C/Y389C stabilizes the NT-CT interaction by forming a disulfide bond that prevents the proton-induced conformational transition. c, Intraventricular administration of a synthetic membrane-permeable peptide NT_{1-20} significantly reduced ischemic brain damage in the mouse MCAO model similarly to the *Asic1a* knockout (KO). The E → A substituted peptide, $NT_{1-20}^{E/A}$, which is unable to bind to ASIC1a CT, failed to show any protection.

long as the acidosis was present, which contrasted sharply to the rapid development and fast inactivation of the transient acid-evoked current mediated by ASIC1a. Furthermore, depletion of the first 20 amino acid residues at NT ($\Delta 1$–20 ASIC1a) totally abolishes NT-CT interaction and frees CP-1 cytotoxic motif. As a result, overexpression of $\Delta 1$–20 ASIC1a in Chinese hamster ovary cells (CHO) caused constitutive cell death at physiological pH in the absence of any acid stimulation. In contrast to $\Delta 1$–20 ASIC1a, overexpression of WT ASIC1a in CHO cells caused no cell death at the physiological pH. Moreover, acidic stimulation also causes N-ethylmaleimide-sensitive fusion ATPase (NSF) to bind to the distal NT of ASIC1a, which prevents the NT from binding back to CT, allowing extended exposure of the CP-1 cytotoxic motif. It was found that the knockdown of NSF by shRNA significantly slowed down the acidosis-induced NT-CT dissociation and inhibited the acidosis-induced neuronal death.

14.3.3 RIPK-1 Plays a Key Role in ASIC1a-Mediated Acidotoxicity

The CP-1 cytotoxic motif induces cell death through binding to RIPK-1 and causing its phosphorylation, which initiates a programmed necrotic cell death pathway, known as necroptosis (Galluzzi et al., 2017; Shan et al., 2018). It has become evident that any manipulation that leads to the exposure of ASIC1a CP-1 motif at the cytoplasmic side can trigger RIPK1 activation and/or cell death. These include the use of membrane penetrating CP-1 peptide, proton activation of ASIC1a to disrupt the autoinhibitory NT-CT interaction, deletion of the distal NT to deprotect its CT, and neutralization of NT glutamates E^6EEE^9 to AAAA to disrupt their interaction with lysines in the CP-1 motif. Conversely, manipulations that promote the NT-CT interaction, such as knockdown or pharmacological inhibition of NSF or treatment with a synthetic membrane penetrating NT peptide representing amino acids 1–20 of ASIC1a, NT_{1-20}. Consistent with the mechanism of blocking CP-1 death motif exposure, the NT_{1-20} peptide also suppressed cell death induced by the incubation with the CP-1 peptide (Wang et al., 2020).

The key role of RIPK1 in acidotoxicity has been demonstrated by knocking down RIPK-1 expression by shRNA and the use of a RIPK-1 specific inhibitor, necrostatin-1 (Nec-1), which both dramatically suppressed acidosis-induced neuronal death to the same extent as the ASIC1a specific antagonist, PcTX1. PcTX1 and Nec-1 also similarly inhibited acidosis-induced RIPK-1 phosphorylation. In addition, Nec-1, but not PcTX1, inhibited constitutive cell death caused by the expression of $\Delta 1$–20 ASIC1a, suggesting that the NT deletion mutant-induced cell death occurs through RIPK1-triggered necroptosis without involving channel opening (Wang et al., 2020). Supporting a necrotic death mechanism of acidotoxicity, the pH 6.0-treated mouse cortical neurons exhibit typical necrotic phenotypes, including plasma membrane rupture, organelle swelling, and cell lysis (Wang et al., 2015). Therefore, all evidence collected so far strongly suggests that acidosis-induced conformational change of ASIC1a initiates necroptosis pathway in an ion conduction-independent manner.

14.3.4 ASIC1a MEDIATES NEURONAL NECROPTOSIS IN ISCHEMIC BRAIN

Necroptosis is a major death pathway of ischemic neuronal demise (Galluzzi et al., 2017; Shan et al., 2018). Ischemic attack causes rapid phosphorylation of RIPK-1, which further phosphorylates RIPK-3. The activated RIPK-3 then phosphorylates Mixed Lineage Kinase Domain-like (MLKL) pseudokinase to cause necroptosis. Nec-1 as well as RIPK-3 inhibitors, like dabrafenib, show strong neuroprotection against brain ischemic injuries in rodent ischemic stroke models such as MCAO (Cruz et al., 2018; Degterev et al., 2005). However, it remains unclear which upstream factor(s) activates RIPK-1 in the ischemic brain. There have been several hypotheses implicating death receptors, toll-like receptors, genotoxic stresses and excitotoxic stresses as the trigger for RIPK-1 activation in cerebral ischemia/reperfusion (Galluzzi et al., 2017; Shan et al., 2018). Recently, we proposed the hypothesis that acidosis-induced conformational changes of ASIC1a activate RIPK1 to initiate ischemic necroptosis (Wang et al., 2015, 2020). We demonstrated the formation of an NSF-ASIC1a-RIPK-1 complex and a significant increase in RIPK-1 phosphorylation in the ischemic hemispheres within a half hour of the onset of MCAO. However, in the brain of global *Asic1a* knockout mice, the same MCAO treatment failed to induce any detectable increase of RIPK-1 phosphorylation in the ischemic hemispheres, suggesting that the ischemia-induced RIPK-1 phosphorylation depends on ASIC1a (Wang et al., 2015). Furthermore, intraventricular administration of the membrane-penetrating NT_{1-20} peptide 30 min before MCAO showed a neuroprotective effect against ischemic brain injury to the similar degree as the *Asic1a* gene knockout, as quantified by brain infarct volumes. Administration a negative control peptide, $NT_{1-20}^{E/A}$, in which the four glutamate residues in NT_{1-20} are substituted by alanines, had no neuroprotective effect (Wang et al., 2015, 2020), suggesting that the NT-CT association of ASIC1a represents the main mechanism that prevents CP-1 death motif exposure and the consequent toxicity mediated by ASIC1a in ischemic brain (Figure 14.1b).

Collectively, the above data strongly argue for an ion conduction-independent mechanism in ASIC1a-mediated neuronal necroptosis in brain ischemia. However, there are still many unanswered questions about the molecular mechanism of ASIC1a conformational change that induces ischemic brain injury. For instance, how is RIPK-1 phosphorylated by the interaction with ASIC1a? Does ASIC1a CT function as a kinase like the CT of TRPM7, or does the exposed ASIC1a CT recruit another kinase to phosphorylate RIPK-1? Answers to these questions will not only greatly improve our understanding of the molecular and cell signaling mechanisms that underlie neuronal acidotoxicity and ischemic brain damage, but also facilitate development of ASIC1a-targeted therapeutic strategies against ischemic stroke and other neurodegenerative disorders.

14.3.5 OTHER POSSIBLE MECHANISMS OF ASIC1a-MEDIATED ISCHEMIC DAMAGE

The demonstration of ASIC1a in ion conduction-independent pathway to mediate acidotoxicity does not exclude the contribution of ASIC1a-mediated ionic currents in ischemic neuronal death. First, a number of factors (e.g. spermine and lactate) can enhance the ion channel function of ASIC1a and hence augment Ca^{2+} influx to levels high enough to induce cell death. In the ischemic brain, the disrupted metabolic activities lead to abnormal generation and accumulation of small molecules such as spermine, lactate, and histamine (Tantini et al., 2006). Spermine profoundly curtails desensitization of ASIC1a by shifting the threshold of steady-state desensitization to more acidic pH. Therefore, in the presence of extracellular spermine, homomeric ASIC1a channels exhibit larger acid-induced currents and are easier to recover from steady-state desensitization than in its absence (Duan et al., 2011). Lactate and histamine can also potentiate ASIC1a currents by increasing the sensitivity of ASIC1a channels to minor pH drops (Biran et al., 2008; Brouns et al., 2008; Immke and McCleskey, 2001; Nagaeva et al., 2016). In addition, during ischemic stroke, activation of NR2B-containing NMDARs triggers CaMKII to phosphorylate ASIC1a at CT residues Ser478 and Ser479, which robustly potentiates ASIC1a currents (Gao et al., 2005). Therefore, in the ischemic brain, the same level of acidification leads to much greater ASIC1a currents and intracellular Ca^{2+} rise than in the healthy brain, which may cause excessive membrane depolarization and Ca^{2+} overload, and in turn downstream death cascade. Plausibly, the elevated extracellular spermine, lactate, and histamine, and/or phosphorylation at Ser478 and Ser479, may also affect ASIC1a regulation of neuronal death through conformational coupling. These possibilities warrant further investigation. Sorting out how various small molecule modulators and post-translational modifications regulate the ion conduction-dependent and -independent functions of ASIC1a and how they contribute to acidosis-induced neuronal death represents a major area of future research and a key to fully understand the roles of ASIC1a in ischemic brain damage.

Second, ASIC1a can form heteromeric channels with other ASIC subunits, like ASIC2b (Wemmie et al., 2013). ASIC1a/2b channels exhibit different channel properties from homomeric ASIC1a channels, which is believed to impact the ion conduction-dependent regulation of acidosis-induced neuronal death (Sherwood et al., 2011). Third, ASIC1a may regulate other ion channels and thereby indirectly affect neuronal death through changes in ion conductance. For example, ASIC1a interacts with the large-conductance Ca^{2+}- and voltage-activated K^+ (BK) channel, which is a key endogenous neuroprotective factor in ischemic stroke (Contet et al., 2016; Petroff et al., 2008). During ischemia, neurons are severely depolarized and overloaded with Ca^{2+}, which provokes many death cascades. Activation of the BK channel by Ca^{2+} and depolarization repolarizes (or even hyperpolarizes) the neuronal plasma membrane, serving as a feedback neuroprotective mechanism to brake Ca^{2+} overload and excitotoxicity (Bentzen et al., 2014). Under physiological conditions, the closed ASIC1a associates with BK and inhibits its current. During ischemia, the activation of ASIC1a disrupts the ASIC1a-BK interaction and relieves BK from inhibition. This disinhibition strongly boosts BK current and facilitates membrane repolarization. Therefore, the rapid activation of ASIC1a at the onset of ischemic stroke may protect neurons from Ca^{2+} overload and excitotoxicity by disinhibiting BK.

Finally, ASIC1a has been found to localize in mitochondrial membrane of mouse cortical neurons (Wang et al., 2013). Mitochondrial ASIC1a (mtASIC1a) contributes to mitochondrial Na^+/Ca^{2+} signaling and metabolic activities. In primary cultures of mouse cortical neurons, even a mild pH shift (from 7.4 to 7.0) in the culture medium is sufficient to activate mtASIC1a in digitonin-permeabilized neurons (Savic Azoulay et al., 2020). Such activity may enhance mitochondrial respiration and evoke an increase in mitochondrial Na^+ concentration. Moreover, a recent study using Tetrahedron DNA-based Nanosensor showed that O_2^- and amyloid β (Aβ)-triggered transitory cytoplasmic acidosis can activate mtASIC1a, which leads to mitochondrial Ca^{2+} overload and abnormal pH alterations (Liu et al., 2018). Deletion of the *Asic1a* gene or blockade with PcTX1 suppresses mtASIC1a-mediated mitochondrial Ca^{2+} overburdening and pH abnormality. These findings suggest that mtASIC1a can be activated by intracellular acidification to regulate mitochondrial Na^+/Ca^{2+} signaling. Furthermore, mtASIC1a physically associates with adenine nucleotide translocase (ANT) and regulates mitochondrial permeability transition (MPT). Deletion of the *Asic1a* gene impairs MPT in mouse cortical neurons. For instance, ASIC1a-null mitochondria display enhanced Ca^{2+} retention capacity (CRC) and mitochondrial Ca^{2+} uptake rate compared to that in WT neurons; hydrogen peroxide treatment also induces less cytochrome C release and smaller inner mitochondrial membrane depolarization in $Asic1a^{-/-}$ neurons than in WT ones. Therefore, $Asic1a^{-/-}$ neurons are more resistant to MPT-depended oxidative neuronal death.

In summary, ischemic attack causes multiple pathological alterations in an affected brain, including extracellular acidosis and oxidative stress. Surface ASIC1a mediates extracellular acidosis-induced neuronal necroptosis by exposing its CP-1 death motif, which in turn activates RIPK-1 and the subsequent necroptotic cascade. This process does not require the ion conduction function of ASIC1a and is facilitated by NSF. In the presence of spermine, lactate, and histamine, activation of surface ASIC1a also contributes to Ca^{2+}-dependent cell demise (Figure 14.2). On the other hand, intracellular mtASIC1a contributes to oxidative stress-induced neuronal death by regulating MPT. Although cell surface ASIC1a and mtASIC1a are seemingly involved in distinct death pathways, there may be cross-talks between them. With the necroptotic cell death, activated MLKL is translocated to mitochondria, where it opens the MPT pore and causes necrotic cell death (Galluzzi et al., 2017; Shan et al., 2018). Therefore, mtASIC1a may contribute to surface ASIC1a-mediated necroptosis by regulating the MPT pore. Moreover, mtASIC1a may also be involved in the production of lactate and other ASIC1a modulators under ischemic conditions, which can enhance the function of surface ASIC1a. Conversely, the surface ASIC1a may regulate mtASIC1a function through affecting intracellular Na^+ signaling near mitochondria. Compared to the extensive knowledge on surface ASIC1a, the study of mtASIC1a is still at its infancy. Neither how mtASIC1a is targeted to mitochondria nor whether it mediates inward or outward current is known. It could be possible that mtASIC1a only functions as a structural scaffold to facilitate the formation or maintenance of the MPT pore through ANT. Better understanding of the mtASIC1a function may shed new light on the molecule mechanisms of ischemic brain damage.

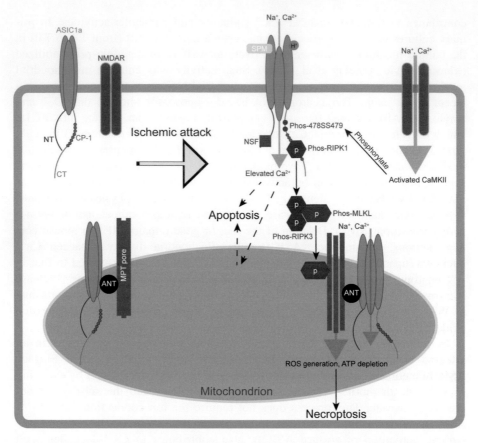

FIGURE 14.2 Distinct roles of surface and mitochondrial ASIC1a in ischemic neuronal death. During ischemic attack, extracellular acidosis activates surface ASIC1a and exposes CP-1 death motif, which causes phosphorylation of RIPK-1. Activated RIPK-1 further phosphorylates RIPK-3 and MLKL. Phosphorylated-MLKL then traffics to mitochondria and binds to MPT pore complex, which leads to increased ROS production and ATP depletion. Since mtASIC1a is a component of MPT pore complex, it may contribute to this process by regulating mitochondrial Na^+ and Ca^{2+} balances, which may accelerate the process of ischemic necroptosis. NMDAR is also activated during ischemic stroke, which motivates CaMKII. Surface ASIC1a can be phosphorylated by CaMKII at S478/S479 sites, which sensitizes the ASIC1a channel. Spermine also sensitizes the ASIC1a channel. The sensitized ASIC1a channel significantly increases Ca^{2+} influx, which leads to apoptosis in ischemic brains.

14.4 CONCLUDING REMARK

Accumulating evidence shows both ionotropic and metabotropic functions can co-exist in a single protein (Desai et al., 2012; Feinshreiber et al., 2010; Gomez-Ospina et al., 2006; Jin et al., 2008; Kaczmarek, 2006; Krapivinsky et al., 2014; Petrovic et al., 2017; Runnels et al., 2001; Wang et al., 2015; Weilinger et al., 2016). In particular, an ion channel can mediate ionic flux and alter cell function through changes in membrane potential and ionic concentrations, or it can affect cell signaling through conformational coupling to other proteins without the involvement of ionic flux. While the signaling cascades triggered by the ionotropic and metabotropic mechanisms may be quite different, the resulting biological consequences are often similar or related. For example, the Zn^{2+}-conducting and alpha-kinase functions of TRPM7 work together to regulate gene expression (Krapivinsky et al., 2014). ASIC1a can mediate acidosis-induced ischemic neuronal death either by conducting Na^+ and Ca^{2+} or by activating RIPK-1 through conformational coupling via the CP-1 death motif located at its CT (Wang et al., 2015; Xiong et al., 2004; Yermolaieva et al., 2004). The multiple actions carried out by the same protein through different pathways for a similar biological outcome may reflect evolutionary convergence whereby different protein modules that acted together to accomplish a biological function were assembled together in the same polypeptide for shared regulation (e.g. ligand sensing) and/or coupling efficiency. This explains why many ion channels can perform ion conduction-independent metabotropic functions. Because of co-evolution, it would be natural for the metabotropic function to interact with the io-notropic function of the ion channel, but it is equally possible that the metabotropic function retains the ability to work independently of the ionotropic actions of the channel for certain biological responses. Ion channels are popular therapeutic targets, including the treatment of ischemic stroke (Bansal et al., 2013; Lai et al., 2014). Many efforts have been and are still being made in developing ion channel blockers. However, the clinical performances of those blockers have not lived up to expectations. Given that the blockers are often identified based on their ability to alter channel gating or channel activity, the largely neglected ion conduction-independent functions may be one possible reason for the failures. Therefore, deepening the un-derstanding of conformational coupling of ion channels in neuronal death associated with ischemic stroke will greatly promote development of new and effective ther-apeutic strategies for stroke treatment.

REFERENCES

Akins, P.T., and Atkinson, R.P. (2002). Glutamate AMPA receptor antagonist treatment for ischaemic stroke. Curr Med Res Opin *18 Suppl 2*, s9–s13.

Aras, M.A., Saadi, R.A., and Aizenman, E. (2009). Zn^{2+} regulates Kv2.1 voltage-dependent gating and localization following ischemia. Eur J Neurosci *30*, 2250–2257.

Bansal, S., Sangha, K.S., and Khatri, P. (2013). Drug treatment of acute ischemic stroke. Am J Cardiovasc Drugs *13*, 57–69.

Bentzen, B.H., Olesen, S.P., Ronn, L.C., and Grunnet, M. (2014). BK channel activators and their therapeutic perspectives. Front Physiol *5*, 389.

Biran, V., Cochois, V., Karroubi, A., Arrang, J.M., Charriaut-Marlangue, C., and Heron, A. (2008). Stroke induces histamine accumulation and mast cell degranulation in the neonatal rat brain. Brain Pathol *18*, 1–9.

Branon, T.C., Bosch, J.A., Sanchez, A.D., Udeshi, N.D., Svinkina, T., Carr, S.A., Feldman, J.L., Perrimon, N., and Ting, A.Y. (2018). Efficient proximity labeling in living cells and organisms with TurboID. Nat Biotechnol *36*, 880–887.

Brouns, R., Sheorajpanday, R., Wauters, A., De Surgeloose, D., Marien, P., and De Deyn, P.P. (2008). Evaluation of lactate as a marker of metabolic stress and cause of sec-ondary damage in acute ischemic stroke or TIA. Clin Chim Acta *397*, 27–31.

Carmichael, S.T. (2012). Brain excitability in stroke: the yin and yang of stroke progression. Arch Neurol *69*, 161–167.

Castro-Rodrigues, A.F., Zhao, Y., Fonseca, F., Gabant, G., Cadene, M., Robertson, G.A., and Morais-Cabral, J.H. (2018). The interaction between the Drosophila EAG potassium channel and the protein kinase CaMKII involves an extensive interface at the active site of the kinase. J Mol Biol *430*, 5029–5049.

Cazares-Ordonez, V., and Pardo, L.A. (2017). Kv10.1 potassium channel: from the brain to the tumors. Biochem Cell Biol *95*, 531–536.

Chen, Y., Sanchez, A., Rubio, M.E., Kohl, T., Pardo, L.A., and Stuhmer, W. (2011). Functional K(v)10.1 channels localize to the inner nuclear membrane. PLoS One *6*, e19257.

Contet, C., Goulding, S.P., Kuljis, D.A., and Barth, A.L. (2016). BK channels in the central nervous system. Int Rev Neurobiol *128*, 281–342.

Cruz, S.A., Qin, Z., Stewart, A.F.R., and Chen, H.H. (2018). Dabrafenib, an inhibitor of RIP3 kinase-dependent necroptosis, reduces ischemic brain injury. Neural Regen Res *13*, 252–256.

De Las Rivas, J., and Fontanillo, C. (2010). Protein-protein interactions essentials: key concepts to building and analyzing interactome networks. PLoS Comput Biol *6*, e1000807.

Degterev, A., Huang, Z., Boyce, M., Li, Y., Jagtap, P., Mizushima, N., Cuny, G.D., Mitchison, T.J., Moskowitz, M.A., and Yuan, J. (2005). Chemical inhibitor of non-apoptotic cell death with therapeutic potential for ischemic brain injury. Nat Chem Biol *1*, 112–119.

Desai, B.N., Krapivinsky, G., Navarro, B., Krapivinsky, L., Carter, B.C., Febvay, S., Delling, M., Penumaka, A., Ramsey, I.S., Manasian, Y., *et al.* (2012). Cleavage of TRPM7 releases the kinase domain from the ion channel and regulates its participation in Fas-induced apoptosis. Dev Cell *22*, 1149–1162.

Duan, B., Wang, Y.Z., Yang, T., Chu, X.P., Yu, Y., Huang, Y., Cao, H., Hansen, J., Simon, R.P., Zhu, M.X., *et al.* (2011). Extracellular spermine exacerbates ischemic neuronal injury through sensitization of ASIC1a channels to extracellular acidosis. J Neurosci *31*, 2101–2112.

Dunham, W.H., Mullin, M., and Gingras, A.C. (2012). Affinity-purification coupled to mass spectrometry: basic principles and strategies. Proteomics *12*, 1576–1590.

Feinshreiber, L., Singer-Lahat, D., Friedrich, R., Matti, U., Sheinin, A., Yizhar, O., Nachman, R., Chikvashvili, D., Rettig, J., Ashery, U., *et al.* (2010). Non-conducting function of the Kv2.1 channel enables it to recruit vesicles for release in neuroendo-crine and nerve cells. J Cell Sci *123*, 1940–1947.

Fox, P.D., Haberkorn, C.J., Akin, E.J., Seel, P.J., Krapf, D., and Tamkun, M.M. (2015). Induction of stable ER-plasma-membrane junctions by Kv2.1 potassium channels. J Cell Sci *128*, 2096–2105.

Galluzzi, L., Kepp, O., Chan, F.K., and Kroemer, G. (2017). Necroptosis: mechanisms and relevance to disease. Annu Rev Pathol *12*, 103–130.

Gao, J., Duan, B., Wang, D.G., Deng, X.H., Zhang, G.Y., Xu, L., and Xu, T.L. (2005). Coupling between NMDA receptor and acid-sensing ion channel contributes to is-chemic neuronal death. Neuron *48*, 635–646.

Ge, Y., Chen, W., Axerio-Cilies, P., and Wang, Y.T. (2020). NMDARs in cell survival and death: implications in stroke pathogenesis and treatment. Trends Mol Med *26*, 533–551.

Gomez-Ospina, N., Tsuruta, F., Barreto-Chang, O., Hu, L., and Dolmetsch, R. (2006). The C terminus of the L-type voltage-gated calcium channel Ca(V)1.2 encodes a transcription factor. Cell *127*, 591–606.

Hegle, A.P., Marble, D.D., and Wilson, G.F. (2006). A voltage-driven switch for ion-independent signaling by ether-a-go-go K^+ channels. Proc Natl Acad Sci USA *103*, 2886–2891.

Hoyte, L., Barber, P.A., Buchan, A.M., and Hill, M.D. (2004). The rise and fall of NMDA antagonists for ischemic stroke. Curr Mol Med *4*, 131–136.

Hughes, F.M., Jr., and Cidlowski, J.A. (1999). Potassium is a critical regulator of apoptotic enzymes in vitro and in vivo. Adv Enzyme Regul *39*, 157–171.

Immke, D.C., and McCleskey, E.W. (2001). Lactate enhances the acid-sensing Na^+ channel on ischemia-sensing neurons. Nat Neurosci *4*, 869–870.

Jin, J., Desai, B.N., Navarro, B., Donovan, A., Andrews, N.C., and Clapham, D.E. (2008). Deletion of Trpm7 disrupts embryonic development and thymopoiesis without altering Mg^{2+} homeostasis. Science *322*, 756–760.

Johnson, B., Leek, A.N., Sole, L., Maverick, E.E., Levine, T.P., and Tamkun, M.M. (2018). Kv2 potassium channels form endoplasmic reticulum/plasma membrane junctions via interaction with VAPA and VAPB. Proc Natl Acad Sci USA *115*, E7331–E7340.

Kaczmarek, L.K. (2006). Non-conducting functions of voltage-gated ion channels. Nat Rev Neurosci *7*, 761–771.

Kahle, K.T., Simard, J.M., Staley, K.J., Nahed, B.V., Jones, P.S., and Sun, D. (2009). Molecular mechanisms of ischemic cerebral edema: role of electroneutral ion transport. Physiology (Bethesda) *24*, 257–265.

Kirmiz, M., Palacio, S., Thapa, P., King, A.N., Sack, J.T., and Trimmer, J.S. (2018). Remodeling neuronal ER-PM junctions is a conserved nonconducting function of Kv2 plasma membrane ion channels. Mol Biol Cell *29*, 2410–2432.

Krapivinsky, G., Krapivinsky, L., Manasian, Y., and Clapham, D.E. (2014). The TRPM7 chanzyme is cleaved to release a chromatin-modifying kinase. Cell *157*, 1061–1072.

Kreple, C.J., Lu, Y., Taugher, R.J., Schwager-Gutman, A.L., Du, J., Stump, M., Wang, Y., Ghobbeh, A., Fan, R., Cosme, C.V., *et al.* (2014). Acid-sensing ion channels contribute to synaptic transmission and inhibit cocaine-evoked plasticity. Nat Neurosci *17*, 1083–1091.

Kruger, L.C., and Isom, L.L. (2016). Voltage-gated Na^+ channels: not just for conduction. Cold Spring Harb Perspect Biol *8*, a029264.

Lai, T.W., Zhang, S., and Wang, Y.T. (2014). Excitotoxicity and stroke: identifying novel targets for neuroprotection. Prog Neurobiol *115*, 157–188.

Lee, A., Fakler, B., Kaczmarek, L.K., and Isom, L.L. (2014). More than a pore: ion channel signaling complexes. J Neurosci *34*, 15159–15169.

Liu, Z., Pei, H., Zhang, L., and Tian, Y. (2018). Mitochondria-targeted DNA nanoprobe for real-time imaging and simultaneous quantification of Ca(2+) and pH in neurons. ACS Nano *12*, 12357–12368.

Misonou, H., Mohapatra, D.P., and Trimmer, J.S. (2005). Kv2.1: a voltage-gated k+ channel critical to dynamic control of neuronal excitability. Neurotoxicology *26*, 743–752.

Misonou, H., Thompson, S.M., and Cai, X. (2008). Dynamic regulation of the Kv2.1 voltage-gated potassium channel during brain ischemia through neuroglial interaction. J Neurosci *28*, 8529–8538.

Nagaeva, E.I., Tikhonova, T.B., Magazanik, L.G., and Tikhonov, D.B. (2016). Histamine selectively potentiates acid-sensing ion channel 1a. Neurosci Lett *632*, 136–140.

Nilius, B., and Owsianik, G. (2011). The transient receptor potential family of ion channels. Genome Biol *12*, 218.

O'Connell, K.M., Loftus, R., and Tamkun, M.M. (2010). Localization-dependent activity of the Kv2.1 delayed-rectifier K+ channel. Proc Natl Acad Sci USA *107*, 12351–12356.

Perkins, J.R., Diboun, I., Dessailly, B.H., Lees, J.G., and Orengo, C. (2010). Transient protein-protein interactions: structural, functional, and network properties. Structure *18*, 1233–1243.

Perraud, A.L., Fleig, A., Dunn, C.A., Bagley, L.A., Launay, P., Schmitz, C., Stokes, A.J., Zhu, Q., Bessman, M.J., Penner, R., *et al.* (2001). ADP-ribose gating of the calcium-permeable LTRPC2 channel revealed by Nudix motif homology. Nature *411*, 595–599.

Petroff, E.Y., Price, M.P., Snitsarev, V., Gong, H., Korovkina, V., Abboud, F.M., and Welsh, M.J. (2008). Acid-sensing ion channels interact with and inhibit BK K+ channels. Proc Natl Acad Sci USA *105*, 3140–3144.

Petrovic, M.M., Viana da Silva, S., Clement, J.P., Vyklicky, L., Mulle, C., Gonzalez-Gonzalez, I.M., and Henley, J.M. (2017). Metabotropic action of postsynaptic kainate receptors triggers hippocampal long-term potentiation. Nat Neurosci *20*, 529–539.

Pignataro, G., Simon, R.P., and Xiong, Z.G. (2007). Prolonged activation of ASIC1a and the time window for neuroprotection in cerebral ischaemia. Brain *130*, 151–158.

Purves, D., Augustine, G.J., Fitzpatrick, D., *et al.*, (eds.)e.a. (2001). Neuroscience. 2nd edition. 2001. Two Families of Postsynaptic Receptors. (Sunderland, MA: Sinauer Associates).

Ramsey, I.S., Delling, M., and Clapham, D.E. (2006). An introduction to TRP channels. Annu Rev Physiol *68*, 619–647.

Riazanova, L.V., Pavur, K.S., Petrov, A.N., Dorovkov, M.V., and Riazanov, A.G. (2001). [Novel type of signaling molecules: protein kinases covalently linked to ion channels]. Mol Biol (Mosk) *35*, 321–332.

Runnels, L.W., Yue, L., and Clapham, D.E. (2001). TRP-PLIK, a bifunctional protein with kinase and ion channel activities. Science *291*, 1043–1047.

Samways, D.S., Harkins, A.B., and Egan, T.M. (2009). Native and recombinant ASIC1a receptors conduct negligible Ca^{2+} entry. Cell Calcium *45*, 319–325.

Sano, Y., Inamura, K., Miyake, A., Mochizuki, S., Yokoi, H., Matsushime, H., and Furuichi, K. (2001). Immunocyte Ca^{2+} influx system mediated by LTRPC2. Science *293*, 1327–1330.

Savic Azoulay, I., Liu, F., Hu, Q., Rozenfeld, M., Ben Kasus Nissim, T., Zhu, M.X., Sekler, I., and Xu, T.L. (2020). ASIC1a channels regulate mitochondrial ion signaling and energy homeostasis in neurons. J Neurochem *153*, 203–215.

Schulien, A.J., Yeh, C.Y., Orange, B.N., Pav, O.J., Hopkins, M.P., Moutal, A., Khanna, R., Sun, D., Justice, J.A., and Aizenman, E. (2020). Targeted disruption of Kv2.1-VAPA association provides neuroprotection against ischemic stroke in mice by declustering Kv2.1 channels. Sci Adv *6*, eaaz8110.

Shah, N.H., Schulien, A.J., Clemens, K., Aizenman, T.D., Hageman, T.M., Wills, Z.P., and Aizenman, E. (2014). Cyclin e1 regulates Kv2.1 channel phosphorylation and localization in neuronal ischemia. J Neurosci *34*, 4326–4331.

Shan, B., Pan, H., Najafov, A., and Yuan, J. (2018). Necroptosis in development and diseases. Genes Dev *32*, 327–340.

Sherwood, T.W., Lee, K.G., Gormley, M.G., and Askwith, C.C. (2011). Heteromeric acid-sensing ion channels (ASICs) composed of ASIC2b and ASIC1a display novel channel properties and contribute to acidosis-induced neuronal death. J Neurosci *31*, 9723–9734.

Stoilova-McPhie, S., Ali, S., and Laezza, F. (2013). Protein-protein interactions as new targets for ion channel drug discovery. Austin J Pharmacol Ther *1*, 5.

Sun, H.S., Jackson, M.F., Martin, L.J., Jansen, K., Teves, L., Cui, H., Kiyonaka, S., Mori, Y., Jones, M., Forder, J.P., *et al.* (2009). Suppression of hippocampal TRPM7 protein prevents delayed neuronal death in brain ischemia. Nat Neurosci *12*, 1300–1307.

Tantini, B., Fiumana, E., Cetrullo, S., Pignatti, C., Bonavita, F., Shantz, L.M., Giordano, E., Muscari, C., Flamigni, F., Guarnieri, C., *et al.* (2006). Involvement of polyamines in apoptosis of cardiac myoblasts in a model of simulated ischemia. J Mol Cell Cardiol *40*, 775–782.

Tombola, F., Pathak, M.M., and Isacoff, E.Y. (2006). How does voltage open an ion channel? Annu Rev Cell Dev Biol *22*, 23–52.

Unwin, N. (1989). The structure of ion channels in membranes of excitable cells. Neuron *3*, 665–676.

Vierra, N.C., Kirmiz, M., van der List, D., Santana, L.F., and Trimmer, J.S. (2019). Kv2.1 mediates spatial and functional coupling of L-type calcium channels and ryanodine receptors in mammalian neurons. Elife *8*, e49953.

Wang, J.J., Liu, F., Yang, F., Wang, Y.Z., Qi, X., Li, Y., Hu, Q., Zhu, M.X., and Xu, T.L. (2020). Disruption of auto-inhibition underlies conformational signaling of ASIC1a to induce neuronal necroptosis. Nat Commun *11*, 475.

Wang, Y.Z., Wang, J.J., Huang, Y., Liu, F., Zeng, W.Z., Li, Y., Xiong, Z.G., Zhu, M.X., and Xu, T.L. (2015). Tissue acidosis induces neuronal necroptosis via ASIC1a channel independent of its ionic conduction. Elife *4*, e05682.

Wang, Y.Z., and Xu, T.L. (2011). Acidosis, acid-sensing ion channels, and neuronal cell death. Mol Neurobiol *44*, 350–358.

Wang, Y.Z., Zeng, W.Z., Xiao, X., Huang, Y., Song, X.L., Yu, Z., Tang, D., Dong, X.P., Zhu, M.X., and Xu, T.L. (2013). Intracellular ASIC1a regulates mitochondrial permeability transition-dependent neuronal death. Cell Death Differ *20*, 1359–1369.

Weilinger, N.L., Lohman, A.W., Rakai, B.D., Ma, E.M., Bialecki, J., Maslieieva, V., Rilea, T., Bandet, M.V., Ikuta, N.T., Scott, L., et al. (2016). Metabotropic NMDA receptor signaling couples Src family kinases to pannexin-1 during excitotoxicity. Nat Neurosci *19*, 432–442.

Weilinger, N.L., Maslieieva, V., Bialecki, J., Sridharan, S.S., Tang, P.L., and Thompson, R.J. (2013). Ionotropic receptors and ion channels in ischemic neuronal death and dysfunction. Acta Pharmacol Sin *34*, 39–48.

Wemmie, J.A., Taugher, R.J., and Kreple, C.J. (2013). Acid-sensing ion channels in pain and disease. Nat Rev Neurosci *14*, 461–471.

Wulff, H., Castle, N.A., and Pardo, L.A. (2009). Voltage-gated potassium channels as therapeutic targets. Nat Rev Drug Discov *8*, 982–1001.

Xia, J., Zhang, X., Staudinger, J., and Huganir, R.L. (1999). Clustering of AMPA receptors by the synaptic PDZ domain-containing protein PICK1. Neuron *22*, 179–187.

Xiong, Z.G., Zhu, X.M., Chu, X.P., Minami, M., Hey, J., Wei, W.L., MacDonald, J.F., Wemmie, J.A., Price, M.P., Welsh, M.J., et al. (2004). Neuroprotection in ischemia: blocking calcium-permeable acid-sensing ion channels. Cell *118*, 687–698.

Yeh, C.Y., Bulas, A.M., Moutal, A., Saloman, J.L., Hartnett, K.A., Anderson, C.T., Tzounopoulos, T., Sun, D., Khanna, R., and Aizenman, E. (2017). Targeting a potassium channel/syntaxin interaction ameliorates cell death in ischemic stroke. J Neurosci *37*, 5648–5658.

Yeh, C.Y., Schulien, A.J., Molyneaux, B.J., and Aizenman, E. (2020). Lessons from recent advances in ischemic stroke management and targeting Kv2.1 for neuroprotection. Int J Mol Sci *21*.

Yeh, C.Y., Ye, Z., Moutal, A., Gaur, S., Henton, A.M., Kouvaros, S., Saloman, J.L., Hartnett-Scott, K.A., Tzounopoulos, T., Khanna, R., et al. (2019). Defining the Kv2.1-syntaxin molecular interaction identifies a first-in-class small molecule neuroprotectant. Proc Natl Acad Sci U S A *116*, 15696–15705.

Yellen, G. (2002). The voltage-gated potassium channels and their relatives. Nature *419*, 35–42.

Yermolaieva, O., Leonard, A.S., Schnizler, M.K., Abboud, F.M., and Welsh, M.J. (2004). Extracellular acidosis increases neuronal cell calcium by activating acid-sensing ion channel 1a. Proc Natl Acad Sci U S A *101*, 6752–6757.

Yu, F.H., Yarov-Yarovoy, V., Gutman, G.A., and Catterall, W.A. (2005). Overview of molecular relationships in the voltage-gated ion channel superfamily. Pharmacol Rev *57*, 387–395.

Yu, S.P. (2003). Regulation and critical role of potassium homeostasis in apoptosis. Prog Neurobiol *70*, 363–386.

15 Nonclassical Ion Channels and Ischemia

Jun Gao and Hui-Xin Zhang

CONTENTS

This chapter discusses mainly three types of ion channels, which can be characterized as nonclassical ion channels. With a combination of data and researches, the nonclassical ion channels, including acid-sensing ion channels, transient receptor potential channels, and sulfonylurea receptor 1-transient receptor potential melastatin 4, will be discussed in detail for further practice and provide more possibilities for drug targets in translation medical research.

Stroke is the second leading cause of death in the world, with high mortality in China, which has nearly 20% of the world's population [1]. Stroke can be classified to mainly two types: hemorrhagic stroke and ischemic stroke, and the latter constitutes about two-thirds of all stroke patients [2]. Ischemic stroke is caused by thrombosis, which reduces blood flow and interrupts blood supply, inducing localized damage in brain-specific tissues under hypoxia, leading to a complex syndrome in the brain. Ischemic stroke survivors suffer different dysfunction in sensory function, motor skills, cognition, and learning and memory. Currently in clinical practice, stroke patients can receive recombination tissue plasminogen activator (tPA) for injection within a therapeutic window of 4.5 hours. This has promising effectiveness but is only suitable for 3%–4% of patients and still has a risk of hemorrhage [3]. With acute ischemia, the ischemic core suffers an irreversible injury and subsequent cell death in minutes, and the ischemic penumbra progressively converts to ischemic core over several hours or days. Understanding the differences between penumbra and core, many treatments have tried to focus on the preservation of the penumbral brain, which can be salvaged and recover its normal function [4]. Besides, along with the development of new therapeutic intervention, more and more researches explore the complex molecular mechanism of ischemia to find potential strategies for treatment.

Generally, ischemic stroke with low cerebral blood flow leads to oxygen and glucose deprivation (OGD) in the brain, continuous with lack of adenosine triphosphate (ATP) supply, which has molecular consequences in neurons, glia, and endothelial cells and affects neuronal function, vascular permeability, and inflammation [5]. Blood-brain barrier dysfunction and release of signaling molecules from astrocytes, microglia, and oligodendrocytes lead to an inflammatory response [6]. With triggers by free radicals and reactive oxygen species (ROS) and reactive nitrogen species (RNS), apoptosis and necrosis results in DNA fragmentation, degradation of cytoskeletal and nuclear proteins, cross-linking of proteins, formation of apoptotic bodies, expression of ligands for phagocytic cell receptors, and finally uptake by phagocytic cells [7]. As for the ligand-gated ion channel, ionotropic glutamate receptors allow rapid ion influx in response to glutamate and generate a neuronal excitotoxicity [8]. The two main subtypes of ionotropic glutamate receptors are N-methyl-D-aspartate (NMDA) receptors (NMDARs) and α-amino-3-hydroxy-5-methylisoxazole-4-propionic acid (AMPA) receptors (AMPARs). NMDARs play a significant part in excitotoxic neuronal death caused in the development of ischemic stroke. When glutamate is released from presynaptic sites, activated AMPARs form a partial depolarization in the post-synaptic membrane sufficient to remove the Mg^{2+} block from NMDARs. excess glutamate release results in over-activation of NMDARs and intracellular calcium influx, eventually leading to excitotoxicity and calcium overload inside the neurons [8,9]. Overload of calcium induces a range of downstream pro-death signaling events such as calpain activation, resulting in cell death programs [10,11].

Since the NMDARs and AMPARs have been investigated as a key gateway to regulate neural death in ischemic stroke, the clinical trials have tested the NMDA and AMPA receptor blockers—even inhalational anesthetics can interact with many ion channels and some of them have neuroprotective effect—but the results turned out to be ineffective and/or had side effects [12]. With the limitation of the glutamate mechanism and the undesirable outcomes, researchers have made efforts to discover the new molecules and pathways for ischemic stroke treatment, including non glutamate-related therapeutic targets, which can also influence the process of ionic imbalance and cell death. Some of these channels include: acid-sensing ion channels (ASICs), transient receptor potential (TRP) channels, non-selective cation channels, hemichannels, volume-regulated anion channels, and sodium-calcium exchangers [13]. This chapter concentrates on the latest researches in ASICs, TRP channels, and sulfonylurea receptor 1-transient receptor potential melastatin 4 in ischemic stroke, and summarizes different types of dysfunction of nonclassical ion channels, attempting to suggest promising ideas for translational stroke research.

15.1   THE ACID-SENSING ION CHANNELS (ASICs)

In ischemic stroke, loss of oxygen causes anaerobic generation of lactate and the accumulation of acid, which disrupts cell ionic homeostasis. Anaerobic hydrolysis increases lactic acid, and the protons from ATP lead to a pH decrease, playing an important role in brain injury during ischemia [14]. Therefore, failure to clear excitatory amino acids, leading to an increase of lactate and proton, is also a critical step for neuronal loss. ASICs are a proton-gated subgroup of the degenerin/epithelial Na^+ channel (DEG/ENaC) family of cation channels, which are expressed on neurons and assemble into homomultimeric and heteromultimeric complexes [15]. ASICs are composed by six ASIC subunits (ASIC1a, ASIC1b, ASIC2a, ASIC2b, ASIC3 and ASIC4) and coded by four genes (*Accn, Accn2, and Accn3*) [15,16]. ASIC1b and ASIC3 are mainly found in sensory neurons while ASIC1a, ASIC2a, ASIC2b, and ASIC4 are located in both sensory and central neurons. In central neurons, ASICs are localized to the cell body, dendrites, and dendritic spines, and can be activated by reductions in extracellular pH and depolarize the membrane. ASIC inhibition has been extensively studied for neuroprotection in stroke treatment, so ASICs may act as an important and challenging therapeutic targets, and one approach is to generate the functional antibodies or peptides against these channels [17,18].

ASIC1a channels were found in the amygdala, which responds to extracellular pH reduction ranging from 6.9 to 5.0 (ischemia, ~pH 6.0–6.5) to generate rapid depolarizing currents [19,20]. In the case of homomeric ASIC1a channels, acid activation also induces Ca^{2+} entry directly through these channels, and the ASIC-mediated membrane depolarization may facilitate the activation of voltage-gated Ca^{2+} channels and NMDA receptor-gated channels, further promoting neuronal excitation and Ca^{2+} accumulation [21–23]. Recently studies have shown that activation of ASIC1a plays a key role in ischemic brain injury. Activation by a solution of low pH in vitro, ASIC1a-specific blockade or deletion of the ASIC1a-encoding gene in mice can rescue the neuronal damage [24]. With intracerebroventricular administration of an ASIC1a blocker in an ischemic stroke model in rodents, the severity of stroke can be reduced by 60% [25]. Surprisingly, the β-estradiol has been reported to decrease acidosis-induced cytotoxicity and attenuated ASIC currents via acid-induced elevation of intracellular Ca^{2+} and the level of ASIC1a protein expression in brain tissues, and the degree of neuroprotection by ASIC1a blockade were lower in female mice, which could be attenuated by ovariectomy. In this pathway the estrogen receptor α was involved, which caused a difference between male and female by suppressing ASIC1a protein expression and channel function with the effect of β-estradiol [26]. ASIC1a is selectively permeable to Ca^{2+}, and activation of ASIC1a-bearing channels promotes Ca^{2+} influx. The activation of ASIC1a is regarded as a trigger membrane depolarization and Ca^{2+} influx directly via ASIC1a homomers or ASIC1a/2b heteromers, voltage-gated Ca^{2+} channels, and NMDARs. The extracellular protons also trigger serine/threonine kinase receptor interaction protein 1 (RIP1) for neuronal necroptosis via ASIC1a. Acid stimulation recruits RIP1 to the ASIC1a C-terminus, causing RIP1 phosphorylation and subsequent neuronal death [27]. ASIC1a could mediate acidic neuronal necroptosis via recruiting receptor-interacting protein kinase 1 (RIPK1) to its C terminus, which interacts with the N-terminus to form an auto-inhibition that prevents RIPK1 recruitment/activation under resting conditions and is disrupted by acidosis [28]. The CA1 region of the hippocampus is particularly vulnerable to ischemic damage, which is also necessary for physiologically induced LTP. ASIC1a, which is expressed in CA1 pyramidal neurons contributing to neuronal death in ischemic stroke, and a- LTP and the delayed increase in the prevalence of CP-AMPARs are dependent on ASIC1a activation during ischemia driving CP-AMPAR plasticity for neuroprotective benefit [29].

In pharmaceutical practice, sophocarpine down-regulates the expression of ASIC1 in the ischemic cortex, which is one of the major alkaloid compounds isolated from *Sophora pachycarpa* and highly valued and important in traditional Chinese medicine. [30]. With the effect of spermine in reducing desensitization of ASIC1a by slowing down desensitization in the open state, shifting steady-state desensitization to more acidic pH, and accelerating recovery between repeated periods of acid stimulation. Spermine-mediated potentiation of ASIC1a activity is occluded by PcTX1 (psalmotoxin 1), a specific ASIC1a inhibitor binding to its extracellular domain, which suggests new neuroprotective strategies for stroke patients via inhibition of polyamine synthesis and subsequent spermine-ASIC interaction [31]. Ginsenoside (GS)-Rd, a mono-compound isolated from traditional Chinese herb panax ginseng, blocks calcium influx after excitotoxic injury, indicating that GS-Rd may act on cation channels with TRPM7, ASIC1a, and ASIC2a [32]. ASIC1a was found to be inhibited by NSAIDs, particularly flurbiprofen and ibuprofen, which contributed to reduce the proteolytic products (SBDPs) caused by ischemic activation of calcium-dependent protease calpain [33].

However, ASIC1 has little effect on neural expression in the hypothalamus during ischemia, and ASIC2 performs an up-regulated expression in the anti-apoptotic pathway with Bcl-2 and Bcl-W. So ASIC2 may take part in preventing apoptosis induced by ischemia, and the function of ASICs may vary in different pathologic region [34]. Moreover, the new discovery about potentiation of ASICs by quinine depends on the presence of the ASIC1a and ASIC2a subunits, and the amino acids in ASIC1a are involved in the modulation of ASICs by pHi [35]. In addition to regulation of pH, the antibodies or inhibitors need to block the transport of cations, including calcium, thereby preventing acid-induced cell death [36]. Hi1a, a disulfide-rich spider venom peptide, is a kind of ASIC1a inhibitor and partly inhibits ASIC1a activation in a pH-independent and slowly reversible way, so Hi1a might be a promising neuroprotective agent for the development of therapeutics to protect the brain from ischemic injury in humans [37]. The ASIC2b subunit is expressed in the brain, which overlaps ASIC1a, and its combination with ASIC1a in Xenopus oocytes results in novel proton-gated currents with properties distinct from ASIC1a homomeric channels. ASIC2b/1a heteromeric channels are inhibited by the nonselective potassium channel blockers tetraethylammonium and barium to play a role in acidosis-induced neuronal death [38,39]. Considering the diverse pharmacological profile, it will be still a long process for clinical application, but with the appropriate physical properties, they have the potential to be a suitable treatment for stroke.

15.2   THE TRANSIENT RECEPTOR POTENTIAL (TRP) CHANNELS

There are 28 mammalian TRP channels separated into 6 sub-families: TRPC (where C denotes canonical), TRPM (melastatin), TRPV (vanilloid), TRPA (ankyrin), TRPP (poly-cystin), and TRPML (mucolipin). They share similar structure, but their functions are diverse and display a different activation mechanism. It has been proved that TRPM, TRPV, TRPC, and TRPA channels are located in glial cells. The mammalian TRPC (canonical) family consists of seven members (TRPC1–7) that are organized into four groups: TRPC1, TRPC2, TRPC3/6/7 and TRPC4/5, because they are with sequence homology and functional similarities [40]. The TRP is a super-family of non-selective cation channels that are widely expressed in mammalian cells [13,41], and they are activated by a multitude of ligands, physical stimuli (e.g. force and temperature), differing local ion concentrations and secondary messengers released via G-protein coupled receptors [42].

TRP channels are involved in various physiological processes, including sensory transduction. The TRP channel TRPV6 mediates calcium uptake in epithelia and its expression is dramatically increased in numerous types of cancer [43]. In ischemia, myelin is damaged in Ca^{2+} involved pathway which is devoted to glutamate release activating NMDA receptors. The H^+-gated $[Ca^{2+}]_i$ elevation is mediated by channels with characteristics of TRPA1, which could be inhibited by ruthenium red, isopentenyl pyrophosphate, HC-030031, A967079 or TRPA1 knockout, suggesting that TRPA1-containing ion channels performs a potential value in white matter ischemia [44]. In edema, ischemic stroke can be prevented by TRPV4, because TRPV4 is activated by body temperature and is enhanced by heating through glutamate receptors [45]. TRPV4 is a calcium-permeable cation channel that is also sensitive to cell swelling, modulating the mitogen-activated protein kinase (MAPK) and phosphatidyl inositol 3 kinase (PI3K)/protein kinase B (Akt) signaling pathways that regulate cell apoptosis when activated [46]. Up-regulation of TRPV4 is mediated through NR2B-NMDAR down-regulating the Akt signaling pathway for neurotoxicity [47]. In the penumbra, peri-infarct depolarizations (PIDs) are accompanied by strong intracellular calcium elevations in astrocytes and neurons, thereby negatively affecting infarct size and clinical outcome. TRPV4 channels contribute to calcium influx into astrocytes and neurons and subsequent extra-cellular glutamate accumulation ameliorate the PID-induced calcium overload during acute stroke [48]. TRPM2 has a high sensitivity to oxidative damage, so the oxidative stress and the release of extracellular Ca^{2+}, hydrogen peroxide, adenosine diphosphate ribose, and nicotinic acid adenine dinucleotide phosphate can change the activity of TRPM2 in the central nervous system and the immune system [49]. As a result, N-acetyl-l-cysteine (NAC) treatment can provide neuroprotection via regulation of TRPM2 [50]. Besides, the suppression of TRPM7 channels reduce delayed neuronal cell death and preserved neuronal functions in global cerebral ischemia [13]. And the TRPA1 is expressed by primary afferent nerve fibers, which is also a 'receptor-operated' channel whose activation downstream of metabotropic receptors elicits inflammatory pain or itch. It functions as a low-threshold sensor for structurally diverse electrophilic irritants, which may act as an attractive target for novel analgesic therapies [51]. Zn^{2+} is transferred from endolysosomal vesicles to the cytosol through the TRPML1 channel, and its sensitivity of Ca^{2+} play critical roles in neuronal function [52]. The TRPC5 ion channel is involved in ischemia which is related to endothelial cell sprouting and angiogenesis. Riluzole, the TRPC5 activator, are tested on ischemic injury regulating nuclear factor of activated T cell isoform c3 and angiopoietin-1 which could provide the mechanism for the angiogenic function of TRPC5 [53].

15.3 THE SULFONYLUREA RECEPTOR 1-TRANSIENT RECEPTOR POTENTIAL MELASTATIN 4 (SUR1-TRPM4)

SUR1 is a molecule with more diverse and critically important functions as a subunit involved in formation of a subset of K(ATP) channels, and the accumulating evidence indicates that SUR1 is newly up-regulated in CNS ischemia and injury in association with different pore-forming subunits. The NC(Ca-ATP) channel, in ischemic astrocytes that is regulated by SUR1, is opened by depletion of ATP and causes cytotoxic edema. SUR1 are expressed in ischemic neurons, astrocytes and capillaries. Up-regulation of SUR1 is linked to activation of the transcription factor Sp1 and is associated with expression of functional NC(Ca-ATP) but not K(ATP) channels, because SUR1-based K(ATP) channels are not obligatory for neuronal pre-conditioning or augmentation of neurodegeneration by 5-hydroxydecanoate [54]. The NC(Ca-ATP) channel is crucially involved in development of cerebral edema, and targets SUR1 as a promising treatment in stroke, trauma and spinal cord injury [55,56]. TRPM4 channel specific antibody M4P could inhibit TRPM4 current and down-regulate TRPM4 surface expression, therefore preventing hypoxia-induced cell swelling, attenuating reperfusion injury in stroke recanalization [57]. In cerebral ischemia/reperfusion, SUR1-TRPM4 channel up-regulates microvascular endothelium with matrix metalloproteinase-9 (MMP-9). The SUR1-TRPM4 is involved in the process that transform the hemorrhage and plasma MMP-9 is induced by protease activated receptor 1 (PAR1)-mediated and Ca^{2+}-dependent phasic secretion [58]. The chronic intermittent hypobaric hypoxia (CIHH) has protective effects against cerebral ischemia via the mitochondrial membrane ATP-sensitive potassium channel (mitoKATP) [59]. Glibenclamide administered during the early stages of stroke might foster neuroprotective microglial activity through ATP-sensitive potassium (K(ATP)) channel blockade. These reactive microglial cells express SUR1, SUR2B, and Kir6.2 proteins that assemble in functional K(ATP) channels, suggesting K(ATP) channels in the control of microglial reactivity are consistent with a microglial effect of glibenclamide into the ischemic brain in the early stages of stroke [60]. Glimepiride shows comparable efficacy with glibenclamide protecting against ischemic injury by blocking the de novo assembled SUR1-TRPM4 channel [61]. Hypoxia-ischemia and ATP depletion are associated with cytotoxic edema of glial cells and the isolated native reactive astrocytes (NRAs) following exposure to NaN3 which depletes cellular ATP, which causes profound and sustained depolarization [62,63]. The mRNA transcription and protein expression of SUR1 but not SUR2 are confirmed in reactive astrocytes, consistent with participation of this channel in cation flux involved in cell swelling [64].

15.4 CONCLUSION

When stroke occurs, cerebral ischemia leads to intracellular Na^+ and Ca^{2+} accumulation, which interrupts the ionic transportation and channel function. The glutamate receptor related channels mediate an acute excitotoxicity and induce a transient change, which is not easy to apply to the limited therapeutic window. Therefore, researches continue to search ASICs, TRP channels, SUR1-TRMP4 channels,

Na+/K+-ATPase, and other channels involved in the pathological process of cerebral ischemic stroke, which have the potential for neuroprotection. ASICs suggest an acid-induced pathway and trigger Ca^{2+} inflex, while TRP channels combine Ca^{2+} with a cell apoptosis pathway in regulation of neural death. The study of SUR1-TRPM4 inhibition in edema provides more evidence for cell swelling under hypoxia. It can be promising to translate the therapeutical targets for treatment of ischemic stroke and other CNS injury diseases with more studies in vitro and in vivo, and it also is worth exploring the mechanisms in neurobiology to explain brain-specific functions under physical and pathological conditions.

REFERENCES

1. Wu, S., et al., Stroke in China: advances and challenges in epidemiology, prevention, and management. *Lancet Neurol*, 2019. **18**(4): p. 394–405.
2. Adams, H.P., Jr., et al., Guidelines for the management of patients with acute ischemic stroke. A statement for healthcare professionals from a special writing group of the Stroke Council, American Heart Association. *Stroke*, 1994. **25**(9): p. 1901–1914.
3. Adibhatla, R.M. and J.F. Hatcher, Tissue plasminogen activator (tPA) and matrix metalloproteinases in the pathogenesis of stroke: therapeutic strategies. *CNS Neurol Disord Drug Targets*, 2008. **7**(3): p. 243–253.
4. Campbell, B.C.V., et al., Ischaemic stroke. *Nat Rev Dis Primers*, 2019. **5**(1): p. 70.
5. George, P.M. and G.K. Steinberg, Novel stroke therapeutics: unraveling stroke pathophysiology and its impact on clinical treatments. *Neuron*, 2015. **87**(2): p. 297–309.
6. Castro, P., E. Azevedo, and F. Sorond, Cerebral autoregulation in stroke. *Curr Atheroscler Rep*, 2018. **20**(8): p. 37.
7. Radak, D., et al., Apoptosis and acute brain ischemia in ischemic stroke. *Curr Vasc Pharmacol*, 2017. **15**(2): p. 115–122.
8. Tymianski, M., et al., Source specificity of early calcium neurotoxicity in cultured embryonic spinal neurons. *J Neurosci*, 1993. **13**(5): p. 2085–2104.
9. Sattler, R. and M. Tymianski, Molecular mechanisms of glutamate receptor-mediated excitotoxic neuronal cell death. *Mol Neurobiol*, 2001. **24**(1-3): p. 107–129.
10. Bano, D. and P. Nicotera, Ca^{2+} signals and neuronal death in brain ischemia. *Stroke*, 2007. **38**(2 Suppl): p. 674–676.
11. Singh, V., et al., Modes of calcium regulation in ischemic neuron. *Indian J Clin Biochem*, 2019. **34**(3): p. 246–253.
12. Lehmke, L., et al., Inhalational anesthetics accelerate desensitization of acid-sensing ion channels. *Neuropharmacology*, 2018. **135**: p. 496–505.
13. Bae, C.Y. and H.S. Sun, TRPM7 in cerebral ischemia and potential target for drug development in stroke. *Acta Pharmacol Sin*, 2011. **32**(6): p. 725–733.
14. Kreple, C.J., et al., Acid-sensing ion channels contribute to synaptic transmission and inhibit cocaine-evoked plasticity. *Nat Neurosci*, 2014. **17**(8): p. 1083–1091.
15. Wemmie, J.A., M.P. Price, and M.J. Welsh, Acid-sensing ion channels: advances, questions and therapeutic opportunities. *Trends Neurosci*, 2006. **29**(10): p. 578–586.
16. Imre, S.G., I. Fekete, and T. Farkas, Increased proportion of docosahexanoic acid and high lipid peroxidation capacity in erythrocytes of stroke patients. *Stroke*, 1994. **25**(12): p. 2416–2420.
17. Hu, H.J. and M. Song, Disrupted ionic homeostasis in ischemic stroke and new therapeutic targets. *J Stroke Cerebrovasc Dis*, 2017. **26**(12): p. 2706–2719.

18. O'Bryant, Z., K.T. Vann, and Z.G. Xiong, Translational strategies for neuroprotection in ischemic stroke—focusing on acid-sensing ion channel 1a. *Transl Stroke Res*, 2014. **5**(1): p. 59–68.

19. Vick, J.S. and C.C. Askwith, ASICs and neuropeptides. *Neuropharmacology*, 2015. **94**: p. 36–41.

20. Zeng, W.Z., D.S. Liu, and T.L. Xu, Acid-sensing ion channels: trafficking and pathophysiology. *Channels (Austin)*, 2014. **8**(6): p. 481–487.

21. Cuomo, O., et al., Ionic homeostasis in brain conditioning. *Front Neurosci*, 2015. **9**: p. 277.

22. Gao, J., et al., Coupling between NMDA receptor and acid-sensing ion channel contributes to ischemic neuronal death. *Neuron*, 2005. **48**(4): p. 635–646.

23. Xiong, Z.G., et al., Neuroprotection in ischemia: blocking calcium-permeable acid-sensing ion channels. *Cell*, 2004. **118**(6): p. 687–698.

24. Li, M., et al., Acid-sensing ion channels in acidosis-induced injury of human brain neurons. *J Cereb Blood Flow Metab*, 2010. **30**(6): p. 1247–1260.

25. Pignataro, G., R.P. Simon, and Z.G. Xiong, Prolonged activation of ASIC1a and the time window for neuroprotection in cerebral ischaemia. *Brain*, 2007. **130**(Pt 1): p. 151–158.

26. Zhou, R., et al., β-Estradiol protects against acidosis-mediated and ischemic neuronal injury by promoting ASIC1a (acid-sensing ion channel 1a) protein degradation. *Stroke*, 2019. **50**(10): p. 2902–2911.

27. Wang, Y.Z., et al., Tissue acidosis induces neuronal necroptosis via ASIC1a channel independent of its ionic conduction. *Elife*, 2015. **4**, e05682.

28. Wang, J.J., et al., Disruption of auto-inhibition underlies conformational signaling of ASIC1a to induce neuronal necroptosis. *Nat Commun*, 2020. **11**(1): p. 475.

29. Quintana, P., et al., Acid-sensing ion channel 1a drives AMPA receptor plasticity following ischaemia and acidosis in hippocampal CA1 neurons. *J Physiol*, 2015. **593**(19): p. 4373–4386.

30. Yifeng, M., et al., Neuroprotective effect of sophocarpine against transient focal cerebral ischemia via down-regulation of the acid-sensing ion channel 1 in rats. *Brain Res*, 2011. **1382**: p. 245–251.

31. Duan, B., et al., Extracellular spermine exacerbates ischemic neuronal injury through sensitization of ASIC1a channels to extracellular acidosis. *J Neurosci*, 2011. **31**(6): p. 2101–2112.

32. Zhang, Y., et al., Ginsenoside-Rd attenuates TRPM7 and ASIC1a but promotes ASIC2a expression in rats after focal cerebral ischemia. *Neurol Sci*, 2012. **33**(5): p. 1125–1131.

33. Mishra, V., R. Verma, and R. Raghubir, Neuroprotective effect of flurbiprofen in focal cerebral ischemia: the possible role of ASIC1a. *Neuropharmacology*, 2010. **59**(7-8): p. 582–588.

34. Liu, S., et al., Acid-sensing ion channels: potential therapeutic targets for neurologic diseases. *Transl Neurodegener*, 2015. **4**: p. 10.

35. Li, M.H., et al., Modulation of acid-sensing ion channel 1a by intracellular pH and its role in ischemic stroke. *J Biol Chem*, 2016. **291**(35): p. 18370–18383.

36. Qiang, M., et al., Selection of an ASIC1a-blocking combinatorial antibody that protects cells from ischemic death. *Proc Natl Acad Sci USA*, 2018. **115**(32): p. E7469–E7477.

37. Ren, Y., et al., Hi1a as a novel neuroprotective agent for ischemic stroke by inhibition of acid-sensing ion channel 1a. *Transl Stroke Res*, 2018. **9**(2): p. 96–98.

38. Sherwood, T.W., et al., Heteromeric acid-sensing ion channels (ASICs) composed of ASIC2b and ASIC1a display novel channel properties and contribute to acidosis-induced neuronal death. *J Neurosci*, 2011. **31**(26): p. 9723–9734.

39. Wu, W.N., et al., Sinomenine protects against ischaemic brain injury: involvement of co-inhibition of acid-sensing ion channel 1a and L-type calcium channels. *Br J Pharmacol*, 2011. **164**(5): p. 1445–1459.
40. Clapham, D.E., L.W. Runnels, and C. Strubing, The TRP ion channel family. *Nat Rev Neurosci*, 2001. **2**(6): p. 387–396.
41. Lin, J. and Z.G. Xiong, TRPM7 is a unique target for therapeutic intervention of stroke. *Int J Physiol Pathophysiol Pharmacol*, 2017, **9**(6): p. 211–216.
42. Cornillot, M., V. Giacco, and N.B. Hamilton, The role of TRP channels in white matter function and ischaemia. *Neurosci Lett*, 2019. **690**: p. 202–209.
43. Singh, A.K., et al., Structural bases of TRP channel TRPV6 allosteric modulation by 2-APB. *Nat Commun*, 2018. **9**(1): p. 2465.
44. Hamilton, N.B., et al., Proton-gated Ca(2+)-permeable TRP channels damage myelin in conditions mimicking ischaemia. *Nature*, 2016. **529**(7587): p. 523–527.
45. Hoshi, Y., et al., Ischemic brain injury leads to brain edema via hyperthermia-induced TRPV4 activation. *J Neurosci*, 2018. **38**(25): p. 5700–5709.
46. Jie, P., et al., Activation of transient receptor potential vanilloid 4 induces apoptosis in hippocampus through downregulating PI3K/Akt and upregulating p38 MAPK signaling pathways. *Cell Death Dis*, 2015. **6**: p. e1775.
47. Tian, Y., et al., Activation of transient receptor potential vanilloid 4 promotes the proliferation of stem cells in the adult hippocampal dentate gyrus. *Mol Neurobiol*, 2017. **54**(8): p. 5768–5779.
48. Rakers, C., M. Schmid, and G.C. Petzold, TRPV4 channels contribute to calcium transients in astrocytes and neurons during peri-infarct depolarizations in a stroke model. *Glia*, 2017. **65**(9): p. 1550–1561.
49. Gelderblom, M., et al., Transient receptor potential melastatin subfamily member 2 cation channel regulates detrimental immune cell invasion in ischemic stroke. *Stroke*, 2014. **45**(11): p. 3395–3402.
50. Hong, D.K., et al., Transient receptor potential melastatin 2 (TRPM2) inhibition by antioxidant, N-acetyl-l-cysteine, reduces global cerebral ischemia-induced neuronal death. *Int J Mol Sci*, 2020. **21**(17), 6026
51. Zhao, J., et al., Irritant-evoked activation and calcium modulation of the TRPA1 receptor. *Nature*, 2020. **585**(7823): p. 141–145.
52. Minckley, T.F., et al., Sub-nanomolar sensitive GZnP3 reveals TRPML1-mediated neuronal Zn2+ signals. *Nat Comm*, 2019. **10**, 4806.
53. Zhu, Y., et al., The TRPC5 channel regulates angiogenesis and promotes recovery from ischemic injury in mice. *J Biol Chem*, 2019. **294**(1): p. 28–37.
54. Munoz, A., et al., Ischemic preconditioning in the hippocampus of a knockout mouse lacking SUR1-based K-ATP channels. *Stroke*, 2003. **34**(1): p. 164–170.
55. Simard, J.M., et al., Drugs acting on SUR1 to treat CNS ischemia and trauma. *Curr Opin Pharmacol*, 2008. **8**(1): p. 42–49.
56. Simard, J.M., et al., Newly expressed SUR1-regulated NC(Ca-ATP) channel mediates cerebral edema after ischemic stroke. *Nat Med*, 2006. **12**(4): p. 433–440.
57. Chen, B., et al., TRPM4-specific blocking antibody attenuates reperfusion injury in a rat model of stroke. *Pflugers Arch*, 2019. **471**(11-12): p. 1455–1466.
58. Gerzanich, V., et al., SUR1-TRPM4 channel activation and phasic secretion of MMP-9 induced by tPA in brain endothelial cells. *PLoS One*, 2018. **13**(4): p. e0195526.
59. Zhang, S., et al., Chronic intermittent hybobaric hypoxia protects against cerebral ischemia via modulation of mitoKATP. *Neurosci Lett*, 2016. **635**: p. 8–16.
60. Ortega, F.J., et al., ATP-dependent potassium channel blockade strengthens microglial neuroprotection after hypoxia-ischemia in rats. *Exp Neurol*, 2012. **235**(1): p. 282–296.
61. Wang, X.Q., et al., Glimepiride and glibenclamide have comparable efficacy in treating acute ischemic stroke in mice. *Neuropharmacology*, 2020. **162**, 107845.

62. Gerzanich, V., et al., SUR1-TRPM4 channel activation and phasic secretion of MMP-9 induced by tPA in brain endothelial cells. *PLoS One*, 2018. **13**(4), e0195526.
63. Woo, S.K., et al., Sequential activation of hypoxia-inducible factor 1 and specificity protein 1 is required for hypoxia-induced transcriptional stimulation of Abcc8. *J Cereb Blood Flow Metab*, 2012. **32**(3): p. 525–536.
64. Simard, J.M. and M.K. Chen, Regulation by sulfanylurea receptor type 1 of a non-selective cation channel involved in cytotoxic edema of reactive astrocytes. *J Neurosurg Anesthesiol*, 2004. **16**(1): p. 98–99.

16 Transient Receptor Potential Channels and Itch

Mahar Fatima, Jingyi Liu, and Bo Duan

CONTENTS

16.1 INTRODUCTION

We have all experienced an itch (pruritus), which is defined as an unpleasant sensation that elicits the desire or reflex to scratch. Itch can be evoked in the skin by chemical mediators, such as histamine, referred as chemical itch, or by physical stimuli, such as innocuous mechanical stimuli, referred as mechanical itch. Itch can be an acute sensation, associated with mosquito bite, fuzzy sweater, or a chronic condition, which raises a major therapeutic problem in many skin diseases, such as atopic dermatitis, and a variety of systemic diseases, including kidney failure, neurological disorders as well as cancers. Based on the underlying causative mechanisms, itch can be primarily categorized into four categories: pruriceptive, neurogenic, neuropathic, and psychogenic (1). Pruriceptive itch or dermatological itch refers to the itch induced by a pruritic mediator, such as histamine, resulting in the activation of peripheral itch neurons. Neurogenic itch is defined as itch that results from a non-diseased central nervous system. Neuropathic itch refers to itch that is caused by a damaged peripheral or central nervous system. Psychogenic itch is considered to be psychiatric in origin where the desire to scratch excessively is present in otherwise normal skin.

Itch signals are received in the skin by pruriceptive skin cells and peripheral afferents of primary sensory neurons, with cell bodies located in dorsal root ganglia (DRG) and trigeminal ganglia (TG) (2–4). The somatic sensation of itch is then converted into electrical signals to be transmitted to the dorsal horn of the spinal cord, and thence to the brain (4–6). It is unclear how the dermal cells interact with the sensory afferent endings to propagate the pruriceptive signals. A growing body of recent evidence reveals transient receptor potential (TRP) channels as key players of both acute and chronic itch elicited by a diversity of pruritogens.

The TRP channel family is a superfamily of non- or weakly-selective cation tetrameric channels. Activation of TRP channels can depolarize the membrane potential, which can lead to the opening or closing of voltage-gated channels. Based on sequence similarity, TRP family can be divided into seven different sub-families: TRP 'Ankyrin' family (TRPA), TRP 'Canonical' family (TRPC), TRP 'Melastatin' family (TRPM), TRP 'Mucolipin' family (TRPML), TRP 'NOMPC' (TRPN), TRP 'Polycystin' family (TRPP), and TRP 'Vanilloid' family (TRPV) (7). TRP channels were initially discovered in the fruit fly *Drosophila melanogaster* when a *trp* mutant showed a 'transient receptor potential' unlike an expected sustained waveform on the electroretinogram (8). Twenty years later, the first *trp* gene was cloned in *Drosophila melanogaster* in 1989 (9), opening avenues to explore the biologically ubiquitous yet unique TRP superfamily.

Functional TRP channels are homo or heterotetramers. The TRP family members have some unifying features, such as six transmembrane segments where the first, second, third, and fourth transmembrane helices constitute the sensor. A pore loop exists between the fifth and sixth transmembrane that forms a conduction pathway on tetramerization. The pore spans the lipid bilayer and assists the hydrophilic molecules to pass across the bilayer. The N-terminal and C-terminal face the cytoplasm. The cytosolic domains contain regulatory elements that control channel opening. Cytosolic diversity among different TRP channels is striking with a low sequence similarity in the N-terminal and C-terminal cytosolic regions across the subfamilies (10–13). TRP channels also carry consensus regions for direct phosphorylation sites for serine, threonine, and tyrosine kinases (14,15).

TRP channels have been shown to play essential roles in various sensory functions, including vision, olfaction, taste, mechanosensation, thermosensation, and pain (12). In the skin, TRP channels are expressed in the sensory neuronal endings, keratinocytes, mast cells, macrophages, Langerhans cells, sebocytes, hair follicles, and melanocytes (16) (Figure 16.1). Recent studies demonstrate that TRP channels are crucially involved in both acute and chronic itch.

16.2 TRPV1 AND ITCH

TRPV1 (also known as VR1) is the founding member of thermosensitive TRP channels that enables primary polymodal nociceptors to detect ambient

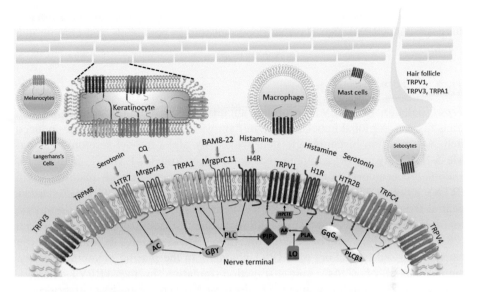

FIGURE 16.1 Schematic showing expression of TRP channels in different cellular sub-types in the skin and the molecular cascades turned on to activate TRP channels to mediate itch. Itch implicated TRP channels are shown in the figure. TRP channels are dominantly expressed in the nerve terminal of sensory neurons, keratinocytes, macrophages, and mast cells. Different subtypes of sensory endings can express a different combination of TRP channels. TRPV1 and TRPV3 are also expressed in the inner root sheath cells of the hair follicle. Itch stimulus is perceived by the dermal cells and the sensory nerve terminal in the periphery. TRPV1 channels are activated by the activation of PLC pathway or phospholipase A2 (PLA2)/Lipoxygenase (LO) pathway. Stimulation of histamine receptor 1R activation PLA2 resulting in accumulation of arachidonic acid which is then acted upon by lipox-ygenase to produce eicosanoids such as 2-hydroperoxyeicosatetraenoic acid (12-HPETE). Activation of histamine receptor H4R activates PLC, which activates TRPV1, which may be occurring through sequestering of phosphatidylinositol 4,5-bisphosphate (PIP2). TRPA1 channels are activated by MrgprC11 via phospholipase C and by MrgrprA3 by $G_{\beta\gamma}$. TRPA1 channels are also activated by serotonin receptor (Htr7) through adenylate cyclase (AC) and $G_{\beta\gamma}$ signaling. TRPC4 channels are activated by activation of PLCβ3/ G_qG_{11} pathway. TRPV1, TRPV4, TRPC4 and TRPA1 are primarily pruritic TRP channels, whereas TRPM8 is anti-pruritic.

temperature changes ($>43°C$) (15,17–20). In addition, TRPV1 can be activated by capsaicin, a principal pungent component of peppers, low pH, and numbers of molecules associated with inflammation and tissue damage, such as bradykinin, prokineticin, prostaglandins, anandamide, and retinoids (20–25). TRPV1 majorly marks a population of unmyelinated, slowly conducting neurons (C-nociceptor) that express neuropeptides substance P, neurokinin A, and calcitonin gene-related peptide and constitute approximately 30%–50% of all sensory neurons situated in the rodent sensory ganglia (19,26). TRPV1 functions as a polymodal receptor for noxious heat, pain, and itch (26,27). Previous researches have extensively investigated the functions of TRPV1 in acute and chronic pain. Current evidence suggests that TRPV1 also plays an important role in both acute and chronic itch conditions.

16.2.1 TRPV1 IN ACUTE ITCH

Trpv1 deletion or TRPV1$^+$ neuronal ablation has differential effects on itch transmission. A global knockout of *Trpv1* shows attenuation in histamine-induced itch, whereas non-histaminergic itch evoked by alpha-methyl-serotonin (alpha-Me-5-HT, a ligand of serotonin receptors) or endothelin-1 (ET-1, a ligand of endothelin receptors) is unaffected. However, pretreatment with resiniferatoxin (RTX), a potent TRPV1 agonist that can effectively ablate the TRPV1$^+$ sensory central terminals, or deletion of TRPV1$^+$ neurons by the expression of diphtheria toxin, can result in attenuation of both histaminergic and non-histaminergic itch (28). Interestingly, a gain of function mutation (*G564S*) in the TRPV1 channel can cause decreased trafficking of TRPV1 channels to the plasma membrane. In the animal model of *G564S* gain-of-function mutation in the TRPV1 channel, histamine-induced itch is dramatically decreased, whereas non-histaminergic itch remains unaffected (29). Hence, histaminergic itch is mediated by TRPV1 channels, while both histaminergic and non-histaminergic itch are dependent on TRPV1$^+$ sensory neurons.

TRPV1 receptor-mediated itch transmission occurs through activation of histamine receptor 1R (H1R). H1R activation switches on phospholipase A2/lipoxygenase pathway to activate TRPV1 receptor to drive calcium influx in the sensory neurons (30) (Figure 16.1). Additionally, a population of DRG neurons sensitive to H4 receptor agonist immepip is also shown to be responsive to capsaicin (31). Subcutaneous injection of immepip evokes robust scratching in mice, which is significantly inhibited by TRPV1 antagonist AMG9810. Immepip-mediated activation of TRPV1 is routed through the phospholipase C (PLC) pathway (31). PLC activates TRPV1 by hydrolyzing and hence inactivating the inhibitor phosphatidylinositol-4,5-bisphosphate (PIP$_2$) (32) (Figure 16.1). Whether PIP$_2$ inhibition-induced TRPV1 activation is also implicated in itch transmission has not been confirmed yet.

Besides being involved in histaminergic and non-histaminergic pathways for itch, TRPV1 is also involved in non-canonical pathways for itch. TRPV1 mediates itch induced via activation of Toll-like receptors 7 (TLR7). Capsaicin-responsive neurons show calcium influx on exposure to imiquimod, a ligand for TLR7 and TLR8 (33). Chemical ablation of TRPV1$^+$ central terminals results in reduction of TLR7/8 ligand imiquimod-induced itching while imiquimod-induced scratching in *Trpv1$^{-/-}$* animals remains unaffected (34). Another group reported a reduction in imiquimod-induced scratching in *Trpv1$^{-/-}$* animals in comparison to control (33).

16.2.2 TRPV1 IN CHRONIC ITCH

Chronic itch represents a debilitating symptom in cholestatic liver disease, causing significant impairment in quality of life. Lysophosphatidic acid (LPA), a ubiquitous biolipid, is known to be a causal factor in chronic cholestatic itch. TRPV1 channels in DRG neurons can be activated by LPA. *Trpv1* knockout mice exhibit reduced scratching in response to LPA-induced itch in comparison to control (35). Cyclic phosphatidic acid (cPA), another naturally occurring unsaturated lipid, can also activate TRPV1 and induce scratching behaviors when injected intradermally whereas in *Trpv1* knockout mice, cPA fails to evoke any scratching behaviors (36).

Patients with chronic renal failure and undergoing hemodialysis suffer from severe uremic pruritus (37). The serum concentration of β2-MG, a subunit of major histocompatibility complex class I antigen, is elevated in patients undergoing hemodialysis (38). Cheek injections of β2-MG in mice evoke scratching responses. Selective blockade of TRPV1 inhibits β2-MG-elicited itch (39), suggesting TRPV1 channels may play an important role in uremic pruritus.

Contact hypersensitivity in allergic contact dermatitis is a cutaneous immune response elicited by hapten sensitization. Squaric acid dibutylester (SADBE), a small molecule hapten, is commonly used in the treatment of alopecia areata and warts but often causes contact hypersensitivity in patients. TRPV1 does not affect SADBE-mediated acute itch. By contrast, TRPV1 is a crucial mediator of chronic itch in the mouse model of SADBE-induced contact hypersensitivity. Genetic deletion of *Trpv1* or pharmacological ablation of TRPV1$^+$ sensory nerves reduces SADBE-induced persistent scratching in mice (40). Strikingly, *Trpv1* deficiency promotes skin inflammation in SADBE-induced contact hypersensitivity. Furthermore, the study shows that TRPV1 modulates skin inflammation by regulating the function of dermal macrophages (40). These results indicate distinct roles of TRPV1 channels in the pathogenesis of skin inflammation and chronic itch.

Barrier damages in the skin can increase inflammatory mediators augmenting atopic dermatitis. TRPV1 antagonist PAC-14028 alleviates the atopic dermatitis symptoms by improving the skin barrier (41,42). Conversely, TRPV1 agonist capsaicin has been an effective antipruritic agent in different conditions characterized by moderate to severe itching such as psoriasis vulgaris, aquagenic pruritus, and prurigo nodularis (42–45).

Alteration in cytokine profiles can result in pruritus. Notably, excessive levels of Interleukin-31 (IL-31), a cytokine produced by TH2 cells, can result in atopic-like dermatitis (46). IL-13 induces calcium influx in TRPV1$^+$ DRG neurons. Administration of IL-31 either cutaneously or intrathecally induces dose-dependent scratching bouts. IL-31 receptor A is expressed by a subset of the TRPV1$^+$ population. Pharmacological ablation of TRPV1$^+$ sensory nerves also culminates in the loss of IL31 receptor A-positive nerve terminals in the dorsal spinal cord. IL-13 induced scratching responses are dramatically reduced in TRPV1$^+$ neuron-ablated or *Trpv1* KO mice (47).

16.2.3 TRPV1-MEDIATED ITCH SUPPRESSION BY PAIN

It is a common experience that the itch sensation can be reduced by painful sensations. Two studies show that VGLUT2-dependent glutamate release from primary

C-nociceptor afferents marked by the expression of the Nav1.8 sodium channel or TRPV1 channel is required to transmit pain and suppress itch in mice (48,49). Conditional deletion of VGLUT2 gene in C-nociceptors leads to enhanced transmission of itch (48,49). Another report supports the hypothesis that the TRPV1$^+$ nociceptors suppress itch by enforcing inhibition on itch transmission (50). Activation of TRPV1$^+$ nociceptors by the exposure of capsaicin leads to increased EPSCs in Bhlhb5$^+$ inhibitory interneurons in the dorsal spinal cord, which present the inhibitory gate on the spinal chemical itch circuit (50).

A capsaicin topical patch has been shown to alleviate itch in patients suffering from neuropathic itch related to small fiber neuropathy and entrapment neuropathies characterized by focally zoned itchy skin areas (51,52). A TRPV1 antagonist, PAC-14028, suppresses the scratching in chemically evoked acute itch and the atopic dermatitis-like model of chronic itch in mice (41). However, a randomized clinical trial conducted on healthy human volunteers using a TRPV1 selective potent antagonist, SB705498, failed to show any relief from histamine or histamine-independent cowhage challenge-evoked scratching (53).

16.3   TRPA1 AND ITCH

TRPA1 (also known as ANKTM1) is the only member of the TRPA subfamily in mammals. TRPA1 channels are characterized by the extensive ankyrin repeats array (16 in humans) (54–56). Ankyrin repeats are 33-amino-acid motifs that are predicted to mediate protein-protein interactions. TRPA1 is selectively expressed in a subset of small-diameter neurons in the dorsal root, trigeminal, and nodose ganglia and keratinocytes (57–60). TRPA1 can be activated by a range of inflammatory agents such as reactive carbonyl species, 4-hydroxynonenal (4-HNE) and 4-oxononenal (4-ONE), and reactive nitrogen species (ROS) (61). Prostaglandin derivative 15d-PGJ2 can exclusively activate TRPA1, but not TRPV1 or TRPM8 (62). Exogenous chemicals that can activate the TRPA1 channel include acrolein (found in tear gas), anesthetics such as isoflurane and lidocaine, allyl isothiocyanate (AITC) found in mustard, cinnamaldehyde (found in cinnamon), and allicin (found in garlic) (61). TRPA1 was initially suggested to contribute to noxious cold sensation (63), which is still controversial (64). Current evidence indicates that TRPA1 plays an important role in pain and itch. Itch-related G protein-coupled receptors (GPCRs), such as Mas-related G protein-coupled receptors A3 (MrgprA3) and MrgprC11, can positively regulate activation of TRPA1 channels. Genetic manipulation and pharmacological blockade studies have shown that TRPA1 functions as the downstream mediator of GPCR signaling in acute and chronic itch.

16.3.1   TRPA1 in Acute Itch

TRPA1 is involved in mediating non-histaminergic acute itch responses to pruritogens like chloroquine and bovine adrenal medulla 8–22 peptide (BAM8–22, simplified as BAM). $Trpa1^{-/-}$ mice fail to respond to chloroquine and BAM-mediated itch (65). Chloroquine and BAM elicits action potential in TRPA1-expressing neurons which remains unaffected by the blockade of TRPV1. TRPA1 acts downstream to MrgrpA3 and MrgprC11 to facilitate itch response to chloroquine and BAM, respectively (65). However, distinct mechanisms are involved in MrgprA3 and MrgprC11 signaling to activate TRPA1. MrgprC11 activates TRPA1 via phospholipase C, whereas $G_{\beta\gamma}$ signaling facilitates MrgprA3-mediated activation of TRPA1 (Figure 16.1).

16.3.2 TRPA1 in Chronic Itch

TRPA1 receptors are extensively reported to be a crucial regulator of induction and maintenance of chronic pruritus. Scratching evoked by an impaired skin barrier is abolished in *Trpa1*-deficient animals in a dry skin mouse model of chronic itch. TRPA1 is required both for transduction of chronic itch and for the dramatic skin changes triggered by dry-skin-evoked itch and scratching. These data suggest that TRPA1 regulates both itch transduction and pathophysiological changes in the skin to promote chronic itch (66).

The cytokine thymic stromal lymphopoietin (TSLP) acts as a master switch that triggers both the initiation and maintenance of atopic dermatitis, which is characterized by chronic debilitating itch (67). Keratinocyte-derived TSLP activates sensory neurons directly to evoke itch in an atopic dermatitis mouse model (68). TSLP-sensitive sensory neurons represent a previously unidentified subset of sensory neurons that require both functional TSLP receptors and TRPA1 channels to promote TSLP-evoked itch behaviors (68).

IL-13, a Th2 cytokine, can also induce atopic dermatitis. In the mouse model of transgenic IL-13 (K5- tTA-IL-13) controlled by the DOX on/off system, there is a dramatic increase in the number of spontaneous scratching events in the animals following the artificial activation of transcription of *Il 13* gene. With an increased level of IL-13 in Tg (K5- tTA-IL-13) mice, the expression level of TRPA1is also increased. A TRPA1 blocker efficiently attenuates the excessive scratching behavior in IL-13 Tg (K5- tTA-IL-13) mice (69), suggesting that TRPA1 may be downstream of IL-13 signaling.

TRPA1 is also involved in initiating and maintaining inflammation in the allergy contact dermatitis caused by haptens such as oxazolone and urushiol (an allergen found in poison ivy). In the mouse model of oxazolone- and/or urushiol-induced allergic contact dermatitis, *Trpa*$^{-/-}$ mice show less severe dermatitis with lowered edema, decreased thickening of the skin and leukocyte infiltration, lowered levels of pruritic and proinflammatory cytokines-IL-4, IL-6, and CXCL-2, and decreased levels of endogenous pruritic mediators such as nerve growth factor (NGF), substance P and serotonin (70). In addition to TRPV1, TRPA1 is also required to produce acute and chronic scratching in SABDE-induced models of contact dermatitis (40).

TRPA1 is regarded as a critical effector of serotonin-evoked acute and chronic itch. Altered serotonin levels have been reported in chronic itch (71–73). Serotonin receptor Htr7 agonist LP44 fails to trigger scratching in *Htr7*- or *Trpa1*-deficient animals. Contrastingly, Akiyama et al. showed acute scratching evoked by 5-HT is unaffected in the *Trpa1*-deficient animals (74). LP44-sensitive sensory neurons also respond to TRPA1 agonist AITC (75). The MC903-induced mouse model of atopic dermatitis shows lesser scratching and reduced severity of lesions in the *Trpa1*$^{-/-}$ or

Htr7⁻/⁻ animal. Further study demonstrates that serotonin receptor Htr7 activates TRPA1 via adenylate cyclase and $G_{\beta\gamma}$ signaling (Figure 16.1).

TRPA1 can be activated by non-conventional ligands such as microRNA to induce acute and chronic itch. microRNA-711 elicits itch in a TRPA1-dependent manner. MiRNA-711 can induce a far more potent inward current in comparison to TRPA1 agonist AITC in the TRPA1-expressing HEK293 cells. Cutaneous T-cell lymphoma-associated chronic itch can be relieved by disrupting the interaction between miRNA-711 and TRPA1 (76).

TRPA1 is suggested to be critically involved in cholestatic itch—a type of systemic itch caused by liver dysfunction. Increased bile acid levels can cause excessive itch through the activation of G-protein–coupled bile acid receptor 1 TGR5. TGR5 is co-expressed with TRPA1 in a subset of the TRPA1 population of sensory neurons. Deletion or blocking of TRPA1 receptors abolishes bile acid- or TGR5-evoked scratching (77). Lysophosphatidic acid (LPA) is another itch mediator found in cholestatic itch patients (35). LPA-induced itch is dependent on activation of TRPA1 and TRPV1 channels in the DRG. Extracellular LPA activates the LPA5 receptor and membrane lipids are metabolized mainly by Phospholipase D (PLD). Activity of PLD and calcium-independent phospholipase A2 (iPLA2) leads to intracellular LPA production, which results in the activation of TRPA1 and TRPV1 channels.

16.3.3   TRPA1-MEDIATED ITCH SUPPRESSION BY PAIN

Although ample evidence exists that activation of TRPA1 is required in itch transmission, a diametrical result by Ross et al. showed that TRPA1⁺ sensory afferents, when activated by agonist AITC, can result in strengthened inhibition on the spinal itch transmission circuit, which results in reduced transmission of itch. AITC application can evoke EPSC in Bhlhb5⁺ itch gating neurons in the dorsal spinal cord. (50). Since TRPA1⁺ sensory neurons are also involved in pain transmission, a sub-population of TRPA1 that mediates pain may provide activation of the inhibitory gate to achieve itch inhibition.

16.4   TRPV3 AND ITCH

TRPV3 is a temperature-responsive non-selective cation channel (78–80). With an increase in the temperature from 22°C to 40°C, hTRPV3 transfected cells have been shown to respond with an increase in intracellular calcium through non-selective cation influx (79). TRPV3 gets activated not only by innocuous warmer temperature but also by synthetic compounds such as 2-aminoethoxy diphenylborate (2-APB) (81), naturally occurring compounds such as camphor and carvacol (82), and endogenous ligands such as farnesyl pyrophosphate (83). TRPV3 is highly expressed in keratinocytes and in the cells neighboring to the hair follicle. The expression of TRPV3 is also reported in endothelial cells, gut epithelium, placenta cells, skeletal muscle, stomach, intestine, adipose, placenta, brain, and sensory neurons (78–80,82).

TRPV3 channels are widely involved in skin barrier function, hair growth, skin inflammation, pain, and itch (84). The *G573S* substitution in the *Trpv3* gene (*Nh* mutation) in DS-*Nh* mice or *G573C* substitution (*Ht* mutation) in the *Trpv3* gene in WBN/Kob-*Ht* rats results in hairlessness and allergic and pruritic dermatitis (85,86). Both *G573S* and *G573C* mutations of murine *Trpv3* gene are constitutively active under normal physiological conditions, which might alter ion homeostasis and membrane potentials of skin keratinocytes, leading to the hairlessness phenotype and dermatitis (87). Missense mutations in the human *Trpv3 gene*, including *G573S*, have been found in patients with Olmsted syndrome, a rare skin disorder characterized by palmo-plantar hyperkeratosis, severe pain, and itch, indicating that TRPV3 may play a role in itch signaling (88). Although *Trpv3* mutants in DS-*Nh* mice, WBN/Kob-*Ht* rats, and OS patients display itch symptoms; it remains unclear whether TRPV3 channels directly sense itch signals or regulate itch transmission through a secondary mechanism resulting due to skin pathogenesis. The *Trpv3* mutant *G573S* mice develop dermatitis spontaneously when the animals are housed in the presence of a specific pathogen, namely *Staphylococcus aureus* (86). Intriguingly, in the absence of this particular pathogen *Staphylococcus aureus*, these mice do not develop any chronic pruritus (86). These results support the hypothesis that itch may arise through an indirect mechanism that involves TRPV3-mediated skin pathogenesis.

Constitutive activation of TRPV3 channels in DS-*Nh* mice is linked with increased release of thymic stromal lymphopoietin (TSLP) from keratinocytes (89). Recent research reports that the PAR2-TRPV3 coupling regulates PAR2-elicited acute itch and chronic itch in the atopic dermatitis model via facilitating TLSP release in keratinocytes (90). By contrast, histamine-dependent or BAM-induced acute itch is TRPV3-independent (90). However, other studies show that antagonists of TRPV3 from Chinese medicinal herbs attenuated histamine-dependent acute itch and persistent spontaneous scratching in chronic itch models (91,92). Existing controversies and inconsistencies may reflect the complexity of animal-based experimental systems. Though a mechanism by which TRPV3 channels promote itch sensation seems to be appealing, it remains to be validated in the future.

16.5   TRPM8 AND ITCH

TRPM8 is a non-selective channel and somatic transducer for sensing temperature less than 28°C. Chemical compounds that elicit cooling sensation, such as menthol, icilin, and eucalyptol, can also activate TRPM8 channels (93–95). TRPM8 is expressed in lung epithelial cells, colon tissue, nasal mucosa, prostrate, scrotal skin, seminiferous tubules, testicles, cornea, Merkel cells, keratinocytes, and primary sensory neurons in the DRG and the TG, consistent with its predominant role in the detection of cool in mice (96,97).

Cooling induced by chemicals or by cooling the temperature is shown to be antipruritic in the case of acute and chronic itch. Human patients with atopic dermatitis or human subjects who were given an acute histamine challenge reported relief when the temperature of the itchy skin area is cooled (98). Emerging evidence shows that TRPM8 channels are involved in cooling-mediated itch suppression.

Pruritogen compound 48/80, chloroquine, and histamine-evoked acute scratching behaviors in mice were dramatically attenuated when the agitated skin area was mildly cooled down to 20°C (99). Innocuous cooling induced decrease in chloroquine-evoked itch is revoked in TRPM8 knockout mice or TRPM8$^+$ neurons ablated animals (99), implying that the antipruritic role of cooling requires M8 channels and TRPM8$^+$ peripheral sensory neurons. Interestingly, a noxious cold temperature of 17°C does not have any effect on the cooling-induced decrease in chloroquine-evoked behaviors in *Trpm8* knockout animals, whereas in the case of ablation of TRPM8$^+$ sensory neuron, even temperature as low as 10°C does not effectively reduce chloroquine-induced itch (99). Hence, the effect of noxious cold on the inhibition of itch transmission is mediated primarily by TRPM8$^+$ sensory neurons and appears to be TRPM8 receptor- independent. Contrastingly, Sanders et al. reported that local skin cooling does not affect chloroquine-elicited itch (100).

Cooling of the skin has contrasting effects on serotonin-evoked itch in comparison to histamine-dependent itch. Local cooling of the skin shortens the onset of serotonin-evoked itch and increases the number of scratching bouts in the first five minutes of serotonin administration (100). Additionally, following administration of TRPM8 antagonist, innocuous cold exposure reduces the number of scratching bouts and the latency for induction of serotonin-mediated itch (100).

Cooling can also attenuate the development of spontaneous scratching in chronic itch models (99,100). Activation of TRPM8 can effectively alleviate the symptoms of chronic and spontaneous itch in psoriasis, a chronic itch condition characterized by immune deregulation and spontaneous scratching. Thymol, a naturally occurring agonist of TRPM8, found primarily in thyme, oregano, and tangerine peel (101), can alleviate immunoquid (IMQ)-evoked itch and the severity of skin lesions in the IMQ-induced model of psoriasis. Thymol can effectively reduce the immune burden in the skin by reducing the numbers of dermal neutrophils and dendritic cells. Following thymol treatment, there is a reduction in proinflammatory cytokines in the serum such as TNF-α, IL-6, IL-1B, IL-17A, and IFN-γ expression induced by IMQ treatment (102). The effect of thymol in successfully reversing IMQ-induced aberrant itch and immune profile is TRPM8 dependent (102).

How does the activation of TRPM8 channels relieve itch? Spinal Bhlhb5$^+$ inhibitory neurons block the chemical itch transmission by preventing the activation of the gastrin-releasing peptide receptor-positive (Grpr$^+$) neurons which are the transmission neurons for chemical itch. Bhlhb5$^+$ neurons can be activated by menthol activation of TRPM8$^+$ sensory primary afferents (50). Menthol treatment results in an increase in EPSC frequency in Bhlhb5$^+$ neurons (50). These findings have led to a hypothesis that glutamate release from TRPM8$^+$ primary sensory neurons activates Bhlhb5$^+$ itch gating neurons to suppress itch in the spinal dorsal horn.

16.6 TRPV4 AND ITCH

TRPV4 is a non-selective cation channel with higher permeability for calcium than for Mg^{2+} and Na^+. TRPV4 is activated by warmer temperatures (27°C–34°C). TRPV4 is expressed in lung epithelium, urothelial cells, osteoblasts, osteoclasts, and chondrocytes, vascular endothelial cells, smooth muscles in the pulmonary arteries and aorta, keratinocytes, epidermal macrophages, hippocampal astrocytes, and CNS and PNS neurons. TRPV4 plays a crucial role in osmoregulation, mechanosensation, thermosensation, pain, vascular tone, and epithelial and endothelial barrier integrity (103,104).

The precise role of TRPV4 is still controversial in acute itch. Chen et al. reported that keratinocytes relied on TRPV4 for calcium influx in response to histaminergic pruritogens. TRPV4 activation in keratinocytes evoked phosphorylation of mitogen-activated protein kinase, ERK, for histaminergic pruritogens (105). These results suggest that TRPV4 functions as a pruriceptor-TRP in skin keratinocytes in histaminergic itch. However, Akiyama et al. showed that *Trpv4* knockout mice displayed a significant reduction in scratching behavior in response to serotonin, but not to histamine, or SLIGRL (74). Even more surprisingly, chloroquine-elicited scratching response increases in these *Trpv4*-null animals (74). Contrastingly, Kim et al. report that chloroquine-mediated itch is decreased in *Trpv4*-null animals (106).

A line of evidence indicates the function of TRPV4 in regulating chronic itch. Cell-specific deletion of *Trpv4* in keratinocytes or skin macrophages can result in reduced severity in allergic and non-allergic chronic itch development. Conditional deletion of *Trpv4* in macrophages can rescue non-allergic dry skin-induced chronic itch but do not affect SADBE-induced allergic chronic itch. Cell-specific depletion of *Trpv4* in keratinocytes attenuates scratching in SADBE-evoked spontaneous itch, whereas the chronic itch-induced spontaneous itch remains unchanged. TRPV4-dependent chronic itch is mediated by platelets but not mast cells. Serotonin signaling is crucial in TRPV4-dependent allergic and non-allergic chronic itch. Interestingly, platelets do not express TRPV4 but are a major source of serotonin that is released in chronic itch mouse models. Absence of TRPV4 may result in lowered serotonin levels and hence reduces the severity of scratching chronic itch conditions (107). TRPV4 can also modulate itch transmission through the interference of temperature change.

Pretreatment with TRPV4 antagonist followed by an increase in the skin temperature from 30°C to 38°C evokes a faster onset of histamine-evoked-itch whereas the same condition of TRPV4 antagonist pretreatment and skin warming decreases the serotonin-induced itch and spontaneous itch in Ova-induced rodent model of atopic dermatitis (100).

16.7 TRPC AND ITCH

Seven members of this family have been identified in mammals. TRPC4, expressed in peptidergic neurons in the DRG, is shown to be a facilitator of serotonergic and histaminergic itch. *Trpc4* knockout animals and TRPC4 inhibitor ML204-treated animals show deficits in serotonin- and histamine-evoked itch (108) whereas chloroquine-induced itch remains unaffected. Blocking or depleting TRPC4 receptors is also shown to be effective in attenuating chronic itch symptoms in the mouse model of psoriasis. Either global genetic deletion, mRNA depletion, or local inhibition of TRPC4 receptors using inhibitor ML204 can help to reduce the symptoms of chronic itch and local inflammation in an IMQ-induced mouse model of psoriasis (108).

Selective serotonin reuptake inhibitors (SSRIs) that are widely used to treat depressive disorders have been commonly associated with adverse side effects including skin rashes and urticaria (hives), and itch. SSRI sertraline-, citalopram-, and fluoxetine-elicited scratching in the mouse cheek model of acute itch was revoked in rodents that were co-administered TRPC4 antagonist ML204. Sertraline-evoked itch is diminished in *Trpc4* knockout mice (109). SSRI or serotonin binds and activates serotonin receptor 2B, a GPCR that is coupled G_q/G_{11} leading to the activation of phospholipase C $\beta3$ which mediates the opening of the TRPC4 channel (109) (Figure 16.1). Another TRPC channel, TRPC3, is highly enriched in non-peptidergic MrgprD$^+$ sensory neurons but does not play any role in MrgprD evoked chemical itch behaviors (110).

Recently, TRPC channels, expressed in the spinal astrocytes, have been shown to play an important role in the development of chronic itch. In the mouse model of contact dermatitis, IL-6 released by the pruriceptive endings induces STAT3 activation in the spinal astrocytes, converting them into reactive astrocytes that produce inflammatory responses contributing to chronic itch neuropathology. Reactive astrocytes in the spinal dorsal horn release inflammatory factor lipocalin-2 (LCN2), which is crucial for chronic itch. To maintain the reactive state of astrocytes, persistent activation of STAT3 is important, which requires IP3R1-TRPC3/7 pathway-evoked Ca^{2+} signals. Inhibitors of TRPC3 and TRPC4 attenuate activation of STAT3 when cultured astrocytes are challenged with IL-6 (111). Hence astrocytic TRPC3/7 channels are crucial for the induction and maintenance of the reactive state of astrocytes that contributes to the neuropathology of chronic itch.

16.8 CONCLUSION

Itch is a spatial integration of complex sensory information interplayed and exchanged between keratinocytes, sensory neuronal endings, and various other dermal cells. TRP channels have been extensively reported to contribute significantly to the pathophysiology of itch disorders. Though our understanding of the entirety or extent of contribution of the TRP superfamily as regulators of itch is still in its infancy, there is convincing evidence that implicates TRP channels as crucial effectors in the cellular and molecular pathway of itch transmission. While the roles of TRPV3 and TRPV4 remain controversial in itch induction; some TRP channels have been clearly established as pruritic TRP channels, such as TRPV1 and TRPA1, and as antipruritic TRPM8. Therapeutic use of antagonists or agonists of these TRPs to treat itch has been effective in some cases and continues to be an area of active investigation to develop more specific activators or inhibitors with no or the least off-target effects (Table 16.1).

TABLE 16.1

Agonist and Antagonist of Itch-Implicated TRP Channels

	Agonist		Antagonist	
	Exogenous	**Endogenous**	**Natural**	**Synthetic**
TRPV1	Capsaicin (20), resiniferatoxin (112), cannabidiol (113) cannabidivarin (113)	12(S)-hydroxyeicosapentae-noic acid (114), 5- and 15-(S)-hydroxyeicosatetrae-noic acids (114) leukotriene B4 (114) Oxytocin (115), Hydrogen sulfide (116)	Adenosine (117), N-arachidonoyl serotonin (118)	BCTC (119), SB-705498 (120), Quinazoline analogs (AMG-517) (121), Imidazole analogs (4,5-biarylimidazole) (122), RRRRWW-NH2 (123),
TRPA1	capsiate (124), eugenol (125), carvacrol (126), 1,4 cineole (95), allyl isothiocyanate (127), cinnamaldehyde (128), allicin (129), nicotine (130), acrolein (131), lidocaine (132),	Methylglyoxal (133), H_2O_2 (134), Arachidonic and docosahexaenoic (DHA) acids (135)	1,8-cineole (95), borneol (136), 2-methylisobor-neol (136), fenchyl alcohol (136), cardamonin (137), resveratrol (138),	HC-030031 (139), AP-18 (140),
TRPM8	Icilin (93)	1,8-cineole (95), 1,4-cineole (95), Camphor (141), Menthol (94),	Sesamin (142)	M8-B (143), AMTB (144), PBMC (145),
TRPV3	Farnesyl pyrophosphate (83) Ethylvanillin (146),	Aminoethoxy diphenylborate (2-APB) (81), eugenol (126), Cannabidiol (147), Carvacrol (82,148), Thymol (82,126), Camphor (149), Borneol (82), Vanillin (126),	17(R)-resolvin D1 (150),	GRC 15133 (151), GRC 17173 (151), GRC 15300 (151),
TRPV4	Cannabigerol (147), cannabigerolic acid (147), cannabigerovarin	Anandamide (155), Epoxyeicosatrienoic acids (5,6-EET) (155), Ntyr (156), N-acyl		GSK2193874 (158), HC-067047 (159), RN-1734 (160), RN-9893 (161)

(*continued on next page*)

TABLE 16.1 (*continued*)

Agonist		Antagonist	
Exogenous	Endogenous	Natural	Synthetic
(147), cannabinoid (147), GSK10116790A (152), phorbol esters (4-PDD) (153), RN-1747 (154)	tryptophan (157), arachidonic acid (155),		
TRPC4 englerin A (162)			HC-070 (163)ML204 (164)Pico145 (HC-608) (165) M084 (166)

REFERENCES

1. Twycross R., Greaves M.W., Handwerker H., Jones E.A., Libretto S.E., et al. 2003. Itch: scratching more than the surface. *QJM* 96:7–26.
2. Dong X., Dong X. 2018. Peripheral and central mechanisms of itch. *Neuron* 98:482–494.
3. Ikoma A., Steinhoff M., Stander S., Yosipovitch G., Schmelz M. 2006. The neurobiology of itch. *Nat Rev Neurosci* 7:535–547.
4. Han L., Dong X. 2014. Itch mechanisms and circuits. *Annu Rev Biophys* 43:331–355
5. Barry D.M., Munanairi A., Chen Z.F. 2018. Spinal mechanisms of itch transmission. *Neurosci Bull* 34:156–164.
6. Duan B., Cheng L., Ma Q. 2018. Spinal circuits transmitting mechanical pain and itch. *Neurosci Bull* 34:186–193.
7. Montell C., Birnbaumer L., Flockerzi V., Bindels R.J., Bruford E.A., et al. 2002. A unified nomenclature for the superfamily of TRP cation channels. *Mol Cell* 9:229–231
8. Cosens D.J., Manning A. 1969. Abnormal electroretinogram from a Drosophila mutant. *Nature* 224:285–287.
9. Montell C., Rubin G.M. 1989. Molecular characterization of the Drosophila trp locus: a putative integral membrane protein required for phototransduction. *Neuron* 2:1313–1323.
10. Nilius B., Owsianik G. 2011. The transient receptor potential family of ion channels. *Genome Biol* 12:218.
11. Li M., Yu Y., Yang J. 2011. Structural biology of TRP channels. *Adv Exp Med Biol* 704:1–23.
12. Venkatachalam K., Montell C. 2007. TRP channels. *Annu Rev Biochem* 76:387–417.
13. Rosasco M.G., Gordon S.E. 2017. TRP Channels: What Do They Look Like? In *Neurobiology of TRP Channels*, ed. T.L.R. Emir, Boca Raton, FL: CRC Press. 1–9 pp.
14. Xu H., Zhao H., Tian W., Yoshida K., Roullet J.B., Cohen D.M. 2003. Regulation of a transient receptor potential (TRP) channel by tyrosine phosphorylation. SRC family kinase-dependent tyrosine phosphorylation of TRPV4 on TYR-253 mediates its response to hypotonic stress. *J Biol Chem* 278:11520–11527.

15. Pedersen S.F., Owsianik G., Nilius B. 2005. TRP channels: an overview. *Cell Calcium* 38:233–252.
16. Ho J.C., Lee C.H. 2015. TRP channels in skin: from physiological implications to clinical significances. *Biophysics (Nagoya-shi)* 11:17–24.
17. Bevan S., Quallo T., Andersson D.A. 2014. Trpv1. *Handb Exp Pharmacol* 222:207–245.
18. Tominaga M., Tominaga T. 2005. Structure and function of TRPV1. *Pflugers Arch* 451:143–150.
19. Kobayashi K., Fukuoka T., Obata K., Yamanaka H., Dai Y., et al. 2005. Distinct expression of TRPM8, TRPA1, and TRPV1 mRNAs in rat primary afferent neurons with adelta/c-fibers and colocalization with trk receptors. *J Comp Neurol* 493:596–606.
20. Caterina M.J., Schumacher M.A., Tominaga M., Rosen T.A., Levine J.D., Julius D. 1997. The capsaicin receptor: a heat-activated ion channel in the pain pathway. *Nature* 389:816–824.
21. Carnevale V., Rohacs T. 2016. TRPV1: a target for rational drug design. *Pharmaceuticals (Basel)* 9, 52.
22. Brederson J.D., Kym P.R., Szallasi A. 2013. Targeting TRP channels for pain relief. *Eur J Pharmacol* 716:61–76.
23. Cortright D.N., Szallasi A. 2009. TRP channels and pain. *Curr Pharm Des* 15:1736–1749.
24. Eid S.R. 2011. Therapeutic targeting of TRP channels—the TR(i)P to pain relief. *Curr Top Med Chem* 11:2118–2130.
25. Yin S., Luo J., Qian A., Du J., Yang Q., et al. 2013. Retinoids activate the irritant receptor TRPV1 and produce sensory hypersensitivity. *J Clin Invest* 123:3941–3951.
26. Tominaga M., Caterina M.J., Malmberg A.B., Rosen T.A., Gilbert H., et al. 1998. The cloned capsaicin receptor integrates multiple pain-producing stimuli. *Neuron* 21:531–543.
27. Willis W.D., Jr. 2009. The role of TRPV1 receptors in pain evoked by noxious thermal and chemical stimuli. *Exp Brain Res* 196:5–11.
28. Imamachi N., Park G.H., Lee H., Anderson D.J., Simon M.I., et al. 2009. TRPV1-expressing primary afferents generate behavioral responses to pruritogens via multiple mechanisms. *Proc Natl Acad Sci U S A* 106:11330–11335.
29. Duo L., Hu L., Tian N., Cheng G., Wang H., et al. 2018. TRPV1 gain-of-function mutation impairs pain and itch sensations in mice. *Mol Pain* 14:1744806918762031.
30. Kajihara Y., Murakami M., Imagawa T., Otsuguro K., Ito S., Ohta T. 2010. Histamine potentiates acid-induced responses mediating transient receptor potential V1 in mouse primary sensory neurons. *Neuroscience* 166:292–304.
31. Jian T., Yang N., Yang Y., Zhu C., Yuan X., et al. 2016. TRPV1 and PLC participate in histamine H4 receptor-induced itch. *Neural Plast* 2016:1682972.
32. Prescott E.D., Julius D. 2003. A modular PIP2 binding site as a determinant of capsaicin receptor sensitivity. *Science* 300:1284–1288.
33. Kim S.J., Park G.H., Kim D., Lee J., Min H., et al. 2011. Analysis of cellular and behavioral responses to imiquimod reveals a unique itch pathway in transient receptor potential vanilloid 1 (TRPV1)-expressing neurons. *Proc Natl Acad Sci USA* 108:3371–3376.
34. Liu T., Xu Z.Z., Park C.K., Berta T., Ji R.R. 2010. Toll-like receptor 7 mediates pruritus. *Nat Neurosci* 13:1460–1462.
35. Kittaka H., Uchida K., Fukuta N., Tominaga M. 2017. Lysophosphatidic acid-induced itch is mediated by signalling of LPA5 receptor, phospholipase D and TRPA1/TRPV1. *J Physiol* 595:2681–2698.
36. Morales-Lazaro S.L., Llorente I., Sierra-Ramirez F., Lopez-Romero A.E., Ortiz-Renteria M., et al. 2016. Inhibition of TRPV1 channels by a naturally occurring omega-9 fatty acid reduces pain and itch. *Nat Commun* 7:13092.

37. Falodun O., Ogunbiyi A., Salako B., George A.K. 2011. Skin changes in patients with chronic renal failure. *Saudi J Kidney Dis Transpl* 22:268–272.

38. Cheung A.K., Greene T., Leypoldt J.K., Yan G., Allon M., et al. 2008. Association between serum 2-microglobulin level and infectious mortality in hemodialysis patients. *Clin J Am Soc Nephrol* 3:69–77.

39. Andoh T., Maki T., Li S., Uta D. 2017. beta2-Microglobulin elicits itch-related responses in mice through the direct activation of primary afferent neurons expressing transient receptor potential vanilloid 1. *Eur J Pharmacol* 810:134–140.

40. Feng J., Yang P., Mack M.R., Dryn D., Luo J., et al. 2017. Sensory TRP channels contribute differentially to skin inflammation and persistent itch. *Nat Commun* 8:980.

41. Yun J.W., Seo J.A., Jang W.H., Koh H.J., Bae I.H., et al. 2011. Antipruritic effects of TRPV1 antagonist in murine atopic dermatitis and itching models. *J Invest Dermatol* 131:1576–1579.

42. Yun J.W., Seo J.A., Jeong Y.S., Bae I.H., Jang W.H., et al. 2011. TRPV1 antagonist can suppress the atopic dermatitis-like symptoms by accelerating skin barrier recovery. *J Dermatol Sci* 62:8–15.

43. Ellis C.N., Berberian B., Sulica V.I., Dodd W.A., Jarratt M.T., et al. 1993. A double-blind evaluation of topical capsaicin in pruritic psoriasis. *J Am Acad Dermatol* 29:438–442.

44. Lotti T., Teofoli P., Tsampau D. 1994. Treatment of aquagenic pruritus with topical capsaicin cream. *J Am Acad Dermatol* 30:232–235.

45. Hautmann G., Teofoli P., Lotti T. 1994. Aquagenic pruritus, PUVA and capsaicin treatments. *Br J Dermatol* 131:920–921.

46. Dillon S.R., Sprecher C., Hammond A., Bilsborough J., Rosenfeld-Franklin M., et al. 2004. Interleukin 31, a cytokine produced by activated T cells, induces dermatitis in mice. *Nat Immunol* 5:752–760.

47. Cevikbas F., Wang X., Akiyama T., Kempkes C., Savinko T., et al. 2014. A sensory neuron-expressed IL-31 receptor mediates T helper cell-dependent itch: involvement of TRPV1 and TRPA1. *J Allergy Clin Immunol* 133:448–460.

48. Liu Y., Abdel Samad O., Zhang L., Duan B., Tong Q., et al. 2010. VGLUT2-dependent glutamate release from nociceptors is required to sense pain and suppress itch. *Neuron* 68:543–556.

49. Lagerstrom M.C., Rogoz K., Abrahamsen B., Persson E., Reinius B., et al. 2010. VGLUT2-dependent sensory neurons in the TRPV1 population regulate pain and itch. *Neuron* 68:529–542.

50. Kardon A.P., Polgar E., Hachisuka J., Snyder L.M., Cameron D., et al. 2014. Dynorphin acts as a neuromodulator to inhibit itch in the dorsal horn of the spinal cord. *Neuron* 82:573–586.

51. Andersen H.H., Sand C., Elberling J. 2016. Considerable variability in the efficacy of 8% capsaicin topical patches in the treatment of chronic pruritus in 3 patients with notalgia paresthetica. *Ann Dermatol* 28:86–89.

52. Misery L., Erfan N., Castela E., Brenaut E., Lanteri-Minet M., et al. 2015. Successful treatment of refractory neuropathic pruritus with capsaicin 8% patch: a bicentric retrospective study with long-term follow-up. *Acta Derm Venereol* 95:864–865.

53. Gibson R.A., Robertson J., Mistry H., McCallum S., Fernando D., et al. 2014. A randomised trial evaluating the effects of the TRPV1 antagonist SB705498 on pruritus induced by histamine, and cowhage challenge in healthy volunteers. *PLoS One* 9:e100610.

54. Paulsen C.E., Armache J.P., Gao Y., Cheng Y., Julius D. 2015. Structure of the TRPA1 ion channel suggests regulatory mechanisms. *Nature* 520:511–517.

55. Paulsen C.E., Armache J.P., Gao Y., Cheng Y., Julius D. 2015. Structure of the TRPA1 ion channel suggests regulatory mechanisms. *Nature* 525:552.

56. Clapham D.E. 2015. Structural biology: pain-sensing TRPA1 channel resolved. *Nature* 520:439–441.
57. Cho H.J., Callaghan B., Bron R., Bravo D.M., Furness J.B. 2014. Identification of enteroendocrine cells that express TRPA1 channels in the mouse intestine. *Cell Tissue Res* 356:77–82.
58. Atoyan R., Shander D., Botchkareva N.V. 2009. Non-neuronal expression of transient receptor potential type A1 (TRPA1) in human skin. *J Invest Dermatol* 129:2312–2315.
59. Cao D.S., Zhong L., Hsieh T.H., Abooj M., Bishnoi M., et al. 2012. Expression of transient receptor potential ankyrin 1 (TRPA1) and its role in insulin release from rat pancreatic beta cells. *PLoS One* 7:e38005.
60. Nilius B., Appendino G., Owsianik G. 2012. The transient receptor potential channel TRPA1: from gene to pathophysiology. *Pflugers Arch* 464:425–458.
61. Bautista D.M., Pellegrino M., Tsunozaki M. 2013. TRPA1: a gatekeeper for inflammation. *Annu Rev Physiol* 75:181–200.
62. Cruz-Orengo L., Dhaka A., Heuermann R.J., Young T.J., Montana M.C., et al. 2008. Cutaneous nociception evoked by 15-delta PGJ2 via activation of ion channel TRPA1. *Mol Pain* 4:30.
63. Story G.M., Peier A.M., Reeve A.J., Eid S.R., Mosbacher J., et al. 2003. ANKTM1, a TRP-like channel expressed in nociceptive neurons, is activated by cold temperatures. *Cell* 112:819–829.
64. Caspani O., Heppenstall P.A. 2009. TRPA1 and cold transduction: an unresolved issue? *J Gen Physiol* 133:245–249.
65. Hojland C.R., Andersen H.H., Poulsen J.N., Arendt-Nielsen L., Gazerani P. 2015. A human surrogate model of itch utilizing the TRPA1 agonist trans-cinnamaldehyde. *Acta Derm Venereol* 95:798–803.
66. Wilson S.R., Nelson A.M., Batia L., Morita T., Estandian D., et al. 2013. The ion channel TRPA1 is required for chronic itch. *J Neurosci* 33:9283–9294.
67. Ziegler S.F., Roan F., Bell B.D., Stoklasek T.A., Kitajima M., Han H. 2013. The biology of thymic stromal lymphopoietin (TSLP). *Adv Pharmacol* 66:129–155.
68. Wilson S.R., The L., Batia L.M., Beattie K., Katibah G.E., et al. 2013. The epithelial cell-derived atopic dermatitis cytokine TSLP activates neurons to induce itch. *Cell* 155:285–295.
69. Oh M.H., Oh S.Y., Lu J., Lou H., Myers A.C., et al. 2013. TRPA1-dependent pruritus in IL-13-induced chronic atopic dermatitis. *J Immunol* 191:5371–5382.
70. Reese R.M., Dourado M., Anderson K., Warming S., Stark K.L., et al. 2020. Behavioral characterization of a CRISPR-generated TRPA1 knockout rat in models of pain, itch, and asthma. *Sci Rep* 10:979.
71. Huang J., Li G., Xiang J., Yin D., Chi R. 2004. Immunohistochemical study of serotonin in lesions of chronic eczema. *Int J Dermatol* 43:723–726.
72. Huang J., Li G., Xiang J., Yin D., Chi R. 2004. Immunohistochemical study of serotonin in lesions of psoriasis. *Int J Dermatol* 43:408–411.
73. Lundeberg L., Liang Y., Sundstrom E., Nordlind K., Verhofstad A., et al. 1999. Serotonin in human allergic contact dermatitis. An immunohistochemical and high-performance liquid chromatographic study. *Arch Dermatol Res* 291:269–274.
74. Akiyama T., Ivanov M., Nagamine M., Davoodi A., Carstens M.I., et al. 2016. Involvement of TRPV4 in serotonin-evoked scratching. *J Invest Dermatol* 136:154–160.
75. Morita T., McClain S.P., Batia L.M., Pellegrino M., Wilson S.R., et al. 2015. HTR7 Mediates serotonergic acute and chronic itch. *Neuron* 87:124–138.
76. Han Q., Liu D., Convertino M., Wang Z., Jiang C., et al. 2018. miRNA-711 binds and activates TRPA1 extracellularly to evoke acute and chronic pruritus. *Neuron* 99:449–463.e6.

77. Lieu T., Jayaweera G., Zhao P., Poole D.P., Jensen D., et al. 2014. The bile acid receptor TGR5 activates the TRPA1 channel to induce itch in mice. *Gastroenterology* 147:1417–1428.
78. Peier A.M., Reeve A.J., Andersson D.A., Moqrich A., Earley T.J., et al. 2002. A heat-sensitive TRP channel expressed in keratinocytes. *Science* 296:2046–2049.
79. Xu H., Ramsey I.S., Kotecha S.A., Moran M.M., Chong J.A., et al. 2002. TRPV3 is a calcium-permeable temperature-sensitive cation channel. *Nature* 418:181–186.
80. Smith G.D., Gunthorpe M.J., Kelsell R.E., Hayes P.D., Reilly P., et al. 2002. TRPV3 is a temperature-sensitive vanilloid receptor-like protein. *Nature* 418:186–190.
81. Hu H.Z., Gu Q., Wang C., Colton C.K., Tang J., et al. 2004. 2-Aminoethoxydiphenyl borate is a common activator of TRPV1, TRPV2, and TRPV3. *J Biol Chem* 279:35741–35748.
82. Vogt-Eisele A.K., Weber K., Sherkheli M.A., Vielhaber G., Panten J., et al. 2007. Monoterpenoid agonists of TRPV3. *Br J Pharmacol* 151:530–540.
83. Bang S., Yoo S., Yang T.J., Cho H., Hwang S.W. 2010. Farnesyl pyrophosphate is a novel pain-producing molecule via specific activation of TRPV3. *J Biol Chem* 285:19362–19371.
84. Wang G., Wang K. 2017. The Ca(2+)-permeable cation transient receptor potential TRPV3 channel: an emerging pivotal target for itch and skin diseases. *Mol Pharmacol* 92:193–200.
85. Asakawa M., Yoshioka T., Hikita I., Matsutani T., Hirasawa T., et al. 2005. WBN/Kob-Ht rats spontaneously develop dermatitis under conventional conditions: another possible model for atopic dermatitis. *Exp Anim* 54:461–465.
86. Asakawa M., Yoshioka T., Matsutani T., Hikita I., Suzuki M., et al. 2006. Association of a mutation in TRPV3 with defective hair growth in rodents. *J Invest Dermatol* 126:2664–2672.
87. Xiao R., Tian J., Tang J., Zhu M.X. 2008. The TRPV3 mutation associated with the hairless phenotype in rodents is constitutively active. *Cell Calcium* 43:334–343.
88. Yadav M., Goswami C. 2017. TRPV3 mutants causing Olmsted Syndrome induce impaired cell adhesion and nonfunctional lysosomes. *Channels (Austin)* 11:196–208.
89. Yamamoto-Kasai E., Yasui K., Shichijo M., Sakata T., Yoshioka T. 2013. Impact of TRPV3 on the development of allergic dermatitis as a dendritic cell modulator. *Exp Dermatol* 22:820–824.
90. Zhao J., Munanairi A., Liu X.Y., Zhang J., Hu L., et al. 2020. PAR2 mediates itch via TRPV3 signaling in keratinocytes. *J Invest Dermatol* 140:1524–1532.
91. Zhang H., Sun X., Qi H., Ma Q., Zhou Q., et al. 2019. Pharmacological inhibition of the temperature-sensitive and Ca(2+)-permeable transient receptor potential vanilloid TRPV3 channel by natural forsythoside B attenuates pruritus and cytotoxicity of keratinocytes. *J Pharmacol Exp Ther* 368:21–31.
92. Sun X.Y., Sun L.L., Qi H., Gao Q., Wang G.X., et al. 2018. Antipruritic effect of natural coumarin osthole through selective inhibition of thermosensitive TRPV3 channel in the skin. *Mol Pharmacol* 94:1164–1173.
93. Chuang H.H., Neuhausser W.M., Julius D. 2004. The super-cooling agent icilin reveals a mechanism of coincidence detection by a temperature-sensitive TRP channel. *Neuron* 43:859–869.
94. Peier A.M., Moqrich A., Hergarden A.C., Reeve A.J., Andersson D.A., et al. 2002. A TRP channel that senses cold stimuli and menthol. *Cell* 108:705–715.
95. Takaishi M., Fujita F., Uchida K., Yamamoto S., Sawada Shimizu M., et al. 2012. 1,8-cineole, a TRPM8 agonist, is a novel natural antagonist of human TRPA1. *Mol Pain* 8:86.
96. Liu Y., Mikrani R., He Y., Faran Ashraf Baig M.M., Abbas M., et al. 2020. TRPM8 channels: a review of distribution and clinical role. *Eur J Pharmacol* 882:173312.

97. Bouvier V., Roudaut Y., Osorio N., Aimonetti J.M., Ribot-Ciscar E., et al. 2018. Merkel cells sense cooling with TRPM8 channels. *J Invest Dermatol* 138:946–956.
98. Fruhstorfer H., Hermanns M., Latzke L. 1986. The effects of thermal stimulation on clinical and experimental itch. *Pain* 24:259–269.
99. Palkar R., Ongun S., Catich E., Li N., Borad N., et al. 2018. Cooling relief of acute and chronic itch requires TRPM8 channels and neurons. *J Invest Dermatol* 138:1391–1399.
100. Sanders K.M., Hashimoto T., Sakai K., Akiyama T. 2018. Modulation of itch by localized skin warming and cooling. *Acta Derm Venereol* 98:855–861.
101. Liang D., Li F., Fu Y., Cao Y., Song X., et al. 2014. Thymol inhibits LPS-stimulated inflammatory response via down-regulation of NF-kappaB and MAPK signaling pathways in mouse mammary epithelial cells. *Inflammation* 37:214–222.
102. Wang W., Wang H., Zhao Z., Huang X., Xiong H., Mei Z. 2020. Thymol activates TRPM8-mediated Ca(2+) influx for its antipruritic effects and alleviates inflammatory response in Imiquimod-induced mice. *Toxicol Appl Pharmacol* 407:115247.
103. Darby W.G., Grace M.S., Baratchi S., McIntyre P. 2016. Modulation of TRPV4 by diverse mechanisms. *Int J Biochem Cell Biol* 78:217–228.
104. Rosenbaum T., Benitez-Angeles M., Sanchez-Hernandez R., Morales-Lazaro S.L., Hiriart M., et al. 2020. TRPV4: a physio and pathophysiologically significant ion channel. *Int J Mol Sci* 21, 3837.
105. Chen Y., Fang Q., Wang Z., Zhang J.Y., MacLeod A.S., et al. 2016. Transient receptor potential vanilloid 4 ion channel functions as a pruriceptor in epidermal keratinocytes to evoke histaminergic itch. *J Biol Chem* 291:10252–10262.
106. Kim S., Barry D.M., Liu X.Y., Yin S., Munanairi A., et al. 2016. Facilitation of TRPV4 by TRPV1 is required for itch transmission in some sensory neuron populations. *Sci Signal* 9:ra71.
107. Luo J., Feng J., Yu G., Yang P., Mack M.R., et al. 2018. Transient receptor potential vanilloid 4-expressing macrophages and keratinocytes contribute differentially to allergic and nonallergic chronic itch. *J Allergy Clin Immunol* 141:608–619.e7.
108. Lee S.H., Tonello R., Choi Y., Jung S.J., Berta T. 2020. Sensory neuron-expressed TRPC4 is a target for the relief of psoriasiform itch and skin inflammation in mice. *J Invest Dermatol* 140:2221–2229.e6.
109. Lee S.H., Cho P.S., Tonello R., Lee H.K., Jang J.H., et al. 2018. Peripheral serotonin receptor 2B and transient receptor potential channel 4 mediate pruritus to serotonergic antidepressants in mice. *J Allergy Clin Immunol* 142:1349–1352.e16.
110. Dong P., Guo C., Huang S., Ma M., Liu Q., Luo W. 2017. TRPC3 is dispensable for beta-alanine triggered acute itch. *Sci Rep* 7:13869.
111. Shiratori-Hayashi M., Yamaguchi C., Eguchi K., Shiraishi Y., Kohno K., et al. 2020. Astrocytic STAT3 activation and chronic itch require IP3R1/TRPC-dependent Ca(2+) signals in mice. *J Allergy Clin Immunol*. In press..
112. Raisinghani M., Pabbidi R.M., Premkumar L.S. 2005. Activation of transient receptor potential vanilloid 1 (TRPV1) by resiniferatoxin. *J Physiol* 567:771–786.
113. Iannotti F.A., Hill C.L., Leo A., Alhusaini A., Soubrane C., et al. 2014. Nonpsychotropic plant cannabinoids, cannabidivarin (CBDV) and cannabidiol (CBD), activate and desensitize transient receptor potential vanilloid 1 (TRPV1) channels in vitro: potential for the treatment of neuronal hyperexcitability. *ACS Chem Neurosci* 5:1131–1141.
114. Hwang S.W., Cho H., Kwak J., Lee S.Y., Kang C.J., et al. 2000. Direct activation of capsaicin receptors by products of lipoxygenases: endogenous capsaicin-like substances. *Proc Natl Acad Sci USA* 97:6155–6160.
115. Nersesyan Y., Demirkhanyan L., Cabezas-Bratesco D., Oakes V., Kusuda R., et al. 2017. Oxytocin modulates nociception as an agonist of pain-sensing TRPV1. *Cell Rep* 21:1681–1691.

116. Trevisani M., Patacchini R., Nicoletti P., Gatti R., Gazzieri D., et al. 2005. Hydrogen sulfide causes vanilloid receptor 1-mediated neurogenic inflammation in the airways. *Br J Pharmacol* 145:1123–1131.

117. Puntambekar P., Van Buren J., Raisinghani M., Premkumar L.S., Ramkumar V. 2004. Direct interaction of adenosine with the TRPV1 channel protein. *J Neurosci* 24:3663–3671.

118. Maione S., De Petrocellis L., de Novellis V., Moriello A.S., Petrosino S., et al. 2007. Analgesic actions of N-arachidonoyl-serotonin, a fatty acid amide hydrolase inhibitor with antagonistic activity at vanilloid TRPV1 receptors. *Br J Pharmacol* 150:766–781.

119. Pomonis J.D., Harrison J.E., Mark L., Bristol D.R., Valenzano K.J., Walker K. 2003. N-(4-Tertiarybutylphenyl)-4-(3-cholorphyridin-2-yl)tetrahydropyrazine -1(2H)-carbox-amide (BCTC), a novel, orally effective vanilloid receptor 1 antagonist with analgesic properties: II. In vivo characterization in rat models of inflammatory and neuropathic pain. *J Pharmacol Exp Ther* 306:387–393.

120. Rami H.K., Thompson M., Stemp G., Fell S., Jerman J.C., et al. 2006. Discovery of SB-705498: a potent, selective and orally bioavailable TRPV1 antagonist suitable for clinical development. *Bioorg Med Chem Lett* 16:3287–3291.

121. Doherty E.M., Fotsch C., Bannon A.W., Bo Y., Chen N., et al. 2007. Novel vanilloid receptor-1 antagonists: 2. Structure-activity relationships of 4-oxopyrimidines leading to the selection of a clinical candidate. *J Med Chem* 50:3515–3527.

122. Gore V.K., Ma V.V., Tamir R., Gavva N.R., Treanor J.J., Norman M.H. 2007. Structure-activity relationship (SAR) investigations of substituted imidazole analogs as TRPV1 antagonists. *Bioorg Med Chem Lett* 17:5825–5830.

123. Himmel H.M., Kiss T., Borvendeg S.J., Gillen C., Illes P. 2002. The arginine-rich hexapeptide R4W2 is a stereoselective antagonist at the vanilloid receptor 1: a Ca^{2+} imaging study in adult rat dorsal root ganglion neurons. *J Pharmacol Exp Ther* 301:981–986.

124. Shintaku K., Uchida K., Suzuki Y., Zhou Y., Fushiki T., et al. 2012. Activation of transient receptor potential A1 by a non-pungent capsaicin-like compound, capsiate. *Br J Pharmacol* 165:1476–1486.

125. Chung G., Im S.T., Kim Y.H., Jung S.J., Rhyu M.R., Oh S.B. 2014. Activation of transient receptor potential ankyrin 1 by eugenol. *Neuroscience* 261:153–160.

126. Xu H., Delling M., Jun J.C., Clapham D.E. 2006. Oregano, thyme and clove-derived flavors and skin sensitizers activate specific TRP channels. *Nat Neurosci* 9:628–635.

127. Jordt S.E., Bautista D.M., Chuang H.H., McKemy D.D., Zygmunt P.M., et al. 2004. Mustard oils and cannabinoids excite sensory nerve fibres through the TRP channel ANKTM1. *Nature* 427:260–265.

128. Namer B., Seifert F., Handwerker H.O., Maihofner C. 2005. TRPA1 and TRPM8 activation in humans: effects of cinnamaldehyde and menthol. *Neuroreport* 16:955–959.

129. Bautista D.M., Movahed P., Hinman A., Axelsson H.E., Sterner O., et al. 2005. Pungent products from garlic activate the sensory ion channel TRPA1. *Proc Natl Acad Sci USA* 102:12248–12252.

130. Talavera K., Gees M., Karashima Y., Meseguer V.M., Vanoirbeek J.A., et al. 2009. Nicotine activates the chemosensory cation channel TRPA1. *Nat Neurosci* 12:1293–1299.

131. Bautista D.M., Jordt S.E., Nikai T., Tsuruda P.R., Read A.J., et al. 2006. TRPA1 mediates the inflammatory actions of environmental irritants and proalgesic agents. *Cell* 124:1269–1282.

132. Leffler A., Lattrell A., Kronewald S., Niedermirtl F., Nau C. 2011. Activation of TRPA1 by membrane permeable local anesthetics. *Mol Pain* 7:62.

133. Eberhardt M.J., Filipovic M.R., Leffler A., de la Roche J., Kistner K., et al. 2012. Methylglyoxal activates nociceptors through transient receptor potential channel A1

(TRPA1): a possible mechanism of metabolic neuropathies. *J Biol Chem* 287:28291–28306.

134. Trevisan G., Hoffmeister C., Rossato M.F., Oliveira S.M., Silva M.A., et al. 2014. TRPA1 receptor stimulation by hydrogen peroxide is critical to trigger hyperalgesia and inflammation in a model of acute gout. *Free Radic Biol Med* 72:200–209.

135. Motter A.L., Ahern G.P. 2012. TRPA1 is a polyunsaturated fatty acid sensor in mammals. *PLoS One* 7:e38439.

136. Takaishi M., Uchida K., Fujita F., Tominaga M. 2014. Inhibitory effects of mono-terpenes on human TRPA1 and the structural basis of their activity. *J Physiol Sci* 64:47–57.

137. Wang S., Zhai C., Zhang Y., Yu Y., Zhang Y., et al. 2016. Cardamonin, a novel antagonist of hTRPA1 cation channel, reveals therapeutic mechanism of pathological pain. *Molecules* 21, 1145.

138. Yu L., Wang S., Kogure Y., Yamamoto S., Noguchi K., Dai Y. 2013. Modulation of TRP channels by resveratrol and other stilbenoids. *Mol Pain* 9:3.

139. Eid S.R., Crown E.D., Moore E.L., Liang H.A., Choong K.C., et al. 2008. HC-030031, a TRPA1 selective antagonist, attenuates inflammatory- and neuropathy-induced mechanical hypersensitivity. *Mol Pain* 4:48.

140. Petrus M., Peier A.M., Bandell M., Hwang S.W., Huynh T., et al. 2007. A role of TRPA1 in mechanical hyperalgesia is revealed by pharmacological inhibition. *Mol Pain* 3:40.

141. Selescu T., Ciobanu A.C., Dobre C., Reid G., Babes A. 2013. Camphor activates and sensitizes transient receptor potential melastatin 8 (TRPM8) to cooling and icilin. *Chem Senses* 38:563–575.

142. Sui Y., Li S., Zhao Y., Liu Q., Qiao Y., et al. 2020. Identification of a natural compound, sesamin, as a novel TRPM8 antagonist with inhibitory effects on prostate adenocarcinoma. *Fitoterapia* 145:104631.

143. Almeida M.C., Hew-Butler T., Soriano R.N., Rao S., Wang W., et al. 2012. Pharmacological blockade of the cold receptor TRPM8 attenuates autonomic and behavioral cold defenses and decreases deep body temperature. *J Neurosci* 32:2086–2099.

144. Lashinger E.S., Steiginga M.S., Hieble J.P., Leon L.A., Gardner S.D., et al. 2008. AMTB, a TRPM8 channel blocker: evidence in rats for activity in overactive bladder and painful bladder syndrome. *Am J Physiol Renal Physiol* 295:F803–F810.

145. Knowlton W.M., Daniels R.L., Palkar R., McCoy D.D., McKemy D.D. 2011. Pharmacological blockade of TRPM8 ion channels alters cold and cold pain responses in mice. *PLoS One* 6:e25894.

146. Masamoto Y., Kawabata F., Fushiki T. 2009. Intragastric administration of TRPV1, TRPV3, TRPM8, and TRPA1 agonists modulates autonomic thermoregulation in different manners in mice. *Biosci Biotechnol Biochem* 73:1021–1027.

147. De Petrocellis L., Orlando P., Moriello A.S., Aviello G., Stott C., et al. 2012. Cannabinoid actions at TRPV channels: effects on TRPV3 and TRPV4 and their potential relevance to gastrointestinal inflammation. *Acta Physiol (Oxf)* 204:255–266.

148. Cui T.T., Wang G.X., Wei N.N., Wang K. 2018. A pivotal role for the activation of TRPV3 channel in itch sensations induced by the natural skin sensitizer carvacrol. *Acta Pharmacol Sin* 39:331–335.

149. Moqrich A., Hwang S.W., Earley T.J., Petrus M.J., Murray A.N., et al. 2005. Impaired thermosensation in mice lacking TRPV3, a heat and camphor sensor in the skin. *Science* 307:1468–1472.

150. Bang S., Yoo S., Yang T.J., Cho H., Hwang S.W. 2012. 17(R)-resolvin D1 specifically inhibits transient receptor potential ion channel vanilloid 3 leading to peripheral antinociception. *Br J Pharmacol* 165:683–692.

151. Reilly R.M., Kym P.R. 2011. Analgesic potential of TRPV3 antagonists. *Curr Top Med Chem* 11:2210–2215.
152. Thorneloe K.S., Sulpizio A.C., Lin Z., Figueroa D.J., Clouse A.K., et al. 2008. N-((1S)-1-{[4-((2S)-2-{[(2,4-dichlorophenyl)sulfonyl]amino}-3-hydroxypropanoyl)-1 -piperazinyl] carbonyl}-3-methylbutyl)-1-benzothiophene-2-carboxamide (GSK1016790A), a novel and potent transient receptor potential vanilloid 4 channel agonist induces urinary bladder contraction and hyperactivity: part I. *J Pharmacol Exp Ther* 326:432–442.
153. Watanabe H., Davis J.B., Smart D., Jerman J.C., Smith G.D., et al. 2002. Activation of TRPV4 channels (hVRL-2/mTRP12) by phorbol derivatives. *J Biol Chem* 277:13569–13577.
154. Pfanzagl B., Pfragner R., Jensen-Jarolim E. 2019. The transient receptor potential vanilloid 4 agonist RN-1747 inhibits the calcium response to histamine. *Pharmacology* 104:166–172.
155. Watanabe H., Vriens J., Prenen J., Droogmans G., Voets T., Nilius B. 2003. Anandamide and arachidonic acid use epoxyeicosatrienoic acids to activate TRPV4 channels. *Nature* 424:434–438.
156. Raboune S., Stuart J.M., Leishman E., Takacs S.M., Rhodes B., et al. 2014. Novel endogenous N-acyl amides activate TRPV1-4 receptors, BV-2 microglia, and are regulated in brain in an acute model of inflammation. *Front Cell Neurosci* 8:195.
157. Muller C., Morales P., Reggio P.H. 2018. Cannabinoid ligands targeting TRP channels. *Front Mol Neurosci* 11:487.
158. Cheung M., Bao W., Behm D.J., Brooks C.A., Bury M.J., et al. 2017. Discovery of GSK2193874: an orally active, potent, and selective blocker of transient receptor potential vanilloid 4. *ACS Med Chem Lett* 8:549–554.
159. Everaerts W., Zhen X., Ghosh D., Vriens J., Gevaert T., et al. 2010. Inhibition of the cation channel TRPV4 improves bladder function in mice and rats with cyclophosphamide-induced cystitis. *Proc Natl Acad Sci USA* 107:19084–19089.
160. Vincent F., Acevedo A., Nguyen M.T., Dourado M., DeFalco J., et al. 2009. Identification and characterization of novel TRPV4 modulators. *Biochem Biophys Res Commun* 389:490–494.
161. Wei Z.L., Nguyen M.T., O'Mahony D.J., Acevedo A., Zipfel S., et al. 2015. Identification of orally-bioavailable antagonists of the TRPV4 ion-channel. *Bioorg Med Chem Lett* 25:4011–4015.
162. Carson C., Raman P., Tullai J., Xu L., Henault M., et al. 2015. Englerin A agonizes the TRPC4/C5 cation channels to inhibit tumor cell line proliferation. *PLoS One* 10:e0127498.
163. Just S., Chenard B.L., Ceci A., Strassmaier T., Chong J.A., et al. 2018. Treatment with HC-070, a potent inhibitor of TRPC4 and TRPC5, leads to anxiolytic and anti-depressant effects in mice. *PLoS One* 13:e0191225.
164. Miller M., Shi J., Zhu Y., Kustov M., Tian J.B., et al. 2011. Identification of ML204, a novel potent antagonist that selectively modulates native TRPC4/C5 ion channels. *J Biol Chem* 286:33436–33446.
165. Rubaiy H.N., Ludlow M.J., Henrot M., Gaunt H.J., Miteva K., et al. 2017. Picomolar, selective, and subtype-specific small-molecule inhibition of TRPC1/4/5 channels. *J Biol Chem* 292:8158–8173.
166. Zhu Y., Lu Y., Qu C., Miller M., Tian J., et al. 2015. Identification and optimization of 2-aminobenzimidazole derivatives as novel inhibitors of TRPC4 and TRPC5 channels. *Br J Pharmacol* 172:3495–3509.

17 Two-Pore Domain Potassium Channels in Pain and Depression

Shaoying Zhang and Huaiyu Yang

CONTENTS

17.1 STRUCTURES OF K_{2P} CHANNELS AND THEIR EFFECTS ON CELL EXCITABILITY

As the largest family of ion channels, potassium (K^+) channels have an extremely broad distribution from viruses and bacteria to plants and animals (Korovkina and England 2002; Prindle et al. 2015; Siotto et al. 2014; Wang et al. 2018). In animals, cells have been classified as excitable and nonexcitable cells according to their functions of K^+ channels (Lewis and Cahalan 1995; Rudy 1988). In excitable cells, K^+ channels play roles in modulating the repolarization of action potential and changing the duration of action potential. In addition, K^+ channels are also involved in regulating the firing rate, neurotransmitter release and presynaptic facilitation (Rudy 1988). In nonexcitable cells, K^+ channels control the membrane potential and regulate the cell volume, as well as interfere with the proliferative activity and differentiation of cells (Lewis and Cahalan 1995). According to their architecture and sequence relatedness, K^+ channels are organized into four distinct classes, namely, the voltage-gated (Kv), calcium-activated (Kca), inwardly rectifier (K_{ir}), and K_{2P} channels (Gutman et al. 2003; Yu et al. 2005). K_{2P} channels, the most recently discovered group of K^+ channels, are located on the cell membranes of various types of cells, and they have a wide distribution from single-cell yeast to plants and higher mammals (Czempinski et al. 1997; Ketchum et al. 1995). Mammalian K_{2P} channels are encoded by 15 genes that can be subdivided into six subfamilies on the basis of their sequence similarity and functional properties: weak inward rectifying channels (TWIKs, tandem of pore domains in a weak inward rectifying K^+ channels), lipid- and mechano-sensitive channels (TREKs, TWIK-related K^+ channels), alkaline-activated channels (TALKs, TWIK-related alkaline pH-activated K^+ channels), acid sensitive channels (TASKs, TWIK-related acid-sensitive K^+ channels), halothane-inhibited channels (THIKs, tandem pore domain halothane-inhibited K^+ channels), and channels expressed in the spinal cord (TRESK, TWIK-related spinal cord K^+ channel) (Enyedi and Czirjak 2010; Goldstein et al. 2005).

In 2012, the crystal structures of K_{2P} channel were solved (Brohawn et al. 2012; Miller and Long 2012). Both channels presented a dimeric form and seemed to be crystallized in the open state. Subsequently, the crystal structures of other K_{2P} channels, such as TREK-1, TREK-2, and TASK-1, have also been reported (Brohawn et al. 2013, 2014, 2019; Dong et al. 2015; Li et al. 2020; Lolicato et al. 2014, 2017, 2020; Lolicato 2017; Pope et al. 2020; Rodstrom 2020). Although the amino acid identity between different K_{2P} subfamilies is low, all members share a similar structural arrangement (Fink et al. 1996; Lesage and Lazdunski 2000). Each subunit of the dimeric form of the K_{2P} channel contains four transmembrane domains (M1-M4), two pore-forming domains (P1 and P2), and two extracellular helices (E1 and E2) (Brohawn et al. 2013, 2014, 2019; Dong et al. 2015; Li et al. 2020; Lolicato et al. 2014, 2017, 2020; Lolicato 2017; Pope et al. 2020; Rodstrom et al. 2020) (Figure 17.1a). The short NH_2-terminus and the long COOH-terminus of the subunit are located in the cytoplasm (Lesage and Lazdunski 2000). Conformational changes in the transmembrane domain have been captured by the crystal structures, especially for M4, which adopts up (active-like) or down (inactive-like) conformations (Brohawn et al. 2013, 2014; Dong et al. 2015; Lolicato et al. 2014). The channel can form two open fenestrations on the side exposed to the inner membrane when M4 is in the down conformation. The crystal structure of TREK-2 cocrystallized with norfluoxetine suggests that these fenestration regions are potential binding sites for small molecules (Dong et al. 2015). The extracellular helices of K_{2P} extend ~35 Å beyond the outer leaflet of the plasma membrane to form a cap domain (Figure 17.1b). This domain is positioned above the outward pore, forming two lateral entrances outside the cell that bifurcate the K^+ conduction pathway (Brohawn et al. 2012; Miller and Long 2012; Zuniga and Zuniga 2016). Owing to this unique extracellular cap structure creating steric hindrance, K_{2P} channels are insensitive to classical K^+ channel blockers, such as tetraethylammonium ions and 4-aminopyridine (Concha et al. 2018; Lesage and Lazdunski 2000; Lotshaw 2007). The pore-forming domain is arranged in tandem and functions as homo- or heterodimeric channels. This highly conserved structural motif forms a single K^+ selective conductance pore (selectivity filter, SF) (Niemeyer et al. 2016). The C terminus region serves as a sensor for channel gating in response to a range of physical and chemical stimulation (Honore 2007). A substantial repositioning of the C-terminal would pull the transmembrane helix that conveys the signal to the SF gate, mediating the open or closed state of the K_{2P} channels (Lolicato et al. 2014; Piechotta et al. 2011; Zhuo et al. 2016). However, the conformations of the intracellular C-terminal region have not been solved in the present crystal structures of K_{2P} channels (Brohawn et al. 2013, 2014, 2019; Dong et al. 2015; Li et al. 2020; Lolicato et al. 2014, 2017, 2020; Pope et al. 2020; Rodstrom et al. 2020). In addition to the conservative topology of their structures,

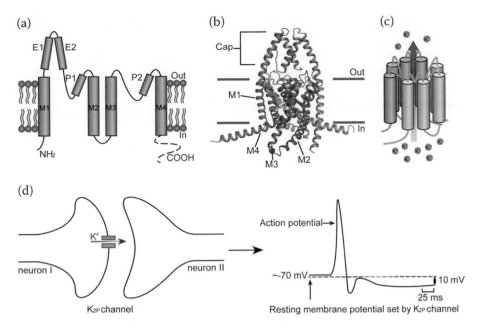

FIGURE 17.1 The structure and function of K_{2P} channels. (a) A cartoon showing that K_{2P} subunits are integral membrane proteins with internal amino (NH_2) and carboxy (COOH) termini, four transmembrane domains, M1-M4, two pore-forming domains, P1 and P2, and two extracellular helices, E1 and E2. (b) The crystal structure of TREK-1 (Lolicato et al. 2017) showing the architecture of the channel including the transmembrane domains and the extracellular cap domain. (c) Two subunits of K_{2P} channels come together as dimers to form a K^+ ion selective channel. Under physiological conditions, K^+ ions move from intracellular to extracellular via channels along their electrochemical gradient. (d) Role of K_{2P} channels in the membrane of neurons. K_{2P} channels leak K^+ ion out of neurons to set resting membrane potential.

the biophysical features of K_{2P} channels are also similar in many aspects. It is well accepted that SF is constitutively open at rest to leak K^+ ions out of cells for maintaining the background (also called baseline or leak) K^+ currents (Honore 2007; Lesage and Lazdunski 2000) (Figure 17.1c, d); thus, this family acts as a background channel in both excitable and nonexcitable cells.

Leak K^+ currents stabilize the hyperpolarized membrane voltages of the cells below the trigger threshold to determine the stimulation magnitude required to initiate an action potential (AP) (Goldstein et al. 2001; Jones 1989) (Figure 17.1d). Through regulating the duration, frequency, and amplitude of action potentials, the leak K^+ currents influence the resting membrane potential and consequently control the excitability of cells (Hodgkin and Huxley 1952; Plant 2012). If the leak K^+ currents are augmented, cells will become less excitable due to the V_m (cellular membrane potential) moving closer to E_k (the Nernst equilibrium potential for cation K) (Fink et al. 1996; Plant 2012). K_{2P} channels are responsible for the background K^+ currents. Since stimuli that can increase the activity of K_{2P} channels dampen cell excitability, inhibitors or blockers of K_{2P} channels increase excitability under physiological conditions (Plant 2012). Imiquimod is an agonist of Toll-like receptor 7, and it can enhance the excitability of dorsal root ganglia (DRG) neurons at least partially by blocking TREK-1 and TRAAK channels (Lee et al. 2012). Moreover, TASK-3 channels not only limit neuronal excitability by increasing resting K^+ conductance but also increase the excitability of neurons by supporting high-frequency firing when the AP threshold is reached (Brickley et al. 2007). These results indicate that the normal function of K_{2P} channels is one of the determinants of cellular excitability. K_{2P} channels are widely expressed in the nervous system (Marsh et al. 2012; Talley et al. 2001, 2003), and their effects on cell excitability may play crucial roles in the process of neural signaling transduction. Schwarz et al. described the current voltage curve of human peripheral nerves (Schwarz et al. 1995). The K^+ currents could be isolated from the compound action potential after blocking Na^+ currents completely by treatment with tetrodotoxin. It was found that the leak currents took a large portion of outward K^+ currents that significantly add the K^+ flux at all phases of the action potential (Schwarz et al. 1995). This current changes the excitability of nerve cells by regulating the resting membrane potential and then controls neuronal activity and signal transduction throughout the nervous system.

At present, many reported studies on K_{2P} channels in the nervous system focus on neurons of the central and peripheral nervous systems, including interneurons and sensory neurons (Cadaveira-Mosquera et al. 2012; La and Gebhart 2011; Leist et al. 2017; Weir et al. 2019). Compared with the number of neurons in the brain, the glia are much more abundant, constituting a greater component of brain cells (~90%), and astrocytes are one of the major types of glial cells (Kimelberg and Nedergaard 2010). It has been reported that some members of K_{2P}, such as TWIK-1, TREK-1, and TASK-1, are expressed in astrocytes (Kindler et al. 2000; Seifert et al. 2009). Further functional studies have shown that K_{2P} channels not only can mediate the passive conductance of astrocytes and improve the ability of astrocytes but also can contribute to the rapid release of glutamate (Hwang et al. 2014; Woo et al. 2012; Zhou et al. 2009).

Consistent with the importance of K_{2P} channels in modulating the fundamental properties of neurons and astrocytes, a variety of evidence suggests that K_{2P} channels are potential therapeutic targets for the treatment of mental disorders. The remainder of this chapter mainly focuses on the research progress into revealing the role of K_{2P} channels in pain and depression.

17.2 K_{2P}: POTENTIAL TARGETS FOR TREATING PAIN

The definition of pain described by the International Association for the Study of Pain is "an unpleasant sensory and emotional experience associated with actual or potential tissue damage, or described in terms of such damage" (International Association for the Study of Pain 1986). There are three main categories of pain: nociceptive pain, inflammatory pain, and neuropathic pain (NP) (Patapoutian et al. 2009). Although distinct in etiology and clinical features, the three may have the same pain signal transmission mode (Costigan et al. 2009). Pain is usually initiated by the activation of sensory neurons, also termed nociceptors, which have a cell body located in the DRG or trigeminal ganglia (TG) (Basbaum et al. 2009; Fenton et al. 2015). These neurons possess a peripheral axon that innervates target tissues such as skin and muscle to react to sensory stimuli through terminals and a central axon that innervates the spinal cord to transfer information to the central nervous system (CNS) (Patapoutian et al. 2009). Pain-related stimuli, such as extreme temperatures, mechanical stress and chemical irritants, can be detected by specific ion channels (i.e. TRPV1, TRPA1, Nav1.8, Nav1.9, and ASIC), which are expressed at peripheral nerve terminals of nociceptors (Basbaum et al. 2009; Dubin and Patapoutian 2010; Hung and Tan 2018; Yekkirala et al. 2017). These stimuli cause the membrane potentials of the peripheral nerve terminals to move to the depolarizing direction (Dubin and Patapoutian 2010). Voltage-dependent Na^+ channels are activated when the membrane potential reaches the threshold that can induce an action potential, converting pain perception into long-ranging electrical signals (Dubin and Patapoutian 2010). These signals are coordinated through different ion channels and eventually propagate from distant parts in the body to the CNS of the spinal cord and brain (Basbaum et al. 2009; Dubin and Patapoutian 2010; Plant 2012). Notably, only noxious stimuli, such as a pinprick, can be detected by nociceptors when they reach the threshold for eliciting pain, but innocuous light touch, such as the brush of a feather, is incapable (Tsunozaki and Bautista 2009). This result suggests that ion channels responding to noxious stimuli are potential targets for analgesia. TRPV1, Nav1.8, and Nav1.9 have been widely studied as potential analgesic targets because of their high expression in nociceptors and their remarkable abilities to control the firing of nociceptors by mediating ion influx in response to noxious stimuli (Dubin and Patapoutian 2010; Yekkirala et al. 2017).

In contrast to TRPV1, Nav1.8, and Nav1.9 channels, which regulate Na^+ or Ca^{2+} influx, K^+ channels facilitate K^+ ion efflux across the plasma membrane, hyperpolarizing the membrane potential of sensory nerve fiber terminals (Li and Toyoda 2015). This hyperpolarization prevents neurons from generating action potentials, which can inhibit the transmission of electrical signals (Li and Toyoda 2015). Moreover, neurons express only a few types of Nav channels, but they express a panoply of K^+ channels, suggesting that differential expression and regulation of K^+ channel subtypes may be associated with dynamic mechanisms that modulate the excitability of cells (Plant 2012; Yekkirala et al. 2017). It is pivotal to identify which K^+ channels participate in pain signal perception and transduction. Since K_{2P} channels were identified in 1995, multiple lines of evidence implicated that this family plays a crucial role at each step along the pain signal conducting pathways, not only acting as the physiological and noxious stimuli receptors but also functioning as the determinants of nociceptor excitability and conductivity (Basbaum et al. 2009; Li and Toyoda 2015; Plant 2012). The reduction of background K^+ currents mediated by K_{2P} channels upregulates the pain sensation, which has been shown by studies on gene knockout (KO) mice (Heurteaux et al. 2004). Existing first-line drugs, from nonsteroidal anti-inflammatory drugs such as aspirin, to opioids such as morphine have been confirmed to be effective in the treatment of chronic pain (Labianca et al. 2012). However, they often cannot completely alleviate pain (including NP), and are also associated with a series of adverse effects, including tolerance and addiction to opioids (Labianca et al. 2012; Mathie 2010). Accumulating evidence has indicated that K_{2P} channels have a close relationship with multiple types of pain, including thermal and mechanical, neuropathic, inflammatory, migraine, and cancer pain, and do not have opioid-like adverse effects (Devilliers et al. 2013; Gada and Plant 2019). Therefore, this family is a potential target for developing new-generation analgesics. Next, we introduce the research progress on the involvement of K_{2P} channels, such as TREK-1, TRESK, and TREK-2 in pain regulation.

TREK-1. TREK-1 channels are related to thermal stimuli and provide beneficial analgesic effects in pain. TREK-1 channels are widely expressed in the CNS with the highest expression level in the striatum, in parts of the cortex (layer IV) and hippocampus (CA2 pyramidal neurons), as well as in small- and medium-sized DRG neurons (Alloui et al. 2006; Fink et al. 1996; Talley et al. 2001). TREK-1 channels are colocalized with TRPV1 channels, which are thermal sensing ion channels. Alloui et al. found that ~40% of the DRG neurons that express TREK-1 also express TRPV1, and correspondingly, ~60% of the neurons that express TRPV1 also express TREK-1 (Alloui et al. 2006). This suggests that TREK-1 may be associated with the detection of thermal stimuli. The functions of TREK-1 channels in thermal perception have been further demonstrated in mice with a disrupted *TREK-1* gene. Compared with wild-type (WT) mice, TREK-1$^{-/-}$ mice were more sensitive to painful heat sensations near the threshold between anxious warmth and painful heat (Alloui et al. 2006). The activity of TREK-1 channels increases with temperature from 22°C and decreases at 42°C (a noxious heat level), which indicated that TREK-1 is a heat-activated K$^+$ channel (Maingret et al. 2000; Noel et al. 2009). This channel is a potential sensor of painful heat stimuli. In contrast, there is no difference in sensitivity to cold induced by a drop of acetone between WT and the gene deletion mice of TREK-1 or TRAAK, but TREK-1$^{-/-}$-TRAAK$^{-/-}$ mice displayed increased sensitivity to cold at 20, 15 and 10°C (Alloui et al. 2006; Noel et al. 2009), suggesting that TREK-1 may act as the regulator of nociceptor activation by cold through working together with TRAAK. In addition to sensing heat pain, TREK-1$^{-/-}$ mice were also subjected to allodynia after a range of mechanical stimuli and showed increased sensitivity to mechanical stimuli. TREK-1$^{-/-}$ mice subjected to intraplantar injection with carrageenan to produce inflammation showed thermal and mechanical hyperalgesia (Alloui et al. 2006). This evidence indicates that this channel plays a crucial role in peripheral sensitization of nociceptors during inflammation. As a key anatomical site, the DRG is important for the development and maintenance of NP. Han et al. confirmed that TREK-1 was upregulated and activated in the DRG of a CCI model (a model mimicking the clinical nerve injury conditions) (Han et al. 2016), suggesting that TREK-1 may have beneficial analgesic effects in NP patients. Recently, Royal et al. reported that TREK-1$^{-/-}$-TREK-2$^{-/-}$ mice showed migraine-like hypersensitivity to mechanical stimuli (Royal et al. 2019). Therefore, the TREK-1 channel is an attractive analgesic target related to polymodal pain perception. Significantly, Devilliers et al. demonstrated that TREK-1 channels contributed to strong analgesic effects induced by morphine in mice without causing opioid-like adverse effects such as constipation, respiratory depression and dependence

(Devilliers et al. 2013). Therefore, TREK-1 may be a better analgesic target than the opioid receptor.

TRESK. TRESK channels are associated with migraines, neuropathic and inflammatory pain. TRESK is a newly identified member of the K_{2P} family that was cloned from the human spinal cord, and its expression was found in the mouse cerebellum and in the rat and human brain (Czirjak et al. 2004; Liu et al. 2004; Sano et al. 2003; Yoo et al. 2009). Through endpoint and quantitative RT-PCR analysis, Dobler et al. and Marsh et al. found that the TRESK channel in mice was mainly expressed in DRGs and was the most abundant K_{2P} channel in the DRG (Dobler et al. 2007; Marsh et al. 2012). TRESK channels were also prominently expressed in the TG (Lafreniere et al. 2010; Weir et al. 2019). As the important conduction regions, DRG and TG are closely related to the neural signal transduction of migraines. Several mutations in TRESK have been identified to be associated with migraines (Lafreniere and Rouleau 2011). K2P18-A34V is related to the occurrence of typical migraines, and the F139WfsX24 (frameshift mutation) is linked to familial migraine with aura. This mutant is characterized by the absence of a two-base pair (2 bp) region at its C-terminus, which leads to TRESK subunits existing as frameshift mutants and to the premature truncation of the protein to 162 residues. The truncated K2P18 mutant channel (TRESK-MT) not only is nonfunctional but also downregulates WT channel activity in a dominant-negative fashion in a heterogenic expression system (Lafreniere et al. 2010). Interestingly, the other missense mutation that causes loss of TRESK function (C110R) is not able to induce TG neuron hyperexcitability and has no correlation with migraine (Andres-Enguix et al. 2012; Guo et al. 2014). The reason for this has been recently explored and explained by Royal et al., who revealed that TRESK-MT (F139WfsX24) functions in regulating migraines through a newly identified mechanism, termed frameshift mutation-induced alternative translation initiation (fsATI). The 2-bp frameshift mutation of TRESK places a new start condone in its open reading frame, which allows the mutant to produce two truncated fragments. One is the N-terminus of TRESK, and the other is the C-terminus of TRESK, and they can coassemble with TREK-1 or TREK-2 to inhibit their currents, leading to an increase in TG excitability. This is dissimilar from TRESK mutation (C110R), which only inhibits the currents of TRESK, not TREK-1 or TREK-2 (Royal et al. 2019).

TRESK channels provided most of the background K^+ currents of DRG neurons at 24°C (Kang and Kim 2006). Compared with WT littermates, the standing outward current IK_{so} was reduced significantly and cellular excitability was augmented significantly in DRG neurons of TRESK (G339R) functional KO mice (Dobler et al. 2007). Overexpression of TRESK channels can effectively reduce excitability and inhibit capsaicin-evoked spikes in small-diameter TG neurons (Guo and Cao 2014). It should be noted that the capsaicin-evoked substance P is related to the peripheral mechanism of NP, which suggests that TRESK may be associated with NP (Zhou et al. 2012). Tulleuda et al. reported that TRESK mRNA transcript levels decreased by approximately 50% in the rat DRG in response to axotomy of peripheral nerves (causing animals to produce behavioral alterations analogous to those seen in human NP patients). Downregulation of TRESK channels after axonal injury led to nociceptor hyperexcitability and increased sensitivity to painful pressure

(Tulleuda et al. 2011). The same result was also observed in TRESK KO mice (Weir et al. 2019). In contrast, enhancing the expression level of TRESK via intrathecal delivery of adenovirus constructs could inhibit capsaicin-evoked substance P released from DRG neurons, and relieve NP syndromes induced by peripheral nerve injury (Zhou et al. 2012, 2013). In terms of NP relief, in addition to regulating the excitability of neurons, TRESK also contributes to the pathogenesis of NP by upregulating gap junctions (synaptic transmission) and activating gliocytes (Zhou et al. 2017). Furthermore, the TRESK mRNA level was decreased in during inflammation induced by intradermal injection with complete Freund's adjuvant (CFA) (Marsh et al. 2012). TG neurons of TRESK KO mice exhibited increased action potential discharge during inflammation (Weir et al. 2019). These studies suggest that TRESK channels may be associated with inflammation-induced pain and may play a key role in nociceptor sensitization by inflammation. In addition, the reduced abundance of TRESK channels sensitized cancer-associated pain. Yang et al. found that the calcineurin inhibitor DSCR1 decreased calcineurin-mediated activation of the transcription factor NFAT, which eventually led to a reduction in the transcription of the *TRESK* gene (Yang et al. 2018). Therefore, using a calcineurin agonist to enhance TRESK channel currents is a potential treatment for pain. Giblin et al. reported that although both human and rat TRESK contain potential anionic phospholipid binding sites located at the membrane-cytosol interface that consist of 14 amino acids in the loop, only the human channel is able to respond to the electrostatic interaction mediated by anionic phospholipids [such as phosphatidylinositol-4,5-bisphosphate (PIP_2)] (Giblin et al. 2019). This implies that this novel regulatory mechanism is a characteristic of the human TRESK channel that is not present in rodent orthologs. Therefore, we should thoroughly examine the conclusions from animal models; further, this discovery may provide a new area for the development of novel channel modulators as analgesics.

TREK-2. TREK-2 channels are distributed in the CNS and are mainly expressed in the cerebellar granule cell layer and the DRG (Kang and Kim 2006; Talley et al. 2001). Single-channel current recording showed that under physiological conditions (37°C), TREK-2 channels provided the most (69%) background K^+ currents in DRG neurons, which was followed by TRESK (16%), and TREK-2 channels were most likely to induce the majority of background K^+ currents in the cell body of small- to medium-sized DRG neurons (Kang and Kim 2006). Therefore, TREK-2 channels play an active role in the regulation of the excitability of DRG neurons. Similar to the TREK-1, TREK-2 is also a temperature-sensitive channel and responds to temperature changes by controlling the firing activity of peripheral sensory C-fibers (Pereira et al. 2014). TREK-2 displays a 20-fold increase in of activity from 24 to 42°C (Kang et al. 2005), especially in the perception of nonaversive warm, between 40°C and 46°C, and moderated ambient cool temperatures, between 20°C and 25°C in mice (Pereira et al. 2014). This evidence suggests that TREK-2 channels may be involved in heat pain signal conduction. A role for TREK-2 in spontaneous pain has also been discovered in recent years. Spontaneous pain is implicated in the frequency of spontaneous firing in intact adult DRG C-fiber nociceptors, and uninjured C-nociceptors had a lower spontaneous firing rate observed in NP models with more hyperpolarized resting membrane potentials (Ems) (Acosta et al. 2014; Djouhri et al. 2006). As a member of the leak K^+ current family, TREK-2 was found to be selectively expressed in IB4-binding rat C-nociceptors. Both small interfering RNA (siRNA) and axotomy can reduce the expression level of TREK-2, which depolarizes Ems in C-neurons by approximately 10 mV and increases spontaneous foot lifting (SFL, a measure of spontaneous pain) behavior (Acosta et al. 2014). Therefore, downregulating the activity of TREK-2 may be a therapeutic method for treating pathological spontaneous pain. Moreover, TREK-2 was also found to be involved in the perception of mechanical pain, osmotic pain after sensitization by prostaglandin E2 and cold allodynia (a neuropathy feature) (Pereira et al. 2014). However, recently, Han et al. found that the TREK-2 protein expression level did not show a significant increasing trend in the DRG of a CCI model and was much weaker than that of TREK-1, which means that more research is needed to investigate the functional role of TREK-2 in terms of NP (Han et al. 2016). Notably, the effects of commonly used analgesics, such as acetaminophen, ibuprofen, nabumetone, and bupropion, inhibit TRESK but have no effect on TREK-2 (Park et al. 2016). This provides a good strategy for the development of selective inhibitors or agonists of TREK-2.

TRAAK. TRAAK channels, which belong to the TREK subfamily, are expressed in the DRG and TG (Marsh et al. 2012; Noel et al. 2011; Tulleuda et al. 2011; Yamamoto et al. 2009). As mentioned above, imiquimod can enhance the excitability of DRG neurons by blocking at least a portion of TREK-1 and TRAAK channels (Lee et al. 2012). Therefore, TRAAK may also be involved in the transmission of pain-related nerve signals. Similar to two other members of the TREK family, TRAAK is also associated with the perception of thermal, mechanical and chemical stimuli (Kang et al. 2005; Maingret et al. 1999). The activity of TRAAK increased progressively when the bash temperature rose from 24 to 42°C (Kang et al. 2005). Furthermore, compared with WT mice, the behavioral performance of TRAAK KO mice is very significantly linked to mechanical hypersensitivity with reduced paw withdrawal thresholds, higher sensitivity to noxious heat and capsaicin of small-diameter DRG neurons, and obvious heat hyperalgesia at temperatures between 46°C and 50°C (Noel et al. 2009). This indicates that TRAAK channels play an active role in the regulation of pain perception. Recently, a study aiming at identifying genetic predictors of persistent postsurgical NP found that only one possible association was reached for one single nucleotide polymorphism (SNP), the rs2286614, which encodes for TRAAK, suggesting that TRAAK may also be linked to NP (Blanc et al. 2019). In brief, TRAAK is an attractive target for mechanical and heat pain treatment.

TASK-1. Many researchers have found that TASK-1 was expressed in mouse and rat DRG neurons at a low level compared with that of TRESK or TREK-1, whereas the highest accumulation of TASK-1 mRNA was seen in the cerebellum and somatic motoneurons (Kindler et al. 2000; Marsh et al. 2012; Talley et al. 2001). Data showed that the mRNA level of TASK-1 was reduced at 4 days of inflammation, and this reduction was related to spontaneous pain behavior (Marsh et al. 2012). Furthermore, by assessing the participation of spinal TASK-1 and TASK-3 in inflammatory (formalin test) and neuropathic (spinal nerve ligation, SNL) pain in rats, Navarro et al. demonstrated that TASK-1 and TASK-3 had a relevant antinociceptive role in inflammatory and neuropathic pain (Garcia et al. 2019). The function of TASK-1 in nociception was also observed in TASK-1 KO animals. The KO mice exhibited enhanced sensitivity to thermal nociception in a hot-plate test but not in a tail-flick test. Moreover, the analgesic, sedative and hypothermic effects of a cannabinoid agonist (WIN55212-2) were reduced in TASK-1 KO mice (Linden et al. 2006). This evidence illustrates that TASK-1 channels are also associated with nociception in pain pathways and therefore present a potential target for treating pain.

TASK-3. TASK-3 channels were found to be much more widely distributed in the CNS, with robust expression in all brain regions, especially in somatic moto-neurons, cerebellar granule neurons, the locus coeruleus, raphe nuclei, and various nuclei of the hypothalamus (Talley et al. 2001). Although the overall expression level of TASK-3 in DRG neurons was low, its mRNA level was reduced during inflammation, similar to that of TASK-1; this decrease was related to spontaneous pain behavior, which showed a positive correlation between ipsilateral SFL duration and ipsilateral TASK-3 mRNA during inflammation (Marsh et al. 2012). Furthermore, TASK-3 also contributed to inflammatory and neuropathic pain an-tinociception. The antinociceptive effect of terbinafine, a TASK-3 activator, was fully blocked by the TASK-3 inhibitor PK-THPP. Spinal TASK-3 activation in-duced by terbinafine was also able to reduce tactile allodynia in rats with spinal nerve injuries (Garcia et al. 2019). This implies that the downregulation of TASK-3 channels is closely related to NP, inflammatory pain, and allodynia. TASK-3 channels also play a crucial role in temperature perception. Recent studies have shown that TASK-3 channels were highly enriched in a subpopulation of sensory neurons, which also express TRPM8, a cold- and menthol-activated ion channel. The thermal threshold of TRPM8-expressing cold neurons was decreased during TASK-3 inhibition and in TASK-3 KO mice, and these mice displayed hy-persensitivity to cold (Morenilla-Palao et al. 2014). In addition, we found that TASK-3 also responded to heat hyperalgesia and participated in temperature no-ciception signal conduction (Liao et al. 2019). As a dominant oncogene, TASK-3 can promote tumor growth and hypoxia resistance (Mu et al. 2003). Two SNPs (rs3780039 and rs11166921) of TASK-3 were linked to the occurrence of pre-operative breast pain (Langford et al. 2014). It seems that TASK-3 is a potential drug target for alleviating cancer pain.

TASK-2. Due to low sequence similarity between TASK-2 and TASK-1 and sensitivity to alkaline pH, TASK-2 was classified as a TALK subfamily member (Enyedi and Czirjak 2010; Niemeyer et al. 2007). Using immunohistochemical methods, Gabriel et al. demonstrated that prominent TASK-2 immunoreactivity was found in the rat hippocampal formation. The localization of TREK-2 proteins was also found in neuronal soma and proximal regions of neurites of cerebellar granule cells, sensory and motoneurons of the spinal cord and small-diameter neurons of DRG (Gabriel et al. 2002). Therefore, these results indicate that TASK-2 channels may participate in central mechanisms for controlling neuronal excitability and peripheral signal transduction. Similar to TRESK, TASK-1, and TASK-3, the mRNA for TASK-2 decreased approximately 3-fold at 4 days of inflammation, and it exhibited the greatest reduction (Marsh et al. 2012). The data suggest that TASK-2 may be linked to inflammation-induced pain. Thus far, no other studies have reported the function of TASK-2 in pain regulation, and further exploration is needed to understand the role of TASK-2 channels in pain signal conduction.

TWIK-1. As the first two-pore family member to be cloned, TWIK-1 is a weakly inward rectifying K^+ channel that is involved in the control of background K^+ membrane conductance and resting membrane potential setting (Lesage et al. 1996). The mRNA of TWIK-1 was found to be highly expressed in the CNS of rats and mice, especially in the cerebellar granule cell layer, thalamic reticular nucleus, and piriform cortex (Talley et al. 2001). A study on the expression level of the TWIK-1 channel in DRG neurons found that this channel showed robust expression in large- and medium-size neurons as well. Furthermore, TWIK-1 expression was much greater than TASK-1 and TASK-3 expression, and it decreased dramatically 1, 2, and 4 weeks after neuropathic injury, suggesting that TWIK-1 may be associated with the maintenance of NP (Pollema-Mays et al. 2013). Evidence supporting this speculation was recently demonstrated. Mao et al. recently illustrated that TWIK-1 could block the development and maintenance of SNL-induced NP by over-expressing its mRNA in the injured DRG (Mao et al. 2017). Systemic injection of the chemotherapy agent paclitaxel led to chemotherapy-induced NP. Recently, it has been found that paclitaxel could induce an increase in DNMT3a, which downregulated TWIK-1 in the DRG by elevating its DNA methylation level (Mao et al. 2019). This evidence implies that the TWIK-1 channel is an attractive target for NP therapy. However, the background K^+ currents produced by TWIK-1 were very small in both heterologous expression systems and native neurons (Feliciangeli et al. 2007; Yarishkin et al. 2014). Thus, the function of TWIK-1 is uncertain because of the difficulty in detecting its currents. It remains challenging to determine the functional role of TWIK-1 channels in pain signaling.

17.3 K_{2P}: POTENTIAL TARGETS FOR TREATING DEPRESSION

Major depression disorder (MDD) is a syndrome that has multiple subtypes and causes, with nearly one in five people experiencing one episode at some point in their lives (Malhi and Mann 2018; Menard et al. 2016). It was predicted that MDD will rank as the first cause of disease burden worldwide by 2030 (Malhi and Mann 2018). The patients, their families and society face a heavy burden because of the functional impairment caused by depression. Moreover, depression can increase suicide rates (Huang et al. 2017). Therefore, it is urgent to elucidate the pathogenesis of depression and develop effective treatment methods. The monoamine hypothesis proposes that the reduced availability of major monoamine neurotransmitters (serotonin [5-HT], norepinephrine [NE], and dopamine [DA]) leads to depression. In addition, endocrine system abnormalities, genetic factors, the influence of environmental stress, immunologic factors, and neurogenesis have a close relationship to the pathogenesis of depression (Jesulola et al. 2018; Menard et al. 2016). Different patients have distinct pathogenesis or pathophysiological conditions, and the same patient can even experience different episodes at different times; therefore, a single model or mechanism that can satisfactorily explain all aspects of depression has not yet been identified (Malhi and Mann 2018). Sequenced Treatment Alternatives to Relieve Depression (STAR*D) is a strategic plan for treatment-resistant depression, providing four steps for antidepressant treatment (Fava et al. 2003; Rush et al. 2006). After the four steps of acute treatment, the overall cumulative remission rate was only 67%, and those people who required more treatment steps had higher relapse rates (Rush et al. 2006). Identifying the neurobiological basis of antidepressant treatment and improving treatment strategies remain a primary challenge that modern psychiatry currently faces. The currently available antidepressants are not used broadly because of their side effects (i.e. produce neurotoxic effects), inadequate treatment and slow response (Hillhouse and Porter 2015; Wang et al. 2013). Many patients either observe no effect after taking the medicine, or they relapse after recovery. A meta-analysis showed that the overall pooled response rate of atypical agent antidepressant treatment was only 44.2% (Nelson and Papakostas 2009). Therefore, the need to develop new antidepressants is crucial. Recently, preclinical studies using experimental mouse models have provided innovative therapeutic ideas. The regions of prefrontal cortex and hippocampus can mediate cognitive aspects of depression (Nestler et al. 2002), and many TREK-1 channels are expressed well (Fink et al. 1996; Heurteaux et al. 2006). Heurteaux et al. found that TREK-1-deficient mice had an increased efficacy of 5-HT neurotransmission and showed a depression-resistant phenotype. Moreover, TREK-1-deficient mice showed similar behavior to that of naive animals treated with fluoxetine (an antidepressant), implying that TREK-1 may be a potential target for treatment of depressant (Heurteaux et al. 2006). Although K_{2P} channels have been found to be involved in the regulation of depression, their specific mechanism is not clear. It may be quite distinct from traditional antidepressant mechanisms. Hence, the following section will introduce the research progress of K_{2P} channels as they relate to the modulation of depression.

TREK-1. The TREK-1 channels are strongly expressed in the prefrontal cortex and the hippocampus regions, where they seem to mediate cognitive aspects of depression, such as memory impairments and feelings of worthlessness, hopelessness, guilt, doom, and suicidality (Heurteaux et al. 2006; Nestler et al. 2002; Talley et al. 2001). As mentioned above, TREK-1 is highly expressed in the striatum, amygdala, and hypothalamus and is related to emotional memory (Maingret et al. 2000; Nestler et al. 2002; Talley et al. 2001). As a result, TREK-1 may be associated with the anhedonia (the main symptom of MDD), anxiety, and reduced motivation that are observed in depression (Nestler et al. 2002). Perlis et al. provided evidence for the role of TREK-1 in MDD using the STAR*D study (Perlis et al. 2008). Through analyzing the association of genetic variants of TREK-1 with treatment response in nonpsychotic subjects with depression in the STAR*D study, they found that allelic variations in four TREK-1 SNP$_S$ were associated with a positive response to antidepressants: the A allele of rs10494996, G allele of rs12136349, C allele of rs2841608, and G allele of rs2841616 (Perlis et al. 2008). This is the first study to examine candidate genes derived from mouse models of response to an antidepressant and validate the usefulness of these models in understanding the genetic basis of the human response to antidepressants. On those bases, Dillon et al. illustrated that individuals possessing rs10494996, rs2841608, and rs2841616 TREK-1 genotypes showed stronger basal ganglia responses to gains relative to individuals in the at-risk groups. Through a whole-brain regression analysis, they also discovered that many protective alleles across these SNPs were positively correlated with neural responses to gains in several regions of the reward network, such as the right caudate and right putamen (Dillon et al. 2010). In addition to these four SNPs, Liou et al. also found that the genotype frequency of TREK-1 rs6686529 differed significantly between MDD patients and controls. Individuals with homozygous genotypes (CC or GG) showed better susceptibility to MDD than those with heterozygous genotypes, suggesting that these gene polymorphisms may result in heterosis effects on MDD (Liou et al. 2009). Accumulated evidence has indicated that certain genotypes of TREK-1 may promote remission from depression by enhancing the relative neurological response.

In terms of animal experiments, TREK-1 KO mice exhibited an antidepressant phenotype in five different depression models: the Porsolt forced swim test (FST), the tail suspension test (TST), the conditioned suppression of motility test (CSMT), the learned helplessness test (LH), and the novelty-suppressed feeding test (NSF). Regardless of whether they received antidepressant treatment, TREK-1 KO mice exhibited behaviors similar to those of naive animals treated with acutely or chronically classical antidepressants such as fluoxetine (Heurteaux et al. 2006). The involvement of the neurotransmitter 5-HT in the pathophysiology of depression and the role of antidepressant therapy have been characterized (Jesulola et al. 2018; Menard et al. 2016). The TREK-1 channel is expressed in midbrain 5-HT neurons located in the dorsal raphe nucleus (DRN). In mice, the deletion of the *TREK-1* gene led to animals having an increased efficacy of 5-HT neurotransmission in DRN neurons and a substantially reduced elevation of corticosterone levels under stress conditions. Compared with WT mice, the firing rate of 5-HT neurons in TREK-1 KO mice was approximately twofold higher. This increase may enhance the release of 5-HT in target structures such as the hippocampus. The activity of 5-HT neurons in TREK-1 mutant mice increased, as did selective 5-HT reuptake inhibitor (SSRI)-treated animals (Heurteaux et al. 2006). This was explained by Gordon and Hen: the TREK-1 channels in 5-HT neurons may participate in 5-HT 1A receptor-dependent negative feedback inhibition. 5-HT fed back onto 5-HT neurons, activating presynaptic 5-HT 1A receptors, which primarily opened GIRK (G-protein-gated inwardly rectifying K^+ channels) channels, hyperpolarizing the cell and reducing firing. 5-HT 1A receptor activation also decreases cAMP levels, which may in turn result in disinhibition of TREK-1 channels (Gordon and Hen 2006). This mechanism explains why combined treatment with fluoxetine (an SSRI) and p-chlorophenyl alanine methyl ester (a tryptophan hydroxylase inhibitor), which depletes 5-HT from nerve terminals, can completely prevent the antidepressant phenotype (Heurteaux et al. 2006). To verify that TREK-1 channels play a role in this pathway instead of other K_{2P} members, Gordon and Hen compared TREK-1 KO mice to mice lacking the TRAAK channel, which is a channel that is closely related to TREK-1 and is also expressed in 5-HT neurons. Although TREK-1 and TRAAK share many of the same regulatory influences, only TREK-1 was inhibited by activated protein kinase A (PKA) and protein kinase C (PKC), potentially linking the TREK-1 channel to G-protein-coupled receptors, such as the 5-HT 1A receptor. Furthermore, TRAAK KO mice did not display an antidepressant-like phenotype (Gordon and Hen 2006). This also argues against the idea that the antidepressant phenotype is simply the result of 5-HT neuron depolarization occurring due to reduced K^+ ion efflux. All these findings suggest that genetic variation in TREK-1 may identify individuals at risk for treatment resistance.

However, the role of 5-HT in depression has been questioned in recent years (Cowen and Browning 2015). To obtain a better understanding of the TREK-1 channel in antidepressant treatment, it is necessary to determine the relationship between TREK-1 and other cells in addition to 5-HT neurons. Hwang et al. reported that TREK-1 formed heterodimeric channels with TWIK-1 in astrocytes. These heterodimers mediated the release of glutamate in astrocytes induced by cannabinoid (Hwang et al. 2014). According to preclinical and clinical trials, glutamatergic system dysfunction has been implicated in the pathophysiology of mood disorders such as bipolar depression and MDD (Henter et al. 2018). Therefore, TREK-1/TWIK-1 heterodimers of astrocytes may be a potential target in the treatment of depression. By generating a Cre-dependent TREK-1 knockdown (Cd-TREK-1 KD) transgenic mouse, Kim et al. found that selectively inhibiting TREK-1 in hippocampal neurons could provide antidepressant effects (Kim et al. 2019). This result implies that Cd-TREK-1 KD mice will be a valuable tool for revealing the antidepressant function of TREK-1 in the brain.

In addition to MDD and acute depression, the function of TREK-1 has also been demonstrated in poststroke depression (PSD). PSD is one of the most common neuropsychiatric complications following stroke (Gaete and Bogousslavsky 2008; Robinson 2003). It has been reported that the expression levels of the TREK-1 protein were increased significantly in the hippocampal CA1, CA3, dentate gyrus, and prefrontal cortex in PSD rats, and TREK-1 mRNA was increased in all the same regions except for hippocampal CA1 (Lin et al. 2015). As a potent antidepressant, mini-spadin prevented PSD by enhancing the activity of TREK-1, as analyzed by FST and NSF tests (Djillani et al. 2017). In summary, modulating TREK-1 channels in the early and chronic phases of stroke could play a crucial role in PSD treatment (Pietri et al. 2019).

TASK-3. As mentioned above, the TASK-3 channel exhibits robust expression in all brain regions (Talley et al. 2001). Therefore, it has been hypothesized that TASK-3 channels may play a role in mood disorders, sleep-wake control and cognition (Borsotto et al. 2015). Through analyzing locomotor activity and circadian rhythm, researchers have discovered that the amplitudes of nocturnal activity bout duration and bout number were augmented in TASK-3 mutant and KO mice well beyond those seen in WT mice, while the light phase activities were similar between the different mice (Gotter et al. 2011; Linden et al. 2007). From the continuous EEG (wake and sleep electroencephalogram)/EMG recordings, a slower progression of TASK-3 KO mice from waking to sleeping states compared with that of WT was observed, and their sleep episodes as well as their rapid eye movement (REM, a hallmark of antidepressant treatment) θ oscillations were more fragmented during the sleeping period (Gotter et al. 2011; Pang et al. 2009; Steiger and Kimura 2010). Such behaviors were also consistent with the observations of despair-related animal models and MDD patients. These phenomena were reversed in WT mice by antidepressant treatments, but they were not reversed in TASK-3 KO mice (Gotter et al. 2011; Steiger and Kimura 2010). TASK-3 KO mice displayed resistance to despair behavior, which was characterized by significant decreases in immobility relative to WT controls in both FST and TST tests (Gotter et al. 2011). The same results were also observed in mice in which the TASK-3 channel was knocked down through small interfering RNA (siRNA) treatment (Fullana et al. 2019). Together, these results identify the TASK-3 channel as a possible therapeutic target for antidepressant action.

17.4 MODULATORS OF K$_{2P}$ CHANNELS

Among K$^+$ channels, the K$_{2P}$ family may have the largest variety of endogenous regulatory factors, including mechanical stress, temperature, membrane depolarization, arachidonic acid (AA) and other polyunsaturated fatty acids (PUFAs), voltage, pH, posttranslational modification, and accessory proteins (i.e. Gαq- and Gαo-coupled receptors and protein kinases such as PKC and PKA) (Mathie and Veale 2015; Patel and Honore 2001). It has been reported that PIP$_2$ can modulate the voltage dependence of TREK-1. The addition of PIP$_2$ to intracellular medium leads to an increase in the current amplitude. In contrast, PIP$_2$ hydrolysis inhibits channel activity and shifts the voltage dependence of the TREK-1 channel (Chemin et al. 2005; Lopes et al. 2005). TREK-1 can also be activated by PUFAs in a dose-dependent manner (Patel et al. 1998). Protein modification is also important in regulating the function of K$_{2P}$ channels. Bockenhauer et al. found that phosphorylation of TREK-1 on Ser348 by PKA converted hippocampal TREK-1 channels from being in the K$^+$ leak mode to having a voltage-dependent phenotype (Bockenhauer et al. 2001). Rajan et al. suggested that the phenomenon of TWIK-1 having low activity in heterologous expression systems (Nematian-Ardestani et al. 2020) was attributed to the occurrence of SUMO modification on TWIK-1 expressed in the plasma membrane that eventually led to the channel function to be "silenced" (Rajan et al. 2005). Plant et al. reported a mature SUMO1$_{97}$ form, which binds to the Lys274 ε-amino group of TWIK-1 via a covalent bond, resulting in the TWIK-1 channel being "silenced," and this phenomenon could be reversed by adding the SUMO protease (SENP-1) (Plant et al. 2010). Glycosylation and alternative translation initiation were also observed in the K$_{2P}$ family (Mant et al. 2013; Royal et al. 2019; Wiedmann et al. 2019).

Although some K$_{2P}$ channels are sensitive to classic K$^+$ channel inhibitors (i.e. Ba^{2+} and quinine), almost all of the K$_{2P}$ members lack selective pharmacological modulators (Lesage et al. 1996; Ma et al. 2011; Zhou et al. 2009). There is substantial interest in identifying pharmacological modulators for K$_{2P}$ that would be useful tools for studying the channels.

Fluoxetine and its active metabolite, norfluoxetine, produced a concentration-dependent inhibition of TREK-1 currents (Kennard et al. 2005). Dong et al. used the crystal structure of the TREK-2 channel in complex with norfluoxetine to find the binding pocket. This pocket was under the P2 helix of the SF and consisted of M2, M3, and M4 transmembrane helices (Dong et al. 2015). Fluoxetine stabilized the down state of TREK channels by binding within the fenestrations. In contrast, AA stabilized TREK channels in the up state (Brohawn et al. 2014; Dong et al. 2015).

The binding of fluoxetine and its derivatives could not interact with the channel when TREK-2 is in the up state upon AA stimulation (McClenaghan et al. 2016).

We discovered the TASK-3 selective agonist, CHET3, using computer-aided drug design (Liao et al. 2019). The binding pocket of this agonist is a transmembrane cavity underneath the SF (Figure 17.2). The same binding cavity is also observed in the TRAAK crystal structure in which a lipid acyl chain binds (Brohawn et al. 2014). Furthermore, this cavity is a traditional inhibitor binding site that is highly conserved in K^+ channels. Usually, compounds binding to this conserved central cavity are not selective modulators (Rinne et al. 2019). Schewe et al. reported a class of negatively charged activators (NCAs) binding in the same cavity, which can activate K_{2P} channels. Their conclusion confirmed that this site can indeed act as an agonist binding

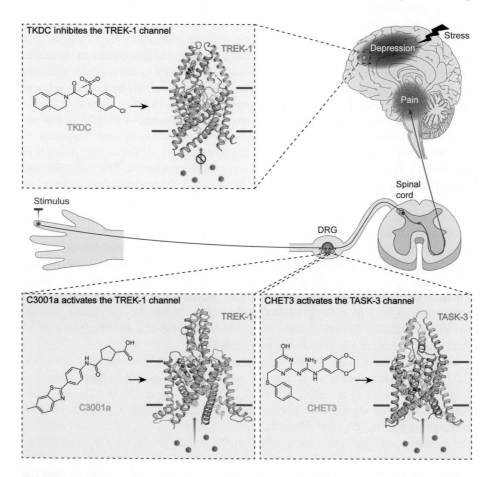

FIGURE 17.2   The schematic of K_{2P} modulators in pain signaling pathway and depression. TKDC is an inhibitor of TREK-1 channel and exhibits the antidepressant effects by inhibiting the activity of TREK-1 channel. C3001a is a selective activator of TREK-1 channel. CHET3 is a selective agonist of TASK-3 channels. C3001a and CHET3 exhibit analgesic effects by enhancing the activity of their target protein in the peripheral nervous system, particularly in DRG.

site. However, NCAs are nonselective agonists that activate K_{2P} channels, voltage-gated hERG (human ether-à-go-go-related gene) channels and calcium (Ca^{2+})-activated big-conductance K^+ (BK)-type channels (Schewe et al. 2019). In contrast, CHET3 serves as a highly selective allosteric activator acting on TASK-3-containing channels, and CHET3 exhibits potent analgesic effects in a variety of acute and chronic pain models of rodents without causing side effects, such as respiratory and cardiovascular malfunction. The analgesic functions of CHET3 can be abolished pharmacologically by treatment with TASK-3 inhibitors or by *TASK-3* gene deletion, indicating that CHET3 indeed acts through interacting with TASK-3 (Liao et al. 2019). Therefore, using CHET3 as a chemical probe, our work suggested that TASK-3 is a promising target for developing analgesic treatment.

The crystal structure of TREK-1 in complex with ML335 or ML402 that was reported by Lolicato et al. identified a new druggable L-shaped pocket in TREK. This pocket was a cryptic binding site located behind the SF formed by the P1 pore helix and M4 transmembrane helix intrasubunit interface. Activators ML335 and ML402 activated the channel by interacting with residues behind the SF. These activators served as molecular wedges that restricted the movement of the inter-domain interface and stimulated the C-type gate directly (Lolicato et al. 2017). We also found a selective activator, C3001a, that enhanced the activity of TREK channels (Figure 17.2). C3001a forms hydrophobic interactions with F149 and W290, similar to the interactions between ML335 and the equivalent residues F134 and W275. However, C3001a forms a hydrogen bond interaction with Y285 through its carboxyl group that is not seen in the crystal structure and is replaced by the interaction between the sulfonyl group of ML335 or its derivatives with H141 and S146 (Qiu et al. 2020). This research not only supports the conclusion that stabilizing the SF by restricting the movement of this pocket is feasible but also indicates that the same cavity can possess various regulatory modes and mechanisms of action. C3001a targets TREK channels in the peripheral nervous system (PNS) and alleviates spontaneous pain, cold hyperalgesia, mechanical allodynia and inflammation in a variety of animal models (Qiu et al. 2020).

The extracellular domains of K_{2P} channels are also regulatory sites. We found the TREK inhibitor TKDC which is bound to the extracellular cap domain (Luo et al. 2017) (Figure 17.2). The TKDC binding pocket, consisting of E1 and E2 helices, acts as an allosteric site. When TKDC binds to the pocket of the cap domain, the E2 helix moves toward the SF and directly interacts with the pore region of the other subunit to block the extracellular ion pathway. The results of electrophysiological experiments show that the half-maximal inhibition concentration (IC_{50}) of 10 μM TKDC to TREK-1 and TREK-2 is 4.9 ± 0.6 μM and 5.2 ± 1.9 μM respectively, but TRAAK is weaker (65.9 ± 7 μM), indicating that TKDC has certain subtype selectivity. In animal behavior tests, TKDC shows a shorter onset time than the classic antidepressant fluoxetine (Luo et al. 2017).

BL-1249 is a member of the fenamate class of nonsteroidal anti-inflammatory drugs, and its extracellular application activated all TREK subfamily members but had no effect on other subfamilies (Pope et al. 2018). BL-1249 exhibited selectivity among the TREK subfamily, with the ability to activate TREK-1 and TREK-2 10-fold more potently than TRAAK. Using mutants and TREK-1/TRAAK chimeras, researchers identified the M2/M3 transmembrane helix interface as a key site for determining the selectivity of BL-1249, and the integrity of the C-terminal tail was crucial for BL-1249 to perform the function (Pope et al. 2018). Therefore, the action mechanism of BL-1249 may involve an SF C-type gate.

The binding sites of several small-molecule modulators of K_{2P} are unclear. These modulators include riluzole, pranlukast, PK-THPP, ML6733, A1899, et al. (Bagriantsev et al. 2013; Chokshi et al. 2015; Cotten et al. 2006; Cunningham et al. 2019; Duprat et al. 2000; Flaherty et al. 2014; Kiper et al. 2015; O'Donohoe et al. 2018; Ramirez et al. 2019; Rodstrom et al. 2020; Streit et al. 2011; Wiedmann et al. 2019; Wright et al. 2017, 2019).

In addition to small molecules, several peptide modulators of K_{2P} were identified. Spadin is the first peptide antidepressant found to block the TREK-1 channel (Moha Ou Maati et al. 2012), and it is a secreted peptide (Ala12-Arg28) derived from the N-terminal propeptide (Gln1-Arg44), which is released through cleaving the mature neurotensin receptor 3 (NTSR3/Sortilin) by furin in the posterior region of the Golgi apparatus (Mazella et al. 2010). Spadin bound to TREK-1 specifically with an affinity of 10 nM and was unable to inhibit currents generated by TREK-2, TRAAK, TASK and TRESK channels (Mazella et al. 2010; Moha Ou Maati et al. 2012). In five behavioral tests for predicting antidepressant response, spadin-treated mice showed the same phenotype as TREK-1 KO mice (Mazella et al. 2010). It is of great importance to develop spadin analogs (Djillani et al. 2017; Ma and Lewis 2020).

Some natural-product compounds can also regulate the activity of K_{2P} channels. Bautista et al. found that the active ingredient in Szechuan peppercorns, hydroxy-α-sanshool, induced depolarization of sensory neurons by inhibiting some K_{2P} channels (TASK-1, TASK-3, and TRESK) to achieve analgesic effects (Bautista et al. 2008). Aristolochic acid (AristA) was found in traditional medicines and has been used to treat pain. Veale and Mathie found that AristA exerted its analgesic effects by enhancing the activity of TREK-1 and TREK-2 and by inhibiting the activity of TRESK. However, the pharmacological effects of aristolochic acid are complex and are related to nephritis and carcinogenesis (Veale and Mathie 2016). Monoterpenes (MT) are terpenes composed of two five-carbon isoprene units. For centuries, MT has been known for its antifungal, antibacterial and analgesic properties (Arazi et al. 2020). Ariza et al. recently discovered that TREK channels were activated by several MTs. In addition, carvacrol and cinnamaldehyde, two types of MT, robustly enhanced currents of alkaline-sensitive TASK-2, whereas carvone, another type of MT, could selectively inhibit the voltage-dependent current of the TASK family (Arazi et al. 2020).

17.5   CHALLENGES AND FUTURE PERSPECTIVES

With the analysis of crystal structures, and the continuous exploration of mechanism of action and functions, we have gained a deeper understanding of the K_{2P} family. The K_{2P} family is widely distributed in the body, and its expression level is even more abundant in the nervous system. Moreover, this family has a unique biophysical property that mediates the background K^+ currents of various types of cells. Therefore, K_{2P} channels play an irreplaceable role in regulating cell excitability and affecting nerve signal transduction. With the maturity of gene knockout technology, in recent years, the functional characteristics of the K_{2P} family have been continuously identified: both TREK-1 and TRAAK KO mice show increased sensitivity to pain. TREK-1 and TASK-3 KO mice also show an antidepressant phenotype. Therefore, researchers found that the K_{2P} family is not only a good analgesic target but also an attractive antidepressant target. There are many endogenous factors regulating the activity of the K_{2P} family members in vivo, suggesting that it is a good candidate for regulation. Therefore, drugs designed for this family may achieve better analgesic or antidepressant effects than what is currently available. Significantly, each member of this family has a unique distribution pattern in the body. Although they are structurally similarities between the family members, their sequence similarity is low. Hence, there is an opportunity for rational drug design for K_{2P} channels.

However, as far as the current research progress is concerned, many questions remain. First, there is a lack of selective regulatory molecules and drugs that can specifically identify K_{2P} channel subtypes. This is affected by the wide distribution of K_{2P} channels in the body and the difficulties in performing high-throughput screening of drugs through ion channels, and moreover, the limited number of crystal structures. The unique secondary structure of K_{2P} channels also prevents inference from the existing structures of Kv channels, Kir channels, or standard K^+ channels (KcsA). It also hampers the understanding of protein regulatory mechanisms. Last but not least, the in vivo functions of many K_{2P} members are still poorly understood. In summary, research on K_{2P} channels and the discovery of selective small molecules and functional exploration in vivo still has a long way to go.

REFERENCES

Acosta, C., L. Djouhri, R. Watkins, C. Berry, K. Bromage, and S. N. Lawson. 2014. TREK2 expressed selectively in IB4-binding C-fiber nociceptors hyperpolarizes their membrane potentials and limits spontaneous pain. *J Neurosci* 34: 1494–1509.

Alloui, A., K. Zimmermann, J. Mamet, F. Duprat, J. Noël, J. Chemin, N. Guy, et al. 2006. TREK-1, a K^+ channel involved in polymodal pain perception. *EMBO J* 25: 2368–2376.

Andres-Enguix, I., L. Shang, P. J. Stansfeld, J. M. Morahan, M. S. Sansom, R. G. Lafrenière, B. Roy, et al. 2012. Functional analysis of missense variants in the TRESK (*KCNK18*) K^+ channel. *Sci Rep* 2: 237.

Arazi, E., G. Blecher, and N. Zilberberg. 2020. Monoterpenes differently regulate acid-sensitive and mechano-gated K_{2P} channels. *Front Pharmacol* 11: 704.

Bagriantsev, S. N., K. H. Ang, A. Gallardo-Godoy, K. A. Clark, M. R. Arkin, A. R. Renslo, and D. L. Minor, Jr. 2013. A high-throughput functional screen identifies small molecule regulators of temperature- and mechano-sensitive K_{2P} channels. *ACS Chem Biol* 8: 1841–1851.

Basbaum, A. I., D. M. Bautista, G. Scherrer, and D. Julius. 2009. Cellular and molecular mechanisms of pain. *Cell* 139: 267–284.

Bautista, D. M., Y. M. Sigal, A. D. Milstein, J. L. Garrison, J. A. Zorn, P. R. Tsuruda, R. A. Nicoll, and D. Julius. 2008. Pungent agents from Szechuan peppers excite sensory neurons by inhibiting two-pore potassium channels. *Nat Neurosci* 11: 772–779.

Blanc, P., E. Génin, B. Jesson, C. Dubray, C. Dualé, and E. DONIS-gene Investigating group. 2019. Genetics and postsurgical neuropathic pain: An ancillary study of a multicentre survey. *Eur J Anaesthesiol* 36: 342–350.

Bockenhauer, D., N. Zilberberg, and S. A. Goldstein. 2001. KCNK2: reversible conversion of a hippocampal potassium leak into a voltage-dependent channel. *Nat Neurosci* 4: 486–491.

Borsotto, M., J. Veyssiere, H. Moha Ou Maati, C. Devader, J. Mazella, and C. Heurteaux. 2015. Targeting two-pore domain K^+ channels TREK-1 and TASK-3 for the treatment of depression: a new therapeutic concept. *Br J Pharmacol* 172: 771–784.

Brickley, S. G., M. I. Aller, C. Sandu, E. L. Veale, F. G. Alder, H. Sambi, A. Mathie, and W. Wisden. 2007. TASK-3 two-pore domain potassium channels enable sustained high-frequency firing in cerebellar granule neurons. *J Neurosci* 27: 9329–9340.

Brohawn, S. G., E. B. Campbell, and R. MacKinnon. 2013. Domain-swapped chain connectivity and gated membrane access in a Fab-mediated crystal of the human TRAAK K^+ channel. *Proc Natl Acad Sci U S A* 110: 2129–2134.

Brohawn, S. G., E. B. Campbell, and R. MacKinnon. 2014. Physical mechanism for gating and mechanosensitivity of the human TRAAK K^+ channel. *Nature* 516: 126–130.

Brohawn, S. G., J. del Mármol, and R. MacKinnon. 2012. Crystal structure of the human K2P TRAAK, a lipid- and mechano-sensitive K^+ ion channel. *Science* 335: 436–441.

Brohawn, S. G., W. Wang, A. Handler, E. B. Campbell, J. R. Schwarz, and R. MacKinnon. 2019. The mechanosensitive ion channel TRAAK is localized to the mammalian node of Ranvier. *Elife* 8: e50403.

Cadaveira-Mosquera, A., M. Pérez, A. Reboreda, P. Rivas-Ramírez, D. Fernández-Fernández, and J. A. Lamas. 2012. Expression of K2P channels in sensory and motor neurons of the autonomic nervous system. *J Mol Neurosci* 48: 86–96.

Chemin, J., A. J. Patel, F. Duprat, I. Lauritzen, M. Lazdunski, and E. Honoré. 2005. A phospholipid sensor controls mechanogating of the K^+ channel TREK-1. *EMBO J* 24: 44–53.

Chokshi, R. H., A. T. Larsen, B. Bhayana, and J. F. Cotten. 2015. Breathing stimulant compounds inhibit TASK-3 potassium channel function likely by binding at a common site in the channel pore. *Mol Pharmacol* 88: 926–934.

Concha, G., D. Bustos, R. Zúñiga, M. A. Catalán, and L. Zúñiga. 2018. The insensitivity of TASK-3 K_2P channels to external tetraethylammonium (TEA) partially depends on the cap structure. *Int J Mol Sci* 19: 2437.

Costigan, M., J. Scholz, and C. J. Woolf. 2009. Neuropathic pain: A maladaptive response of the nervous system to damage. *Annu Rev Neurosci* 32: 1–32.

Cotten, J. F., B. Keshavaprasad, M. J. Laster, E. I. Eger, 2nd, and C. S. Yost. 2006. The ventilatory stimulant doxapram inhibits TASK tandem pore (K_2P) potassium channel function but does not affect minimum alveolar anesthetic concentration. *Anesth Analg* 102: 779–785.

Cowen, P. J., and M. Browning. 2015. What has serotonin to do with depression? *World Psychiatry* 14: 158–160.

Cunningham, K. P., R. G. Holden, P. M. Escribano-Subias, A. Cogolludo, E. L. Veale, and A. Mathie. 2019. Characterization and regulation of wild-type and mutant TASK-1 two pore domain potassium channels indicated in pulmonary arterial hypertension. *J Physiol* 597: 1087–1101.

Czempinski, K., S. Zimmermann, T. Ehrhardt, and B. Müller-Röber. 1997. New structure and function in plant K^+ channels: KCO1, an outward rectifier with a steep Ca^{2+} dependency. *EMBO J* 16: 2565–2575.

Czirják, G., Z. E. Tóth, and P. Enyedi. 2004. The two-pore domain K^+ channel, TRESK, is activated by the cytoplasmic calcium signal through calcineurin. *J Biol Chem* 279: 18550–18558.

Devilliers, M., J. Busserolles, S. Lolignier, E. Deval, V. Pereira, A. Alloui, M. Christin, *et al.* 2013. Activation of TREK-1 by morphine results in analgesia without adverse side effects. *Nat Commun* 4: 2941.

Dillon, D. G., R. Bogdan, J. Fagerness, A. J. Holmes, R. H. Perlis, and D. A. Pizzagalli. 2010. Variation in TREK1 gene linked to depression-resistant phenotype is associated with potentiated neural responses to rewards in humans. *Hum Brain Mapp* 31: 210–221.

Djillani, A., M. Pietri, S. Moreno, C. Heurteaux, J. Mazella, and M. Borsotto. 2017. Shortened spadin analogs display better TREK-1 inhibition, *in vivo* stability and antidepressant activity. *Front Pharmacol* 8: 643.

Djouhri, L., S. Koutsikou, X. Fang, S. McMullan, and S. N. Lawson. 2006. Spontaneous pain, both neuropathic and inflammatory, is related to frequency of spontaneous firing in intact C-fiber nociceptors. *J Neurosci* 26: 1281–1292.

Dobler, T., A. Springauf, S. Tovornik, M. Weber, A. Schmitt, R. Sedlmeier, E. Wischmeyer, and F. Döring. 2007. TRESK two-pore-domain K⁺ channels constitute a significant component of background potassium currents in murine dorsal root ganglion neurones. *J Physiol* 585: 867–879.

Dong, Y. Y., A. C. Pike, A. Mackenzie, C. McClenaghan, P. Aryal, L. Dong, A. Quigley, *et al.* 2015. K2P channel gating mechanisms revealed by structures of TREK-2 and a complex with Prozac. *Science* 347: 1256–1259.

Dubin, A. E., and A. Patapoutian. 2010. Nociceptors: The sensors of the pain pathway. *J Clin Invest* 120: 3760–3772.

Duprat, F., F. Lesage, A. J. Patel, M. Fink, G. Romey, and M. Lazdunski. 2000. The neuroprotective agent riluzole activates the two P domain K⁺ channels TREK-1 and TRAAK. *Mol Pharmacol* 57: 906–912.

Enyedi, P., and G. Czirják. 2010. Molecular background of leak K⁺ currents: Two-pore domain potassium channels. *Physiol Rev* 90: 559–605.

Fava, M., A. J. Rush, M. H. Trivedi, A. A. Nierenberg, M. E. Thase, H. A. Sackeim, F. M. Quitkin, *et al.* 2003. Background and rationale for the sequenced treatment alternatives to relieve depression (STAR*D) study. *Psychiatr Clin North Am* 26: 457–494, x.

Feliciangeli, S., S. Bendahhou, G. Sandoz, P. Gounon, M. Reichold, R. Warth, M. Lazdunski, J. Barhanin, and F. Lesage. 2007. Does sumoylation control K2P1/TWIK1 background K⁺ channels? *Cell* 130: 563–569.

Fenton, B. W., E. Shih, and J. Zolton. 2015. The neurobiology of pain perception in normal and persistent pain. *Pain Manag* 5: 297–317.

Fink, M., F. Duprat, F. Lesage, R. Reyes, G. Romey, C. Heurteaux, and M. Lazdunski. 1996. Cloning, functional expression and brain localization of a novel unconventional outward rectifier K⁺ channel. *EMBO J* 15: 6854–6862.

Flaherty, D. P., D. S. Simpson, M. Miller, B. E. Maki, B. Zou, J. Shi, M. Wu, *et al.* 2014. Potent and selective inhibitors of the TASK-1 potassium channel through chemical optimization of a bis-amide scaffold. *Bioorg Med Chem Lett* 24: 3968–3973.

Fullana, M. N., A. Ferrés-Coy, J. E. Ortega, E. Ruiz-Bronchal, V. Paz, J. J. Meana, F. Artigas, and A. Bortolozzi. 2019. Selective knockdown of TASK3 potassium channel in monoamine neurons: A new therapeutic approach for depression. *Mol Neurobiol* 56: 3038–3052.

Gabriel, A., M. Abdallah, C. S. Yost, B. D. Winegar, and C. H. Kindler. 2002. Localization of the tandem pore domain K⁺ channel KCNK5 (TASK-2) in the rat central nervous system. *Brain Res Mol Brain Res* 98: 153–163.

Gada, K., and L. D. Plant. 2019. Two-pore domain potassium channels: Emerging targets for novel analgesic drugs: IUPHAR review 26. *Br J Pharmacol* 176: 256–266.

Gaete, J. M., and J. Bogousslavsky. 2008. Post-stroke depression. *Expert Rev Neurother* 8: 75–92.

García, G., R. Noriega-Navarro, V. A. Martínez-Rojas, E. J. Gutiérrez-Lara, N. Oviedo, and J. Murbartián. 2019. Spinal TASK-1 and TASK-3 modulate inflammatory and neuropathic pain. *Eur J Pharmacol* 862: 172631.

Giblin, J. P., I. Etayo, A. Castellanos, A. Andres-Bilbe, and X. Gasull. 2019. Anionic phospholipids bind to and modulate the activity of human TRESK background K^+ channel. *Mol Neurobiol* 56: 2524–2541.

Goldstein, S. A., D. A. Bayliss, D. Kim, F. Lesage, L. D. Plant, and S. Rajan. 2005. International Union of Pharmacology. LV. Nomenclature and molecular relationships of two-P potassium channels. *Pharmacol Rev* 57: 527–540.

Goldstein, S. A., D. Bockenhauer, I. O'Kelly, and N. Zilberberg. 2001. Potassium leak channels and the KCNK family of two-p-domain subunits. *Nat Rev Neurosci* 2: 175–184.

Gordon, J. A., and R. Hen. 2006. TREKing toward new antidepressants. *Nat Neurosci* 9: 1081–1083.

Gotter, A. L., V. P. Santarelli, S. M. Doran, P. L. Tannenbaum, R. L. Kraus, T. W. Rosahl, H. Meziane, *et al.* 2011. TASK-3 as a potential antidepressant target. *Brain Res* 1416: 69–79.

Guo, Z., and Y. Q. Cao. 2014. Over-expression of TRESK K^+ channels reduces the excitability of trigeminal ganglion nociceptors. *PLoS One* 9: e87029.

Guo, Z., P. Liu, F. Ren, and Y. Q. Cao. 2014. Nonmigraine-associated TRESK K^+ channel variant C110R does not increase the excitability of trigeminal ganglion neurons. *J Neurophysiol* 112: 568–579.

Gutman, G. A., K. G. Chandy, J. P. Adelman, J. Aiyar, D. A. Bayliss, D. E. Clapham, M. Covarrubias, *et al.* 2003. International Union of Pharmacology. XLI. Compendium of voltage-gated ion channels: Potassium channels. *Pharmacol Rev* 55: 583–586.

Han, H. J., S. W. Lee, G. T. Kim, E. J. Kim, B. Kwon, D. Kang, H. J. Kim, and K. S. Seo. 2016. Enhanced expression of TREK-1 is related with chronic constriction injury of neuropathic pain mouse model in dorsal root ganglion. *Biomol Ther (Seoul)* 24: 252–259.

Henter, I. D., R. T. de Sousa, and C. A. Zarate, Jr. 2018. Glutamatergic modulators in depression. *Harv Rev Psychiatry* 26: 307–319.

Heurteaux, C., N. Guy, C. Laigle, N. Blondeau, F. Duprat, M. Mazzuca, L. Lang-Lazdunski, *et al.* 2004. TREK-1, a K^+ channel involved in neuroprotection and general anesthesia. *EMBO J* 23: 2684–2695.

Heurteaux, C., G. Lucas, N. Guy, M. El Yacoubi, S. Thümmler, X. D. Peng, F. Noble, *et al.* 2006. Deletion of the background potassium channel TREK-1 results in a depression-resistant phenotype. *Nat Neurosci* 9: 1134–1141.

Hillhouse, T. M., and J. H. Porter. 2015. A brief history of the development of antidepressant drugs: From monoamines to glutamate. *Exp Clin Psychopharmacol* 23: 1–21.

Hodgkin, A. L., and A. F. Huxley. 1952. A quantitative description of membrane current and its application to conduction and excitation in nerve. *J Physiol* 117: 500–544.

Honoré, E. 2007. The neuronal background K_{2P} channels: Focus on TREK1. *Nat Rev Neurosci* 8: 251–261.

Huang, Y. J., H. Y. Lane, and C. H. Lin. 2017. New treatment strategies of depression: Based on mechanisms related to neuroplasticity. *Neural Plast* 2017: 4605971.

Hung, C. Y., and C. H. Tan. 2018. TRP channels in nociception and pathological pain. *Adv Exp Med Biol* 1099: 13–27.

Hwang, E. M., E. Kim, O. Yarishkin, D. H. Woo, K. S. Han, N. Park, Y. Bae, *et al.* 2014. A disulphide-linked heterodimer of TWIK-1 and TREK-1 mediates passive conductance in astrocytes. *Nat Commun* 5: 3227.

International Association for the Study of Pain. 1986. Classification of chronic pain: Descriptors of chronic pain syndromes and definition of pain terms. In *Pain Suppl*, ed. H. Merskey and N. Bogduk, S1–226. United States of America: IASP Press.

Jesulola, E., P. Micalos, and I. J. Baguley. 2018. Understanding the pathophysiology of depression: From monoamines to the neurogenesis hypothesis model—Are we there yet? *Behav Brain Res* 341: 79–90.

Jones, S. W. 1989. On the resting potential of isolated frog sympathetic neurons. *Neuron* 3: 153–161.

Kang, D., C. Choe, and D. Kim. 2005. Thermosensitivity of the two-pore domain K^+ channels TREK-2 and TRAAK. *J Physiol* 564: 103–116.

Kang, D., and D. Kim. 2006. TREK-2 ($K_{2P}10.1$) and TRESK ($K_{2P}18.1$) are major background K^+ channels in dorsal root ganglion neurons. *Am J Physiol Cell Physiol* 291: C138–C146.

Kennard, L. E., J. R. Chumbley, K. M. Ranatunga, S. J. Armstrong, E. L. Veale, and A. Mathie. 2005. Inhibition of the human two-pore domain potassium channel, TREK-1, by fluoxetine and its metabolite norfluoxetine. *Br J Pharmacol* 144: 821–829.

Ketchum, K. A., W. J. Joiner, A. J. Sellers, L. K. Kaczmarek, and S. A. Goldstein. 1995. A new family of outwardly rectifying potassium channel proteins with two pore domains in tandem. *Nature* 376: 690–695.

Kim, A., H. G. Jung, Y. E. Kim, S. C. Kim, J. Y. Park, S. G. Lee, and E. M. Hwang. 2019. The knockdown of TREK-1 in hippocampal neurons attenuate lipopolysaccharide-induced depressive-like behavior in mice. *Int J Mol Sci* 20: 5902.

Kimelberg, H. K., and M. Nedergaard. 2010. Functions of astrocytes and their potential as therapeutic targets. *Neurotherapeutics* 7: 338–353.

Kindler, C. H., C. Pietruck, C. S. Yost, E. R. Sampson, and A. T. Gray. 2000. Localization of the tandem pore domain K^+ channel TASK-1 in the rat central nervous system. *Brain Res Mol Brain Res* 80: 99–108.

Kiper, A. K., S. Rinné, C. Rolfes, D. Ramírez, G. Seebohm, M. F. Netter, W. González, and N. Decher. 2015. Kv1.5 blockers preferentially inhibit TASK-1 channels: TASK-1 as a target against atrial fibrillation and obstructive sleep apnea? *Pflugers Arch* 467: 1081–1090.

Korovkina, V. P., and S. K. England. 2002. Detection and implications of potassium channel alterations. *Vascul Pharmacol* 38: 3–12.

La, J. H., and G. F. Gebhart. 2011. Colitis decreases mechanosensitive K_{2P} channel expression and function in mouse colon sensory neurons. *Am J Physiol Gastrointest Liver Physiol* 301: G165–G174.

Labianca, R., P. Sarzi-Puttini, S. M. Zuccaro, P. Cherubino, R. Vellucci, and D. Fornasari. 2012. Adverse effects associated with non-opioid and opioid treatment in patients with chronic pain. *Clin Drug Investig* 32 Suppl 1: 53–63.

Lafrenière, R. G., M. Z. Cader, J. F. Poulin, I. Andres-Enguix, M. Simoneau, N. Gupta, K. Boisvert, *et al.* 2010. A dominant-negative mutation in the TRESK potassium channel is linked to familial migraine with aura. *Nat Med* 16: 1157–1160.

Lafrenière, R. G., and G. A. Rouleau. 2011. Migraine: Role of the TRESK two-pore potassium channel. *Int J Biochem Cell Biol* 43: 1533–1536.

Langford, D. J., C. West, C. Elboim, B. A. Cooper, G. Abrams, S. M. Paul, B. L. Schmidt, *et al.* 2014. Variations in potassium channel genes are associated with breast pain in women prior to breast cancer surgery. *J Neurogenet* 28: 122–135.

Lee, J., T. Kim, J. Hong, J. Woo, H. Min, E. Hwang, S. J. Lee, and C. J. Lee. 2012. Imiquimod enhances excitability of dorsal root ganglion neurons by inhibiting background (K_{2P}) and voltage-gated ($K_v1.1$ and $K_v1.2$) potassium channels. *Mol Pain* 8: 2.

Leist, M., S. Rinné, M. Datunashvili, A. Aissaoui, H. C. Pape, N. Decher, S. G. Meuth, and T. Budde. 2017. Acetylcholine-dependent upregulation of TASK-1 channels in thalamic interneurons by a smooth muscle-like signalling pathway. *J Physiol* 595: 5875–5893.

Lesage, F., E. Guillemare, M. Fink, F. Duprat, M. Lazdunski, G. Romey, and J. Barhanin. 1996. TWIK-1, a ubiquitous human weakly inward rectifying K$^+$ channel with a novel structure. *EMBO J* 15: 1004–1011.

Lesage, F., and M. Lazdunski. 2000. Molecular and functional properties of two-pore-domain potassium channels. *Am J Physiol Renal Physiol* 279: F793–F801.

Lewis, R. S., and M. D. Cahalan. 1995. Potassium and calcium channels in lymphocytes. *Annu Rev Immunol* 13: 623–653.

Li, X. Y., and H. Toyoda. 2015. Role of leak potassium channels in pain signaling. *Brain Res Bull* 119: 73–79.

Li, B., R. A.Rietmeijer, andS. G. Brohawn. 2020. Structural basis for pH gating of the two-pore domain K+ channel TASK2. Nature 586: 457–462.

Lin, D. H., X. R. Zhang, D. Q. Ye, G. J. Xi, J. J. Hui, S. S. Liu, L. J. Li, and Z. J. Zhang. 2015. The role of the two-pore domain potassium channel TREK-1 in the therapeutic effects of escitalopram in a rat model of poststroke depression. *CNS Neurosci Ther* 21: 504–512.

Liao, P., Y. Qiu, Y. Mo, J. Fu, Z. Song, L. Huang, S. Bai, *et al.* 2019. Selective activation of TWIK-related acid-sensitive K$^+$ 3 subunit-containing channels is analgesic in rodent models. *Sci Transl Med* 11: eaaw8434.

Linden, A. M., M. I. Aller, E. Leppä, O. Vekovischeva, T. Aitta-Aho, E. L. Veale, A. Mathie, *et al.* 2006. The in vivo contributions of TASK-1-containing channels to the actions of inhalation anesthetics, the α_2 adrenergic sedative dexmedetomidine, and cannabinoid agonists. *J Pharmacol Exp Ther* 317: 615–626.

Linden, A. M., C. Sandu, M. I. Aller, O. Y. Vekovischeva, P. H. Rosenberg, W. Wisden, and E. R. Korpi. 2007. TASK-3 knockout mice exhibit exaggerated nocturnal activity, impairments in cognitive functions, and reduced sensitivity to inhalation anesthetics. *J Pharmacol Exp Ther* 323: 924–934.

Liou, Y. J., T. J. Chen, S. J. Tsai, Y. W. Yu, C. Y. Cheng, and C. J. Hong. 2009. Support for the involvement of the *KCNK2* gene in major depressive disorder and response to antidepressant treatment. *Pharmacogenet Genomics* 19: 735–741.

Liu, C., J. D. Au, H. L. Zou, J. F. Cotten, and C. S. Yost. 2004. Potent activation of the human tandem pore domain K channel TRESK with clinical concentrations of volatile anesthetics. *Anesth Analg* 99: 1715–1722, table of contents.

Lolicato, M., C. Arrigoni, T. Mori, Y. Sekioka, C. Bryant, K. A. Clark, and D. L. Minor, Jr. 2017. K$_{2P}$2.1 (TREK-1): activator complexes reveal a cryptic selectivity filter binding site. *Nature* 547: 364–368.

Lolicato, M., P. M.Riegelhaupt, C. Arrigoni, K. A.Clark, and D. L. Minor, Jr. 2014.Transmembrane helix straightening and buckling underlies activation of mechanosensitive and thermosensitive K$_{2P}$ channels. *Neuron* 84: 1198–1212.

Lolicato, M., A. M., Natale, F. Abderemane-Ali, S. Capponi, R. Duman, A. Qagner, *et al.* 2020. K2P channel C-type gating involves asymmetric selectivity filter order-disorder transitions. *Sci Adv* 6: eabc9174.

Lopes, C. M., T. Rohács, G. Czirják, T. Balla, P. Enyedi, and D. E. Logothetis. 2005. PIP$_2$ hydrolysis underlies agonist-induced inhibition and regulates voltage gating of two-pore domain K$^+$ channels. *J Physiol* 564: 117–129.

Lotshaw, D. P. 2007. Biophysical, pharmacological, and functional characteristics of cloned and native mammalian two-pore domain K$^+$ channels. *Cell Biochem Biophys* 47: 209–256.

Luo, Q., L. Chen, X. Cheng, Y. Ma, X. Li, B. Zhang, L. Li, *et al.* 2017. An allosteric ligand-binding site in the extracellular cap of K2P channels. *Nat Commun* 8: 378.

Ma, R., and A. Lewis. 2020. Spadin selectively antagonizes arachidonic acid activation of TREK-1 channels. *Front Pharmacol* 11: 434.

Ma, X. Y., J. M. Yu, S. Z. Zhang, X. Y. Liu, B. H. Wu, X. L. Wei, J. Q. Yan, *et al.* 2011. External Ba^{2+} block of the two-pore domain potassium channel TREK-1 defines conformational transition in its selectivity filter. *J Biol Chem* 286: 39813–39822.

Maingret, F., M. Fosset, F. Lesage, M. Lazdunski, and E. Honoré. 1999. TRAAK is a mammalian neuronal mechano-gated K^+ channel. *J Biol Chem* 274: 1381–1387.

Maingret, F., I. Lauritzen, A. J. Patel, C. Heurteaux, R. Reyes, F. Lesage, M. Lazdunski, and E. Honore. 2000. TREK-1 is a heat-activated background K^+ channel. *EMBO J* 19: 2483–2491.

Malhi, G. S., and J. J. Mann. 2018. Depression. *Lancet* 392: 2299–2312.

Mant, A., S. Williams, L. Roncoroni, E. Lowry, D. Johnson, and I. O'Kelly. 2013. N-glycosylation-dependent control of functional expression of background potassium channels $K_{2P}3.1$ and $K_{2P}9.1$. *J Biol Chem* 288: 3251–3264.

Mao, Q., S. Wu, X. Gu, S. Du, K. Mo, L. Sun, J. Cao, *et al.* 2019. DNMT3a-triggered downregulation of K_{2p} 1.1 gene in primary sensory neurons contributes to paclitaxel-induced neuropathic pain. *Int J Cancer* 145: 2122–2134.

Mao, Q., J. Yuan, M. Xiong, S. Wu, L. Chen, A. Bekker, Y. X. Tao, and T. Yang. 2017. Role of dorsal root ganglion $K_{2P}1.1$ in peripheral nerve injury-induced neuropathic pain. *Mol Pain* 13: 1744806917701135.

Marsh, B., C. Acosta, L. Djouhri, and S. N. Lawson. 2012. Leak K^+ channel mRNAs in dorsal root ganglia: Relation to inflammation and spontaneous pain behaviour. *Mol Cell Neurosci* 49: 375–386.

Mathie, A. 2010. Ion channels as novel therapeutic targets in the treatment of pain. *J Pharm Pharmacol* 62: 1089–1095.

Mathie, A., and E. L. Veale. 2015. Two-pore domain potassium channels: Potential therapeutic targets for the treatment of pain. *Pflugers Arch* 467: 931–943.

Mazella, J., O. Pétrault, G. Lucas, E. Deval, S. Béraud-Dufour, C. Gandin, M. El-Yacoubi, *et al.* 2010. Spadin, a sortilin-derived peptide, targeting rodent TREK-1 channels: A new concept in the antidepressant drug design. *PLoS Biol* 8: e1000355.

McClenaghan, C., M. Schewe, P. Aryal, E. P. Carpenter, T. Baukrowitz, and S. J. Tucker. 2016. Polymodal activation of the TREK-2 K2P channel produces structurally distinct open states. *J Gen Physiol* 147: 497–505.

Ménard, C., G. E. Hodes, and S. J. Russo. 2016. Pathogenesis of depression: Insights from human and rodent studies. *Neuroscience* 321: 138–162.

Miller, A. N., and S. B. Long. 2012. Crystal structure of the human two-pore domain potassium channel K2P1. *Science* 335: 432–436.

Moha Ou Maati, H., J. Veyssiere, F. Labbal, T. Coppola, C. Gandin, C. Widmann, J. Mazella, C. Heurteaux, and M. Borsotto. 2012. Spadin as a new antidepressant: Absence of TREK-1-related side effects. *Neuropharmacology* 62: 278–288.

Morenilla-Palao, C., E. Luis, C. Fernandez-Peña, E. Quintero, J. L. Weaver, D. A. Bayliss, and F. Viana. 2014. Ion channel profile of TRPM8 cold receptors reveals a role of TASK-3 potassium channels in thermosensation. *Cell Rep* 8: 1571–1582.

Mu, D., L. Chen, X. Zhang, L. H. See, C. M. Koch, C. Yen, J. J. Tong, *et al.* 2003. Genomic amplification and oncogenic properties of the *KCNK9* potassium channel gene. *Cancer Cell* 3: 297–302.

Nelson, J. C., and G. I. Papakostas. 2009. Atypical antipsychotic augmentation in major depressive disorder: A meta-analysis of placebo-controlled randomized trials. *Am J Psychiatry* 166: 980–991.

Nematian-Ardestani, E., F. Abd-Wahab, F. C. Chatelain, H. Sun, M. Schewe, T. Baukrowitz, and S. J. Tucker. 2020. Selectivity filter instability dominates the low intrinsic activity of the TWIK-1 K2P K^+ channel. *J Biol Chem* 295: 610–618.

Nestler, E. J., M. Barrot, R. J. DiLeone, A. J. Eisch, S. J. Gold, and L. M. Monteggia. 2002. Neurobiology of depression. *Neuron* 34: 13–25.

Niemeyer, M. I., L. P. Cid, W. González, and F. V. Sepúlveda. 2016. Gating, regulation, and structure in K_{2P} K^+ channels: In varietate concordia? *Mol Pharmacol* 90: 309–317.

Niemeyer, M. I., F. D. González-Nilo, L. Zúñiga, W. González, L. P. Cid, and F. V. Sepúlveda. 2007. Neutralization of a single arginine residue gates open a two-pore domain, alkali-activated K^+ channel. *Proc Natl Acad Sci USA* 104: 666–671.

Noël, J., G. Sandoz, and F. Lesage. 2011. Molecular regulations governing TREK and TRAAK channel functions. *Channels (Austin)* 5: 402–409.

Noël, J., K. Zimmermann, J. Busserolles, E. Deval, A. Alloui, S. Diochot, N. Guy, *et al.* 2009. The mechano-activated K^+ channels TRAAK and TREK-1 control both warm and cold perception. *EMBO J* 28: 1308–1318.

O'Donohoe, P. B., N. Huskens, P. J. Turner, J. J. Pandit, and K. J. Buckler. 2018. A1899, PK-THPP, ML365, and doxapram inhibit endogenous TASK channels and excite calcium signaling in carotid body type-1 cells. *Physiol Rep* 6: e13876.

Pang, D. S., C. J. Robledo, D. R. Carr, T. C. Gent, A. L. Vyssotski, A. Caley, A. Y. Zecharia, *et al.* 2009. An unexpected role for TASK-3 potassium channels in network oscillations with implications for sleep mechanisms and anesthetic action. *Proc Natl Acad Sci USA* 106: 17546–17551.

Park, H., E. J. Kim, J. Han, J. Han, and D. Kang. 2016. Effects of analgesics and anti-depressants on TREK-2 and TRESK currents. *Korean J Physiol Pharmacol* 20: 379–385.

Patapoutian, A., S. Tate, and C. J. Woolf. 2009. Transient receptor potential channels: Targeting pain at the source. *Nat Rev Drug Discov* 8: 55–68.

Patel, A. J., and E. Honoré. 2001. Properties and modulation of mammalian 2P domain K^+ channels. *Trends Neurosci* 24: 339–346.

Patel, A. J., E. Honoré, F. Maingret, F. Lesage, M. Fink, F. Duprat, and M. Lazdunski. 1998. A mammalian two pore domain mechano-gated S-like K^+ channel. *EMBO J* 17: 4283–4290.

Pereira, V., J. Busserolles, M. Christin, M. Devilliers, L. Poupon, W. Legha, A. Alloui, *et al.* 2014. Role of the TREK2 potassium channel in cold and warm thermosensation and in pain perception. *Pain* 155: 2534–2544.

Perlis, R. H., P. Moorjani, J. Fagerness, S. Purcell, M. H. Trivedi, M. Fava, A. J. Rush, and J. W. Smoller. 2008. Pharmacogenetic analysis of genes implicated in rodent models of antidepressant response: Association of *TREK1* and treatment resistance in the STAR*D study. *Neuropsychopharmacology* 33: 2810–2819.

Piechotta, P. L., M. Rapedius, P. J. Stansfeld, M. K. Bollepalli, G. Ehrlich, I. Andres-Enguix, H. Fritzenschaft, *et al.* 2011. The pore structure and gating mechanism of K2P channels. *EMBO J* 30: 3607–3619.

Pietri, M., A. Djillani, J. Mazella, M. Borsotto, and C. Heurteaux. 2019. First evidence of protective effects on stroke recovery and post-stroke depression induced by sortilin-derived peptides. *Neuropharmacology* 158: 107715.

Plant, L. D. 2012. A role for K2P channels in the operation of somatosensory nociceptors. *Front Mol Neurosci* 5: 21.

Plant, L. D., I. S. Dementieva, A. Kollewe, S. Olikara, J. D. Marks, and S. A. Goldstein. 2010. One SUMO is sufficient to silence the dimeric potassium channel K2P1. *Proc Natl Acad Sci USA* 107: 10743–10748.

Pollema-Mays, S. L., M. V. Centeno, C. J. Ashford, A. V. Apkarian, and M. Martina. 2013. Expression of background potassium channels in rat DRG is cell-specific and down-regulated in a neuropathic pain model. *Mol Cell Neurosci* 57: 1–9.

Pope, L., C. Arrigoni, H. Lou, C. Bryant, A. Gallardo-Godoy, A. R. Renslo, and D. L. Minor, Jr. 2018. Protein and chemical determinants of BL-1249 action and selectivity for K_{2P} channels. *ACS Chem Neurosci* 9: 3153–3165.

Pope, L., M. Lolicato, and D. L. Minor, Jr. 2020. Polynuclear ruthenium amines inhibit K_{2P} channels via a "finger in the dam" mechanism. *Cell Chem Biol* 27: 511–524 e4.

Prindle, A., J. Liu, M. Asally, S. Ly, J. Garcia-Ojalvo, and G. M. Süel. 2015. Ion channels enable electrical communication in bacterial communities. *Nature* 527: 59–63.

Qiu, Y., L. Huang, J. Fu, C. Han, J. Fang, P. Liao, Z. Chen, *et al.* 2020. TREK channel family activator with a well-defined structure-activation relationship for pain and neurogenic inflammation. *J Med Chem* 63: 3665–3677.

Rajan, S., L. D. Plant, M. L. Rabin, M. H. Butler, and S. A. Goldstein. 2005. Sumoylation silences the plasma membrane leak K^+ channel K2P1. *Cell* 121: 37–47.

Ramirez, D., G. Concha, B. Arevalo, L. Prent-Penaloza, L. Zuniga, A. K. Kiper, S. Rinne, *et al.* 2019. Discovery of novel TASK-3 channel blockers using a pharmacophore-based virtual screening. *Int J Mol Sci* 20: 4014.

Rinné, S., A. K. Kiper, K. S. Vowinkel, D. Ramírez, M. Schewe, M. Bedoya, D. Aser, *et al.* 2019. The molecular basis for an allosteric inhibition of K^+-flux gating in K_{2P} channels. *Elife* 8.

Robinson, R. G. 2003. Poststroke depression: Prevalence, diagnosis, treatment, and disease progression. *Biol Psychiatry* 54: 376–387.

Rödstrom, K. E. J., A. K. Kiper, W. Zhang, S. Rinné, A. C. W. Pike, M. Goldstein, L. J. Conrad, *et al.* 2020. A lower X-gate in TASK channels traps inhibitors within the vestibule. *Nature* 582: 443–447.

Royal, P., A. Andres-Bilbe, P. Á. Prado, C. Verkest, B. Wdziekonski, S. Schaub, A. Baron, *et al.* 2019. Migraine-associated TRESK mutations increase neuronal excitability through alternative translation initiation and inhibition of TREK. *Neuron* 101: 232–45 e6.

Rudy, B. 1988. Diversity and ubiquity of K channels. *Neuroscience* 25: 729–749.

Rush, A. J., M. H. Trivedi, S. R. Wisniewski, A. A. Nierenberg, J. W. Stewart, D. Warden, G. Niederehe, *et al.* 2006. Acute and longer-term outcomes in depressed outpatients requiring one or several treatment steps: A STAR*D report. *Am J Psychiatry* 163: 1905–1917.

Sano, Y., K. Inamura, A. Miyake, S. Mochizuki, C. Kitada, H. Yokoi, K. Nozawa, *et al.* 2003. A novel two-pore domain K^+ channel, TRESK, Is localized in the spinal cord. *J Biol Chem* 278: 27406–27412.

Schewe, M., H. Sun, Ü. Mert, A. Mackenzie, A. C. W. Pike, F. Schulz, C. Constantin, *et al.* 2019. A pharmacological master key mechanism that unlocks the selectivity filter gate in K^+ channels. *Science* 363: 875–880.

Schwarz, J. R., G. Reid, and H. Bostock. 1995. Action potentials and membrane currents in the human node of Ranvier. *Pflugers Arch* 430: 283–292.

Seifert, G., K. Hüttmann, D. K. Binder, C. Hartmann, A. Wyczynski, C. Neusch, and C. Steinhäuser. 2009. Analysis of astroglial K^+ channel expression in the developing hippocampus reveals a predominant role of the Kir4.1 subunit. *J Neurosci* 29: 7474–7488.

Siotto, F., C. Martin, O. Rauh, J. L. Van Etten, I. Schroeder, A. Moroni, and G. Thiel. 2014. Viruses infecting marine picoplancton encode functional potassium ion channels. *Virology* 466–467: 103–111.

Steiger, A., and M. Kimura. 2010. Wake and sleep EEG provide biomarkers in depression. *J Psychiatr Res* 44: 242–252.

Streit, A. K., M. F. Netter, F. Kempf, M. Walecki, S. Rinné, M. K. Bollepalli, R. Preisig-Müller, *et al.* 2011. A specific two-pore domain potassium channel blocker defines the structure of the TASK-1 open pore. *J Biol Chem* 286: 13977–13984.

Talley, E. M., J. E. Sirois, Q. Lei, and D. A. Bayliss. 2003. Two-pore-domain (KCNk) potassium channels: Dynamic roles in neuronal function. *Neuroscientist* 9: 46–56.

Talley, E. M., G. Solórzano, Q. Lei, D. Kim, and D. A. Bayliss. 2001. CNS distribution of members of the two-pore-domain (KCNK) potassium channel family. *J Neurosci* 21: 7491–7505.

Tsunozaki, M., and D. M. Bautista. 2009. Mammalian somatosensory mechanotransduction. *Curr Opin Neurobiol* 19: 362–369.

Tulleuda, A., B. Cokic, G. Callejo, B. Saiani, J. Serra, and X. Gasull. 2011. TRESK channel contribution to nociceptive sensory neurons excitability: Modulation by nerve injury. *Mol Pain* 7: 30.

Veale, E. L., and A. Mathie. 2016. Aristolochic acid, a plant extract used in the treatment of pain and linked to Balkan endemic nephropathy, is a regulator of K2P channels. *Br J Pharmacol* 173: 1639–1652.

Wang, C., D. Zheng, J. Xu, W. Lam, and D. T. Yew. 2013. Brain damages in ketamine addicts as revealed by magnetic resonance imaging. *Front Neuroanat* 7: 23.

Wang, S., M. Song, J. Guo, Y. Huang, F. Zhang, C. Xu, Y. Xiao, and L. Zhang. 2018. The potassium channel *FaTPK1* plays a critical role in fruit quality formation in strawberry (*Fragaria* x *ananassa*). *Plant Biotechnol J* 16: 737–748.

Weir, G. A., P. Pettingill, Y. Wu, G. Duggal, A. S. Ilie, C. J. Akerman, and M. Z. Cader. 2019. The role of TRESK in discrete sensory neuron populations and somatosensory processing. *Front Mol Neurosci* 12: 170.

Wiedmann, F., A. K. Kiper, M. Bedoya, A. Ratte, S. Rinné, M. Kraft, M. Waibel, *et al.* 2019. Identification of the A293 (AVE1231) binding site in the cardiac two-pore-domain potassium channel TASK-1: A common low affinity antiarrhythmic drug binding site. *Cell Physiol Biochem* 52: 1223–1235.

Wiedmann, F., D. Schlund, F. Faustino, M. Kraft, A. Ratte, D. Thomas, H. A. Katus, and C. Schmidt. 2019. *N*-glycosylation of TREK-1/hK$_{2P}$2.1 two-pore-domain potassium (K$_{2P}$) channels. *Int J Mol Sci* 20: 5193.

Woo, D. H., K. S. Han, J. W. Shim, B. E. Yoon, E. Kim, J. Y. Bae, S. J. Oh, *et al.* 2012. TREK-1 and Best1 channels mediate fast and slow glutamate release in astrocytes upon GPCR activation. *Cell* 151: 25–40.

Wright, P. D., D. McCoull, Y. Walsh, J. M. Large, B. W. Hadrys, E. Gaurilcikaite, L. Byrom, *et al.* 2019. Pranlukast is a novel small molecule activator of the two-pore domain potassium channel TREK2. *Biochem Biophys Res Commun* 520: 35–40.

Wright, P. D., E. L. Veale, D. McCoull, D. C. Tickle, J. M. Large, E. Ococks, G. Gothard, *et al.* 2017. Terbinafine is a novel and selective activator of the two-pore domain potassium channel TASK3. *Biochem Biophys Res Commun* 493: 444–450.

Yamamoto, Y., T. Hatakeyama, and K. Taniguchi. 2009. Immunohistochemical colocalization of TREK-1, TREK-2 and TRAAK with TRP channels in the trigeminal ganglion cells. *Neurosci Lett* 454: 129–133.

Yang, Y., S. Li, Z. R. Jin, H. B. Jing, H. Y. Zhao, B. H. Liu, Y. J. Liang, *et al.* 2018. Decreased abundance of TRESK two-pore domain potassium channels in sensory neurons underlies the pain associated with bone metastasis. *Sci Signal* 11: eaao5150.

Yarishkin, O., D. Y. Lee, E. Kim, C. H. Cho, J. H. Choi, C. J. Lee, E. M. Hwang, and J. Y. Park. 2014. TWIK-1 contributes to the intrinsic excitability of dentate granule cells in mouse hippocampus. *Mol Brain* 7: 80.

Yekkirala, A. S., D. P. Roberson, B. P. Bean, and C. J. Woolf. 2017. Breaking barriers to novel analgesic drug development. *Nat Rev Drug Discov* 16: 545–564.

Yoo, S., J. Liu, M. Sabbadini, P. Au, G. X. Xie, and C. S. Yost. 2009. Regional expression of the anesthetic-activated potassium channel TRESK in the rat nervous system. *Neurosci Lett* 465: 79–84.

Yu, F. H., V. Yarov-Yarovoy, G. A. Gutman, and W. A. Catterall. 2005. Overview of molecular relationships in the voltage-gated ion channel superfamily. *Pharmacol Rev* 57: 387–395.

Zhou, J., H. Chen, C. Yang, J. Zhong, W. He, and Q. Xiong. 2017. Reversal of TRESK downregulation alleviates neuropathic pain by inhibiting activation of gliocytes in the spinal cord. *Neurochem Res* 42: 1288–1298.

Zhou, J., C. X. Yang, J. Y. Zhong, and H. B. Wang. 2013. Intrathecal *TRESK* gene recombinant adenovirus attenuates spared nerve injury-induced neuropathic pain in rats. *Neuroreport* 24: 131–136.

Zhou, J., S. L. Yao, C. X. Yang, J. Y. Zhong, H. B. Wang, and Y. Zhang. 2012. TRESK gene recombinant adenovirus vector inhibits capsaicin-mediated substance P release from cultured rat dorsal root ganglion neurons. *Mol Med Rep* 5: 1049–1052.

Zhou, M., G. Xu, M. Xie, X. Zhang, G. P. Schools, L. Ma, H. K. Kimelberg, and H. Chen. 2009. TWIK-1 and TREK-1 are potassium channels contributing significantly to astrocyte passive conductance in rat hippocampal slices. *J Neurosci* 29: 8551–8564.

Zhuo, R. G., P. Peng, X. Y. Liu, H. T. Yan, J. P. Xu, J. Q. Zheng, X. L. Wei, and X. Y. Ma. 2016. Allosteric coupling between proximal C-terminus and selectivity filter is facilitated by the movement of transmembrane segment 4 in TREK-2 channel. *Sci Rep* 6: 21248.

Zúñiga, L., and R. Zúñiga. 2016. Understanding the cap structure in K2P channels. *Front Physiol* 7: 228.

18 Lysosomal Ion Channels and Human Diseases

Peng Huang, Mengnan Xu, Yi Wu, and Xian-Ping Dong

CONTENTS

18.1 INTRODUCTION

Lysosomes are membrane-bound organelles that are discovered and named by Belgian biologist Christian de Duve in 1955. A mammalian cell normally has several hundred lysosomes that constitute up to 5% of cellular volume and display great heterogeneity in size, morphology and function (Alberts et al., 2014; Luzio et al., 2007; Xu and Ren, 2015). In general, they function as the recycling centers of the cell, where both materials taken up from outside the cell and obsolete components of the cell itself are degraded by more than 60 specific acid hydrolases. Lysosomes are distinguished from other intracellular organelles not only by acid hydrolases but also by numerous lysosomal transmembrane proteins (LMPs) including lysosomal ion channels, transporters, and structural proteins. Deficiency in either the enzymes or the LMPs often results in a group of inherited metabolic disorders termed Lysosomal Storage Diseases (LSDs) (Lloyd-Evans and Platt, 2011; Luzio et al., 2007, 2000; Xu and Ren, 2015). There are approximately 60 LSDs (Kiselyov et al., 2010; Lloyd-Evans and Platt, 2011; Luzio et al., 2000, 2007; Morgan et al., 2011; Platt et al., 2012; Samie and Xu, 2014) that are generally divided into two groups. One is attributed to deficiency of acid hydrolases, and the other results from defective LMPs which are essential for metabolite transports and for creating an optimal environment for lysosomal enzymes. LSDs affect most body organs, but severe neurodegeneration and motor deterioration are the prominent pathological hallmark of most LSDs. Although each type of LSD is rare, collectively they are estimated to affect about 1 in 7,700 births, making them a relatively common and significant health problem. While lysosomal enzymes have been known for quite some time, the understanding of LMPs, particularly lysosomal ion channels, has been impeded due to the limitation of traditional approaches. In this chapter, we aim to provide an updated overview of lysosomal ion channels and their physiopathological implications.

18.2   LYSOSOME ION HOMEOSTASIS

The lysosome has been suggested to be a reservoir of many ions, including proton (H^+), sodium (Na^+), potassium (K^+), calcium (Ca^{2+}), and chloride (Cl^-) (Cang et al., 2013, 2014b; Chakraborty et al., 2017; Li et al., 2019; Lloyd-Evans and Platt, 2011; Luzio et al., 2000, 2007; Mindell, 2012; Saha et al., 2015; Xu and Ren, 2015) (Figure 18.1). Lysosomal H^+ concentration is estimated to be 10–31.6 μM (pH of ~4.5–5.0), that is established by the vacuolar-type H^+-ATPase, a well-studied H^+ transporter present in the lysosomal membrane. Compared with H^+, the concentrations of other lysosomal ions are much less understood. Recent studies have suggested that intralysosomal Ca^{2+} concentration is ~0.5 mM (Christensen et al., 2002; Wang et al., 2012) that is likely established by an unidentified Ca^{2+}/H^+ exchanger or Ca^{2+} transporter (Garrity et al., 2016; Li et al., 2019; Melchionda et al., 2016; Morgan et al., 2011; Narayanaswamy et al., 2019; Wang et al., 2017; Xu and Ren, 2015). The concentrations of intralysosomal Na^+ and K^+ have been estimated to be ~100 mM and ~10 mM, respectively (Steinberg et al., 2010; Wang et al., 2012). Considering the concentration of H^+ (63 nM, pH 7.2), Ca^{2+} (100 nM), Na^+ (~12 mM) and K^+ (~150 mM) in the cytosol, there are large gradients across the lysosomal membrane for H^+, Ca^{2+}, Na^+, and K^+. In addition to cations, the lysosome also stores high levels of anions such as Cl^- (> 80 mM) that is ~10 fold higher compared with Cl^- concentration in the cytosol (Chakraborty et al., 2017; Mindell, 2012; Saito et al., 2007; Stauber and Jentsch, 2013). Moreover, the lysosome accumulates several heavy trace metals such as iron (Fe^{2+}) and zinc (Zn^{2+}).

Almost all the lysosomal functions are regulated by ion transport across the lysosomal membrane. At rest, basal ion transport helps maintain a stable ion concentration in the lysosome, which provides an optimal milieu for the activity of lysosomal enzymes, catabolite transport and other lysosomal functions. For example, the intraluminal H^+ is essential for the functions of most lysosomal hydrolases (Lloyd-Evans and Platt, 2011; Luzio et al., 2000, 2007; Mindell, 2012); some lysosomal enzymes also require adequate levels of other ions such Fe^{2+} and Zn^{2+} for their activities (Blaby-Haas and Merchant, 2014; Dong et al., 2008; Terman and Kurz, 2013; Xiong and Zhu, 2016); basal H^+ and K^+ transport are important for stabilizing the lysosomal membrane potential (Li et al., 2019; Xu and Ren, 2015). Upon stimulation, the transient opening of lysosomal ion channels alters both intralysosomal and cytoplasmic ion concentration, helping the cell adapt

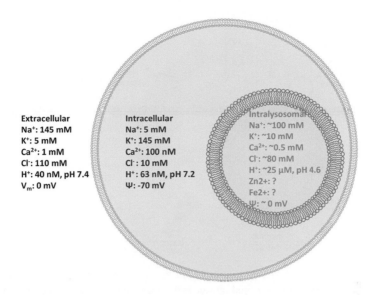

Extracellular
Na$^+$: 145 mM
K$^+$: 5 mM
Ca^{2+}: 1 mM
Cl$^-$: 110 mM
H$^+$: 40 nM, pH 7.4
V$_m$: 0 mV

Intracellular
Na$^+$: 5 mM
K$^+$: 145 mM
Ca^{2+}: 100 nM
Cl$^-$: 10 mM
H$^+$: 63 nM, pH 7.2
Ψ: -70 mV

Intralysosomal
Na$^+$: ~100 mM
K$^+$: ~10 mM
Ca^{2+}: ~0.5 mM
Cl$^-$: ~80 mM
H$^+$: ~25 μM, pH 4.6
Zn2+: ?
Fe2+: ?
Ψ: ~ 0 mV

FIGURE 18.1 Lysosomal ionic homeostasis. Although the ionic composition of extracellular space and cytosol have been well established, the ion composition in the lysosome remain uncertain. Compared with the cytosol, the lysosome contains higher concentrations of Na$^+$, Ca^{2+}, Cl$^-$, and H$^+$, but lower concentrations of K$^+$. The resting membrane potential of the lysosome (ψ_{lyso}, defined as V$_{cytosol}$ − V$_{lumen}$; V$_{lumen}$ set at 0 mV) is close to 0 mV. Lysosomal ions and ψ_{lyso} are important for lysosome functions.

to environmental changes. The spatiotemporal changes of intralysosomal and cytoplasmic ion concentrations are particularly essential for lysosome-associated intracellular signaling and membrane trafficking (Li et al., 2019; Luzio et al., 2007; Morgan et al., 2011; Saftig and Klumperman, 2009; Xiong and Zhu, 2016).

18.3 LYSOSOMAL ION CHANNELS

The ion gradients across the lysosomal membrane suggest the existence of ion channels in the lysosomal membrane. Compared with ion channels in the Plasma membrane (PM), ion channels in the lysosomal membrane are poorly understood. Recent advances in some modern techniques such as lysosomal patch-clamp recording, together with some classical genetic and biochemical approaches, have allowed us to identify and characterize a plethora of ion channels in the lysosomal membrane (Cang et al., 2013, 2014b; Cao et al., 2015b; Dong et al., 2008; Huang et al., 2014; Li et al., 2019; Wang et al., 2012) (Figure 18.2). These include non-selective cation channels such as Transient Receptor Potential Mucolipin 1–3 (TRPML1–3) (Cheng et al., 2010; Dong et al., 2008, 2010; Shen et al., 2011), Transient Receptor Potential Melastatin 2 (TRPM2) (Lange et al., 2009; Sumoza-Toledo et al., 2011) and P2X4 purinoceptor (Huang et al., 2014; Qureshi et al., 2007); Na$^+$ selective channels including Two-Pore Channel 1 (TPC1) (Brailoiu et al., 2009; Cang et al., 2014b) and TPC2 (Calcraft et al., 2009; Cang et al., 2013;

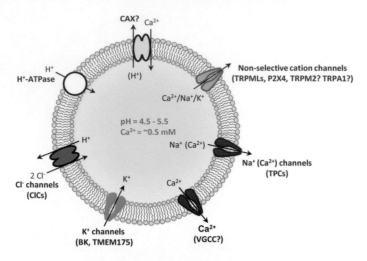

FIGURE 18.2 Ion channels and transporters in the lysosome. Lysosomes have a limiting membrane composed of a single-lipid bilayer and integral proteins and an acidic lumen that contains >60 hydrolytic enzymes. Lysosomal enzymes are responsible for the degradation of intracellular material, whereas the lysosomal membrane proteins participate in metabolite and ion transport, lysosomal trafficking and signaling. Ion transport across the lysosomal membrane is mediated by ion channels and transporters including H^+-ATPase, non-selective cation channels (TRPML1–3, TRPM2, TRPA1, and P2X4), Na^+ or Na^+/Ca^{2+}-selective two-pore channels (TPC1–3), voltage-gated Ca^{2+} channels (VGCC), K^+-selective channels (BK and TMEM175), and Cl^- channels (ClC3–7) that exchange cytosolic Cl^- for lysosomal H^+. Putative lysosomal Ca^{2+} transport protein or Ca^{2+}/H^+ exchanger (CAX) mediates lysosomal uptake of Ca^{2+} from the cytosol and the ER. Note that some of the channels are predominant endosomal channels such as TRPML2–3, TPC1 and ClC3–6, and many of them are also in the PM such as P2X4, TPC3 and BK. TRPM2, TRPA1 and VGCC have not been confirmed as lysosomal ion channels by lysosomal patch-clamp recording. Electrophysiology analysis has also suggested the presence of other molecularly unidentified ion channels and transporters in the lysosome.

Gerndt et al., 2020; Ruas et al., 2015; Wang et al., 2012; Zhang et al., 2019); K^+ selective channels such as Big conductance Ca^{2+}-activated K^+ channel (BK, KCa1.1, MaxiK) (Cao et al., 2015b; Wang et al., 2017) and Transmembrane Protein 175 (TMEM175) (Cang et al., 2015). A chloride channel, ClC7 that is later demonstrated to be a Cl^-/H^+ exchanger, has also been suggested to be expressed in the lysosomal membrane (Graves et al., 2008; Jentsch, 2007; Schieder et al., 2010a; Weinert et al., 2010; Xiong and Zhu, 2016). By controlling ion flux across the lysosomal membrane, these ion channels regulate enzyme activity, metabolite transport, lysosomal membrane potential, cellular signaling, and membrane trafficking (fusion and fission) (Li et al., 2019; Luzio et al., 2007; Morgan et al., 2011; Saftig and Klumperman, 2009; Xiong and Zhu, 2016). Dysfunction of lysosomal ion channels has been implicated in numerous human diseases including LSDs (Bassi et al., 2000; Cao et al., 2015b; Sun et al., 2000; Zhong et al., 2016), classical forms of neurodegenerative diseases (Bae et al., 2014; Coen et al., 2012; Funk and

Kuret, 2012; Hui et al., 2019; Jeyakumar et al., 2005; Neefjes and van der Kant, 2014; Pan et al., 2008; Tsunemi et al., 2019; Wang et al., 2013; Zhang et al., 2009, 2017), muscular dystrophy (Cheng et al., 2014; Yu et al., 2020), phagocytosis of large particles (Samie et al., 2013; Sun et al., 2020), bacterial infection (Capurro et al., 2019; Dayam et al., 2015; Hu et al., 2019a; Miao et al., 2015; Plesch et al., 2018; Sun et al., 2015), and cancer (Hu et al., 2019b; Jung et al., 2019; Kasitinon et al., 2019; Morelli et al., 2016, 2019; Xu et al., 2019; Yin et al., 2019).

18.3.1 LYSOSOMAL Ca^{2+} AND Na^+ CHANNELS

Lysosomal Ca^{2+} and Na^+ channels are a group of proteins whose activation allows both Ca^{2+} and Na^+ to flow out of the lysosome. TRPML1–3 (Cheng et al., 2010; Dong et al., 2008, 2010; Shen et al., 2011), TRPM2 (Lange et al., 2009; Sumoza-Toledo et al., 2011), P2X4 purinoceptor (Huang et al., 2014; Qureshi et al., 2007), and TPC1–3 (Brailoiu et al., 2009; Cai and Patel, 2010; Calcraft et al., 2009; Cang et al., 2013, 2014a,b; Gerndt et al., 2020; Ogunbayo et al., 2015; Ramos et al., 2014; Ruas et al., 2015; Shimomura and Kubo, 2019; Wang et al., 2012; Zhang et al., 2019) belong to this subfamily. By releasing lysosomal Ca^{2+}, these channels participate in intracellular Ca^{2+} signaling and membrane trafficking [i.e. fusion of lysosomes with other cellular membranes including endosomes, autophagosomes, phagosomes, and the PM (Cheng et al., 2010; Li et al., 2019; Luzio et al., 2007; Morgan et al., 2011; Samie and Xu, 2014; Venkatachalam et al., 2015; Xu and Ren, 2015) and lysosomal fission (Cao et al., 2017; Li et al., 2016c; Pryor et al., 2000; Treusch et al., 2004)] by activating downstream Ca^{2+} sensor proteins (Cao et al., 2017; Li et al., 2016c, 2019; Peters and Mayer, 1998; Pryor et al., 2000; Xu and Ren, 2015). In the meanwhile, lysosomal Na^+ release through these channels facilitates lysosomal membrane trafficking by regulating lysosomal membrane potential and pH (Cang et al., 2013) that are important for lysosomal Ca^{2+} homeostasis (Christensen et al., 2002; Melchionda et al., 2016). Dysregulation of these lysosomal Ca^{2+} and Na^+ channels normally leads to impaired membrane trafficking and subsequent accumulation of damaged macromolecules, damaged organelles and intracellular storage (Kiselyov et al., 2010; Lloyd-Evans and Platt, 2011; Morgan et al., 2011; Samie and Xu, 2014; Shen et al., 2012; Venkatachalam et al., 2014).

18.3.1.1 TRPMLs

The TRPML subfamily comprises three channels (TRPML1–3, encoded by *MCOLN1–3* genes) in mammals (Cheng et al., 2010; Li et al., 2019; Venkatachalam et al., 2015; Xu and Ren, 2015). TRPML1 is ubiquitously expressed in all cell types, whereas TRPML2 and TRPML3 are more restricted to specific cell types, with high TRPML3 levels in melanocytes, hair cells of the inner ear, neonatal enterocytes, and bladder epithelial cells (Castiglioni et al., 2011; Cuajungco et al., 2007; Cuajungco and Samie, 2008; Di Palma et al., 2002; Grimm et al., 2010; Kim et al., 2007, 2009; Miao et al., 2015; Nagata et al., 2008; Samie et al., 2009; Xu et al., 2007) and high TRPML2 levels in the thymus, spleen, and kidney (Cheng et al., 2010; Samie et al., 2009; Sun et al., 2015; Venkatachalam et al., 2014). At the cellular level, TRPML1 is primarily localized in the late endocytic pathway (i.e. the late endosome and the lysosome), whereas TRPML2 and TRPML3 (Cheng et al., 2010; Karacsonyi et al.,

2007; Martina et al., 2009; Venkatachalam et al., 2006, 2015) predominantly reside at the earlier compartments of the endocytic pathway [i.e. the recycling endosome for TRPML2 (Karacsonyi et al., 2007; Plesch et al., 2018; Venkatachalam et al., 2006) and the early endosome for TRPML3) (Kim et al., 2009; Martina et al., 2009; Venkatachalam et al., 2006]. Although TRPMLs are primarily intracellular, they are also found in the PM where they are either inactivated (for TRPML1 and TRPML2) (Grimm et al., 2010; Plesch et al., 2018; Zhang et al., 2012) or less active (for TRPML3) (Grimm et al., 2010; Plesch et al., 2018; Xu et al., 2007) compared with intracellular TRPMLs (Chen et al., 2017; Grimm et al., 2010, 2012; Zeevi et al., 2009).

18.3.1.1.1 Characterization of TRPMLs

Like all the other TRP channels (Nilius et al., 2007; Ramsey et al., 2006), TRPML proteins form tetramers, and each pore-forming subunit contains six transmembrane segments (S1-S6) or six transmembrane domains (TM1-TM6) that are separated into an S1-S4 voltage-sensing domain (VSD) and an S5-S6 pore region (P). Distinct from other TRP channels, TRPMLs are characterized by a large extracellular (or intraluminal) loop between S1 and S2 and intracellular endolysosomal localization that is determined by dileucine motifs and their heteromultimerization (Venkatachalam et al., 2006; Vergarajauregui and Puertollano, 2006). Functionally, TRPMLs are inwardly (cation flowing from the lumen to the cytosol) rectifying channels permeable to both Ca^{2+} and Na^+ (Chen et al., 2017; Dong et al., 2008; Plesch et al., 2018). Given the presumed topology of TRPML proteins at the endolysosomal membrane and the electrical properties of the lysosome, TRPML opening leads to Ca^{2+} and Na^+ release from the endolysosome to the cytosol (Chen et al., 2017; Dong et al., 2008, 2010) (Figure 18.3). Additionally, TRPML1 and TRPML2 are also permeable to heavy trace metals such as Fe^{2+} and Zn^{2+} (Dong et al., 2008; Eichelsdoerfer et al., 2010).

In the lysosome, TRPML1 is specifically activated by phosphatidylinositol 3,5-bisphosphate [PI(3,5)P2] (Dong et al., 2010), the major form of phosphoinositide on the lysosomal membrane, whereas in the PM it is inhibited by the PM-specific phosphoinositides PI(4,5)P2, PI(3,4,5)P3, and PI(3,4)P2 (Zhang et al., 2012). Although TRPML2 and TRPML3 are also activated by PI(3,5)P2, it is unclear whether they are sensitive to the PM-specific phosphoinositides (Dong et al., 2010; Plesch et al., 2018). TRPML activity is also regulated by luminal pH, with low pH activating TRPML1 (Chen et al., 2014; Li et al., 2017a; Xu et al., 2007) but high pH activating TRPML2 (Plesch et al., 2018) and TRPML3 (Dong et al., 2008; Xu et al., 2007). The pH-dependent activation profiles of TRPMLs are in accordance with their intracellular localization along the endocytic pathway, i.e. TRPML1 resides in more acidic lysosomes whereas TRPML2 and TRPML3 reside in less acidic recycling endosomes and early endosomes, respectively. Therefore, TRPMLs are tightly controlled by compartment-specific phosphoinositides and pH during membrane trafficking.

TRPMLs are also regulated by additional mechanisms, which may favor TRPML functional diversity. TRPML1 is phosphorylated by Protein kinase A (PKA) on S557 and S559 (Vergarajauregui et al., 2008b), and by mammalian target of rapamycin complex 1 (mTORC1) kinase, a master regulator of cell growth and metabolism

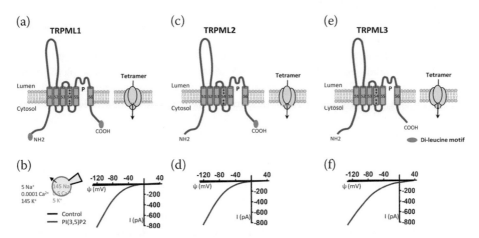

FIGURE 18.3 TRPML ion channels. (A, C, E) TRPML pore-forming subunit contains six transmembrane segments (S1–S6) and a putative pore region (P), with presumably cytosolic N- and C-termini. Distinct from other TRP channels, TRPMLs are characterized by a large extracellular (or intraluminal) loop between S1 and S2. There are dileucine motifs in TRPML1 and TRPML2 at their C-and/or N-termini to determine their intracellular endolysosomal localization. The endolysosomal localization of TRPML3 is determined by its heteromultimerization with other TRPMLs. Like canonical voltage-gated cation channels, TRPML has a voltage sensor domain (VSD) formed by the four transmembrane segments (S1–S4) and a pore domain (S5–S6). Functional TRPMLs are tetramers. (B, D, F) TRPMLs currents measured using lysosome-patch-clamp method. Physiological asymmetric solutions are adopted. The bath solution (cytosolic) contains (in mM) 145 K^+, 5 Na^+, 0.0001 Ca^{2+}, and pH 7.2. Pipette solution (luminal) contains (in mM) 145 Na^+, 5 K^+, 0.5 Ca^{2+}, and pH 4.6. TRPMLs are activated by cytosolic PI(3,5)P2 (1 µM). TRPMLs are inwardly (cation flowing from the lumen to the cytosol) rectifying channels permeable to cations. Given the topology of the TRPML proteins at the endolysosomal membrane and the electrical properties of the lysosome, TRPML opening leads to Ca^{2+} and Na^+ release from the endolysosome to the cytosol.

(Onyenwoke et al., 2015; Sun et al., 2018). The phosphorylation of TRPML1 by mTORC1 is particularly important for the cell to sense nutrient availability. mTORC1 phosphorylates and inhibits TRPML1 when nutrient is abundant. Upon starvation, a reduction of mTORC1 activity stimulates TRPML1 and subsequent lysosomal regeneration that is essentially required for maintaining autophagic flux. TRPML1 also undergoes proteolytic cleavage mediated by cathepsin B that inactivates the channel (Kiselyov et al., 2005). Additionally, TRPML1 (Vergarajauregui and Puertollano, 2006) and TRPML3 (Kim et al., 2019) are regulated by palmitoylation at C-terminal region, controlling their internalization; TRPML2 (Grimm et al., 2012) and TRPML3 (Kim et al., 2007, 2008) in the PM are activated by lowering the extracellular (intraluminal) Na^+ concentration, although the physiological significance remains unclear for this Na^+-dependent activation. Finally, TRPMLs can interact with each other to form heteromultimers, thereby modulating channel functions. The loss of any TRPML leads to lysosomal storage that can be rescued by any other, suggesting the functional overlap of mammalian TRPMLs (Curcio-Morelli et al., 2010b; Zeevi et al., 2009; Zeevi et al., 2010).

In addition to these endogenous factors regulating TRPML property and function, several small-molecule agonists and antagonists for TRPMLs have been identified by using Ca^{2+} imaging-based high-throughput screening method. The agonists belong to different chemotypes, including benzenesulfonamides (e.g. SN-1- or SF-21-type), thiophenesulfonamides (e.g. SF-22-type, including MK6–83), isoindolediones (e.g. SF-51-type, including mucolipin-specific synthetic agonists or ML-SAs), isoxazolines (e.g. SN-2-type), and others (Cao et al., 2015b; Chen et al., 2014; Grimm et al., 2010, 2012; Shen et al., 2012; Wang et al., 2015). Although many of them act at TRPMLs without specificity, SN-2 is selective for TRPML3 (Grimm et al., 2010; Plesch et al., 2018). Through systematic chemical modification of SN-2, TRPML2 selective agonist ML2-SA1 is generated (Plesch et al., 2018). On the other hand, TRPMLs can be inhibited by a variety of small chemical compounds called mucolipin-specific synthetic inhibitors (ML-SIs) (Cheng et al., 2014; Li et al., 2016c; Sahoo et al., 2017; Samie et al., 2013; Wang et al., 2015; Yu et al., 2020; Zhang et al., 2016). These compounds potentially provide important tools to study the functions of TRPMLs and to develop new therapeutic strategies for TRPML-related human diseases.

18.3.1.1.2 TRPMLs in Membrane Trafficking

Lysosomes frequently fuse with intracellular organelles such as endosomes, autophagosomes, and phagosomes to perform their degradation functions (Luzio et al., 2007; Xu and Ren, 2015). Lysosomes also constitutively undergo fission (reformation or biogenesis) after completing their functions, thereby maintaining lysosome homeostasis (Luzio et al., 2007; Saftig and Klumperman, 2009). As with the synaptic vesicle fusion with the PM and endocytic membrane fission, lysosome fusion and fission are also Ca^{2+}-dependent, and the ubiquitously expressed lysosome-associated Ca^{2+} binding proteins synaptotagmin 7 (Syt7), apoptosis-linked gene-2 (ALG-2), and calmodulin (CaM) may serve as Ca^{2+} sensors to trigger fusion and/or fission (Campbell and Fares, 2010; Cao et al., 2017; Lloyd-Evans and Platt, 2011; Miller et al., 2015; Morgan et al., 2011; Peters and Mayer, 1998; Piper and Luzio, 2004; Pryor et al., 2000; Treusch et al., 2004). Mounting evidence has suggested that TRPML1 is the key lysosomal Ca^{2+} release channel regulating both fusion and fission. For example, TRPML1 promotes the fusion between autophagosomes and lysosomes by activating the Ca^{2+}-binding protein ALG-2 to facilitate centripetal movement of lysosomes to the perinuclear area where they fuse with autophagosomes (Li et al., 2016c). TRPML1 also increases phagosome-lysosome fusion upon bacterial infection (Dayam et al., 2015). In the meanwhile, new evidence suggests that TRPML1 facilitates lysosomal fission by activating the CaM-mTORC1 complex (Campbell and Fares, 2010; Cao et al., 2017; Miller et al., 2015; Sun et al., 2018; Treusch et al., 2004). This is critical for maintaining efficient autophagic flux (Yu et al., 2010), maybe also for lysosomal exocytosis (Chen et al., 1998; Shen et al., 2012) and retrograde transport of lactosylceramide to the Golgi compartment (Chen et al., 1998; Pryor et al., 2006). The role of TRPML1 in fission is in accordance with earlier reports showing that (1) TRPML1 mutant cells display enlarged lysosomes and impaired lysosome biogenesis (Cao et al., 2017; Cheng et al., 2010; Dong et al., 2010; Miller et al., 2015; Treusch et al., 2004; Venkatachalam et al., 2015); (2) mTORC1 is required for autophagic lysosomal reformation (Michaillat et al., 2012; Roczniak-Ferguson et al., 2012; Settembre et al., 2012; Sun et al., 2018; Yu et al., 2010), entotic vacuole fission and yeast vacuole (lysosome) fission (Krajcovic et al., 2013; Michaillat et al., 2012); (3) an increase in PI(3,5)P2 promotes vacuole fission (Efe et al., 2005), while the deficiency of PI(3,5)P2 causes vacuole enlargement in both yeast (Efe et al., 2005; Rudge et al., 2004) and mammalian cells (Dong et al., 2010). Thus, TRPML1 plays a dual role in membrane fusion and fission in the late endocytic pathway and the PM, depending on the context.

Lysosomal exocytosis, a Ca^{2+}-dependent process in which lysosomes constitutively exocytose to fuse with the PM and release intraluminal contents to the extracellular environment (Jaiswal et al., 2002; Martinez et al., 2000; Pryor et al., 2000; Reddy et al., 2001; Rodriguez et al., 1997; Samie and Xu, 2014), is particularly important. It has been implicated in cellular waste elimination, membrane repair, neurite outgrowth, and neurotransmitter release (Arantes and Andrews, 2006; Medina et al., 2011; Reddy et al., 2001; Saftig and Klumperman, 2009; Samie and Xu, 2014; Zhang et al., 2007). Emerging evidence suggests that by releasing lysosomal Ca^{2+} TRPML1 is essential for lysosomal exocytosis. The deficiency of TRPML1 results in impaired lysosomal exocytosis whereas TRPML1 activation promotes lysosomal exocytosis (Cheng et al., 2014; Dong et al., 2009; LaPlante et al., 2006; Samie et al., 2013; Samie and Xu, 2014). By regulating lysosomal exocytosis, TRPML1 is implicated in cellular clearance of lysosomal storage in various in vitro and in vivo models of LSDs (Cao et al., 2015b; Medina et al., 2011; Palmieri et al., 2011; Shen et al., 2012; Zhong et al., 2016). Lysosomal exocytosis is a multi-step process including antegrade transport of lysosomes along microtubules, lysosomal docking at and fusion with the PM. TRPML1 may regulate any of these steps to control lysosomal exocytosis. Because the lysosomal membrane protein Syt7 plays a crucial role in Ca^{2+} binding and subsequent tethering of lysosomal and plasma membrane SNAREs preceding fusion (LaPlante et al., 2006; Martinez et al., 2000), it is conceivable that TRPML1 may activate Syt7 to favor membrane fusion. However, the lack of interaction between TRPML1 and Syt7 argues against this (Cao et al., 2017). Alternatively, because of its role in lysosomal fission, TRPML1 may control lysosomal exocytosis by providing small lysosomes that are energy-efficient for transport along cytoskeleton (Campbell and Fares, 2010; Cao et al., 2017; Li et al., 2016b,c; Miller et al., 2015; Sun et al., 2018; Treusch et al., 2004). By increasing lysosomal exocytosis, TRPML1 plays an important role in large-particle phagocytosis (Samie et al., 2013), membrane repair (Cheng et al., 2014), lipid turnover (Cui et al., 2020), and storage clearance (Li et al., 2019; Medina et al., 2011; Shen et al., 2012; Tsunemi et al., 2019; Xu and Ren, 2015; Zhong et al., 2016).

Compared with TRPML1, the role of TRPML2 and TRPML3 in membrane trafficking is poorly studied. TRPML2 has been suggested to play a role in the traffic of cargo through the early endosome and recycling endosome pathway (Karacsonyi et al., 2007; Plesch et al., 2018; Venkatachalam et al., 2006). In murine macrophages, TRPML2 also regulates the exocytosis of recycling endosomes but not lysosomes (Plesch et al., 2018). TRPML3 has been suggested to regulate the early and late endocytic pathway. In the retinal epithelial cell line ARPE19 (Martina et al., 2009), HEK293, and HeLa cells (Kim et al., 2009), overexpression of TRPML3 causes the enlargement of endosomes and delayed degradation of epidermal growth factor receptor (EGFR), whereas the depletion of endogenous TRPML3 enhances EGFR degradation. However, whether TRPML3 has effect on endocytosis is debated (Kim et al., 2009; Martina et al., 2009). Emerging evidence also suggests that TRPML3 can detect lysosomal alkalization to initiate lysosomal exocytosis (Miao et al., 2015).

In summary, all three TRPML channels can regulate intracellular membrane trafficking while the mechanisms by which they affect trafficking might differ. TRPML2 primarily regulates membrane trafficking at the level of early endosomes and recycling endosomes; TRPML3 regulates membrane trafficking at the level of early endosomes and late endosomes; TRPML1 regulates membrane trafficking at the level of late endosomes and lysosomes.

18.3.1.1.3 TRPMLs in Autophagy

Autophagy is a lysosome-dependent cellular process for the turnover of organelles and molecules. It is predominantly a cell-survival mechanism, playing a critical role in cell growth, cell proliferation and differentiation, and tissue homeostasis and development (Klionsky et al., 2016; Levine and Kroemer, 2008; Mizushima and Levine, 2010). Cellular stress conditions such as nutrient and energy deprivation and diseases elevate autophagy. During autophagy, cytoplasmic components are engulfed by autophagosomes that fuse with lysosomes where cytoplasmic components are degraded by lysosomal hydrolases and then recycled. Autophagic flux is regulated by a series of events, including autophagy initiation, autophagosome-lysosome fusion (autophagosome maturation), autolysosome degradation, and lysosome reformation (Klionsky et al., 2016; Yu et al., 2010). Disruption of any of them could lead to impaired autophagy that has been associated with abnormalities such as LSDs, neurodegeneration, cancer, and inflammatory, infectious and autoimmune conditions (Fulda and Kogel, 2015; Hara et al., 2006; Komatsu et al., 2006; Levine and Kroemer, 2008; Levine et al., 2011). Basal autophagy is especially important in postmitotic cells such as neurons and muscles, where the accumulation of aggregated proteins and damaged organelles often results in cell death (Hara et al., 2006; Komatsu et al., 2006; Levine and Kroemer, 2008). Indeed, the suppression of basal autophagy causes neurodegeneration (Hara et al., 2006; Komatsu et al., 2006) and myofiber degeneration and weakness (Masiero et al., 2009; Raben et al., 2008).

Previous studies using different models, including human fibroblasts derived from TRPML1 deficient patients (Jennings et al., 2006; Vergarajauregui et al., 2008a), TRPML1 deficient mouse model (Curcio-Morelli et al., 2010a; Micsenyi et al., 2009), *drosophila* TRPML1 mutants (Venkatachalam et al., 2008), and cup-5 (the *C. elegans* ortholog of the mammalian TRPML1) null mutant (Sun et al., 2011), have suggested that the loss of TRPML1 impairs autophagic flux with characteristics of accumulated autophagosomes (Curcio-Morelli et al., 2010a; Sun et al., 2011; Venkatachalam et al., 2008; Vergarajauregui et al., 2008a; Wong et al., 2012). The accumulation of autophagosomes is likely due to a defect in proteolytic degradation of cargo in autolysosomes following the fusion of autophagosomes and lysosomes (Schaheen et al., 2006; Sun et al., 2011; Venkatachalam et al., 2008). Growing evidence suggests that TRPML1 regulating autophagy is much more complicated, and TRPML1 plays a multifaceted role in autophagy. Sun et al. (2018) suggest that TRPML1 and mTORC1 form a complex acting as a sensor of cellular stresses to control multiple steps of autophagy. Under normal condition, high levels of nutrients activate mTORC1 which phosphorylates and inactivates TRPML1. In the same time, mTORC1 phosphorylates TFEB and inhibits TFEB-dependent transcription of autophagy and lysosomal genes (Settembre et al., 2011). When nutrients are depleted, TRPML1 is activated due to mTORC1 reduction. This initiates several downstream Ca^{2+}-dependent events to facilitate autophagy. First, TRPML1 releases Ca^{2+} to rapidly activate the CaM-dependent protein kinase kinase β (CaMKKβ) and AMP-activated protein kinase (AMPK) pathway to initiate autophagosome formation (Scotto Rosato et al., 2019). Second, TRPML1 promotes Ca^{2+}-dependent centripetal movement of lysosomes toward the perinuclear region to fuse with autophagosomes by activating ALG-2 (Li et al., 2016c). Third, TRPML1 facilitates autophagic flux by increasing proteolytic degradation (Sun et al., 2011). Fourth, TRPML1 reactivates mTORC1 via CaM to promote lysosome reformation and protein synthesis during starvation (Sun et al., 2018; Zoncu et al., 2011). Fifth, TRPML1 activates Syt7-dependent lysosomal exocytosis to remove garbage (Samie and Xu, 2014; Shen et al., 2012). Sixth, TRPML1 activates TFEB-dependent transcription to continuously supply lysosomal and autophagy proteins (Medina et al., 2015; Wang et al., 2015; Zhang et al., 2016). Therefore, TRPML1 orchestrates all these cellular events to help cells maintain high autophagic flux, adapting to environmental changes.

TRPML3 also regulates the autophagic pathway. TRPML3 overexpression increases autophagosome accumulation whereas TRPML3 down-regulation reduces autophagosomes (Kim et al., 2009; Martina et al., 2009; Scotto Rosato et al., 2019). Mechanistically, upon nutrient starvation, TRPML3 may interact with GATE16, a mammalian ATG8 homologue, but not LC3B to facilitate autophagosome formation (Choi and Kim, 2014). TRPML3 regulating autophagy is also under the control of palmitoylation (Kim et al., 2019). Although no direct evidence suggesting a role of TRPML2 in autophagy, TRPML2-mediated Ca^{2+} release may activate TFEB-dependent gene expression to regulate autophagy pathway in some specific types of cells (Ma et al., 2018).

18.3.1.1.4 TRPMLs in Cellular Signaling

As a newly identified Ca^{2+} store (Christensen et al., 2002; Wang et al., 2012), the lysosome in intracellular Ca^{2+} signaling has attracted attention in recent years. By releasing Ca^{2+}, the lysosome not only activates local Ca^{2+} signaling events but also, along with the endoplasmic reticulum (ER), modulates global cytosolic Ca^{2+} signaling events using a mechanism similar to the Ca^{2+}-induced Ca^{2+} release (CICR). This further causes Ca^{2+} entry from the extracellular space due to the depletion of the ER Ca^{2+} store, evoking global Ca^{2+} signals (Kilpatrick et al., 2013). TRPML1 has been implicated in both local and global Ca^{2+}-dependent signaling pathways. On the one hand, by releasing lysosomal Ca^{2+}, TRPML1 activates local Ca^{2+}-dependent signaling pathways including the aforementioned CaM-mTORC1 signaling pathway (Sun et al., 2018) and calcineurin (CaN)-TFEB signaling pathway (Medina et al., 2015) to adapt to nutrient starvation. On the other hand, TRPML1 can evoke global Ca^{2+} signals by inducing ER-dependent Ca^{2+} release, probably through lysosome-ER membrane contact sites. This further induces subsequent Ca^{2+} entry from the extracellular space due to the ER Ca^{2+} depletion (Atakpa et al., 2018; Kilpatrick et al., 2013, 2016, 2017; Morgan et al., 2013; Patel and Brailoiu, 2012; Penny et al., 2015). Moreover, new studies have revealed mitochondria-lysosome membrane contact sites (Wong et al., 2018), at which TRPML1 can directly transfer Ca^{2+} from lysosomes to mitochondria (Peng et al., 2020). Thus, TRPML1 provides an additional mechanism in modulating intracellular Ca^{2+} dynamics.

Emerging evidence suggests that TRPMLs also mediate several other signaling pathways to adapt to environmental stresses. For example, both TRPML1 (Zhang et al., 2016) and TRPML2 (Ma et al., 2018) act as sensors of cellular reactive oxygen species (ROS), and play a role in ROS signaling in the cell. By promoting lysosomal exocytosis, TRPML1 may facilitate the release of intralysosomal signaling molecules such as ATP (~1 mM) (Cao et al., 2014; Huang et al., 2014; Zhang et al., 2007) and adenosine (~1 mM) (Zhong et al., 2017) to the extracellular space where they activate some signaling pathways. Indeed, TRPML1 increases intraluminal ATP release to promote TNBC progression, possibly by activating ATP receptor on the PM or by acting on the extracellular matrix (Xu et al., 2019).

18.3.1.1.5 Other Functions of TRPMLs

In addition to the aforementioned physiological functions, TRPMLs also regulate endolysosomal pH and degradation activity. For example, TRPML1 activity is required for lysosomal acidification (Bach et al., 1999; Bae et al., 2014; Khan et al., 2019) and degradation (Fares and Greenwald, 2001; Schaheen et al., 2006; Sun et al., 2011; Venkatachalam et al., 2008). TRPML3 regulates both endosomal (Lelouvier and Puertollano, 2011; Martina et al., 2009) and lysosomal pH (Hu et al., 2019a; Miao et al., 2015). However, TRPML3 could facilitate (Hu et al., 2019a; Lelouvier and Puertollano, 2011) or hinder acidification (Martina et al., 2009; Miao et al., 2015) depending on the status of the cell.

18.3.1.1.6 TRPMLs and Human Diseases

Because lysosomal Ca^{2+} release plays a role in several important cellular processes such as membrane trafficking, cellular signaling, and autophagy (Cheng et al., 2010; Lloyd-Evans and Platt, 2011; Luzio et al., 2007; Morgan et al., 2011; Venkatachalam et al., 2015; Xu and Ren, 2015), TRPMLs have been implicated in many pathological conditions.

First, TRPMLs have a role in neurodegenerative diseases. Due to its important role in lysosomal Ca^{2+} release and membrane trafficking, mutations in human TRPML1 cause Mucolipidosis type IV (ML-IV), a LSD with neurodegeneration and motor defects (Bassi et al., 2000; Cheng et al., 2010; Dong et al., 2008; Puertollano and Kiselyov, 2009; Sun et al., 2000; Venkatachalam et al., 2008, 2014). Impaired TRPML1 has also been implicated in several other LSDs including Niemann-Pick disease C1 (NPC1), NPA, NPB, and Fabry disease (Cao et al., 2015b; Kiselyov et al., 2010; Shen et al., 2012), and classical forms of neurodegenerative diseases including

Alzheimer's disease (Bae et al., 2014; Coen et al., 2012; Hui et al., 2019; Zhang et al., 2017) and Parkinson's disease (Tsunemi et al., 2019). Emerging evidence further indicates that promoting TRPML1-mediated lysosomal exocytosis represents a promising therapeutic approach for LSDs (Chen et al., 2014; Medina et al., 2011; Samie and Xu, 2014; Shen et al., 2012). Compared with TRPML1, no clinically mutations in TRPML2 and TRPML3 have been reported.

Second, TRPMLs are associated with both innate and adaptive immune responses. For example, TRPML1 up-regulation favors the clearance of *Helicobacter pylori* (*H. pylori*) in *AGS* gastric cells (Capurro et al., 2019), the clearance of exogenous particles in macrophages (Dayam et al., 2015; Samie et al., 2013; Sun et al., 2020), and B-cell survival associated with severe combined immunodeficiency diseases (Zhong et al., 2017). Mechanistically, bacterial infection activates TRPML1 to mediate phagosome-lysosome fusion (Dayam et al., 2015, 2017) and to stimulate the TFEB signaling pathway, thereby enhancing lysosome-based proteolysis and killing of subsequently phagocytosed pathogens (Gray et al., 2016). TRPML1 activation also promotes the arrival of dendritic cells to lymph nodes for antigen presentation to T-cells (Bretou et al., 2017) and the remodeling of secretory lysosomes during natural killer (*NK*) education (Goodridge et al., 2019). In addition to a role in bacterial clearance, TRPML1 also plays a role in viral infection. For example, HIV-1 transactivator of transcription (Tat) is essential for HIV-1 replication and appears to play an important role in the pathogenesis of HIV-associated neurological complications. After being secreted from infected cells, Tat can be taken up via receptor-mediated endocytosis. Following endocytosis and internalization into endolysosomes, Tat must be released into the cytoplasm in order to activate the HIV-1 Toll-like receptor (TLR) promoter and facilitate HIV-1 viral replication in the nucleus. TRPML1 activation enhances the Tat protein degradation by acidifying lysosomes, thereby inhibiting Tat from activating HIV-1 replication and preventing disease progression (Bae et al., 2014; Khan et al., 2019). TRPML1 also participates in stimulating anti-viral adaptive immune responses by enhancing TLR7 and TLR9-mediated responses (Kawai and Akira, 2010; Li et al., 2015). Recent studies further suggest a role of TRPML2 in both bacterial and viral infection. TRPML2 expression in immune cells is dramatically increased by TLR4 activation (Cuajungco et al., 2016; Sun et al., 2015). TRPML2 up-regulation increases the recruitment of peripheral macrophages in response to bacterial infection by facilitating the release of the chemokine CCL2 via the exocytosis of early and recycling endosomes (Plesch et al., 2018; Sun et al., 2015). However, TRPML2 enhances viral infection by facilitating viral entry and replication (Rinkenberger and Schoggins, 2018; Schoggins et al., 2011). TRPML3 has also been involved in bacterial clearance using autophagy as a cell-autonomous defense mechanism. *Uropathogenic E. coli* (UPEC) in bladder epithelial cells and *H. pylori* in gastric epithelial cells are targeted by autophagy but avoid the degradation because of their capacity to neutralize lysosomal pH. In bladder epithelial cells, increased lysosomal pH by UPEC activates lysosomal TRPML3 to initiate lysosomal exocytosis, resulting in the expulsion of exosome-encased bacteria. However, in gastric epithelial cells *H. pylori* is sequestered and survives in autophagosomes. TRPML3 activation acidifies lysosomes, reactivating the autolysosomal degradation function to combat bacterial infection (Hu et al., 2019a). Similar to TRPML2, TRPML3 may regulate the infection of influenza A virus (Rinkenberger and Schoggins, 2018), dengue virus and zika virus (Xia et al., 2020).

Third, TRPMLs regulate cancer progression (Fehrenbacher and Jaattela, 2005; Fulda and Kogel, 2015). Jung et al. (2019) suggest that TRPML1 expression is significantly elevated in HRAS-positive tumors and inversely correlated with patient prognosis. TRPML1 is necessary for the proliferation of cancer cells that bear activating mutations in HRAS. In the meanwhile, Xu et al. (2019) show that TRPML1 expression is specifically elevated in TNBC which favors cancer development by increasing mTORC1 activity and lysosomal ATP release. TRPML1 down-regulation inhibits TNBC cell proliferation in vitro and tumorigenesis and metastasis in vivo (Xu et al., 2019). Distinct from TNBC, TRPML1 promotes melanoma development by inhibiting mTORC1 signaling (Kasitinon et al., 2019). A correlation between TRPML1 expression level and poor clinical characteristics has also been reported in pancreatic ductal adenocarcinoma (PDAC) (Hu et al., 2019b) and non-small-cell lung cancer (NSCLC) (Yin et al., 2019). The links between TRPML2 and glioma (Morelli et al., 2016), breast cancer (Huang et al., 2013), lymphoblastic leukemia (Almamun et al., 2015), colorectal cancer (Perez-Riesgo et al., 2017) have also been suggested. For example, TRPML2 expression increases with glioblastoma (GBM) progression, and knockdown of TRPML2 inhibits the proliferation and viability of glioma cells (Morelli et al., 2016). The role of TRPML3 in cancer has been poorly investigated. Data from the TCGA database reveal that TRPML3 expression is down-regulated in a number of cancers, and it is identified as a protective gene (Wu et al., 2019).

In addition to these conditions mentioned above, TRPML1 has also been suggested to mitigate muscular dystrophy by increasing sarcolemma repair (Cheng et al., 2014; Yu et al., 2020) and Lowe syndrome by enhancing autophagosome-lysosome fusion (De Leo et al., 2016). TRPML3 may be associated with hearing loss (Grimm et al., 2007; Wiwatpanit et al., 2018).

18.3.1.2   TPCs

TPCs comprise a subfamily (TPC1–3, encoded by *TPCN* genes) of cation channels that belong to the voltage-gated ion channel superfamily (Cang et al., 2013, 2014a,b; Grimm et al., 2017; Li et al., 2019; Wang et al., 2012). While TPC1 and TPC2 are ubiquitously expressed in animals, TPC3 is only found in some animals such as sea urchins, fish, reptiles, amphibians, birds, cats and dogs, but not in humans or mice (Cai and Patel, 2010; Cang et al., 2014a,b; Ramos et al., 2014; Zhu et al., 2010). While TPC1 and TPC2 proteins are targeted to the membranes of endosomes and lysosomes (Calcraft et al., 2009; Grimm et al., 2017; Morgan et al., 2011) by their conserved dileucine-based motifs (Brailoiu et al., 2010a; Larisch et al., 2012; Li et al., 2019), TPC3 is localized in both the PM and the endolysosomal membrane (Cang et al., 2014a; Ogunbayo et al., 2015; Shimomura and Kubo, 2019). In addition, TPC1 channels also reside in the vacuolar membrane of plants (Furuichi et al., 2001; Guo et al., 2016; Peiter et al., 2005).

18.3.1.2.1 Characterization of TPCs

The TPC channels form a homodimer, in which each subunit contains two homologous repeats of six-transmembrane helices (IS1-IS6 and IIS1-IIS6) (Calcraft et al., 2009; Guo et al., 2016; Jin et al., 2020; Kintzer and Stroud, 2016; Peiter et al., 2005). As in the superfamily of voltage-gated cation channels, each of the repeats has a voltage sensor domain (VSD) formed by four transmembrane segments (S1-S4) and a pore domain (S5-S6). Although the S4 segments of all three TPC channels contain positively charged amino acid residues, distinct from TPC1 and TPC3 that are voltage-gated, TPC2 lacks such voltage dependence (Cang et al., 2013, 2014a,b; Jha et al., 2014; She et al., 2019; Wang et al., 2012) (Figure 18.4).

The activation mechanism and ion selectivity of TPCs have been debated for a long time. Early studies suggest that TPC1 and TPC2 are implicated in lysosomal Ca^{2+} release in response to nicotinic acid adenine dinucleotide phosphate (NAADP) (Brailoiu et al., 2009; Brailoiu et al., 2010a; Calcraft et al., 2009; Grimm et al., 2017; Jha et al., 2014; Patel et al., 2017; Pitt et al., 2010; Ruas et al., 2010, 2015; Rybalchenko et al., 2012; Schieder et al., 2010a). However, later studies suggest that TPCs are Na^+-selective channels activated by PI(3,5)P2 (Bellono et al., 2016; Cang et al., 2013, 2014b; Gerndt et al., 2020; Guo et al., 2017; Jha et al., 2014; Kirsch et al.,

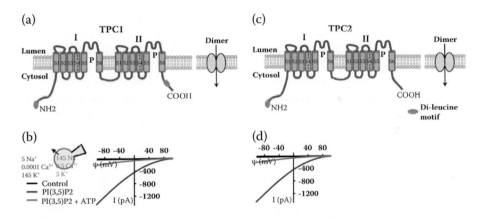

FIGURE 18.4 TPC ion channels. (A) Transmembrane topology of TPC1. Each TPC1 subunit contains two homologous repeats of six-transmembrane helices (IS1-IS6 and IIS1-IIS6), and each of the repeats has a voltage sensor domain formed by the four transmembrane segments (S1–S4) and a pore domain (S5–S6). The S4 segments contain positively-charged amino acid residues. The TPC1 forms a homodimeric channel. (B) TPC1 is activated by cytosolic PI(3,5)P2 (1 μM). PI(3,5)P2-induced TPC1 activity is inhibited by cytosolic ATP (1 mM). The bath solution (cytosolic) contains (in mM) 145 K^+, 5 Na^+, 0.0001 Ca^{2+}, and pH 7.2. Pipette solution (luminal) contains (in mM) 145 Na^+, 5 K^+, 0.5 Ca^{2+}, and pH 4.6. (C) Schematic of TPC2 membrane topology. TPC2 has a similar topology of TPC1. Although the S4 segments of TPC2 channels contain positively-charged amino acid residues, distinct from TPC1 and TPC3 that are voltage-gated, TPC2 lacks such voltage dependence. (D) TPC2 is sensitive to both PI(3,5)P2 (1 μM) and ATP (1 mM) as in TPC1. TPC2 currents are measured using physiological asymmetric solutions as in B. Note that TPCs can also be activated by NAADP to release lysosomal Ca^{2+}.

2018; She et al., 2018, 2019; Wang et al., 2012; Zhang et al., 2019). Although both views have received support in the proceeding years (Gerndt et al., 2020; Grimm et al., 2014; Guo et al., 2017; Jha et al., 2014; Kirsch et al., 2018; Lagostena et al., 2017; Moccia et al., 2020; Ogunbayo et al., 2018; Patel et al., 2017; Ruas et al., 2015; Rybalchenko et al., 2012), increasing evidence suggests that NAADP may indirectly activate TPC1/2 via an accessory protein (Krogsaeter et al., 2019; Lin-Moshier et al., 2012; Pitt et al., 2010; Ruas et al., 2015; She et al., 2019; Walseth et al., 2012; Wang et al., 2012; Xu and Ren, 2015). High-resolution structural studies combined with functional analysis have suggested that the opening of animal TPC1 is dependent on both ligand PI(3,5)P2 and voltage, and the VSD from the second 6-TM domain confers voltage dependence on TPC1 (She et al., 2018). In contrast, TPC2 is simply a PI(3,5)P2-activated channel (She et al., 2019). The binding site of PI(3,5)P2 is located at the first 6-TM domain for both TPC1 (She et al., 2018) and TPC2 (She et al., 2019). Similar to TPC1, the second VSD of TPC3 is involved in sensing changes in the membrane potential to open the channel (Dickinson et al., 2020). Distinct from animal TPC1 and TPC2, animal TPC3 is sensitive to PI(3,4)P2 but not PI(3,5)P2 when recorded at extracellular pH of 7.4, suggesting that TPC3 may function as a PI (3,4)P2-sensitive Na^+ channel in the PM (Cang et al., 2014a; Dickinson et al., 2020; Shimomura and Kubo, 2019). In spite of the high sequence similarity in the filter region, plant TPC1 functions as a nonselective cation channel on the vacuole membrane, with higher selectivity for Ca^{2+} over Na^+, but without selectivity among monovalent cations (Li^+, Na^+, and K^+) (Guo et al., 2017). Plant TPC1 activation requires both voltage and cytosolic Ca^{2+} (Guo et al., 2016) but not PI(3,5)P2 (Boccaccio et al., 2014).

To resolve the conflicting reports on the permeability and gating properties of TPCs, two groups recently suggest that the ion selectivity and gating properties of TPCs are likely determined by the activating ligands. By performing a Ca^{2+} imaging-based high-throughput screen, Gerndt et al. (2020) identified two lipophilic and structurally distinct TPC2 agonists TPC2-A1-N and TPC2-A1-P. TPC2-A1-N evokes robust Ca^{2+}-signals and non-selective cation currents, whereas TPC2-A1-P induces Na^+-selective currents with weaker Ca^{2+}-signals. These properties are mirrored by the Ca^{2+}-mobilizing messenger, NAADP and the phosphoinositide, PI(3,5)P2, respectively. Interestingly, TPC2-A1-N but not TPC2-A1-P also renders the channel H^+-permeable. In the other study, Zhang et al. (2019) reported that five chemically closely related classes of dibenzazepine type tricyclic antidepressants (TCAs, i.e. clomipramine, desipramine, imipramine, amitriptyline, and nortriptyline that are named LyNa-VA1.1 to LyNa-VA1.5) and phenothiazine (i.e. chlorpromazine and triflupromazine that are named LyNa-VA2.1 and LyNa-VA2.2)-based antidepressants induce strong inwardly rectifying TPC currents. However, Riluzole (referred to as Lysosomal Na^+ channel Agonist 1 [LyNA1]), an FDA-approved amyotrophic lateral sclerosis drug that is known to modulate voltage-gated Na^+ channels, evokes TPC2 currents independent of voltage as in the case for PI(3,5)P2. In contrast to TPC2-A1-N and TPC2-A1-P that alter cation permeability in an agonist-dependent manner, LyNa-VA and LyNA do not change ion selectivity of the channel, that is, low Ca^{2+} but high Na^+

permeability. Furthermore, these compounds act on TPC1 and TPC2 differently. TPC2-A1-N and TPC2-A1-P only activate TPC2 but neither inhibits nor activates TPC1 (Gerndt et al., 2020); LyNa-VA1.1 and LyNa-VA1.2 activate both TPC2 and TPC1 in a voltage-dependent manner, while LyNa-VA2.1 and Riluzole inhibit TPC1 (Zhang et al., 2019). Overall, TPCs function as either NAADP-activated Ca^{2+} release channels (Brailoiu et al., 2010a; Grimm et al., 2014; Pitt et al., 2010; Ruas et al., 2015; Schieder et al., 2010b) or PI(3,5)P2-gated Na^+ channels (Cang et al., 2013; Guo et al., 2017; Wang et al., 2012; Zhang et al., 2019). The dual activation mechanism allows TPCs to mediate diverse cellular functions in response to various environmental stimuli.

Pharmacologically, in addition to TPC2-A1-N, TPC2-A1-P, LyNa-VA and LyNA, all three TPCs are insensitive to the voltage-gated Na^+ channel blocker tetrodotoxin (TTX) but sensitive to voltage-gated Ca^{2+} channel blockers including verapamil, Cd^{2+}, and nifedipine (Cang et al., 2013, 2014a,b; Wang et al., 2012). Sphingosines were also reported to induce TPC1-mediated Ca^{2+} release from the lysosome (Hoglinger et al., 2015). However, electrophysiological assay suggests that sphingosine may activate TPCs indirectly (Li et al., 2019; Zhang et al., 2019).

18.3.1.2.2 Functions of TPCs

Similar to TRPML1, by mediating Ca^{2+} release from lysosomal Ca^{2+} stores, the NAADP-TPC pathway regulates both local and global Ca^{2+} signaling events (Calcraft et al., 2009; Churchill and Galione, 2001; Gerasimenko et al., 2015). This contributes to many cellular signaling pathways in response to cues such as acetylcholine (Brailoiu et al., 2010b), histamine (Esposito et al., 2011) and cholecystokinin (Cancela et al., 1999; Gerasimenko et al., 2015; Yamasaki et al., 2004).

TPCs play an important role in endolysosomal membrane trafficking Castonguay et al. suggest that TPC1 channel interacts with syntaxins to regulate the fusion of intracellular vesicles, and TPC1 deficiency impairs efficient protein processing through early and recycling endosomes (Castonguay et al., 2017). Grimm et al. (2014) suggest that cells lacking TPC2 display a profound impairment of low-density lipoprotein (LDL)-cholesterol and epithelial growth factor (EGF)/EGF-receptor trafficking, likely due to a defective fusion between the late endosome and the lysosome. It is also suggested that TPC2 ablation disturbs integrin trafficking in the endolysosomal system (Nguyen et al., 2017). In contrast to LDL-cholesterol and EGF/EGF-receptors that are accumulated in late endosomes, integrins are enriched in early endosomes. TPC2 deletion also causes impaired membrane trafficking of melanosomes, lysosome-related organelles (Ambrosio et al., 2016).

Although NAADP has been suggested to induce alkalinization of lysosomes (Cosker et al., 2010; Morgan and Galione, 2007), the role of TPC2 in regulating lysosomal pH is conflicting. Some studies suggest that the loss of TPC2 leads to an increased lysosomal pH and a reduced lysosomal protease activity (Ambrosio et al., 2016; Cang et al., 2013; Lin et al., 2015), whereas others suggest a normal lysosomal pH in TPC mutant cells (Grimm et al., 2014; Ruas et al., 2015). These conflicting results could be attributed to the experimental conditions because TPC2 agonists differentially affect lysosomal pH and some agonists may render the channel H^+-permeable (Gerndt et al., 2020).

Cang et al. (2013) suggest that TPC1 and TPC2 form a complex with ATP/mTOR to detect the nutrient status. TPCs are inhibited when mTOR are associated with lysosomes in fed conditions. Energy depletion and nutrient starvation cause mTOR falling off from the lysosomal membrane, and this leads to TPC activation to maintain lysosomal membrane potential, pH stability and amino acid homeostasis. Hence, TPCs couple the metabolic status of the cell with lysosome function. Although they did not observe obvious defect in autophagy in the TPC1/TPC2 double knockout mice (Cang et al., 2013), later studies from others suggest that TPC2 contributes to autophagy termination by facilitating mTOR reactivation (Garcia-Rua et al., 2016; Lin et al., 2015).

18.3.1.2.3 TPCs and Human Diseases

The clinical relevance of TPCs is mainly associated with immune response to pathogens. TPCs have been involved in various viral and bacterial infections. Ebolavirus binds to cell surface proteins to initiate its entry into host cells through micropinocytosis. After internalization, the virus is released into the cytoplasm of host cells where replication begins. New evidence suggests that Ebolavirus entry into host cells requires TPCs. Disrupting TPC function prevents Ebolavirus infection (Penny et al., 2019; Sakurai et al., 2015). TPCs also have a role in HIV-1 infection. TPCs are suggested to be required for Tat endolysosome escape and subsequent TLR transactivation. Similarly, Middle East Respiratory Syndrome coronavirus (MERS-CoV) (Gunaratne et al., 2018) and Severe Acute Respiratory Syndrome coronavirus 2 (*SARS-CoV-2*) (Ou et al., 2020) may also adopt the TPC-dependent mechanism to enter the cytoplasm of host cells through the endolysosomal system. Currently, our understanding of TPCs in bacterial infection is very limited. Emerging evidence suggests that the phagocytic uptake of live bacteria requires TPC1 and TPC2 in macrophages (Davis et al., 2020). Therefore, TPC proteins may be effective targets for antiviral and antibacterial therapy.

Emerging evidence has also linked TPCs to cancers including melanoma (Kocarnik et al., 2015; Kosiniak-Kamysz et al., 2014), breast cancer (Jahidin et al., 2016; Nguyen et al., 2017), and colorectal cancer (Faris et al., 2019). Mechanistically, Nguyen et al. (2017) suggest that disruption of TPC function abrogates cancer cell migration in vitro and metastasis in vivo, resulting from disturbed integrin trafficking in the endolysosomal system, reduced adhesion and reduced formation of the leading edge. NAADP-TPC2 may also regulate cancer progression by affecting VEGF-induced angiogenesis (Favia et al., 2014; Pafumi et al., 2017). These studies designate TPCs as promising targets for cancer treatment.

A link between NAADP-TPCs and diabetes has been long appreciated (Fan et al., 2016; Solberg Woods et al., 2012; Tsaih et al., 2014). In response to glucose uptake and metabolism, intracellular Ca^{2+} concentration increases in pancreatic β cells, which triggers insulin release and promotes the absorption of glucose from the blood into liver, fat, and skeletal muscle cells. It is suggested that, in addition to the activation of the voltage-gated calcium channel (VGCC) on the PM, glucose challenge also increases NAADP-dependent Ca^{2+} signals in pancreatic β cells (Masgrau et al., 2003). In agreement with this, interfering with NAADP-TPCs reduces glucose-evoked Ca^{2+} signals and insulin secretion both in vitro and in vivo (Arredouani et al., 2015; Masgrau et al., 2003). Paradoxically, although insulin response to a glucose challenge is decreased, fasting glucose levels in TPC2 knockout mice are also reduced (Tsaih et al., 2014). Additionally, animals with β-cell specific TPC2 knockout have normal glucose-evoked Ca^{2+} signals, insulin secretion, or glucose tolerance (Cane et al., 2016). One possibility to explain these conflicting results is that TPC knockout is compensated for in vivo, or that reduced glucagon secretion from α-cells of the pancreas offsets the effect of insulin on blood sugar levels because TPC2 knockout reduces glucagon secretion (Hamilton et al., 2018).

In addition to the role in infection, cancer, and diabetes, TPCs have also been associated with neurodegenerative diseases such as Parkinson's disease (Hockey et al., 2015) and NPC1 (Hoglinger et al., 2015), non-alcoholic fatty liver disease (NAFLD) (Grimm et al., 2014), ischemia (Davidson et al., 2015; Djerada et al., 2013; Khalaf and Babiker, 2016), mature-onset obesity (Lear et al., 2015), and pigmentation (Ambrosio et al., 2016; Bellono et al., 2016; Chao et al., 2017; Lin-Moshier et al., 2012; Sulem et al., 2008).

18.3.1.3 P2X4

The P2X4 receptor (encoded by the *P2RX4* gene) belongs to the family of purinoceptors that opens in response to ATP binding at the extracellular side (Khakh and North, 2012). It is a trimeric 2-TM channel permeable to both Na^+ and Ca^{2+} when activated by ATP (Figure 18.5). P2X4 is highly expressed in the PM of various tissues and involved in many cellular processes. Recent studies suggest that P2X4 receptors are stored in the lysosomal membrane and brought to the cell surface or to phagosomes in response to a variety of stimuli (Qureshi et al., 2007). Lysosomal P2X4 is activated by luminal ATP in a pH-dependent manner, i.e. it is minimally activated at acidic luminal pH, whereas lysosome alkalization dramatically increases its activity. Physiologically, P2X4 functions as a lysosomal Ca^{2+} channel which activation facilitates homotypic lysosome fusion using a CaM-

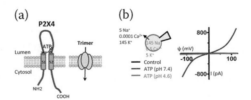

FIGURE 18.5 P2X4 receptors in lysosomes. (A) Schematic representation showing the membrane topology of P2X4 receptor which contains two transmembrane domains. P2X4 forms a trimer that is activated by intraluminal ATP. (B) Intraluminal pH dependent activation of P2X4 in the lysosome. A P2X4 current induced by luminal ATP (0.1 mM) in pH 7.4 Tyrode but not by luminal ATP (0.1 mM) in pH 4.6 Tyrode in a lysosome expressing rP2X4-GFP. The bath solution (cytosolic) contains (in mM) 145 K^+, 5 Na^+, 0.0001 Ca^{2+}, and pH 7.2. Pipette solution (luminal) contains (in mM) 145 Na^+, 5 K^+, 0.5 Ca^{2+}, 0.1 mM ATP, and pH 4.6 or pH 7.4.

dependent mechanism (Cao et al., 2015a). P2X4 is also expressed in lysosome-related vesicles such as Lamellar Bodies, large secretory lysosomes that store lung surfactant in alveolar type II epithelial cells and is inserted into the PM following lysosomal exocytosis. New evidence suggests that the activation of vesicular P2X4 receptors facilitates the secretion of pulmonary surfactant in pulmonary alveoli (Fois et al., 2018; Miklavc et al., 2011; Thompson et al., 2013).

18.3.1.4   Other Lysosomal Ca^{2+} and/or Na^+ Channels

Some other non-selective cation channels have been suggested to function in the lysosomal membrane. However, their activities in the lysosomal membrane have not been confirmed by direct lysosomal patch-clamp. These include TRPM2, TRPA1, and VGCC.

TRPM2 is a ubiquitously expressed ROS-sensitive nonselective cation channel in the TRP family (Nilius et al., 2007; Ramsey et al., 2006). TRPM2 in the PM has been involved in insulin secretion in pancreatic β-cells (Togashi et al., 2006) or immune response in monocytes (Yamamoto et al., 2008). Emerging evidence suggests that TRPM2 may also function as a lysosomal Ca^{2+}-release channel activated by intracellular ADP-ribose in dendritic cells (Sumoza-Toledo et al., 2011) and pancreatic β-cells (Lange et al., 2009). In dendritic cells it regulates cell maturation and chemotaxis (Sumoza-Toledo et al., 2011), whereas in β-cells it plays an important role in H_2O_2-induced β-cell death (Lange et al., 2009). Later studies suggest that TRPM2-mediated lysosomal Zn^{2+} release seems to play a primary role in cell functions. For example, the H_2O_2-mediated activation of TRPM2 channels increases migration of HeLa and prostate cancer (PC)-3 cells by releasing lysosomal Zn^{2+} but not Ca^{2+} to the leading edge of migrating cells, promoting filopodia formation (Li et al., 2016a). In pancreatic β-cell, death induced by the H_2O_2-mediated activation of TRPM2 channels is attributed to an increase in the cytosolic levels of both Ca^{2+} and Zn^{2+}. TRPM2-mediated Zn^{2+} release from lysosomes plays a primary role in ROS-induced β-cell death, whereas extracellular Ca^{2+} entry only potentiates lysosomal Zn^{2+} release through TRPM2 (Li and Jiang, 2019; Manna et al., 2015). Similarly, Aβ$_{42}$ may activate lysosomal ROS-sensitive TRPM2 channel, releasing lysosomal Zn^{2+} to induce neurotoxicity in mouse hippocampal neurons (Li and Jiang, 2018; Li et al., 2017b).

Pharmacological and immunocytochemical analyses revealed the presence of TRPA1 channels in lysosomes of dorsal root ganglion (DRG) neurons (Shang et al., 2016). By inducing Ca^{2+} release from lysosome-like organelles, TRPA1 activation triggers vesicle release in sensory neurons. Interestingly, Fabry disease, a common lysosomal storage disease caused by a deficiency of the lysosomal enzyme α-galactosidase, displays a severe neuropathic pain. This seems to involve TRPA1 because in Fabry rat sensory neurons TRPA1 is sensitized and TRPA1 antagonism reversed the behavioral mechanical sensitization (Miller et al., 2018). Therefore, lysosomal TRPA1 may have a role in pain sensation. However, this was later questioned by Gebhardt et al. (2020). Thus, further investigation is needed to clarify the function of TRPA1 in lysosomes.

In a forward mosaic screen in *Drosophila* designed to identify genes essential for neuronal function and maintenance, a *Drosophila* VGCC *cacophony (cac)* was identified in lysosomes where it regulates the fusion of autophagosomes with lysosomes. The role of VGCC in autophagosome-lysosome fusion is evolutionarily conserved, as the loss of either *cac or* the mouse homologues, CACNA1A and CACNA2D2 leads to autophagic defects in mice (Tian et al., 2015).

18.3.2 LYSOSOMAL K$^+$ CHANNELS

K$^+$ channels are widely distributed in living organisms. More than 80 members of K$^+$ channels have been identified in the PM, and they are classified into four major groups, i.e. voltage-gated K$^+$ channels (K$_v$), Ca^{2+}-activated K$^+$ channels (K$_{Ca}$), inwardly rectifying K$^+$ channels (K$_{ir}$), and tandem pore domain K$^+$ channels (K2P) (Wulff et al., 2009; Yellen, 2002). Compared to the K$^+$ channels in the PM, K$^+$-selective channels in intracellular organelles such as lysosomes are much less understood. By recording the lysosomal membrane, two proteins have been identified as K$^+$-selective ion channels in lysosomes. These include BK (Cao et al., 2015b; Wang et al., 2017) and TMEM175 (Cang et al., 2015).

18.3.2.1 BK

18.3.2.1.1 Characterization of Lysosomal BK

BK (encoded by the *KCNMA1* gene) is a ubiquitously expressed K$^+$ channel consisting of four pore-forming α-subunits. The BK channel α-subunit shares homology with all other voltage-sensitive K$^+$ channels containing six transmembrane segments (S1–S6). Uniquely, it has an additional transmembrane segment, S0, and thus the N-terminus is extracellular. Segments S1-S4 form the VSD, and the pore region is located between S5 and S6. The cytoplasmic C-terminal domain of BK channel encompasses two domains termed Regulators of Conductance for K$^+$ (RCK): the proximal RCK1 and the distal RCK2. Each RCK contains a high affinity Ca^{2+}-binding site. Compared with other K$^+$ channels, it is characterized by a large K$^+$ conductance with dual activation by membrane depolarization and elevated cytosolic Ca^{2+} (Berkefeld et al., 2006; Fakler and Adelman, 2008; Salkoff et al., 2006) (Figure 18.6A).

Interestingly, the BK channel is recently reported to be highly expressed in the lysosomal membrane (Cao et al., 2015b; Wang et al., 2017). The lysosomal localization of the BK channel is determined by two dileucine motifs, D(485)ACLI and D(731)PLLI, located in the RCK domain in the large cytoplasmic C-terminus (Cao et al., 2015b). In agreement with what was known for BK channel on the PM, an increase in the cytosolic Ca^{2+} concentration dramatically increases BK currents and shifts the voltage dependence of channel activation to more negative potentials (Cao et al., 2015b; Wang et al., 2017). This shift of voltage dependence makes BK easily activated in response to a local Ca^{2+} increase upon the opening of lysosomal Ca^{2+} release channels (Berkefeld et al., 2010; Fakler and Adelman, 2008; Salkoff et al., 2006). Like the BK channel in the PM, the lysosomal BK channel is inhibited by paxilline, quinidine, clofilium and iberiotoxin (Garrity et al., 2016; Wang et al., 2017) but activated by NS1619 (Bentzen et al., 2014; Olesen et al., 1994) and isopimaric acid (Garrity et al., 2016; Wang et al., 2017; Yamamura et al., 2001).

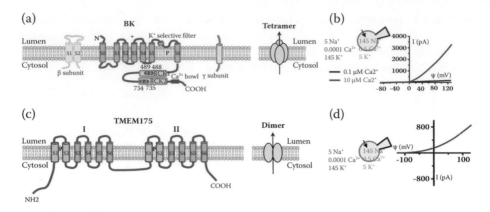

FIGURE 18.6 Lysosomal K^+ channels. (A) Topology and domain structure of human BK channel. BK has seven transmembrane segments, with a luminal N-terminus and a cytoplasmic C-terminus. The cytoplasmic domain of BK channel contains four primary binding sites for Ca^{2+}, called "Ca^{2+} bowls," the Ca^{2+}-sensing domain. There are two dileucine motifs, D(485)ACLI and D(731)PLLI, in the regulator of conductance for K^+ (RCK) domains of the large cytoplasmic C-terminus. BK channels form a tetramer that is regulated by auxiliary β or γ subunit. The arrow indicates K^+ influx into the lysosome. (B) BK currents are induced by Ca^{2+} (10 μM in cytosol) in an enlarged lysosome. The bath solution (cytosolic) contains (in mM) 145 K^+, 5 Na^+, 0.0001 or 0.01 Ca^{2+}, and pH 7.2. Pipette solution (luminal) contains (in mM) 145 Na^+, 5 K^+, 0.5 Ca^{2+}, 0.1 mM ATP, and pH 4.6. (C) Topology and domain structure of hTMEM175 channel. hTMEM175 has a two-repeat structure (I and II), with each repeat containing 6 TMs (S1–S6). IS1 and IIS1 are channel pore-lining helices. hTMEM175 proteins form a dimer. (D) TMEM175 forms a voltage-independent K^+ channel. A TEME175 current recorded using physiological asymmetric solutions. The bath solution (cytosolic) contains (in mM) 145 K^+, 5 Na^+, 0.0001 or 0.01 Ca^{2+}, and pH 7.2. Pipette solution (luminal) contains (in mM) 145 Na^+, 5 K^+, 0.5 Ca^{2+}, 0.1 mM ATP, and pH 4.6.

These data indicate that the BK channel in lysosomes functions in the same manner as its PM counterpart.

18.3.2.1.2 BK in Lysosome Functions

On excitable membranes, the BK channel is activated by Ca^{2+} influx through VGCCs. This leads to an efflux of K^+ from the cell and subsequent cell membrane hyperpolarization, prohibiting excessive Ca^{2+} influx and over-excitation (Berkefeld et al., 2006, 2010; Cui et al., 2009; Fakler and Adelman, 2008; Lee and Cui, 2010; Salkoff et al., 2006). Similarly, BK channel forms physical and functional coupling with the lysosomal Ca^{2+} release channel, TRPML1. In this coupling, Ca^{2+} release via TRPML1 activates BK, which in turn facilitates TRPML1-mediated lysosomal Ca^{2+} release by providing a counterion shunt to dissipate the transmembrane potential generated by TRPML1-mediated Ca^{2+} (Na^+) release (Cao et al., 2015b; Wang et al., 2017). Thus, TRPML1 and BK form a positive feedback loop to ensure efficient Ca^{2+} release. After being released, lysosomal Ca^{2+} stores can be refilled by ER via the presumed formation of ER-lysosome membrane contact sites (Atakpa et al., 2018;

Eden, 2016; Garrity et al., 2016; Haller et al., 1996; Wang et al., 2017). Recently, Wang et al. (2017) showed that genetic ablation or pharmacological inhibition of lysosomal BK, or abolition of its Ca^{2+} sensitivity, blocks lysosomal Ca^{2+} refilling. Therefore, lysosomal BK regulates intracellular Ca^{2+} signaling by controlling both the release and uptake of lysosomal Ca^{2+}. Because efficient Ca^{2+} mobilization via TRPML1 is important for lysosomal trafficking (Cheng et al., 2010; Venkatachalam et al., 2015; Xu and Ren, 2015), by affecting TRPML1-mediated lysosomal Ca^{2+} release, BK also regulates lysosomal membrane trafficking. BK deficiency induces enlarged lysosomes with the buildup of membranous and electron-dense inclusions and lipofuscin (Cao et al., 2015b), closely resembling those found in cells from ML-IV and many other LSDs (Lloyd-Evans and Platt, 2011). By modulating lysosomal Ca^{2+} and membrane potential (ψ_{lyso}, defined as $V_{cytosol}$ - V_{lumen}; V_{lumen} set at 0 mV) (Sun et al., 2011; Venkatachalam et al., 2008; Xu and Ren, 2015), BK is also necessary for proteolytic activity of lysosomes (Wang et al., 2017).

18.3.2.1.3 BK in Human Diseases

Impaired membrane trafficking is common in LSDs (Lloyd-Evans and Platt, 2011; Parenti et al., 2015; Xu and Ren, 2015). Because promoting membrane trafficking, particularly lysosomal exocytosis, represents a promising therapeutic approach for LSDs (Chen et al., 2014; Medina et al., 2011; Samie and Xu, 2014; Shen et al., 2012), by promoting TRPML1-mediated Ca^{2+} release and lysosomal trafficking, BK up-regulation mitigates defects in several LSD patients including NPC1, NPA, mild cases of ML-IV (i.e. TRPML1-F408Δ) and Fabry diseases (Cao et al., 2015b; Shen et al., 2012; Zhong et al., 2016). Therefore, enhancing BK function represents another plausible approach to treat some LSDs.

Lysosomal BK may also regulate the progression of many other disease conditions, particularly in those involving TRPML1. For example, by promoting TRPML1 function, BK up-regulation may help clear *H. pylori* (Capurro et al., 2019), prevent reactivation of latent HIV-1 infection and HIV-1 comorbidities (Bae et al., 2014; Khan et al., 2019), inhibit low-density lipoprotein-induced amyloidogenesis (Hui et al., 2019), and protect dopaminergic neurons from α-synuclein toxicity (Tsunemi et al., 2019). Because TRPML1 also plays a role in cancer progression (Jung et al., 2019; Kasitinon et al., 2019; Xu et al., 2019), severe combined immunodeficiency diseases (Zhong et al., 2017), large particle phagocytosis (Samie et al., 2013; Sun et al., 2020) and membrane repair (Cheng et al., 2014), lysosomal BK may also play a role in these conditions (Khaitan et al., 2009; Liu et al., 2002; Sun et al., 2020).

18.3.2.2 TMEM175

18.3.2.2.1 Characterization of Lysosomal TMEM175

Human TMEM175 (hTMEM175, encoded by the *TMEM175* gene) is widely expressed in all tissues. Using proteomic analyses of enriched lysosome preparation (Chapel et al., 2013; Saftig et al., 2010), TMEM175 was identified as an lysosomal membrane protein with unknown function. Most recent studies have suggested that TMEM175 acts as a K^+-selective ion channel expressed in the lysosomal membrane (Brunner et al., 2020; Cang et al., 2015; Lee et al., 2017; Oh et al., 2020). Mammalian TMEM175 is predicted to form a dimer. Unlike any of the ~80 K^+

channels in the PM, each pore-forming subunit of TMEM175 has two repeats of six-transmembrane-spanning segments (IS1-IS6 and IIS1-IIS6). However, the S4 segment of TMEM175 has no VSD characterized by the presence of charged residues (Figure 18.6B). The conserved TVGYG selectivity filter motif found in all the known K^+ channels (Doyle et al., 1998; Jiang et al., 2003; Long et al., 2005; MacKinnon and Miller, 1989; Papazian et al., 1987; Tao et al., 2017) is also lacking in TMEM175 (Cang et al., 2015). Structural analysis suggests that TMEM175 adopts a distinct structure and K^+ selectivity mechanism from classical K^+ channels (Brunner et al., 2020; Lee et al., 2017; Oh et al., 2020), in which all 6-TM helices are tightly packed within each subunit without undergoing domain swapping. Distinct from canonical K^+ channel, the highly conserved S1 helix but not the S6 forms the pore-lining inner helix at the center of the channel. Three layers of hydrophobic residues on the C-terminal half of the S1 helices form a bottleneck along the ion conduction pathway and serve as the selectivity filter of the channel, and the first layer of the highly conserved isoleucine residues in the filter is primarily responsible for channel selectivity (Long et al., 2005, 2007). Moreover, the conserved isoleucine residues in the center of the pore serve as both the gate and ion selectivity filter (Oh et al., 2020).

The properties of TMEM175 proteins were recently characterized by Cang et al. (2015). TMEM175s show little inactivation or rectification. Like many canonical K^+ channels, TMEM175s are Rb^+ permeable but minimally permeable to Na^+ and Ca^{2+}. Distinct from canonical K^+ channels, mammalian TMEM175s are not inhibited by Cs^+, Ba^{2+}, tetraethylammonium (TEA), or quinine at concentrations commonly used to block canonical K^+ channels. However, Zn^{2+} and 4-aminopyridine (4-AP) inhibit hTMEM175.

18.3.2.2.2 Functions of Lysosomal TMEM175

Physiologically TMEM175 acts as a K^+-permeable leak-like channel, setting resting ψ_{lyso}. Lysosomes lacking TMEM175 exhibit a considerably depolarized membrane potential, subsequently leading to accelerated fusion of autophagosomes to lysosomes (Cang et al., 2015). Because K^+ serves as a counterion to maintain lysosome acidification (Mindell, 2012; Steinberg et al., 2010), TMEM175 knockout compromises lysosomal pH stability and reduces proteolytic activity in lysosomes (Cang et al., 2015; Jinn et al., 2017).

18.3.2.2.3 Pathological Implications of Lysosomal TMEM175

Pathologically, several genome-wide association studies (GWAS) have linked TMEM175 to Parkinson's disease (Chang et al., 2017; Jinn et al., 2017, 2019; Krohn et al., 2020; Lill et al., 2015; Nalls et al., 2014). In line with this, neurons with TMEM175 deficiency are more susceptible to exogenous α-synuclein fibrils (Jinn et al., 2017), potentially due to unstable lysosomal pH and impaired autophagy-mediated clearance of damaged organelles. This is also in agreement with multiple lines of evidence that underscores the importance of the autophagy-lysosomal pathway in Parkinson's disease (Malik et al., 2019; Senkevich and Gan-Or, 2019; Tsunemi et al., 2019; Vidyadhara et al., 2019; Wallings et al., 2019a,b; Ysselstein et al., 2019). Therefore, TMEM175 may play a critical role in the pathogenesis of Parkinson's disease by regulating the autophagy-lysosomal pathway, highlighting TMEM175 as a potential therapeutic target for Parkinson's Disease (PD) treatment.

18.3.3 Lysosomal Cl Channels

Endosomes and lysosomes are highly enriched in Cl⁻ that may control their trafficking and degradative functions (Chakraborty et al., 2017; Jentsch and Pusch, 2018; Saha et al., 2015; Xu and Ren, 2015). A subgroup of ClC proteins (ClC3 through ClC7, encoded by *CLCN3–7* genes) resides in the intracellular membranes of the endolysosomal pathway. Due to their intracellular localization, biophysical characterization of the intracellular ClCs only becomes possible when they are heterologously expressed in the PM. All these channels in the PM yield outwardly rectifying anion currents [ClC3 (Guzman et al., 2015; Li et al., 2002), ClC4 (Friedrich et al., 1999; Mohammad-Panah et al., 2003), ClC5 (Friedrich et al., 1999; Steinmeyer et al., 1995), ClC6 (Buyse et al., 1997; Neagoe et al., 2010), and ClC7 (Leisle et al., 2011)] that are inhibited by low extracellular pH (Friedrich et al., 1999; Leisle et al., 2011). However, the electrophysiological properties and physiological roles of endolysosomal ClCs are far from being understood. They were initially believed to be Cl⁻ channels like the other four mammalian ClC channels in the PM (i.e. ClC1, ClC2, ClCKa, and ClCKb). However, later studies suggest that ClC3-ClC7 are coupled $2Cl^-/1H^+$ antiporters with an anion conductance sequence of $Cl^- > Br^- > I^-$ (Chadda et al., 2016; Graves et al., 2008; Jentsch and Pusch, 2018; Leisle et al., 2011; Picollo and Pusch, 2005; Scheel et al., 2005; Weinert et al., 2010). They all have a conserved 'gating glutamate' that participate in coupling H^+ to Cl⁻ transport. Neutralization of the 'gating glutamate' results in uncoupled Cl⁻ passive conductance and eliminates voltage dependence (Jentsch and Pusch, 2018; Leisle et al., 2011; Novarino 2010; Picollo and Pusch, 2005; Scheel et al., 2005). Endolysosomal ClC proteins form homo- or heteromeric dimers, in which each monomer contains an ion conductance pathway. The physiological roles of endolysosomal ClCs are believed to provide countercurrents to allow efficient endolysosomal acidification by the V-ATPase (Jentsch and Pusch, 2018).

Among intracellular ClC channels, only ClC5 and ClC7 have been associated with human diseases. ClC5 is highly expressed in proximal tubular cells of the kidney where it regulates the uptake of albumin and low-molecular-weight proteins. Dysfunction of ClC5 leads to Dent disease, a rare X-linked recessive nephropathy characterized by excessive urinary loss of small proteins, increased levels of Ca^{2+} in the urine, kidney calcifications and kidney stones (Fisher et al., 1994; Steinmeyer et al., 1995; Wrong et al., 1994). Subcellularly, ClC5 is located in early endosomes of proximal tubular cells where it plays a role in endocytosis by supporting efficient endosomal acidification by a mechanism independent of endosomal acidification (Novarino et al., 2010; Smith and Lippiat, 2010).

ClC7 is the only lysosomal ClC protein (Graves et al., 2008; Kasper et al., 2005; Kornak et al., 2001; Wartosch et al., 2009) whose stability and activity require a accessory β-subunit, osteoclastogenesis asssociated transmembrane protein 1 (OSTM1) (Lange et al., 2006). Its lysosomal targeting is determined by N-terminal leucine-based sorting motifs (Stauber and Jentsch, 2010). Lysosomal CLC7 is believed to provide a shunting conductance to allow efficient lysosomal acidification by the V-ATPase (Graves et al., 2008). However, later studies indicate that lysosomal pH is unchanged in the absence of CLC7 (Kasper et al., 2005; Lange et al., 2006; Weinert et al., 2010, 2014). This is in agreement with the finding by Steinberg et al. (2010) showing that cations but not anions support lysosomal acidification. It is likely that the major role of ClC7 (and other endolysosomal CLCs) is to use the pH gradient created by the V-ATPase to increase the luminal Cl^- concentration (Graves et al., 2008; Jentsch, 2007; Stauber and Jentsch, 2013; Weinert et al., 2010) that is important for the activity of lysosomal enzymes (Chakraborty et al., 2017; Wartosch et al., 2009). ClC7 is also expressed in the acid-secreting ruffled border of bone-resorbing osteoclasts where it may shunt H^+-ATPase currents (Kornak et al., 2001) and regulate osteoclast-mediated extracellular acidification and bone regeneration (Kasper et al., 2005; Kornak et al., 2001; Pangrazio et al., 2010). ClC7 mutations lead to osteopetrosis (Kornak et al., 2001; Pangrazio et al., 2010), lysosomal storage disease and neurological disorders (Chakraborty et al., 2017; Kasper et al., 2005; Park et al., 2019; Wartosch et al., 2009; Weinert et al., 2010).

Other three endolysosomal ClC channels, i.e. ClC3, ClC4, and ClC6, are less understood. Although their physiological functions are largely uncertain and they have not been linked directly to any human diseases, cell models and mouse models have suggested that they play important roles in endolysosomal physiology. ClC3 is widely expressed in most tissues (Kawasaki et al., 1994; Stobrawa et al., 2001), and it mainly resides in endosomes (Suzuki et al., 2006) and synaptic vesicles (Stobrawa et al., 2001). ClC3 deletion in mice results in neurodegeneration in the retina and brain (Dickerson et al., 2002; Stobrawa et al., 2001), with mild lysosomal storage (Yoshikawa et al., 2002). In supporting its role in acidification of intracellular compartments, acidification and Cl^- accumulation were reduced in early and late endosomes in ClC3 knockout mice (Hara-Chikuma et al., 2005). ClC4 is broadly expressed in various tissues with species-specific expression patterns (Jentsch and Pusch, 2018; Mohammad-Panah et al., 2003). It is predominantly localized at the endosomal membrane, and a small portion heterologous ClC4 is also found in the PM (Mohammad-Panah et al., 2003; Suzuki et al., 2006). However, the loss of ClC4 in mice does not show defects in proximal tubular endocytosis (Rickheit et al., 2010). ClC6 protein is almost exclusively expressed in neurons of the central and peripheral nervous systems, with a particularly high expression in dorsal root ganglia. It mainly localizes at the membranes of late endosomes and lysosomes (Poet et al., 2006). ClC6 ablation in mice causes reduced pain sensitivity and moderate behavioral abnormalities. Late in life, ClC6 mutant mice display lysosomal storage at proximal axons without affecting lysosomal steady-state pH (Poet et al., 2006).

18.4 SUMMARY AND PERSPECTIVES

Compared with mitochondria and ERs, lysosomes only occupy a small fraction of the intracellular space in many cell types (Alberts et al., 2014; Holtzman, 1989; Luzio et al., 2007; Xu and Ren, 2015). However, they play an essential role in cell functions by altering their position, number, size, composition, activity, and inter-action with other organelles upon environmental changes. Like the PM, the lyso-somal membrane also contains many ion channels that control almost all lysosomal functions. However, our knowledge about lysosomal ion channels is still sparse. Currently, only eight proteins have been definitely established as lysosomal ion channels. These ion channels not only regulate the intralysosomal environment to meet specific needs such as lysosomal degradation and membrane trafficking in the endocytic, phagocytic, and autophagic pathways but also control the cytosolic en-vironment to activate a variety of intracellular signaling pathways such as Ca^{2+} signaling and nutrient sensing pathways, thereby maintaining cellular homeostasis. Despite some progress in our understanding of lysosomal ion channels, many questions remain to be addressed in more detail. For example, how are they regulated by environmental factors and cellular cues? How do they regulate cell functions and organellular homeostasis? Do they function in lysosome-related or-ganelles in specialized cells, and if so how? Do they contribute to the commu-nication of lysosomes with other cellular structures and if so how? How are they involved in human diseases?

In addition to the known ion channels, previous studies have suggested various unidentified conductances on the lysosomal membrane that are not mediated by the aforementioned channels. These include a H^+ "leak" conductance (Cang et al., 2013, 2014b), several Cl^- conductances (Cang et al., 2014b; Li et al., 2019), and presumably water channels and osmo-sensitive channels. Therefore, another future direction would be to identify molecular determinants responsible for these con-ductances. Because lysosomes exchange membranes with the PM via both en-docytic and exocytic pathways and because lysosomes interact with other organelles, some ion channels in the PM and other organelles may also reside in the lysosomal membrane and play a role in lysosomes. For example, in addition to P2X4 and BK which have been found in both the PM and the lysosomal membrane, the recently identified H^+-activated Cl^- channel Proton-activated chloride channel (PAC) could also be expressed in the endolysosome to regulate organellular functions (Ullrich et al., 2019; Yang et al., 2019).

Since dys-homeostasis of lysosomal ions and deficiency in lysosomal ion channels have been implicated in a variety of human diseases such as LSD, classical neurodegenerative disease, bacterial and viral infection, and cancer, developing therapeutic strategies targeting on lysosomal ion channels could be a direction for drug discovery. As our understanding of lysosomal ion channel deepens, we are expecting that some patients with these diseases will benefit from these studies in near the future.

ACKNOWLEDGMENTS

We apologize to colleagues whose works are not cited due to space limitations. This work was supported by Shanghai Municipal Health Commission grant (201740161) to P.H. and CIHR project grant (PJT-156102) to X.D.

REFERENCES

Alberts B.J.A., Lewis J., Morgan D., Raff M., Roberts K., and Walter P. (2014). Molecular Biology of the Cell, Sixth edn (Garland Science).

Almamun, M., Levinson, B.T., van Swaay, A.C., Johnson, N.T., McKay, S.D., Arthur, G.L., Davis, J.W., and Taylor, K.H. (2015). Integrated methylome and transcriptome analysis reveals novel regulatory elements in pediatric acute lymphoblastic leukemia. Epigenetics *10*, 882–890.

Ambrosio, A.L., Boyle, J.A., Aradi, A.E., Christian, K.A., and Di Pietro, S.M. (2016). TPC2 controls pigmentation by regulating melanosome pH and size. Proceedings of the National Academy of Sciences of the United States of America *113*, 5622–5627.

Arantes, R.M., and Andrews, N.W. (2006). A role for synaptotagmin VII-regulated exocytosis of lysosomes in neurite outgrowth from primary sympathetic neurons. The Journal of Neuroscience: The Official Journal of the Society for Neuroscience *26*, 4630–4637.

Arredouani, A., Ruas, M., Collins, S.C., Parkesh, R., Clough, F., Pillinger, T., Coltart, G., Rietdorf, K., Royle, A., Johnson, P., *et al.* (2015). Nicotinic acid adenine dinucleotide phosphate (NAADP) and endolysosomal two-pore channels modulate membrane excitability and stimulus-secretion coupling in mouse pancreatic beta cells. The Journal of Biological Chemistry *290*, 21376–21392.

Atakpa, P., Thillaiappan, N.B., Mataragka, S., Prole, D.L., and Taylor, C.W. (2018). IP3 receptors preferentially associate with ER-lysosome contact sites and selectively deliver Ca^{2+} to lysosomes. Cell Reports *25*, 3180–3193 e3187.

Bach, G., Chen, C.S., and Pagano, R.E. (1999). Elevated lysosomal pH in Mucolipidosis type IV cells. Clinica Chimica Acta; International Journal of Clinical Chemistry *280*, 173–179.

Bae, M., Patel, N., Xu, H., Lee, M., Tominaga-Yamanaka, K., Nath, A., Geiger, J., Gorospe, M., Mattson, M.P., and Haughey, N.J. (2014). Activation of TRPML1 clears intraneuronal Abeta in preclinical models of HIV infection. The Journal of Neuroscience: The Official Journal of the Society for Neuroscience *34*, 11485–11503.

Bassi, M.T., Manzoni, M., Monti, E., Pizzo, M.T., Ballabio, A., and Borsani, G. (2000). Cloning of the gene encoding a novel integral membrane protein, mucolipidin-and identification of the two major founder mutations causing mucolipidosis type IV. American Journal of Human Genetics *67*, 1110–1120.

Bellono, N.W., Escobar, I.E., and Oancea, E. (2016). A melanosomal two-pore sodium channel regulates pigmentation. Scientific Reports *6*, 26570.

Bentzen, B.H., Olesen, S.P., Ronn, L.C., and Grunnet, M. (2014). BK channel activators and their therapeutic perspectives. Frontiers in Physiology *5*, 389.

Berkefeld, H., Fakler, B., and Schulte, U. (2010). Ca^{2+}-activated K^+ channels: from protein complexes to function. Physiological Reviews *90*, 1437–1459.

Berkefeld, H., Sailer, C.A., Bildl, W., Rohde, V., Thumfart, J.O., Eble, S., Klugbauer, N., Reisinger, E., Bischofberger, J., Oliver, D., *et al.* (2006). BKCa-Cav channel complexes mediate rapid and localized Ca^{2+}-activated K^+ signaling. Science *314*, 615–620.

Blaby-Haas, C.E., and Merchant, S.S. (2014). Lysosome-related organelles as mediators of metal homeostasis. The Journal of Biological Chemistry *289*, 28129–28136.

Boccaccio, A., Scholz-Starke, J., Hamamoto, S., Larisch, N., Festa, M., Gutla, P.V., Costa, A., Dietrich, P., Uozumi, N., and Carpaneto, A. (2014). The phosphoinositide PI(3,5)P (2) mediates activation of mammalian but not plant TPC proteins: functional expression of endolysosomal channels in yeast and plant cells. Cellular and Molecular Life Sciences: CMLS *71*, 4275–4283.

Brailoiu, E., Churamani, D., Cai, X., Schrlau, M.G., Brailoiu, G.C., Gao, X., Hooper, R., Boulware, M.J., Dun, N.J., Marchant, J.S., *et al.* (2009). Essential requirement for two-pore channel 1 in NAADP-mediated calcium signaling. The Journal of Cell Biology *186*, 201–209.

Brailoiu, E., Rahman, T., Churamani, D., Prole, D.L., Brailoiu, G.C., Hooper, R., Taylor, C.W., and Patel, S. (2010a). An NAADP-gated two-pore channel targeted to the plasma membrane uncouples triggering from amplifying Ca^{2+} signals. The Journal of Biological Chemistry *285*, 38511–38516.

Brailoiu, G.C., Gurzu, B., Gao, X., Parkesh, R., Aley, P.K., Trifa, D.I., Galione, A., Dun, N.J., Madesh, M., Patel, S., et al. (2010b). Acidic NAADP-sensitive calcium stores in the endothelium: agonist-specific recruitment and role in regulating blood pressure. The Journal of Biological Chemistry *285*, 37133–37137.

Brunner, J.D., Jakob, R.P., Schulze, T., Neldner, Y., Moroni, A., Thiel, G., Maier, T., and Schenck, S. (2020). Structural basis for ion selectivity in TMEM175 K^+ channels. eLife *9*, e53683.

Bretou, M., Saez, P.J., Sanseau, D., Maurin, M., Lankar, D., Chabaud, M., Spampanato, C., Malbec, O., Barbier, L., Muallem, S., et al. (2017). Lysosome signaling controls the migration of dendritic cells. Sci Immunol *2*, eaak9573.

Buyse, G., Voets, T., Tytgat, J., De Greef, C., Droogmans, G., Nilius, B., and Eggermont, J. (1997). Expression of human pICln and ClC-6 in Xenopus oocytes induces an identical endogenous chloride conductance. The Journal of Biological Chemistry *272*, 3615–3621.

Cai, X., and Patel, S. (2010). Degeneration of an intracellular ion channel in the primate lineage by relaxation of selective constraints. Mol Biol Evol *27*, 2352–2359.

Calcraft, P.J., Ruas, M., Pan, Z., Cheng, X., Arredouani, A., Hao, X., Tang, J., Rietdorf, K., Teboul, L., Chuang, K.T., et al. (2009). NAADP mobilizes calcium from acidic organelles through two-pore channels. Nature *459*, 596–600.

Campbell, E.M., and Fares, H. (2010). Roles of CUP-5, the *Caenorhabditis elegans* orthologue of human TRPML1, in lysosome and gut granule biogenesis. BMC Cell Biology *11*, 40.

Cancela, J.M., Churchill, G.C., and Galione, A. (1999). Coordination of agonist-induced Ca^{2+}-signalling patterns by NAADP in pancreatic acinar cells. Nature *398*, 74–76.

Cane, M.C., Parrington, J., Rorsman, P., Galione, A., and Rutter, G.A. (2016). The two pore channel TPC2 is dispensable in pancreatic beta-cells for normal Ca^{2+} dynamics and insulin secretion. Cell Calcium *59*, 32–40.

Cang, C., Aranda, K., and Ren, D. (2014a). A non-inactivating high-voltage-activated two-pore Na^+ channel that supports ultra-long action potentials and membrane bistability. Nature Communications *5*, 5015.

Cang, C., Bekele, B., and Ren, D. (2014b). The voltage-gated sodium channel TPC1 confers endolysosomal excitability. Nature Chemical Biology *10*, 463–469.

Cang, C., Aranda, K., Seo, Y.J., Gasnier, B., and Ren, D. (2015). TMEM175 is an organelle K^+ channel regulating lysosomal function. Cell *162*, 1101–1112.

Cang, C., Zhou, Y., Navarro, B., Seo, Y.J., Aranda, K., Shi, L., Battaglia-Hsu, S., Nissim, I., Clapham, D.E., and Ren, D. (2013). mTOR regulates lysosomal ATP-sensitive two-pore Na^+ channels to adapt to metabolic state. Cell *152*, 778–790.

Cao, Q., Yang, Y., Zhong, X.Z., and Dong, X.P. (2017). The lysosomal Ca^{2+} release channel TRPML1 regulates lysosome size by activating calmodulin. The Journal of Biological Chemistry *292*, 8424–8435.

Cao, Q., Zhao, K., Zhong, X.Z., Zou, Y., Yu, H., Huang, P., Xu, T.L., and Dong, X.P. (2014). SLC17A9 protein functions as a lysosomal ATP transporter and regulates cell viability. The Journal of Biological Chemistry *289*, 23189–23199.

Cao, Q., Zhong, X.Z., Zou, Y., Murrell-Lagnado, R., Zhu, M.X., and Dong, X.P. (2015a). Calcium release through P2X4 activates calmodulin to promote endolysosomal membrane fusion. The Journal of Cell Biology *209*, 879–894.

Cao, Q., Zhong, X.Z., Zou, Y., Zhang, Z., Toro, L., and Dong, X.P. (2015b). BK channels alleviate lysosomal storage diseases by providing positive feedback regulation of lysosomal Ca^{2+} release. Developmental Cell *33*, 427–441.

Capurro, M.I., Greenfield, L.K., Prashar, A., Xia, S., Abdullah, M., Wong, H., Zhong, X.Z., Bertaux-Skeirik, N., Chakrabarti, J., Siddiqui, I., *et al*. (2019). VacA generates a protective intracellular reservoir for *Helicobacter pylori* that is eliminated by activation of the lysosomal calcium channel TRPML1. Nature Microbiology *4*, 1411–1423.

Castiglioni, A.J., Remis, N.N., Flores, E.N., and Garcia-Anoveros, J. (2011). Expression and vesicular localization of mouse Trpml3 in stria vascularis, hair cells, and vomeronasal and olfactory receptor neurons. The Journal of Comparative Neurology *519*, 1095–1114.

Castonguay, J., Orth, J.H.C., Muller, T., Sleman, F., Grimm, C., Wahl-Schott, C., Biel, M., Mallmann, R.T., Bildl, W., Schulte, U., *et al*. (2017). The two-pore channel TPC1 is required for efficient protein processing through early and recycling endosomes. Scientific Reports *7*, 10038.

Chadda, R., Krishnamani, V., Mersch, K., Wong, J., Brimberry, M., Chadda, A., Kolmakova-Partensky, L., Friedman, L.J., Gelles, J., and Robertson, J.L. (2016). The dimerization equilibrium of a ClC Cl^-/H^+ antiporter in lipid bilayers. eLife *5*, e17438.

Chakraborty, K., Leung, K., and Krishnan, Y. (2017). High lumenal chloride in the lysosome is critical for lysosome function. eLife *6*, e28862.

Chang, D., Nalls, M.A., Hallgrimsdottir, I.B., Hunkapiller, J., van der Brug, M., Cai, F., Kerchner, G.A., Ayalon, G., Bingol, B., Sheng, M., *et al*. (2017). A meta-analysis of genome-wide association studies identifies 17 new Parkinson's disease risk loci. Nature Genetics *49*, 1511–1516.

Chao, Y.K., Schludi, V., Chen, C.C., Butz, E., Nguyen, O.N.P., Muller, M., Kruger, J., Kammerbauer, C., Ben-Johny, M., Vollmar, A.M., *et al*. (2017). TPC2 polymorphisms associated with a hair pigmentation phenotype in humans result in gain of channel function by independent mechanisms. Proceedings of the National Academy of Sciences of the United States of America *114*, E8595–E8602.

Chapel, A., Kieffer-Jaquinod, S., Sagne, C., Verdon, Q., Ivaldi, C., Mellal, M., Thirion, J., Jadot, M., Bruley, C., Garin, J., *et al*. (2013). An extended proteome map of the lysosomal membrane reveals novel potential transporters. Molecular and Cellular Proteomics *12*, 1572–1588.

Chen, C.C., Butz, E.S., Chao, Y.K., Grishchuk, Y., Becker, L., Heller, S., Slaugenhaupt, S.A., Biel, M., Wahl-Schott, C., and Grimm, C. (2017). Small molecules for early endosome-specific patch clamping. Cell Chemical Biology *24*, 907–916 e904.

Chen, C.C., Keller, M., Hess, M., Schiffmann, R., Urban, N., Wolfgardt, A., Schaefer, M., Bracher, F., Biel, M., Wahl-Schott, C., *et al*. (2014). A small molecule restores function to TRPML1 mutant isoforms responsible for mucolipidosis type IV. Nature Communications *5*, 4681.

Chen, C.S., Bach, G., and Pagano, R.E. (1998). Abnormal transport along the lysosomal pathway in mucolipidosis, type IV disease. Proceedings of the National Academy of Sciences of the United States of America *95*, 6373–6378.

Cheng, X., Shen, D., Samie, M., and Xu, H. (2010). Mucolipins: intracellular TRPML1-3 channels. FEBS Letters *584*, 2013–2021.

Cheng, X., Zhang, X., Gao, Q., Ali Samie, M., Azar, M., Tsang, W.L., Dong, L., Sahoo, N., Li, X., Zhuo, Y., *et al*. (2014). The intracellular Ca^{2+} channel MCOLN1 is required for sarcolemma repair to prevent muscular dystrophy. Nature Medicine *20*, 1187–1192.

Choi, S., and Kim, H.J. (2014). The Ca^{2+} channel TRPML3 specifically interacts with the mammalian ATG8 homologue GATE16 to regulate autophagy. Biochemical and Biophysical Research Communications *443*, 56–61.

Christensen, K.A., Myers, J.T., and Swanson, J.A. (2002). pH-dependent regulation of lysosomal calcium in macrophages. Journal of Cell Science *115*, 599–607.

Churchill, G.C., and Galione, A. (2001). NAADP induces Ca^{2+} oscillations via a two-pool mechanism by priming IP3- and cADPR-sensitive Ca^{2+} stores. The EMBO Journal *20*, 2666–2671.

Coen, K., Flannagan, R.S., Baron, S., Carraro-Lacroix, L.R., Wang, D., Vermeire, W., Michiels, C., Munck, S., Baert, V., Sugita, S., *et al.* (2012). Lysosomal calcium homeostasis defects, not proton pump defects, cause endo-lysosomal dysfunction in PSEN-deficient cells. The Journal of Cell Biology *198*, 23–35.

Cosker, F., Cheviron, N., Yamasaki, M., Menteyne, A., Lund, F.E., Moutin, M.J., Galione, A., and Cancela, J.M. (2010). The ecto-enzyme CD38 is a nicotinic acid adenine dinucleotide phosphate (NAADP) synthase that couples receptor activation to Ca^{2+} mobilization from lysosomes in pancreatic acinar cells. The Journal of Biological Chemistry *285*, 38251–38259.

Cuajungco, M.P., Grimm, C., and Heller, S. (2007). TRP channels as candidates for hearing and balance abnormalities in vertebrates. Biochimica et Biophysica Acta *1772*, 1022–1027.

Cuajungco, M.P., and Samie, M.A. (2008). The varitint-waddler mouse phenotypes and the TRPML3 ion channel mutation: cause and consequence. Pflugers Archiv: European Journal of Physiology *457*, 463–473.

Cuajungco, M.P., Silva, J., Habibi, A., and Valadez, J.A. (2016). The mucolipin-2 (TRPML2) ion channel: a tissue-specific protein crucial to normal cell function. Pflugers Archiv: European Journal of Physiology *468*, 177–192.

Cui, J., Yang, H., and Lee, U.S. (2009). Molecular mechanisms of BK channel activation. Cellular and Molecular Life Sciences: CMLS *66*, 852–875.

Cui, W., Sathyanarayan, A., Lopresti, M., Aghajan, M., Chen, C., and Mashek, D.G. (2020). Lipophagy-derived fatty acids undergo extracellular efflux via lysosomal exocytosis. Autophagy, 1–16.

Curcio-Morelli, C., Charles, F.A., Micsenyi, M.C., Cao, Y., Venugopal, B., Browning, M.F., Dobrenis, K., Cotman, S.L., Walkley, S.U., and Slaugenhaupt, S.A. (2010a). Macroautophagy is defective in mucolipin-1-deficient mouse neurons. Neurobiology of Disease *40*, 370–377.

Curcio-Morelli, C., Zhang, P., Venugopal, B., Charles, F.A., Browning, M.F., Cantiello, H.F., and Slaugenhaupt, S.A. (2010b). Functional multimerization of mucolipin channel proteins. Journal of Cellular Physiology *222*, 328–335.

Davidson, S.M., Foote, K., Kunuthur, S., Gosain, R., Tan, N., Tyser, R., Zhao, Y.J., Graeff, R., Ganesan, A., Duchen, M.R., *et al.* (2015). Inhibition of NAADP signalling on reperfusion protects the heart by preventing lethal calcium oscillations via two-pore channel 1 and opening of the mitochondrial permeability transition pore. Cardiovascular Research *108*, 357–366.

Davis, L.C., Morgan, A.J., and Galione, A. (2020). NAADP-regulated two-pore channels drive phagocytosis through endo-lysosomal Ca^{2+} nanodomains, calcineurin and dynamin. The EMBO Journal *39*, e104058.

Dayam, R.M., Saric, A., Shilliday, R.E., and Botelho, R.J. (2015). The phosphoinositide-gated lysosomal Ca^{2+} channel, TRPML1, is required for phagosome maturation. Traffic *16*, 1010–1026.

Dayam, R.M., Sun, C.X., Choy, C.H., Mancuso, G., Glogauer, M., and Botelho, R.J. (2017). The lipid kinase PIKfyve coordinates the neutrophil immune response through the activation of the Rac GTPase. Journal of Immunology *199*, 2096–2105.

De Leo, M.G., Staiano, L., Vicinanza, M., Luciani, A., Carissimo, A., Mutarelli, M., Di Campli, A., Polishchuk, E., Di Tullio, G., Morra, V., *et al.* (2016). Autophagosome-lysosome fusion triggers a lysosomal response mediated by TLR9 and controlled by OCRL. Nature Cell Biology *18*, 839–850.

Di Palma, F., Belyantseva, I.A., Kim, H.J., Vogt, T.F., Kachar, B., and Noben-Trauth, K. (2002). Mutations in Mcoln3 associated with deafness and pigmentation defects in varitint-waddler (Va) mice. Proceedings of the National Academy of Sciences of the United States of America *99*, 14994–14999.

Dickerson, L.W., Bonthius, D.J., Schutte, B.C., Yang, B., Barna, T.J., Bailey, M.C., Nehrke, K., Williamson, R.A., and Lamb, F.S. (2002). Altered GABAergic function accompanies hippocampal degeneration in mice lacking ClC-3 voltage-gated chloride channels. Brain Research *958*, 227–250.

Dickinson, M.S., Myasnikov, A., Eriksen, J., Poweleit, N., and Stroud, R.M. (2020). Resting state structure of the hyperdepolarization activated two-pore channel 3. Proceedings of the National Academy of Sciences of the United States of America *117*, 1988–1993.

Djerada, Z., Peyret, H., Dukic, S., and Millart, H. (2013). Extracellular NAADP affords cardioprotection against ischemia and reperfusion injury and involves the P2Y11-like receptor. Biochemical and Biophysical Research Communications *434*, 428–433.

Dong, X.P., Cheng, X., Mills, E., Delling, M., Wang, F., Kurz, T., and Xu, H. (2008). The type IV mucolipidosis-associated protein TRPML1 is an endolysosomal iron release channel. Nature *455*, 992–996.

Dong, X.P., Shen, D., Wang, X., Dawson, T., Li, X., Zhang, Q., Cheng, X., Zhang, Y., Weisman, L.S., Delling, M., *et al.* (2010). PI(3,5)P(2) controls membrane trafficking by direct activation of mucolipin Ca^{2+} release channels in the endolysosome. Nature Communications *1*, 38.

Dong, X.P., Wang, X., Shen, D., Chen, S., Liu, M., Wang, Y., Mills, E., Cheng, X., Delling, M., and Xu, H. (2009). Activating mutations of the TRPML1 channel revealed by proline-scanning mutagenesis. The Journal of Biological Chemistry *284*, 32040–32052.

Doyle, D.A., Morais Cabral, J., Pfuetzner, R.A., Kuo, A., Gulbis, J.M., Cohen, S.L., Chait, B.T., and MacKinnon, R. (1998). The structure of the potassium channel: molecular basis of K^+ conduction and selectivity. Science *280*, 69–77.

Eden, E.R. (2016). The formation and function of ER-endosome membrane contact sites. Biochimica et Biophysica Acta *1861*, 874–879.

Efe, J.A., Botelho, R.J., and Emr, S.D. (2005). The Fab1 phosphatidylinositol kinase pathway in the regulation of vacuole morphology. Current Opinion in Cell Biology *17*, 402–408.

Eichelsdoerfer, J.L., Evans, J.A., Slaugenhaupt, S.A., and Cuajungco, M.P. (2010). Zinc dyshomeostasis is linked with the loss of mucolipidosis IV-associated TRPML1 ion channel. The Journal of Biological Chemistry *285*, 34304–34308.

Esposito, B., Gambara, G., Lewis, A.M., Palombi, F., D'Alessio, A., Taylor, L.X., Genazzani, A.A., Ziparo, E., Galione, A., Churchill, G.C., *et al.* (2011). NAADP links histamine H1 receptors to secretion of von Willebrand factor in human endothelial cells. Blood *117*, 4968–4977.

Fakler, B., and Adelman, J.P. (2008). Control of K(Ca) channels by calcium nano/microdomains. Neuron *59*, 873–881.

Fan, Y., Li, X., Zhang, Y., Fan, X., Zhang, N., Zheng, H., Song, Y., Shen, C., Shen, J., Ren, F., *et al.* (2016). Genetic variants of TPCN2 associated with Type 2 diabetes risk in the Chinese population. PLoS One *11*, e0149614.

Fares, H., and Greenwald, I. (2001). Regulation of endocytosis by CUP-5, the *Caenorhabditis elegans* mucolipin-1 homolog. Nature Genetics *28*, 64–68.

Faris, P., Pellavio, G., Ferulli, F., Di Nezza, F., Shekha, M., Lim, D., Maestri, M., Guerra, G., Ambrosone, L., Pedrazzoli, P., *et al.* (2019). Nicotinic acid adenine dinucleotide phosphate (NAADP) induces intracellular Ca^{2+} release through the two-pore channel TPC1 in metastatic colorectal cancer cells. Cancers *11*, 542.

Favia, A., Desideri, M., Gambara, G., D'Alessio, A., Ruas, M., Esposito, B., Del Bufalo, D., Parrington, J., Ziparo, E., Palombi, F., et al. (2014). VEGF-induced neoangiogenesis is mediated by NAADP and two-pore channel-2-dependent Ca^{2+} signaling. Proceedings of the National Academy of Sciences of the United States of America 111, E4706–E4715.

Fehrenbacher, N., and Jaattela, M. (2005). Lysosomes as targets for cancer therapy. Cancer Research 65, 2993–2995.

Fisher, S.E., Black, G.C., Lloyd, S.E., Hatchwell, E., Wrong, O., Thakker, R.V., and Craig, I.W. (1994). Isolation and partial characterization of a chloride channel gene which is expressed in kidney and is a candidate for Dent's disease (an X-linked hereditary nephrolithiasis). Human Molecular Genetics 3, 2053–2059.

Fois, G., Winkelmann, V.E., Bareis, L., Staudenmaier, L., Hecht, E., Ziller, C., Ehinger, K., Schymeinsky, J., Kranz, C., and Frick, M. (2018). ATP is stored in lamellar bodies to activate vesicular P2X4 in an autocrine fashion upon exocytosis. The Journal of General Physiology 150, 277–291.

Friedrich, T., Breiderhoff, T., and Jentsch, T.J. (1999). Mutational analysis demonstrates that ClC-4 and ClC-5 directly mediate plasma membrane currents. The Journal of Biological Chemistry 274, 896–902.

Fulda, S., and Kogel, D. (2015). Cell death by autophagy: emerging molecular mechanisms and implications for cancer therapy. Oncogene 34, 5105-13.

Funk, K.E., and Kuret, J. (2012). Lysosomal fusion dysfunction as a unifying hypothesis for Alzheimer's disease pathology. International Journal of Alzheimer's Disease 2012, 752894.

Furuichi, T., Cunningham, K.W., and Muto, S. (2001). A putative two pore channel AtTPC1 mediates Ca^{2+} flux in Arabidopsis leaf cells. Plant and Cell Physiology 42, 900–905.

Garcia-Rua, V., Feijoo-Bandin, S., Rodriguez-Penas, D., Mosquera-Leal, A., Abu-Assi, E., Beiras, A., Maria Seoane, L., Lear, P., Parrington, J., Portoles, M., et al. (2016). Endolysosomal two-pore channels regulate autophagy in cardiomyocytes. The Journal of Physiology 594, 3061–3077.

Garrity, A.G., Wang, W., Collier, C.M., Levey, S.A., Gao, Q., and Xu, H. (2016). The endoplasmic reticulum, not the pH gradient, drives calcium refilling of lysosomes. eLife 5, e15887.

Gebhardt, L.A., Kichko, T.I., Fischer, M.J.M., and Reeh, P.W. (2020). TRPA1-dependent calcium transients and CGRP release in DRG neurons require extracellular calcium. The Journal of Cell Biology 219, e201702151.

Gerasimenko, J.V., Charlesworth, R.M., Sherwood, M.W., Ferdek, P.E., Mikoshiba, K., Parrington, J., Petersen, O.H., and Gerasimenko, O.V. (2015). Both RyRs and TPCs are required for NAADP-induced intracellular Ca^{2+} release. Cell Calcium 58, 237–245.

Gerndt, S., Chen, C.C., Chao, Y.K., Yuan, Y., Burgstaller, S., Scotto Rosato, A., Krogsaeter, E., Urban, N., Jacob, K., Nguyen, O.N.P., et al. (2020). Agonist-mediated switching of ion selectivity in TPC2 differentially promotes lysosomal function. eLife 9, e54712.

Goodridge, J.P., Jacobs, B., Saetersmoen, M.L., Clement, D., Hammer, Q., Clancy, T., Skarpen, E., Brech, A., Landskron, J., Grimm, C., et al. (2019). Remodeling of secretory lysosomes during education tunes functional potential in NK cells. Nature Communications 10, 514.

Graves, A.R., Curran, P.K., Smith, C.L., and Mindell, J.A. (2008). The Cl^-/H^+ antiporter ClC-7 is the primary chloride permeation pathway in lysosomes. Nature 453, 788–792.

Gray, M.A., Choy, C.H., Dayam, R.M., Ospina-Escobar, E., Somerville, A., Xiao, X., Ferguson, S.M., and Botelho, R.J. (2016). Phagocytosis enhances lysosomal and bactericidal properties by activating the transcription factor TFEB. Current Biology: CB 26, 1955–1964.

Grimm, C., Butz, E., Chen, C.C., Wahl-Schott, C., and Biel, M. (2017). From mucolipidosis type IV to Ebola: TRPML and two-pore channels at the crossroads of endo-lysosomal trafficking and disease. Cell Calcium 67, 148–155.

Grimm, C., Cuajungco, M.P., van Aken, A.F., Schnee, M., Jors, S., Kros, C.J., Ricci, A.J., and Heller, S. (2007). A helix-breaking mutation in TRPML3 leads to constitutive activity underlying deafness in the varitint-waddler mouse. Proceedings of the National Academy of Sciences of the United States of America *104*, 19583–19588.

Grimm, C., Holdt, L.M., Chen, C.C., Hassan, S., Muller, C., Jors, S., Cuny, H., Kissing, S., Schroder, B., Butz, E., et al. (2014). High susceptibility to fatty liver disease in two-pore channel 2-deficient mice. Nature Communications *5*, 4699.

Grimm, C., Jors, S., Guo, Z., Obukhov, A.G., and Heller, S. (2012). Constitutive activity of TRPML2 and TRPML3 channels versus activation by low extracellular sodium and small molecules. The Journal of Biological Chemistry *287*, 22701–22708.

Grimm, C., Jors, S., Saldanha, S.A., Obukhov, A.G., Pan, B., Oshima, K., Cuajungco, M.P., Chase, P., Hodder, P., and Heller, S. (2010). Small molecule activators of TRPML3. Chemistry & Biology *17*, 135–148.

Gunaratne, G.S., Yang, Y., Li, F., Walseth, T.F., and Marchant, J.S. (2018). NAADP-dependent Ca^{2+} signaling regulates Middle East respiratory syndrome-coronavirus pseudovirus translocation through the endolysosomal system. Cell Calcium *75*, 30–41.

Guo, J., Zeng, W., Chen, Q., Lee, C., Chen, L., Yang, Y., Cang, C., Ren, D., and Jiang, Y. (2016). Structure of the voltage-gated two-pore channel TPC1 from *Arabidopsis thaliana*. Nature *531*, 196–201.

Guo, J., Zeng, W., and Jiang, Y. (2017). Tuning the ion selectivity of two-pore channels. Proceedings of the National Academy of Sciences of the United States of America *114*, 1009–1014.

Guzman, R.E., Miranda-Laferte, E., Franzen, A., and Fahlke, C. (2015). Neuronal ClC-3 splice variants differ in subcellular localizations, but mediate identical transport functions. The Journal of Biological Chemistry *290*, 25851–25862.

Haller, T., Dietl, P., Deetjen, P., and Volkl, H. (1996). The lysosomal compartment as intracellular calcium store in MDCK cells: a possible involvement in InsP3-mediated Ca^{2+} release. Cell Calcium *19*, 157–165.

Hamilton, A., Zhang, Q., Salehi, A., Willems, M., Knudsen, J.G., Ringgaard, A.K., Chapman, C.E., Gonzalez-Alvarez, A., Surdo, N.C., Zaccolo, M., et al. (2018). Adrenaline stimulates glucagon secretion by Tpc2-dependent Ca^{2+} mobilization from acidic stores in pancreatic alpha-cells. Diabetes *67*, 1128–1139.

Hara-Chikuma, M., Yang, B., Sonawane, N.D., Sasaki, S., Uchida, S., and Verkman, A.S. (2005). ClC-3 chloride channels facilitate endosomal acidification and chloride accumulation. The Journal of Biological Chemistry *280*, 1241–1247.

Hara, T., Nakamura, K., Matsui, M., Yamamoto, A., Nakahara, Y., Suzuki-Migishima, R., Yokoyama, M., Mishima, K., Saito, I., Okano, H., et al. (2006). Suppression of basal autophagy in neural cells causes neurodegenerative disease in mice. Nature *441*, 885–889.

Hockey, L.N., Kilpatrick, B.S., Eden, E.R., Lin-Moshier, Y., Brailoiu, G.C., Brailoiu, E., Futter, C.E., Schapira, A.H., Marchant, J.S., and Patel, S. (2015). Dysregulation of lysosomal morphology by pathogenic LRRK2 is corrected by TPC2 inhibition. Journal of Cell Science *128*, 232–238.

Hoglinger, D., Haberkant, P., Aguilera-Romero, A., Riezman, H., Porter, F.D., Platt, F.M., Galione, A., and Schultz, C. (2015). Intracellular sphingosine releases calcium from lysosomes. eLife *4*, e10616.

Holtzman, E. (1989). Lysosomes. In Cellular Organclles Plenum Press, New York.

Hu, W., Zhang, L., Li, M.X., Shen, J., Liu, X.D., Xiao, Z.G., Wu, D.L., Ho, I.H.T., Wu, J.C.Y., Cheung, C.K.Y., et al. (2019a). Vitamin D3 activates the autolysosomal degradation function against *Helicobacter pylori* through the PDIA3 receptor in gastric epithelial cells. Autophagy *15*, 707–725.

Hu, Z.D., Yan, J., Cao, K.Y., Yin, Z.Q., Xin, W.W., and Zhang, M.F. (2019b). MCOLN1 promotes proliferation and predicts poor survival of patients with pancreatic ductal adenocarcinoma. Disease Markers *2019*, 9436047.

Huang, C.C., Tu, S.H., Lien, H.H., Jeng, J.Y., Huang, C.S., Huang, C.J., Lai, L.C., and Chuang, E.Y. (2013). Concurrent gene signatures for Han Chinese breast cancers. PLoS One *8*, e76421.

Huang, P., Zou, Y., Zhong, X.Z., Cao, Q., Zhao, K., Zhu, M.X., Murell-Lagnado, R., and Dong, X.P. (2014). P2X4 forms functional ATP-activated cation channels on lysosomal membranes regulated by luminal pH. The Journal of Biological Chemistry 289, 17658-67.

Hui, L., Soliman, M.L., Geiger, N.H., Miller, N.M., Afghah, Z., Lakpa, K.L., Chen, X., and Geiger, J.D. (2019). Acidifying endolysosomes prevented low-density lipoprotein-induced amyloidogenesis. Journal of Alzheimer's Disease: JAD *67*, 393–410.

Jahidin, A.H., Stewart, T.A., Thompson, E.W., Roberts-Thomson, S.J., and Monteith, G.R. (2016). Differential effects of two-pore channel protein 1 and 2 silencing in MDA-MB-468 breast cancer cells. Biochemical and Biophysical Research Communications *477*, 731–736.

Jaiswal, J.K., Andrews, N.W., and Simon, S.M. (2002). Membrane proximal lysosomes are the major vesicles responsible for calcium-dependent exocytosis in nonsecretory cells. The Journal of Cell Biology *159*, 625–635.

Jennings, J.J. Jr., Zhu, J.H., Rbaibi, Y., Luo, X., Chu, C.T., and Kiselyov, K. (2006). Mitochondrial aberrations in mucolipidosis Type IV. The Journal of Biological Chemistry *281*, 39041–39050.

Jentsch, T.J. (2007). Chloride and the endosomal-lysosomal pathway: emerging roles of CLC chloride transporters. The Journal of Physiology *578*, 633–640.

Jentsch, T.J., and Pusch, M. (2018). CLC chloride channels and transporters: structure, function, physiology, and disease. Physiological Reviews *98*, 1493–1590.

Jeyakumar, M., Dwek, R.A., Butters, T.D., and Platt, F.M. (2005). Storage solutions: treating lysosomal disorders of the brain. Nature Reviews Neuroscience *6*, 713–725.

Jha, A., Ahuja, M., Patel, S., Brailoiu, E., and Muallem, S. (2014). Convergent regulation of the lysosomal two-pore channel-2 by Mg^{2+}, NAADP, $PI(3,5)P_2$ and multiple protein kinases. The EMBO Journal *33*, 501–511.

Jiang, Y., Lee, A., Chen, J., Ruta, V., Cadene, M., Chait, B.T., and MacKinnon, R. (2003). X-ray structure of a voltage-dependent K+ channel. Nature *423*, 33–41.

Jin, X., Zhang, Y., Alharbi, A., Hanbashi, A., Alhoshani, A., and Parrington, J. (2020). Targeting two-pore channels: current progress and future challenges. Trends in Pharmacological Sciences *41*, 582–594.

Jinn, S., Blauwendraat, C., Toolan, D., Gretzula, C.A., Drolet, R.E., Smith, S., Nalls, M.A., Marcus, J., Singleton, A.B., and Stone, D.J. (2019). Functionalization of the TMEM175 p.M393T variant as a risk factor for Parkinson disease. Human Molecular Genetics *28*, 3244–3254.

Jinn, S., Drolet, R.E., Cramer, P.E., Wong, A.H., Toolan, D.M., Gretzula, C.A., Voleti, B., Vassileva, G., Disa, J., Tadin-Strapps, M., et al. (2017). TMEM175 deficiency impairs lysosomal and mitochondrial function and increases alpha-synuclein aggregation. Proceedings of the National Academy of Sciences of the United States of America *114*, 2389–2394.

Jung, J., Cho, K.J., Naji, A.K., Clemons, K.N., Wong, C.O., Villanueva, M., Gregory, S., Karagas, N.E., Tan, L., Liang, H., et al. (2019). HRAS-driven cancer cells are vulnerable to TRPML1 inhibition. EMBO Reports *20*, e46685.

Karacsonyi, C., Miguel, A.S., and Puertollano, R. (2007). Mucolipin-2 localizes to the Arf6-associated pathway and regulates recycling of GPI-APs. Traffic *8*, 1404–1414.

Kasitinon, S.Y., Eskiocak, U., Martin, M., Bezwada, D., Khivansara, V., Tasdogan, A., Zhao, Z., Mathews, T., Aurora, A.B., and Morrison, S.J. (2019). TRPML1 promotes protein homeostasis in melanoma cells by negatively regulating MAPK and mTORC1 signaling. Cell Reports *28*, 2293–2305 e2299.

Kasper, D., Planells-Cases, R., Fuhrmann, J.C., Scheel, O., Zeitz, O., Ruether, K., Schmitt, A., Poet, M., Steinfeld, R., Schweizer, M., *et al.* (2005). Loss of the chloride channel ClC-7 leads to lysosomal storage disease and neurodegeneration. The EMBO Journal *24*, 1079–1091.

Kawai, T., and Akira, S. (2010). The role of pattern-recognition receptors in innate immunity: update on Toll-like receptors. Nature Immunology *11*, 373–384.

Kawasaki, M., Uchida, S., Monkawa, T., Miyawaki, A., Mikoshiba, K., Marumo, F., and Sasaki, S. (1994). Cloning and expression of a protein kinase C-regulated chloride channel abundantly expressed in rat brain neuronal cells. Neuron *12*, 597–604.

Khaitan, D., Sankpal, U.T., Weksler, B., Meister, E.A., Romero, I.A., Couraud, P.O., and Ningaraj, N.S. (2009). Role of KCNMA1 gene in breast cancer invasion and metastasis to brain. BMC Cancer *9*, 258.

Khakh, B.S., and North, R.A. (2012). Neuromodulation by extracellular ATP and P2X receptors in the CNS. Neuron *76*, 51–69.

Khalaf, A., and Babiker, F. (2016). Discrepancy in calcium release from the sarcoplasmic reticulum and intracellular acidic stores for the protection of the heart against ischemia/reperfusion injury. Journal of Physiology and Biochemistry *72*, 495–508.

Khan, N., Lakpa, K.L., Halcrow, P.W., Afghah, Z., Miller, N.M., Geiger, J.D., and Chen, X. (2019). BK channels regulate extracellular Tat-mediated HIV-1 LTR transactivation. Scientific Reports *9*, 12285.

Kilpatrick, B.S., Eden, E.R., Hockey, L.N., Yates, E., Futter, C.E., and Patel, S. (2017). An endosomal NAADP-Sensitive two-pore Ca^{2+} channel regulates ER-endosome membrane contact sites to control growth factor signaling. Cell Reports *18*, 1636–1645.

Kilpatrick, B.S., Eden, E.R., Schapira, A.H., Futter, C.E., and Patel, S. (2013). Direct mobilisation of lysosomal Ca^{2+} triggers complex Ca^{2+} signals. Journal of Cell Science *126*, 60–66.

Kilpatrick, B.S., Yates, E., Grimm, C., Schapira, A.H., and Patel, S. (2016). Endo-lysosomal TRP mucolipin-1 channels trigger global ER Ca^{2+} release and Ca^{2+} influx. Journal of Cell Science *129*, 3859–3867.

Kim, H.J., Li, Q., Tjon-Kon-Sang, S., So, I., Kiselyov, K., and Muallem, S. (2007). Gain-of-function mutation in TRPML3 causes the mouse varitint-waddler phenotype. The Journal of Biological Chemistry *282*, 36138–36142.

Kim, H.J., Li, Q., Tjon-Kon-Sang, S., So, I., Kiselyov, K., Soyombo, A.A., and Muallem, S. (2008). A novel mode of TRPML3 regulation by extracytosolic pH absent in the varitint-waddler phenotype. The EMBO Journal *27*, 1197–1205.

Kim, H.J., Soyombo, A.A., Tjon-Kon-Sang, S., So, I., and Muallem, S. (2009). The Ca^{2+} channel TRPML3 regulates membrane trafficking and autophagy. Traffic *10*, 1157–1167.

Kim, S.W., Kim, D.H., Park, K.S., Kim, M.K., Park, Y.M., Muallem, S., So, I., and Kim, H.J. (2019). Palmitoylation controls trafficking of the intracellular Ca^{2+} channel MCOLN3/TRPML3 to regulate autophagy. Autophagy *15*, 327–340.

Kintzer, A.F., and Stroud, R.M. (2016). Structure, inhibition and regulation of two-pore channel TPC1 from *Arabidopsis thaliana*. Nature *531*, 258–262.

Kirsch, S.A., Kugemann, A., Carpaneto, A., Bockmann, R.A., and Dietrich, P. (2018). Phosphatidylinositol-3,5-bisphosphate lipid-binding-induced activation of the human two-pore channel 2. Cellular and Molecular Life Sciences: CMLS *75*, 3803–3815.

Kiselyov, K., Chen, J., Rbaibi, Y., Oberdick, D., Tjon-Kon-Sang, S., Shcheynikov, N., Muallem, S., and Soyombo, A. (2005). TRP-ML1 is a lysosomal monovalent cation channel that undergoes proteolytic cleavage. The Journal of Biological Chemistry *280*, 43218–43223.

Kiselyov, K., Yamaguchi, S., Lyons, C.W., and Muallem, S. (2010). Aberrant Ca^{2+} handling in lysosomal storage disorders. Cell Calcium *47*, 103–111.

Klionsky, D.J., Abdelmohsen, K., Abe, A., Abedin, M.J., Abeliovich, H., Acevedo Arozena, A., Adachi, H., Adams, C.M., Adams, P.D., Adeli, K., *et al.* (2016). Guidelines for the use and interpretation of assays for monitoring autophagy (3rd edition). Autophagy *12*, 1–222.

Kocarnik, J.M., Park, S.L., Han, J., Dumitrescu, L., Cheng, I., Wilkens, L.R., Schumacher, F.R., Kolonel, L., Carlson, C.S., Crawford, D.C., *et al.* (2015). Pleiotropic and sex-specific effects of cancer GWAS SNPs on melanoma risk in the population architecture using genomics and epidemiology (PAGE) study. PLoS One *10*, e0120491.

Komatsu, M., Waguri, S., Chiba, T., Murata, S., Iwata, J., Tanida, I., Ueno, T., Koike, M., Uchiyama, Y., Kominami, E., *et al.* (2006). Loss of autophagy in the central nervous system causes neurodegeneration in mice. Nature *441*, 880–884.

Kornak, U., Kasper, D., Bosl, M.R., Kaiser, E., Schweizer, M., Schulz, A., Friedrich, W., Delling, G., and Jentsch, T.J. (2001). Loss of the ClC-7 chloride channel leads to osteopetrosis in mice and man. Cell *104*, 205–215.

Kosiniak-Kamysz, A., Marczakiewicz-Lustig, A., Marcinska, M., Skowron, M., Wojas-Pelc, A., Pospiech, E., and Branicki, W. (2014). Increased risk of developing cutaneous malignant melanoma is associated with variation in pigmentation genes and VDR, and may involve epistatic effects. Melanoma Research *24*, 388–396.

Krajcovic, M., Krishna, S., Akkari, L., Joyce, J.A., and Overholtzer, M. (2013). mTOR regulates phagosome and entotic vacuole fission. Molecular Biology of the Cell *24*, 3736–3745.

Krogsaeter, E.K., Biel, M., Wahl-Schott, C., and Grimm, C. (2019). The protein interaction networks of mucolipins and two-pore channels. Biochimica et Biophysica Acta—Molecular Cell Research *1866*, 1111–1123.

Krohn, L., Ozturk, T.N., Vanderperre, B., Ouled Amar Bencheikh, B., Ruskey, J.A., Laurent, S.B., Spiegelman, D., Postuma, R.B., Arnulf, I., Hu, M.T.M., *et al.* (2020). Genetic, structural, and functional evidence link TMEM175 to Synucleinopathies. Annals of Neurology *87*, 139–153.

Lagostena, L., Festa, M., Pusch, M., and Carpaneto, A. (2017). The human two-pore channel 1 is modulated by cytosolic and luminal calcium. Scientific Reports *7*, 43900.

Lange, I., Yamamoto, S., Partida-Sanchez, S., Mori, Y., Fleig, A., and Penner, R. (2009). TRPM2 functions as a lysosomal Ca^{2+}-release channel in beta cells. Science Signaling *2*, ra23.

Lange, P.F., Wartosch, L., Jentsch, T.J., and Fuhrmann, J.C. (2006). ClC-7 requires Ostm1 as a beta-subunit to support bone resorption and lysosomal function. Nature *440*, 220–223.

LaPlante, J.M., Sun, M., Falardeau, J., Dai, D., Brown, E.M., Slaugenhaupt, S.A., and Vassilev, P.M. (2006). Lysosomal exocytosis is impaired in mucolipidosis type IV. Molecular Genetics and Metabolism *89*, 339–348.

Larisch, N., Schulze, C., Galione, A., and Dietrich, P. (2012). An N-terminal dileucine motif directs two-pore channels to the tonoplast of plant cells. Traffic *13*, 1012–1022.

Lear, P.V., Gonzalez-Touceda, D., Porteiro Couto, B., Viano, P., Guymer, V., Remzova, E., Tunn, R., Chalasani, A., Garcia-Caballero, T., Hargreaves, I.P., *et al.* (2015). Absence of intracellular ion channels TPC1 and TPC2 leads to mature-onset obesity in male mice, due to impaired lipid availability for thermogenesis in brown adipose tissue. Endocrinology *156*, 975–986.

Lee, C., Guo, J., Zeng, W., Kim, S., She, J., Cang, C., Ren, D., and Jiang, Y. (2017). The lysosomal potassium channel TMEM175 adopts a novel tetrameric architecture. Nature *547*, 472–475.

Lee, U.S., and Cui, J. (2010). BK channel activation: structural and functional insights. Trends in Neurosciences *33*, 415–423.

Leisle, L., Ludwig, C.F., Wagner, F.A., Jentsch, T.J., and Stauber, T. (2011). ClC-7 is a slowly voltage-gated 2Cl⁻/1H⁺-exchanger and requires Ostm1 for transport activity. The EMBO Journal *30*, 2140–2152.

Lelouvier, B., and Puertollano, R. (2011). Mucolipin-3 regulates luminal calcium, acidification, and membrane fusion in the endosomal pathway. The Journal of Biological Chemistry *286*, 9826–9832.

Levine, B., and Kroemer, G. (2008). Autophagy in the pathogenesis of disease. Cell *132*, 27–42.

Levine, B., Mizushima, N., and Virgin, H.W. (2011). Autophagy in immunity and inflammation. Nature *469*, 323–335.

Li, F., Abuarab, N., and Sivaprasadarao, A. (2016a). Reciprocal regulation of actin cytoskeleton remodelling and cell migration by Ca^{2+} and Zn^{2+}: role of TRPM2 channels. Journal of Cell Science *129*, 2016–2029.

Li, R.J., Xu, J., Fu, C., Zhang, J., Zheng, Y.G., Jia, H., and Liu, J.O. (2016b). Regulation of mTORC1 by lysosomal calcium and calmodulin. eLife *5*, e19360.

Li, X., Rydzewski, N., Hider, A., Zhang, X., Yang, J., Wang, W., Gao, Q., Cheng, X., and Xu, H. (2016c). A molecular mechanism to regulate lysosome motility for lysosome positioning and tubulation. Nature Cell Biology 18, 404–17.

Li, M., Zhang, W.K., Benvin, N.M., Zhou, X., Su, D., Li, H., Wang, S., Michailidis, I.E., Tong, L., Li, X., *et al.* (2017a). Structural basis of dual Ca^{2+}/pH regulation of the endolysosomal TRPML1 channel. Nature Structural & Molecular Biology *24*, 205–213.

Li, X., Yang, W., and Jiang, L.H. (2017b). Alteration in intracellular Zn^{2+} homeostasis as a result of TRPM2 channel activation contributes to ROS-induced hippocampal neuronal death. Frontiers in Molecular Neuroscience *10*, 414.

Li, P., Gu, M., and Xu, H. (2019). Lysosomal ion channels as decoders of cellular signals. Trends in Biochemical Sciences *44*, 110–124.

Li, X., and Jiang, L.H. (2018). Multiple molecular mechanisms form a positive feedback loop driving amyloid beta42 peptide-induced neurotoxicity via activation of the TRPM2 channel in hippocampal neurons. Cell Death and Disease *9*, 195.

Li, X., and Jiang, L.H. (2019). A critical role of the transient receptor potential melastatin 2 channel in a positive feedback mechanism for reactive oxygen species-induced delayed cell death. Journal of Cellular Physiology *234*, 3647–3660.

Li, X., Saitoh, S., Shibata, T., Tanimura, N., Fukui, R., and Miyake, K. (2015). Mucolipin 1 positively regulates TLR7 responses in dendritic cells by facilitating RNA transportation to lysosomes. International Immunology *27*, 83–94.

Li, X., Wang, T., Zhao, Z., and Weinman, S.A. (2002). The ClC-3 chloride channel promotes acidification of lysosomes in CHO-K1 and Huh-7 cells. American Journal of Physiology Cell Physiology *282*, C1483–C1491.

Lill, C.M., Hansen, J., Olsen, J.H., Binder, H., Ritz, B., and Bertram, L. (2015). Impact of Parkinson's disease risk loci on age at onset. Movement Disorders: Official Journal of the Movement Disorder Society *30*, 847–850.

Lin-Moshier, Y., Walseth, T.F., Churamani, D., Davidson, S.M., Slama, J.T., Hooper, R., Brailoiu, E., Patel, S., and Marchant, J.S. (2012). Photoaffinity labeling of nicotinic acid adenine dinucleotide phosphate (NAADP) targets in mammalian cells. The Journal of Biological Chemistry *287*, 2296–2307.

Lin, P.H., Duann, P., Komazaki, S., Park, K.H., Li, H., Sun, M., Sermersheim, M., Gumpper, K., Parrington, J., Galione, A., *et al.* (2015). Lysosomal two-pore channel subtype 2

(TPC2) regulates skeletal muscle autophagic signaling. The Journal of Biological Chemistry *290*, 3377–3389.

Liu, X., Chang, Y., Reinhart, P.H., and Sontheimer, H. (2002). Cloning and characterization of glioma BK, a novel BK channel isoform highly expressed in human glioma cells. The Journal of Neuroscience: The Official Journal of the Society for Neuroscience *22*, 1840–1849.

Lloyd-Evans, E., and Platt, F.M. (2011). Lysosomal Ca^{2+} homeostasis: role in pathogenesis of lysosomal storage diseases. Cell Calcium *50*, 200–205.

Long, S.B., Campbell, E.B., and Mackinnon, R. (2005). Crystal structure of a mammalian voltage-dependent Shaker family K^+ channel. Science *309*, 897–903.

Long, S.B., Tao, X., Campbell, E.B., and MacKinnon, R. (2007). Atomic structure of a voltage-dependent K^+ channel in a lipid membrane-like environment. Nature *450*, 376–382.

Luzio, J.P., Pryor, P.R., and Bright, N.A. (2007). Lysosomes: fusion and function. Nature Reviews Molecular Cell Biology *8*, 622–632.

Luzio, J.P., Rous, B.A., Bright, N.A., Pryor, P.R., Mullock, B.M., and Piper, R.C. (2000). Lysosome-endosome fusion and lysosome biogenesis. Journal of Cell Science *113* (Pt 9), 1515–1524.

Ma, J., Wei, K., Zhang, H., Tang, K., Li, F., Zhang, T., Liu, J., Xu, P., Yu, Y., Sun, W., *et al.* (2018). Mechanisms by which dendritic cells present tumor microparticle antigens to $CD8^+$ T cells. Cancer Immunology Research *6*, 1057–1068.

MacKinnon, R., and Miller, C. (1989). Mutant potassium channels with altered binding of charybdotoxin, a pore blocking peptide inhibitor. Science *245*, 1382–1385.

Malik, B.R., Maddison, D.C., Smith, G.A., and Peters, O.M. (2019). Autophagic and endo-lysosomal dysfunction in neurodegenerative disease. Molecular Brain *12*, 100.

Manna, P.T., Munsey, T.S., Abuarab, N., Li, F., Asipu, A., Howell, G., Sedo, A., Yang, W., Naylor, J., Beech, D.J., *et al.* (2015). TRPM2-mediated intracellular Zn^{2+} release triggers pancreatic beta-cell death. The Biochemical Journal *466*, 537–546.

Martina, J.A., Lelouvier, B., and Puertollano, R. (2009). The calcium channel mucolipin-3 is a novel regulator of trafficking along the endosomal pathway. Traffic *10*, 1143–1156.

Martinez, I., Chakrabarti, S., Hellevik, T., Morehead, J., Fowler, K., and Andrews, N.W. (2000). Synaptotagmin VII regulates Ca^{2+}-dependent exocytosis of lysosomes in fibroblasts. The Journal of Cell Biology *148*, 1141–1149.

Masgrau, R., Churchill, G.C., Morgan, A.J., Ashcroft, S.J., and Galione, A. (2003). NAADP: a new second messenger for glucose-induced Ca^{2+} responses in clonal pancreatic beta cells. Current Biology: CB *13*, 247–251.

Masiero, E., Agatea, L., Mammucari, C., Blaauw, B., Loro, E., Komatsu, M., Metzger, D., Reggiani, C., Schiaffino, S., and Sandri, M. (2009). Autophagy is required to maintain muscle mass. Cell Metabolism *10*, 507–515.

Medina, D.L., Di Paola, S., Peluso, I., Armani, A., De Stefani, D., Venditti, R., Montefusco, S., Scotto-Rosato, A., Prezioso, C., Forrester, A., *et al.* (2015). Lysosomal calcium signalling regulates autophagy through calcineurin and TFEB. Nature Cell Biology *17*, 288–299.

Medina, D.L., Fraldi, A., Bouche, V., Annunziata, F., Mansueto, G., Spampanato, C., Puri, C., Pignata, A., Martina, J.A., Sardiello, M., *et al.* (2011). Transcriptional activation of lysosomal exocytosis promotes cellular clearance. Developmental Cell *21*, 421–430.

Melchionda, M., Pittman, J.K., Mayor, R., and Patel, S. (2016). Ca^{2+}/H^+ exchange by acidic organelles regulates cell migration in vivo. The Journal of Cell Biology *212*, 803–813.

Miao, Y., Li, G., Zhang, X., Xu, H., and Abraham, S.N. (2015). A TRP channel senses lysosome neutralization by pathogens to trigger their expulsion. Cell *161*, 1306–1319.

Michaillat, L., Baars, T.L., and Mayer, A. (2012). Cell-free reconstitution of vacuole membrane fragmentation reveals regulation of vacuole size and number by TORC1. Molecular Biology of the Cell *23*, 881–895.

Micsenyi, M.C., Dobrenis, K., Stephney, G., Pickel, J., Vanier, M.T., Slaugenhaupt, S.A., and Walkley, S.U. (2009). Neuropathology of the Mcoln1$^{-/-}$ knockout mouse model of mucolipidosis type IV. Journal of Neuropathology and Experimental Neurology *68*, 125–135.

Miklavc, P., Mair, N., Wittekindt, O.H., Haller, T., Dietl, P., Felder, E., Timmler, M., and Frick, M. (2011). Fusion-activated Ca^{2+} entry via vesicular P2X4 receptors promotes fusion pore opening and exocytotic content release in pneumocytes. Proceedings of the National Academy of Sciences of the United States of America *108*, 14503–14508.

Miller, A., Schafer, J., Upchurch, C., Spooner, E., Huynh, J., Hernandez, S., McLaughlin, B., Oden, L., and Fares, H. (2015). Mucolipidosis type IV protein TRPML1-dependent lysosome formation. Traffic *16*, 284–297.

Miller, J.J., Aoki, K., Moehring, F., Murphy, C.A., O'Hara, C.L., Tiemeyer, M., Stucky, C.L., and Dahms, N.M. (2018). Neuropathic pain in a Fabry disease rat model. JCI Insight *3*, e99171.

Mindell, J.A. (2012). Lysosomal acidification mechanisms. Annual Review of Physiology *74*, 69–86.

Mizushima, N., and Levine, B. (2010). Autophagy in mammalian development and differentiation. Nature Cell Biology *12*, 823–830.

Moccia, F., Zuccolo, E., Di Nezza, F., Pellavio, G., Faris, P.S., Negri, S., De Luca, A., Laforenza, U., Ambrosone, L., Rosti, V., et al. (2020). Nicotinic acid adenine dinucleotide phosphate activates two-pore channel TPC1 to mediate lysosomal Ca^{2+} release in endothelial colony-forming cells. Journal of Cellular Physiology *236*, 688–705.

Mohammad-Panah, R., Harrison, R., Dhani, S., Ackerley, C., Huan, L.J., Wang, Y., and Bear, C.E. (2003). The chloride channel ClC-4 contributes to endosomal acidification and trafficking. The Journal of Biological Chemistry *278*, 29267–29277.

Morelli, M.B., Amantini, C., Tomassoni, D., Nabissi, M., Arcella, A., and Santoni, G. (2019). Transient receptor potential Mucolipin-1 channels in glioblastoma: role in patient's survival. Cancers *11*, 525.

Morelli, M.B., Nabissi, M., Amantini, C., Tomassoni, D., Rossi, F., Cardinali, C., Santoni, M., Arcella, A., Oliva, M.A., Santoni, A., et al. (2016). Overexpression of transient receptor potential mucolipin-2 ion channels in gliomas: role in tumor growth and progression. Oncotarget *7*, 43654–43668.

Morgan, A.J., Davis, L.C., Wagner, S.K., Lewis, A.M., Parrington, J., Churchill, G.C., and Galione, A. (2013). Bidirectional Ca^{2+} signaling occurs between the endoplasmic reticulum and acidic organelles. The Journal of Cell Biology *200*, 789–805.

Morgan, A.J., and Galione, A. (2007). NAADP induces pH changes in the lumen of acidic Ca^{2+} stores. The Biochemical Journal *402*, 301–310.

Morgan, A.J., Platt, F.M., Lloyd-Evans, E., and Galione, A. (2011). Molecular mechanisms of endolysosomal Ca^{2+} signalling in health and disease. The Biochemical Journal *439*, 349–374.

Nagata, K., Zheng, L., Madathany, T., Castiglioni, A.J., Bartles, J.R., and Garcia-Anoveros, J. (2008). The varitint-waddler (Va) deafness mutation in TRPML3 generates constitutive, inward rectifying currents and causes cell degeneration. Proceedings of the National Academy of Sciences of the United States of America *105*, 353–358.

Nalls, M.A., Pankratz, N., Lill, C.M., Do, C.B., Hernandez, D.G., Saad, M., DeStefano, A.L., Kara, E., Bras, J., Sharma, M., et al. (2014). Large-scale meta-analysis of genome-wide association data identifies six new risk loci for Parkinson's disease. Nature Genetics *46*, 989–993.

Narayanaswamy, N., Chakraborty, K., Saminathan, A., Zeichner, E., Leung, K., Devany, J., and Krishnan, Y. (2019). A pH-correctable, DNA-based fluorescent reporter for organellar calcium. Nature Methods 16, 95–102.

Neagoe, I., Stauber, T., Fidzinski, P., Bergsdorf, E.Y., and Jentsch, T.J. (2010). The late endosomal ClC-6 mediates proton/chloride countertransport in heterologous plasma membrane expression. The Journal of Biological Chemistry 285, 21689–21697.

Neefjes, J., and van der Kant, R. (2014). Stuck in traffic: an emerging theme in diseases of the nervous system. Trends in Neurosciences 37, 66–76.

Nguyen, O.N., Grimm, C., Schneider, L.S., Chao, Y.K., Atzberger, C., Bartel, K., Watermann, A., Ulrich, M., Mayr, D., Wahl-Schott, C., et al. (2017). Two-pore channel function is crucial for the migration of invasive cancer cells. Cancer Research 77, 1427–1438.

Nilius, B., Owsianik, G., Voets, T., and Peters, J.A. (2007). Transient receptor potential cation channels in disease. Physiological Reviews 87, 165–217.

Novarino, G., Weinert, S., Rickheit, G., and Jentsch, T.J. (2010). Endosomal chloride-proton exchange rather than chloride conductance is crucial for renal endocytosis. Science 328, 1398–1401.

Ogunbayo, O.A., Duan, J., Xiong, J., Wang, Q., Feng, X., Ma, J., Zhu, M.X., and Evans, A.M. (2018). mTORC1 controls lysosomal Ca^{2+} release through the two-pore channel TPC2. Science Signaling 11, eaao5775.

Ogunbayo, O.A., Zhu, Y., Shen, B., Agbani, E., Li, J., Ma, J., Zhu, M.X., and Evans, A.M. (2015). Organelle-specific subunit interactions of the vertebrate two-pore channel family. The Journal of Biological Chemistry 290, 1086–1095.

Oh, S., Paknejad, N., and Hite, R.K. (2020). Gating and selectivity mechanisms for the lysosomal K^+ channel TMEM175. eLife 9, e53430.

Olesen, S.P., Munch, E., Moldt, P., and Drejer, J. (1994). Selective activation of Ca^{2+}-dependent K^+ channels by novel benzimidazolone. European Journal of Pharmacology 251, 53–59.

Onyenwoke, R.U., Sexton, J.Z., Yan, F., Diaz, M.C., Forsberg, L.J., Major, M.B., and Brenman, J.E. (2015). The mucolipidosis IV Ca^{2+} channel TRPML1 (MCOLN1) is regulated by the TOR kinase. The Biochemical Journal 470, 331–342.

Ou, X., Liu, Y., Lei, X., Li, P., Mi, D., Ren, L., Guo, L., Guo, R., Chen, T., Hu, J., et al. (2020). Characterization of spike glycoprotein of SARS-CoV-2 on virus entry and its immune cross-reactivity with SARS-CoV. Nature Communications 11, 1620.

Pafumi, I., Festa, M., Papacci, F., Lagostena, L., Giunta, C., Gutla, V., Cornara, L., Favia, A., Palombi, F., Gambale, F., et al. (2017). Naringenin impairs two-pore channel 2 activity and inhibits VEGF-induced angiogenesis. Scientific Reports 7, 5121.

Palmieri, M., Impey, S., Kang, H., di Ronza, A., Pelz, C., Sardiello, M., and Ballabio, A. (2011). Characterization of the CLEAR network reveals an integrated control of cellular clearance pathways. Human Molecular Genetics 20, 3852–3866.

Pan, T., Kondo, S., Le, W., and Jankovic, J. (2008). The role of autophagy-lysosome pathway in neurodegeneration associated with Parkinson's disease. Brain: A Journal of Neurology 131, 1969–1978.

Pangrazio, A., Pusch, M., Caldana, E., Frattini, A., Lanino, E., Tamhankar, P.M., Phadke, S., Lopez, A.G., Orchard, P., Mihci, E., et al. (2010). Molecular and clinical heterogeneity in CLCN7-dependent osteopetrosis: report of 20 novel mutations. Human Mutation 31, E1071–E1080.

Papazian, D.M., Schwarz, T.L., Tempel, B.L., Jan, Y.N., and Jan, L.Y. (1987). Cloning of genomic and complementary DNA from Shaker, a putative potassium channel gene from Drosophila. Science 237, 749–753.

Parenti, G., Andria, G., and Ballabio, A. (2015). Lysosomal storage diseases: from pathophysiology to therapy. Annual Review of Medicine 66, 471–486.

Park, S.H., Hyun, J.Y., and Shin, I. (2019). A lysosomal chloride ion-selective fluorescent probe for biological applications. Chemical Science *10*, 56–66.

Patel, S., and Brailoiu, E. (2012). Triggering of Ca^{2+} signals by NAADP-gated two-pore channels: a role for membrane contact sites? Biochemical Society Transactions *40*, 153–157.

Patel, S., Churamani, D., and Brailoiu, E. (2017). NAADP-evoked Ca^{2+} signals through two-pore channel-1 require arginine residues in the first S4-S5 linker. Cell Calcium *68*, 1–4.

Peiter, E., Maathuis, F.J., Mills, L.N., Knight, H., Pelloux, J., Hetherington, A.M., and Sanders, D. (2005). The vacuolar Ca^{2+}-activated channel TPC1 regulates germination and stomatal movement. Nature *434*, 404–408.

Peng, W., Wong, Y.C., and Krainc, D. (2020). Mitochondria-lysosome contacts regulate mitochondrial Ca^{2+} dynamics via lysosomal TRPML1. Proceedings of the National Academy of Sciences of the United States of America *117*, 19266–19275.

Penny, C.J., Kilpatrick, B.S., Eden, E.R., and Patel, S. (2015). Coupling acidic organelles with the ER through Ca^{2+} microdomains at membrane contact sites. Cell Calcium *58*, 387–396.

Penny, C.J., Vassileva, K., Jha, A., Yuan, Y., Chee, X., Yates, E., Mazzon, M., Kilpatrick, B.S., Muallem, S., Marsh, M., *et al.* (2019). Mining of Ebola virus entry inhibitors identifies approved drugs as two-pore channel pore blockers. Biochim Biophys Acta Mol Cell Res *1866*, 1151–1161.

Perez-Riesgo, E., Gutierrez, L.G., Ubierna, D., Acedo, A., Moyer, M.P., Nunez, L., and Villalobos, C. (2017). Transcriptomic analysis of calcium remodeling in colorectal cancer. International Journal of Molecular Sciences *18*, 922.

Peters, C., and Mayer, A. (1998). Ca^{2+}/calmodulin signals the completion of docking and triggers a late step of vacuole fusion. Nature *396*, 575–580.

Picollo, A., and Pusch, M. (2005). Chloride/proton antiporter activity of mammalian CLC proteins ClC-4 and ClC-5. Nature *436*, 420–423.

Piper, R.C., and Luzio, J.P. (2004). CUPpling calcium to lysosomal biogenesis. Trends in Cell Biology *14*, 471–473.

Pitt, S.J., Funnell, T.M., Sitsapesan, M., Venturi, E., Rietdorf, K., Ruas, M., Ganesan, A., Gosain, R., Churchill, G.C., Zhu, M.X., *et al.* (2010). TPC2 is a novel NAADP-sensitive Ca^{2+} release channel, operating as a dual sensor of luminal pH and Ca^{2+}. The Journal of Biological Chemistry *285*, 35039–35046.

Platt, F.M., Boland, B., and van der Spoel, A.C. (2012). The cell biology of disease: lysosomal storage disorders: the cellular impact of lysosomal dysfunction. Journal of Cell Biology *199*, 723–734.

Plesch, E., Chen, C.C., Butz, E., Scotto Rosato, A., Krogsaeter, E.K., Yinan, H., Bartel, K., Keller, M., Robaa, D., Teupser, D., *et al.* (2018). Selective agonist of TRPML2 reveals direct role in chemokine release from innate immune cells. eLife *7*, e39720.

Poet, M., Kornak, U., Schweizer, M., Zdebik, A.A., Scheel, O., Hoelter, S., Wurst, W., Schmitt, A., Fuhrmann, J.C., Planells-Cases, R., *et al.* (2006). Lysosomal storage disease upon disruption of the neuronal chloride transport protein ClC-6. Proceedings of the National Academy of Sciences of the United States of America *103*, 13854–13859.

Pryor, P.R., Mullock, B.M., Bright, N.A., Gray, S.R., and Luzio, J.P. (2000). The role of intraorganellar Ca^{2+} in late endosome-lysosome heterotypic fusion and in the re-formation of lysosomes from hybrid organelles. The Journal of Cell Biology *149*, 1053–1062.

Pryor, P.R., Reimann, F., Gribble, F.M., and Luzio, J.P. (2006). Mucolipin-1 is a lysosomal membrane protein required for intracellular lactosylceramide traffic. Traffic *7*, 1388–1398.

Puertollano, R., and Kiselyov, K. (2009). TRPMLs: in sickness and in health. American Journal of Physiology Renal Physiology *296*, F1245–F1254.

Qureshi, O.S., Paramasivam, A., Yu, J.C., and Murrell-Lagnado, R.D. (2007). Regulation of P2X4 receptors by lysosomal targeting, glycan protection and exocytosis. Journal of Cell Science *120*, 3838–3849.

Raben, N., Hill, V., Shea, L., Takikita, S., Baum, R., Mizushima, N., Ralston, E., and Plotz, P. (2008). Suppression of autophagy in skeletal muscle uncovers the accumulation of ubiquitinated proteins and their potential role in muscle damage in Pompe disease. Human Molecular Genetics *17*, 3897–3908.

Ramos, I., Reich, A., and Wessel, G.M. (2014). Two-pore channels function in calcium regulation in sea star oocytes and embryos. Development *141*, 4598–4609.

Ramsey, I.S., Delling, M., and Clapham, D.E. (2006). An introduction to TRP channels. Annual Review of Physiology *68*, 619–647.

Reddy, A., Caler, E.V., and Andrews, N.W. (2001). Plasma membrane repair is mediated by Ca^{2+}-regulated exocytosis of lysosomes. Cell *106*, 157–169.

Rickheit, G., Wartosch, L., Schaffer, S., Stobrawa, S.M., Novarino, G., Weinert, S., and Jentsch, T.J. (2010). Role of ClC-5 in renal endocytosis is unique among ClC exchangers and does not require PY-motif-dependent ubiquitylation. The Journal of Biological Chemistry *285*, 17595–17603.

Rinkenberger, N., and Schoggins, J.W. (2018). Mucolipin-2 cation channel increases trafficking efficiency of endocytosed viruses. mBio *9*, e02314-17.

Roczniak-Ferguson, A., Petit, C.S., Froehlich, F., Qian, S., Ky, J., Angarola, B., Walther, T.C., and Ferguson, S.M. (2012). The transcription factor TFEB links mTORC1 signaling to transcriptional control of lysosome homeostasis. Science Signaling *5*, ra42.

Rodriguez, A., Webster, P., Ortego, J., and Andrews, N.W. (1997). Lysosomes behave as Ca^{2+}-regulated exocytic vesicles in fibroblasts and epithelial cells. The Journal of Cell Biology *137*, 93–104.

Ruas, M., Davis, L.C., Chen, C.C., Morgan, A.J., Chuang, K.T., Walseth, T.F., Grimm, C., Garnham, C., Powell, T., Platt, N., et al. (2015). Expression of Ca^{2+}-permeable two-pore channels rescues NAADP signalling in TPC-deficient cells. The EMBO Journal *34*, 1743–1758.

Ruas, M., Rietdorf, K., Arredouani, A., Davis, L.C., Lloyd-Evans, E., Koegel, H., Funnell, T.M., Morgan, A.J., Ward, J.A., Watanabe, K., et al. (2010). Purified TPC isoforms form NAADP receptors with distinct roles for Ca^{2+} signaling and endolysosomal trafficking. Current Biology: CB *20*, 703–709.

Rudge, S.A., Anderson, D.M., and Emr, S.D. (2004). Vacuole size control: regulation of PtdIns(3,5)P2 levels by the vacuole-associated Vac14-Fig4 complex, a PtdIns(3,5)P2-specific phosphatase. Molecular Biology of the Cell *15*, 24–36.

Rybalchenko, V., Ahuja, M., Coblentz, J., Churamani, D., Patel, S., Kiselyov, K., and Muallem, S. (2012). Membrane potential regulates nicotinic acid adenine dinucleotide phosphate (NAADP) dependence of the pH- and Ca^{2+}-sensitive organellar two-pore channel TPC1. The Journal of Biological Chemistry *287*, 20407–20416.

Saftig, P., and Klumperman, J. (2009). Lysosome biogenesis and lysosomal membrane proteins: trafficking meets function. Nature Reviews Molecular Cell Biology *10*, 623–635.

Saftig, P., Schroder, B., and Blanz, J. (2010). Lysosomal membrane proteins: life between acid and neutral conditions. Biochemical Society Transactions *38*, 1420–1423.

Saha, S., Prakash, V., Halder, S., Chakraborty, K., and Krishnan, Y. (2015). A pH-independent DNA nanodevice for quantifying chloride transport in organelles of living cells. Nature Nanotechnology *10*, 645–651.

Sahoo, N., Gu, M., Zhang, X., Raval, N., Yang, J., Bekier, M., Calvo, R., Patnaik, S., Wang, W., King, G., *et al.* (2017). Gastric acid secretion from parietal cells is mediated by a Ca^{2+} efflux channel in the tubulovesicle. Developmental Cell *41*, 262–273 e266.

Saito, M., Hanson, P.I., and Schlesinger, P. (2007). Luminal chloride-dependent activation of endosome calcium channels: patch clamp study of enlarged endosomes. The Journal of Biological Chemistry *282*, 27327–27333.

Sakurai, Y., Kolokoltsov, A.A., Chen, C.C., Tidwell, M.W., Bauta, W.E., Klugbauer, N., Grimm, C., Wahl-Schott, C., Biel, M., and Davey, R.A. (2015). Ebola virus. Two-pore channels control Ebola virus host cell entry and are drug targets for disease treatment. Science *347*, 995–998.

Salkoff, L., Butler, A., Ferreira, G., Santi, C., and Wei, A. (2006). High-conductance potassium channels of the SLO family. Nature Reviews Neuroscience *7*, 921–931.

Samie, M., Wang, X., Zhang, X., Goschka, A., Li, X., Cheng, X., Gregg, E., Azar, M., Zhuo, Y., Garrity, A.G., *et al.* (2013). A TRP channel in the lysosome regulates large particle phagocytosis via focal exocytosis. Developmental Cell *26*, 511–524.

Samie, M.A., Grimm, C., Evans, J.A., Curcio-Morelli, C., Heller, S., Slaugenhaupt, S.A., and Cuajungco, M.P. (2009). The tissue-specific expression of TRPML2 (MCOLN-2) gene is influenced by the presence of TRPML1. Pflugers Archiv: European Journal of Physiology *459*, 79–91.

Samie, M.A., and Xu, H. (2014). Lysosomal exocytosis and lipid storage disorders. Journal of Lipid Research *55*, 995–1009.

Schaheen, L., Dang, H., and Fares, H. (2006). Basis of lethality in *C. elegans* lacking CUP-5, the Mucolipidosis Type IV orthologue. Developmental Biology *293*, 382–391.

Scheel, O., Zdebik, A.A., Lourdel, S., and Jentsch, T.J. (2005). Voltage-dependent electrogenic chloride/proton exchange by endosomal CLC proteins. Nature *436*, 424–427.

Schieder, M., Rotzer, K., Bruggemann, A., Biel, M., and Wahl-Schott, C. (2010a). Planar patch clamp approach to characterize ionic currents from intact lysosomes. Science Signaling *3*, pl3.

Schieder, M., Rotzer, K., Bruggemann, A., Biel, M., and Wahl-Schott, C.A. (2010b). Characterization of two-pore channel 2 (TPCN2)-mediated Ca^{2+} currents in isolated lysosomes. The Journal of Biological Chemistry *285*, 21219–21222.

Schoggins, J.W., Wilson, S.J., Panis, M., Murphy, M.Y., Jones, C.T., Bieniasz, P., and Rice, C.M. (2011). A diverse range of gene products are effectors of the type I interferon antiviral response. Nature *472*, 481–485.

Scotto Rosato, A., Montefusco, S., Soldati, C., Di Paola, S., Capuozzo, A., Monfregola, J., Polishchuk, E., Amabile, A., Grimm, C., Lombardo, A., *et al.* (2019). TRPML1 links lysosomal calcium to autophagosome biogenesis through the activation of the CaMKKbeta/VPS34 pathway. Nature Communications *10*, 5630.

Senkevich, K., and Gan-Or, Z. (2019). Autophagy lysosomal pathway dysfunction in Parkinson's disease; evidence from human genetics. Parkinsonism & Related Disorders 73, 60–71.

Settembre, C., Di Malta, C., Polito, V.A., Garcia Arencibia, M., Vetrini, F., Erdin, S., Erdin, S.U., Huynh, T., Medina, D., Colella, P., *et al.* (2011). TFEB links autophagy to lysosomal biogenesis. Science *332*, 1429–1433.

Settembre, C., Zoncu, R., Medina, D.L., Vetrini, F., Erdin, S., Erdin, S., Huynh, T., Ferron, M., Karsenty, G., Vellard, M.C., *et al.* (2012). A lysosome-to-nucleus signalling mechanism senses and regulates the lysosome via mTOR and TFEB. The EMBO Journal *31*, 1095–1108.

Shang, S., Zhu, F., Liu, B., Chai, Z., Wu, Q., Hu, M., Wang, Y., Huang, R., Zhang, X., Wu, X., *et al.* (2016). Intracellular TRPA1 mediates Ca^{2+} release from lysosomes in dorsal root ganglion neurons. The Journal of Cell Biology 215, 369–381.

She, J., Guo, J., Chen, Q., Zeng, W., Jiang, Y., and Bai, X.C. (2018). Structural insights into the voltage and phospholipid activation of the mammalian TPC1 channel. Nature *556*, 130–134.

She, J., Zeng, W., Guo, J., Chen, Q., Bai, X.C., and Jiang, Y. (2019). Structural mechanisms of phospholipid activation of the human TPC2 channel. eLife *8*, e45222.

Shen, D., Wang, X., Li, X., Zhang, X., Yao, Z., Dibble, S., Dong, X.P., Yu, T., Lieberman, A.P., Showalter, H.D., *et al.* (2012). Lipid storage disorders block lysosomal trafficking by inhibiting a TRP channel and lysosomal calcium release. Nature Communications *3*, 731.

Shen, D., Wang, X., and Xu, H. (2011). Pairing phosphoinositides with calcium ions in endolysosomal dynamics: phosphoinositides control the direction and specificity of membrane trafficking by regulating the activity of calcium channels in the endolysosomes. BioEssays: News and Reviews in Molecular, Cellular and Developmental Biology *33*, 448–457.

Shimomura, T., and Kubo, Y. (2019). Phosphoinositides modulate the voltage dependence of two-pore channel 3. The Journal of General Physiology *151*, 986–1006.

Smith, A.J., and Lippiat, J.D. (2010). Direct endosomal acidification by the outwardly rectifying CLC-5 Cl^-/H^+ exchanger. The Journal of Physiology *588*, 2033–2045.

Solberg Woods, L.C., Holl, K.L., Oreper, D., Xie, Y., Tsaih, S.W., and Valdar, W. (2012). Fine-mapping diabetes-related traits, including insulin resistance, in heterogeneous stock rats. Physiological Genomics *44*, 1013–1026.

Stauber, T., and Jentsch, T.J. (2010). Sorting motifs of the endosomal/lysosomal CLC chloride transporters. The Journal of Biological Chemistry *285*, 34537–34548.

Stauber, T., and Jentsch, T.J. (2013). Chloride in vesicular trafficking and function. Annual Review of Physiology *75*, 453–477.

Steinberg, B.E., Huynh, K.K., Brodovitch, A., Jabs, S., Stauber, T., Jentsch, T.J., and Grinstein, S. (2010). A cation counterflux supports lysosomal acidification. The Journal of Cell Biology *189*, 1171–1186.

Steinmeyer, K., Schwappach, B., Bens, M., Vandewalle, A., and Jentsch, T.J. (1995). Cloning and functional expression of rat ClC-5, a chloride channel related to kidney disease. The Journal of Biological Chemistry *270*, 31172–31177.

Stobrawa, S.M., Breiderhoff, T., Takamori, S., Engel, D., Schweizer, M., Zdebik, A.A., Bosl, M.R., Ruether, K., Jahn, H., Draguhn, A., *et al.* (2001). Disruption of ClC-3, a chloride channel expressed on synaptic vesicles, leads to a loss of the hippocampus. Neuron *29*, 185–196.

Sulem, P., Gudbjartsson, D.F., Stacey, S.N., Helgason, A., Rafnar, T., Jakobsdottir, M., Steinberg, S., Gudjonsson, S.A., Palsson, A., Thorleifsson, G., *et al.* (2008). Two newly identified genetic determinants of pigmentation in Europeans. Nature Genetics *40*, 835–837.

Sumoza-Toledo, A., Lange, I., Cortado, H., Bhagat, H., Mori, Y., Fleig, A., Penner, R., and Partida-Sanchez, S. (2011). Dendritic cell maturation and chemotaxis is regulated by TRPM2-mediated lysosomal Ca^{2+} release. FASEB Journal: Official Publication of the Federation of American Societies for Experimental Biology *25*, 3529–3542.

Sun, L., Hua, Y., Vergarajauregui, S., Diab, H.I., and Puertollano, R. (2015). Novel Role of TRPML2 in the Regulation of the Innate Immune Response. Journal of Immunology *195*, 4922–4932.

Sun, M., Goldin, E., Stahl, S., Falardeau, J.L., Kennedy, J.C., Acierno, J.S. Jr., Bove, C., Kaneski, C.R., Nagle, J., Bromley, M.C., *et al.* (2000). Mucolipidosis type IV is caused by mutations in a gene encoding a novel transient receptor potential channel. Human Molecular Genetics *9*, 2471–2478.

Sun, T., Wang, X., Lu, Q., Ren, H., and Zhang, H. (2011). CUP-5, the *C. elegans* ortholog of the mammalian lysosomal channel protein MLN1/TRPML1, is required for proteolytic degradation in autolysosomes. Autophagy 7, 1308–1315.

Sun, X., Xu, M., Cao, Q., Huang, P., Zhu, X., and Dong, X.P. (2020). A lysosomal K$^+$ channel regulates large particle phagocytosis by facilitating lysosome Ca^{2+} release. Scientific Reports *10*, 1038.

Sun, X., Yang, Y., Zhong, X.Z., Cao, Q., Zhu, X.H., Zhu, X., and Dong, X.P. (2018). A negative feedback regulation of MTORC1 activity by the lysosomal Ca^{2+} channel MCOLN1 (mucolipin 1) using a CALM (calmodulin)-dependent mechanism. Autophagy *14*, 38–52.

Suzuki, T., Rai, T., Hayama, A., Sohara, E., Suda, S., Itoh, T., Sasaki, S., and Uchida, S. (2006). Intracellular localization of ClC chloride channels and their ability to form hetero-oligomers. Journal of Cellular Physiology *206*, 792–798.

Tao, X., Hite, R.K., and MacKinnon, R. (2017). Cryo-EM structure of the open high-conductance Ca^{2+}-activated K$^+$ channel. Nature *541*, 46–51.

Terman, A., and Kurz, T. (2013). Lysosomal iron, iron chelation, and cell death. Antioxidants & Redox Signaling *18*, 888–898.

Thompson, K.E., Korbmacher, J.P., Hecht, E., Hobi, N., Wittekindt, O.H., Dietl, P., Kranz, C., and Frick, M. (2013). Fusion-activated cation entry (FACE) via P2X(4) couples surfactant secretion and alveolar fluid transport. FASEB Journal: Official Publication of the Federation of American Societies for Experimental Biology *27*, 1772–1783.

Tian, X., Gala, U., Zhang, Y., Shang, W., Nagarkar Jaiswal, S., di Ronza, A., Jaiswal, M., Yamamoto, S., Sandoval, H., Duraine, L., *et al.* (2015). A voltage-gated calcium channel regulates lysosomal fusion with endosomes and autophagosomes and is required for neuronal homeostasis. PLoS Biology *13*, e1002103.

Togashi, K., Hara, Y., Tominaga, T., Higashi, T., Konishi, Y., Mori, Y., and Tominaga, M. (2006). TRPM2 activation by cyclic ADP-ribose at body temperature is involved in insulin secretion. The EMBO Journal *25*, 1804–1815.

Treusch, S., Knuth, S., Slaugenhaupt, S.A., Goldin, E., Grant, B.D., and Fares, H. (2004). *Caenorhabditis elegans* functional orthologue of human protein h-mucolipin-1 is required for lysosome biogenesis. Proceedings of the National Academy of Sciences of the United States of America *101*, 4483–4488.

Tsaih, S.W., Holl, K., Jia, S., Kaldunski, M., Tschannen, M., He, H., Andrae, J.W., Li, S.H., Stoddard, A., Wiederhold, A., *et al.* (2014). Identification of a novel gene for diabetic traits in rats, mice, and humans. Genetics *198*, 17–29.

Tsunemi, T., Perez-Rosello, T., Ishiguro, Y., Yoroisaka, A., Jeon, S., Hamada, K., Rammonhan, M., Wong, Y.C., Xie, Z., Akamatsu, W., *et al.* (2019). Increased lysosomal exocytosis induced by lysosomal Ca^{2+} channel agonists protects human dopaminergic neurons from alpha-synuclein toxicity. The Journal of Neuroscience: The Official Journal of the Society for Neuroscience *39*, 5760–5772.

Ullrich, F., Blin, S., Lazarow, K., Daubitz, T., von Kries, J.P., and Jentsch, T.J. (2019). Identification of TMEM206 proteins as pore of PAORAC/ASOR acid-sensitive chloride channels. eLife*8*, e49187.

Venkatachalam, K., Hofmann, T., and Montell, C. (2006). Lysosomal localization of TRPML3 depends on TRPML2 and the mucolipidosis-associated protein TRPML1. The Journal of Biological Chemistry *281*, 17517–17527.

Venkatachalam, K., Long, A.A., Elsaesser, R., Nikolaeva, D., Broadie, K., and Montell, C. (2008). Motor deficit in a Drosophila model of mucolipidosis type IV due to defective clearance of apoptotic cells. Cell *135*, 838–851.

Venkatachalam, K., Wong, C.O., and Zhu, M.X. (2014). The role of TRPMLs in endolysosomal trafficking and function. Cell Calcium *58*, 48–56.

Venkatachalam, K., Wong, C.O., and Zhu, M.X. (2015). The role of TRPMLs in endolysosomal trafficking and function. Cell Calcium 58, 48–56.

Vergarajauregui, S., Connelly, P.S., Daniels, M.P., and Puertollano, R. (2008a). Autophagic dysfunction in mucolipidosis type IV patients. Human Molecular Genetics 17, 2723–2737.

Vergarajauregui, S., Oberdick, R., Kiselyov, K., and Puertollano, R. (2008b). Mucolipin 1 channel activity is regulated by protein kinase A-mediated phosphorylation. The Biochemical Journal 410, 417–425.

Vergarajauregui, S., and Puertollano, R. (2006). Two di-leucine motifs regulate trafficking of mucolipin-1 to lysosomes. Traffic 7, 337–353.

Vidyadhara, D.J., Lee, J.E., and Chandra, S.S. (2019). Role of the endolysosomal system in Parkinson's disease. Journal of Neurochemistry 150, 487–506.

Wallings, R., Connor-Robson, N., and Wade-Martins, R. (2019a). LRRK2 interacts with the vacuolar-type H+-ATPase pump a1 subunit to regulate lysosomal function. Human Molecular Genetics 28, 2696–2710.

Wallings, R.L., Humble, S.W., Ward, M.E., and Wade-Martins, R. (2019b). Lysosomal dysfunction at the centre of Parkinson's disease and frontotemporal dementia/amyotrophic lateral sclerosis. Trends in Neurosciences 42, 899–912.

Walseth, T.F., Lin-Moshier, Y., Weber, K., Marchant, J.S., Slama, J.T., and Guse, A.H. (2012). Nicotinic acid adenine dinucleotide 2′-phosphate (NAADP) binding proteins in T-lymphocytes. Messenger (Los Angel) 1, 86–94.

Wang, D., Chan, C.C., Cherry, S., and Hiesinger, P.R. (2013). Membrane trafficking in neuronal maintenance and degeneration. Cellular and Molecular Life Sciences: CMLS 70, 2919–2934.

Wang, W., Gao, Q., Yang, M., Zhang, X., Yu, L., Lawas, M., Li, X., Bryant-Genevier, M., Southall, N.T., Marugan, J., et al. (2015). Up-regulation of lysosomal TRPML1 channels is essential for lysosomal adaptation to nutrient starvation. Proceedings of the National Academy of Sciences of the United States of America 112, E1373–E1381.

Wang, W., Zhang, X., Gao, Q., Lawas, M., Yu, L., Cheng, X., Gu, M., Sahoo, N., Li, X., Li, P., et al. (2017). A voltage-dependent K$^+$ channel in the lysosome is required for refilling lysosomal Ca^{2+} stores. The Journal of Cell Biology 216, 1715–1730.

Wang, X., Zhang, X., Dong, X.P., Samie, M., Li, X., Cheng, X., Goschka, A., Shen, D., Zhou, Y., Harlow, J., et al. (2012). TPC proteins are phosphoinositide-activated sodium-selective ion channels in endosomes and lysosomes. Cell 151, 372–383.

Wartosch, L., Fuhrmann, J.C., Schweizer, M., Stauber, T., and Jentsch, T.J. (2009). Lysosomal degradation of endocytosed proteins depends on the chloride transport protein ClC-7. FASEB Journal: Official Publication of the Federation of American Societies for Experimental Biology 23, 4056–4068.

Weinert, S., Jabs, S., Hohensee, S., Chan, W.L., Kornak, U., and Jentsch, T.J. (2014). Transport activity and presence of ClC-7/Ostm1 complex account for different cellular functions. EMBO Reports 15, 784–791.

Weinert, S., Jabs, S., Supanchart, C., Schweizer, M., Gimber, N., Richter, M., Rademann, J., Stauber, T., Kornak, U., and Jentsch, T.J. (2010). Lysosomal pathology and osteopetrosis upon loss of H$^+$-driven lysosomal Cl$^-$ accumulation. Science 328, 1401–1403.

Wiwatpanit, T., Remis, N.N., Ahmad, A., Zhou, Y., Clancy, J.C., Cheatham, M.A., and Garcia-Anoveros, J. (2018). Codeficiency of lysosomal mucolipins 3 and 1 in cochlear hair cells diminishes outer hair cell longevity and accelerates age-related hearing loss. The Journal of Neuroscience: The Official Journal of the Society for Neuroscience 38, 3177–3189.

Wong, C.O., Li, R., Montell, C., and Venkatachalam, K. (2012). Drosophila TRPML is required for TORC1 activation. Current Biology: CB 22, 1616–1621.

Wong, Y.C., Ysselstein, D., and Krainc, D. (2018). Mitochondria-lysosome contacts regulate mitochondrial fission via RAB7 GTP hydrolysis. Nature *554*, 382–386.

Wrong, O.M., Norden, A.G., and Feest, T.G. (1994). Dent's disease; a familial proximal renal tubular syndrome with low-molecular-weight proteinuria, hypercalciuria, nephrocalcinosis, metabolic bone disease, progressive renal failure and a marked male predominance. QJM *87*, 473–493.

Wu, M., Li, X., Zhang, T., Liu, Z., and Zhao, Y. (2019). Identification of a nine-gene signature and establishment of a prognostic nomogram predicting overall survival of pancreatic cancer. Frontiers in Oncology *9*, 996.

Wulff, H., Castle, N.A., and Pardo, L.A. (2009). Voltage-gated potassium channels as therapeutic targets. Nature Reviews Drug Discovery *8*, 982–1001.

Xia, Z., Wang, L., Li, S., Tang, W., Sun, F., Wu, Y., Miao, L., and Cao, Z. (2020). ML-SA1, a selective TRPML agonist, inhibits DENV2 and ZIKV by promoting lysosomal acidification and protease activity. Antiviral Research *182*, 104922.

Xiong, J., and Zhu, M.X. (2016). Regulation of lysosomal ion homeostasis by channels and transporters. Science China Life Sciences *59*, 777–791.

Xu, H., Delling, M., Li, L., Dong, X., and Clapham, D.E. (2007). Activating mutation in a mucolipin transient receptor potential channel leads to melanocyte loss in varitint-waddler mice. Proceedings of the National Academy of Sciences of the United States of America *104*, 18321–18326.

Xu, H., and Ren, D. (2015). Lysosomal physiology. Annual Review of Physiology *77*, 57–80.

Xu, M., Almasi, S., Yang, Y., Yan, C., Sterea, A.M., Rizvi Syeda, A.K., Shen, B., Richard Derek, C., Huang, P., Gujar, S., et al. (2019). The lysosomal TRPML1 channel regulates triple negative breast cancer development by promoting mTORC1 and purinergic signaling pathways. Cell Calcium *79*, 80–88.

Yamamoto, S., Shimizu, S., Kiyonaka, S., Takahashi, N., Wajima, T., Hara, Y., Negoro, T., Hiroi, T., Kiuchi, Y., Okada, T., et al. (2008). TRPM2-mediated Ca^{2+} influx induces chemokine production in monocytes that aggravates inflammatory neutrophil infiltration. Nature Medicine *14*, 738–747.

Yamamura, H., Ohi, Y., Muraki, K., Watanabe, M., and Imaizumi, Y. (2001). BK channel activation by NS-1619 is partially mediated by intracellular Ca^{2+} release in smooth muscle cells of porcine coronary artery. British Journal of Pharmacology *132*, 828–834.

Yamasaki, M., Masgrau, R., Morgan, A.J., Churchill, G.C., Patel, S., Ashcroft, S.J., and Galione, A. (2004). Organelle selection determines agonist-specific Ca^{2+} signals in pancreatic acinar and beta cells. The Journal of Biological Chemistry *279*, 7234–7240.

Yang, J., Chen, J., Del Carmen Vitery, M., Osei-Owusu, J., Chu, J., Yu, H., Sun, S., and Qiu, Z. (2019). PAC, an evolutionarily conserved membrane protein, is a proton-activated chloride channel. Science *364*, 395–399.

Yellen, G. (2002). The voltage-gated potassium channels and their relatives. Nature *419*, 35–42.

Yin, C., Zhang, H., Liu, X., Zhang, Y., Bai, X., Wang, L., Li, H., Li, X., Zhang, S., and Zhang, L. (2019). Downregulated MCOLN1 attenuates the progression of non-small-cell lung cancer by inhibiting lysosome-autophagy. Cancer Management and Research *11*, 8607–8617.

Yoshikawa, M., Uchida, S., Ezaki, J., Rai, T., Hayama, A., Kobayashi, K., Kida, Y., Noda, M., Koike, M., Uchiyama, Y., et al. (2002). CLC-3 deficiency leads to phenotypes similar to human neuronal ceroid lipofuscinosis. Genes to Cells: Devoted to Molecular & Cellular Mechanisms *7*, 597–605.

Ysselstein, D., Nguyen, M., Young, T.J., Severino, A., Schwake, M., Merchant, K., and Krainc, D. (2019). LRRK2 kinase activity regulates lysosomal glucocerebrosidase in neurons derived from Parkinson's disease patients. Nature Communications 10, 5570.

Yu, L., McPhee, C.K., Zheng, L., Mardones, G.A., Rong, Y., Peng, J., Mi, N., Zhao, Y., Liu, Z., Wan, F., et al. (2010). Termination of autophagy and reformation of lysosomes regulated by mTOR. Nature 465, 942–946.

Yu, L., Zhang, X., Yang, Y., Li, D., Tang, K., Zhao, Z., He, W., Wang, C., Sahoo, N., Converso-Baran, K., et al. (2020). Small-molecule activation of lysosomal TRP channels ameliorates Duchenne muscular dystrophy in mouse models. Science Advances 6, eaaz2736.

Zeevi, D.A., Frumkin, A., Offen-Glasner, V., Kogot-Levin, A., and Bach, G. (2009). A potentially dynamic lysosomal role for the endogenous TRPML proteins. The Journal of Pathology 219, 153–162.

Zeevi, D.A., Lev, S., Frumkin, A., Minke, B., and Bach, G. (2010). Heteromultimeric TRPML channel assemblies play a crucial role in the regulation of cell viability models and starvation-induced autophagy. Journal of Cell Science 123, 3112–3124.

Zhang, L., Fang, Y., Cheng, X., Lian, Y., Xu, H., Zeng, Z., and Zhu, H. (2017). TRPML1 Participates in the progression of Alzheimer's disease by regulating the PPARgamma/AMPK/Mtor signalling pathway. Cellular Physiology and Biochemistry 43, 2446–2456.

Zhang, L., Sheng, R., and Qin, Z. (2009). The lysosome and neurodegenerative diseases. Acta Biochimica et Biophysica Sinica 41, 437–445.

Zhang, X., Chen, W., Li, P., Calvo, R., Southall, N., Hu, X., Bryant-Genevier, M., Feng, X., Geng, Q., Gao, C., et al. (2019). Agonist-specific voltage-dependent gating of lysosomal two-pore Na^+ channels. eLife 8, e51423.

Zhang, X., Cheng, X., Yu, L., Yang, J., Calvo, R., Patnaik, S., Hu, X., Gao, Q., Yang, M., Lawas, M., et al. (2016). MCOLN1 is a ROS sensor in lysosomes that regulates autophagy. Nature Communications 7, 12109.

Zhang, X., Li, X., and Xu, H. (2012). Phosphoinositide isoforms determine compartment-specific ion channel activity. Proceedings of the National Academy of Sciences of the United States of America 109, 11384–11389.

Zhang, Z., Chen, G., Zhou, W., Song, A., Xu, T., Luo, Q., Wang, W., Gu, X.S., and Duan, S. (2007). Regulated ATP release from astrocytes through lysosome exocytosis. Nature Cell Biology 9, 945–953.

Zhong, X.Z., Sun, X., Cao, Q., Dong, G., Schiffmann, R., and Dong, X.P. (2016). BK channel agonist represents a potential therapeutic approach for lysosomal storage diseases. Scientific Reports 6, 33684.

Zhong, X.Z., Zou, Y., Sun, X., Dong, G. Sr., Cao, Q., Pandey, A., Rainey, J.K., Zhu, X., and Dong, X.P. (2017). Inhibition of TRPML1 by lysosomal adenosine involved in severe combined immunodeficiency diseases. The Journal of Biological Chemistry 292, 3445–3455.

Zhu, M.X., Ma, J., Parrington, J., Calcraft, P.J., Galione, A., and Evans, A.M. (2010). Calcium signaling via two-pore channels: local or global, that is the question. American Journal of Physiology Cell Physiology 298, C430–C441.

Zoncu, R., Bar-Peled, L., Efeyan, A., Wang, S., Sancak, Y., and Sabatini, D.M. (2011). mTORC1 senses lysosomal amino acids through an inside-out mechanism that requires the vacuolar H^+-ATPase. Science 334, 678–683.

19 Microglial Voltage-Gated Proton Channel Hv1 in Neurological Disorders

Madhuvika Murugan and Long-Jun Wu

CONTENTS

19.1 INTRODUCTION

The voltage-gated proton channel, Hv1, is a unique ion channel with high selectivity for protons. The proton extrusion through the channel is gated via voltage- and pH-dependent mechanisms. The Hv1 channel is mainly expressed in immune cells such as macrophages, neutrophils, and eosinophils (Ramsey et al. 2006, Sasaki, Takagi, and Okamura 2006). As the primary immune cells in the central nervous system (CNS), microglia are known to highly express the Hv1 channel.

Voltage-gated proton currents were characterized in culture microglia many years ago (Eder and DeCoursey 2001), but its molecular identity as the voltage-gated proton channel Hv1 was only recently elucidated (Ramsey et al. 2006, Sasaki, Takagi, and Okamura 2006). Indeed, Hv1 is selectively and functionally expressed in microglia but not neurons in the mouse brain (Wu et al. 2012, Wu 2014a). Microglial Hv1 regulates intracellular pH and promotes nicotinamide adenine dinucleotide phosphate oxidase (NOX)-dependent generation of reactive oxygen species (ROS). In a mouse model of middle cerebral artery occlusion (MCAO), Hv1 knockout (KO) mice were protected from ischemic damage, showing reduced NOX-dependent ROS production, microglia activation and neuronal cell death (Wu et al. 2012, Wu 2014b, Tian et al. 2016). A similar neuroprotective phenotype was noted in experimental models of spinal cord injury (SCI) and multiple sclerosis (MS), wherein, Hv1 KO mice exhibited white matter sparing and improved motor recovery compared to the wild-type (WT) mice (Liu et al. 2015, Murugan et al. 2020, Li et al. 2020, Li et al. 2021). Interestingly, functional Hv1 expression was also identified in human glioblastoma multiforme (GBM), the most common and lethal brain tumor with possible implications in cancer metastasis (Ribeiro-Silva et al. 2016, Fernandez et al. 2016, Bare et al. 2020). These studies illuminate a critical role for Hv1, particularly microglial Hv1 in neurological disorders, providing a strong rationale for targeting Hv1 for therapeutic benefit. This chapter provides a comprehensive overview of our current understanding of microglial Hv1 in normal and pathological states, and recent progress in the pharmaceutical development of small molecules for Hv1 inhibition.

19.2   VOLTAGE-GATED PROTON CHANNEL Hv1: DISCOVERY, STRUCTURE AND GATING

Voltage-gated proton currents have been observed in immune cells and were believed to be a component of NOX (DeCoursey 2003). However, the discovery of Hv1 proton channel demonstrated that voltage-gated proton current is mediated by its own ion channel. In 2006, two independent groups reported that the *HVCN1* gene encodes an ion channel with voltage-dependent, proton selective flux (Ramsey et al. 2006, Sasaki, Takagi, and Okamura 2006). By comparing the sequence homology to the voltage sensor domain of voltage-gated cation channels, the gene *HVCN1* encoding the voltage-gated proton channel from human genome was identified by the Clapham lab and named proton channel Hv1 to denote the first member in the family. Concurrently, Okamura lab used a similar strategy and identified mouse voltage-gated proton channel, mVSOP (mouse voltage-sensor domain-only protein) (Sasaki, Takagi, and Okamura 2006). It is notable that mVSOP is a homologue of the voltage-sensitive lipid phosphatase found in *Ciona intestinalis* (Murata et al. 2005). The Hv1 channel was later identified in coccolithophores (Taylor et al. 2011) and dinoflagellate (Smith et al. 2011). In addition, genes that are homologous to *HVCN1* are present in many species, from green alga, zebrafish, to monkey (DeCoursey 2013).

   The Hv1 proton channel belongs to the voltage-gated ion channel superfamily with unique biophysical properties (Figure 19.1). It contains 273 amino acids with intracellular N- and C-terminals. Hv1 has four transmembrane domains (S1-S4) and lacks a typical S5-S6 pore domain required for ion conduction in many other voltage-gated ion channels. Three arginine residues in the S4 domain are critical for voltage gating, and two histidine residues are required for extracellular inhibition of Hv1 by Zn^{2+} (Ramsey et al. 2006). The crystal structure of mouse Hv1 in resting state (closed configuration) was performed at a resolution of 3.45 Å, and shows a "closed umbrella" shape with a long helix consisting of the cytoplasmic coiled-coil and the voltage-sensing helix, S4, and features a wide inner-accessible vestibule (Takeshita et al. 2014). Two out of three arginines in S4 were located below the phenylalanine constituting the gating charge-transfer center. Much of the investigations on Hv1 channel mechanisms are based on this study, however crystal structures of Hv1 at higher resolution and of other gating states (of open state) is necessary to reveal detailed operating mechanisms of Hv1 and general principles of the voltage-sensing domain that are shared among voltage-gated ion channels. A more recent study used a combination of techniques to refine the known location and function of the Hv1 selectivity filter (Chamberlin et al. 2015). Using a string simulation method, the study showed that the gating transition from the closed to open state in Hv1 channel involves two distinct transitions. Moreover, the presumed closed structure of mouse Hv1 defined by Takeshita et al was more similar to the intermediary state (Chamberlin et al. 2015). This was confirmed by an independent study that used nuclear magnetic resonance (NMR) spectroscopy to elucidate the solution structure of an N- and C-terminally truncated human Hv1 in the resting state.

Hv1 is the most selective ion channel known, showing no detectable permeability to other ions. The ion selectivity filter of Hv1 channel is believed to be comprised of aspartate 112 in S1 and arginine 211 in S4 (Berger and Isacoff 2011, Musset et al. 2011). These studies confirmed that Asp when substituted with Ser, Ala, or His, it results in anion permeability in both human Hv1, as well as, Hv1 from a dinoflagellate, *Karlodinium veneficum*. These effects noted in two diverse

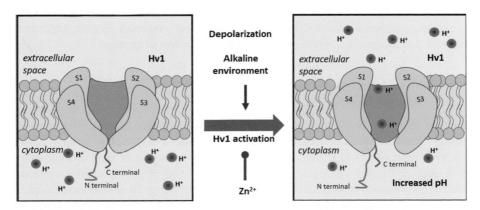

FIGURE 19.1   Hv1 proton channel activation. Schematic diagram shows the structure of Hv1 proton channel made up of four transmembrane domains (S1-S4) with intracellular N- and C-terminals. Hv1 channel activation is triggered by depolarization and an alkaline environment, whereas the presence of zinc inhibits the channel activation.

animal species strongly suggests that the Hv1 selectivity mechanism is evolutionarily conserved (Berger and Isacoff 2011, Musset et al. 2011). With the availability of crystal structure, several studies have attempted to unravel the selectivity mechanism of Hv1 channel (Morgan et al. 2013, Dudev et al. 2015, Lee et al. 2018). Using a reduced quantum model, a study showed that the acidic and basic residues of Asp-Arg form favorable interactions with hydronium but unfavorable interactions with anions/cations (Dudev et al. 2015). Molecular dynamics simulations reveal water molecules in the central crevice of Hv1 model structures but not in homologous voltage-sensor domain structures (Ramsey et al. 2010). These results indicate that the channel forms an internal water wire for selective proton transfer and that interactions between water molecules and S4 arginines may underlie coupling between voltage- and pH-gradient sensing (Ramsey et al. 2010). Taken together, these studies highlight the importance of quantum effects (charge transfer and polarization) in proton selectivity. However, these molecular dynamics simulations are based on the closed structure of Hv1 and do not account for interactions between non-selection filter residues and kinetic barriers, features that may influence proton selectivity.

Hv1 is a dimer and the cytosolic domain of the channel is necessary and sufficient for dimerization (Lee, Letts, and Mackinnon 2008). Interestingly, each subunit of the Hv1 dimer is functional and has a separate permeation pathway with its own pore and voltage sensor (Tombola, Ulbrich, and Isacoff 2008). Recent studies showed that the opening of the two pathways in Hv1 channels is highly cooperative, which involves interactions between the two subunits of the Hv1 dimer to initiate the conformational change during activation (Lee, Letts, and Mackinnon 2008, Qiu et al. 2013). This was further confirmed by voltage- and patch-clamp fluorometry studies, which showed that motion of S1 and S4 are each influenced by residues on the other helix, thus suggesting a dynamic interaction between the two domains (Mony, Berger, and Isacoff 2015). These findings also indicate that the S1 of Hv1 is able to function as part of the channel's gate (Mony, Berger, and Isacoff 2015). Despite the progress on understanding the Hv1 gating and proton permeation mechanisms, additional structural information combined with modeling are needed to unravel mechanisms underlying the efficient conduction of protons.

19.3 MICROGLIAL Hv1: SELECTIVE EXPRESSION AND DEVELOPMENTAL REGULATION

Voltage-gated proton currents in microglia were first recorded in primary microglial culture from the murine neonatal brain (Eder et al. 1995). Later, voltage-gated currents were observed in cultured rat and human microglia as well as microglial cell lines (Eder and DeCoursey 2001). The existence of proton currents *in situ* microglia remained elusive until recently. The discrepancy from two studies, with one study noting proton currents in microglia in P5–9 mouse brain slices, whereas, the other study showing lack of currents in microglia in P21 rat brain slices, further fueled the debate (Schilling and Eder 2007, De Simoni, Allen, and Attwell 2008). The controversy on whether microglia *in situ* express voltage-gated proton current was resolved by recording microglia in both rat and mouse brain slices under the same conditions (Wu et al. 2012). Surprisingly, voltage-gated proton currents in mouse microglia are 10 times larger than those from rat microglia. Considering the close sequence homology between mouse and rat Hv1 proton channel, future studies are needed to investigate the mechanism underlying the species difference in microglial voltage-gated proton current. The species differences also raise the concern of the existence of voltage-gated proton currents in human brain tissue *in situ*, though voltage-gated proton currents are observed in cultured human microglia (McLarnon et al. 1997, Wu et al. 2012).

To test whether the Hv1 channel mediated voltage-gated proton currents in microglia from mouse brain slices, multidisciplinary methods including biochemical, electrophysiological, pharmacological and genetic approaches were used (Wu et al. 2012). (1) Quantitative PCR results showed that *HVCN1* mRNA is expressed in the brain and cultured microglia. However, Hv1 protein could not be detected in whole-brain lysate because of relatively limited number of microglia in the brain. (2) Whole-cell recordings in microglia in mouse brain slices demonstrated the typical voltage-gated proton currents. (3) The voltage-gated proton current in microglia is largely reduced by Zn^{2+}, the known inhibitor of Hv1. (4) More importantly, the voltage-gated proton current in microglia is completely abolished in microglia from Hv1 KO mice (Figure 19.2A). Thus, these results conclude that Hv1 solely mediates voltage-gated proton currents in microglia in the mouse brain.

During development, microglial proton currents are not noticeably regulated. The microglial proton currents persist in both neonatal (P0–2) and adult mouse microglia (P21–23) (Figure 19.2A). In contrast to the electrophysiology results, a study using immunostaining indicated that the protein expression of Hv1 proton channel is absent in the neonatal mouse (P9) brain but present in adult microglia (Okochi et al. 2009). It was argued that Hv1 expression may be regulated in a stage-specific manner, or alternatively, Hv1 could be confined to intracellular compartments, and not abundantly expressed on the cell surface. In the light of the above, the developmental regulation and role of Hv1 remains elusive and warrants further investigation.

In addition to microglia, other immune cells that express Hv1 include macrophages, neutrophils, dendritic cells, T cells and B cells (Sasaki, Takagi, and Okamura 2006, Okochi et al. 2009, El Chemaly et al. 2010). Interestingly, Hv1 activation in B cells modulated B cell antigen receptor (BCR) signaling via generation of ROS (Capasso et al. 2010). Since the CNS is an immune-privileged site, Hv1 expression is limited to microglia in normal physiological conditions. However, in pathological conditions in which the blood-brain barrier is compromised, Hv1 expressed in infiltrating cells might also contribute to the pathological milieu (See Section 19.5).

19.4 MICROGLIAL Hv1: CELLULAR FUNCTIONS

Resting microglia have diverse functions in the brain during normal physiological conditions, including but not limited to—monitoring synapses with dynamic extension and retraction of their highly ramified cell processes (Wake et al. 2009,

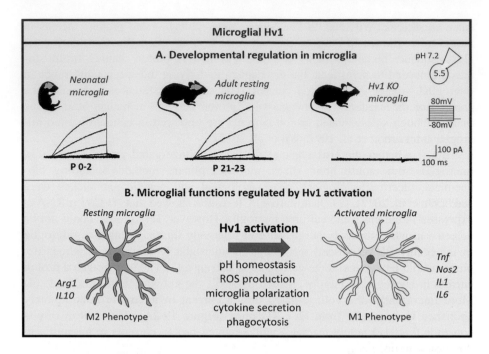

FIGURE 19.2   Microglial Hv1 in development and cellular functions. (A) Representative whole-cell voltage-gated proton currents in mouse microglia from acute hippocampal brain slices. Proton currents in response to voltage steps from −80 to +80 mV (shown in inset) were recorded in microglia under conventional whole-cell voltage-clamp mode with extracellular pH at 7.2 and intracellular pH at 5.5. Holding potential was −60 mV. Proton currents were comparable in P0–2 and P21–23 mouse microglia, but were completely abolished in Hv1 KO mouse microglia. (B) Schematic shows physiological functions of resting and activated microglia regulated by Hv1 such as maintenance of cellular pH, ROS production, microglial polarization and activation, cytokine secretion, and phagocytosis.

Tremblay, Lowery, and Majewska 2010) and phagocytosis of developing (Marin-Teva et al. 2004) and adult-born neurons (Sierra et al. 2010, Paolicelli et al. 2011, Schafer et al. 2012, Eyo and Wu 2013, Hong, Dissing-Olesen, and Stevens 2016). The Hv1 channel is highly expressed in microglia and is one of the major ion channels in resting microglia, indicating a key role for the channel in physiological functions. In this section, we review the cellular functions of microglia that are particularly regulated by Hv1 channel activation such as maintenance of pH, ROS production, microglial polarization, and phagocytosis (Figure 19.2B).

19.4.1   pH Homeostasis

The Hv1 proton channel is activated by membrane depolarization and mediates outward proton currents. Therefore, there are two consequences to Hv1 activation: one is the hyperpolarization of the plasma membrane, and the other is the alkalization of intracellular cytosol. The Hv1-mediated membrane hyperpolarization might be important for Ca^{2+} entry through store-operated Ca^{2+} channels (El

Chemaly et al. 2010). Similarly, microglial K channel-dependent hyperpolarization was shown to promote P2X4-mediated Ca^{2+} elevation (Nguyen et al. 2020). The Hv1-mediated intracellular alkalization potentially regulates pH homeostasis and has been extensively investigated in neutrophils (Morgan et al. 2009, El Chemaly et al. 2010).

In brain microglia, it has been shown that the activation of Hv1 drastically maintains intracellular acidification under certain conditions such as during depolarization (Wu et al. 2012). Since Hv1 activation requires membrane depolarization, it seems that the sodium-proton exchanger (NHE1) but not Hv1 channels maintain basal intracellular pH during resting conditions (Liu et al. 2010, Wu et al. 2012). In the event of a dramatic drop in the intracellular pH (~6.5), Hv1 channels may not be activated to rectify acidosis due to the high voltage threshold for their activation. Microglial membrane depolarization, however, could activate Hv1 and promptly relieve the intracellular acidosis. Compared to NHE1-dependent proton extrusion, Hv1-mediated pH recovery is several times faster in microglia. Therefore, microglial Hv1 is an efficient pH regulator under conditions of strong membrane depolarization, such as induced by increased extracellular K^+ concentration or NOX activation. Considering Hv1 activation requires massive depolarization, it is still unknown whether Hv1 indeed regulates pH homeostasis under physiological conditions *in vivo*. Consistent with this notion, Hv1 channels do not participate in ATP-induced chemotaxis in microglia (Wu et al. 2012), although NHE1-mediated intracellular alkalization is known to be critical for fibroblast and neutrophil migration (Casey, Grinstein, and Orlowski 2010). This may be because ATP activates P2Y receptor-coupled K^+ channel and thus hyperpolarize microglia (Boucsein et al. 2003, Wu, Vadakkan, and Zhuo 2007, Wu and Zhuo 2008). In addition to regulating intracellular pH in physiological conditions, microglial Hv1-dependnent proton extrusion is also reported to mediate tissue acidosis in pathological conditions such as SCI (Li et al. 2021) and discussed in Section 5.

19.4.2 ROS PRODUCTION

Proton currents were proposed to function cooperatively with NOX activities during respiratory burst, which is associated with both intracellular acidosis and membrane depolarization (DeCoursey 2003): (1) NOX activation induces electron transfer across the membrane, which depolarizes the membrane, with an estimated membrane potential of +58 mV during the respiratory burst. The depolarization could reach up to +190 mV within 20 ms in the absence of a charge compensatory mechanism (DeCoursey, Morgan, and Cherny 2003); (2) during NOX activation, protons are accumulated when electrons are transported across the cell membrane. Consistently, NOX activation induces a dramatic decrease in cytoplasmic pH in human neutrophils during phagocytosis (Morgan et al. 2009). Hence, the combined effect of membrane depolarization and intracellular acidosis during NOX activation is sufficient to open Hv1. On the other hand, NOX activity is inhibited by membrane depolarization or intracellular acidosis (DeCoursey 2003). Therefore, under constant NOX activation such as respiratory oxidation, there must be mechanisms to compensate for the charge transfer and to relieve intracellular acidosis. Hv1 is ideally suited for the dual functions in cooperation with NOX. Consistent with this idea, studies have shown that Hv1 is coupled to NOX-dependent pH regulation, membrane depolarization, and ROS production in neutrophils (El Chemaly et al. 2010), B cells (Capasso et al. 2010), and eosinophils (Zhu, Mose, and Zimmermann 2013).

Both Hv1 and NOX (particularly NOX2) are highly expressed in brain microglia. Compared with WT microglia, Hv1 KO microglia exhibited more intracellular acidosis when NOX is activated by phorbol myristate acetate (PMA), suggesting that NOX and Hv1 are cooperatively activated (Wu et al. 2012). In addition, Hv1 is required for NOX-dependent ROS production and Hv1 KO microglia accumulate significantly less ROS compared with that of WT microglia in brain slices. Microglial Hv1 is also critical for NOX-dependent ROS production in ischemia conditions both *in vitro* and *in vivo*. Therefore, similar to that in other immune cells, the Hv1 proton channel in microglia aids in NOX activation and subsequent ROS production. However, it is important to keep in mind that NOX-dependent ROS production is not completely abolished in Hv1 KO, suggesting that NHE1 and other ion channels may serve as alternative mechanisms to compensate for NOX activities, even though the Hv1 channel is the ideal channel for NOX coupling and activation.

19.4.3 MICROGLIAL ACTIVATION AND POLARIZATION

There is a strong correlation between Hv1 and microglial activation, however, whether and how Hv1 activation influences microglial activation state remains largely unknown. Activation of microglia is accompanied by changes in morphology, proliferation, up-regulation of immune surface antigens and the production of cytotoxic or neurotrophic molecules (Ransohoff and Perry 2009). Under normal conditions, Hv1 KO microglia seem to have similar morphology compared with WT microglia. However, the enlarged cell bodies after stroke were attenuated in Hv1 KO microglia compared with WT microglia, suggesting that Hv1 contributes to microglial activation after ischemia (Wu et al. 2012). In line with this, microglial outward rectifier K^+ currents were significantly larger at 7 d after SCI in WT mice compared to Hv1 KO mice (Murugan et al. 2020). These currents typical of voltage-gated Kv1.3 channel activation are required for microglial pro-inflammatory activation in vivo (Di Lucente et al. 2018) and emphasize a role for Hv1 in accentuating microglial activation. In the neutrophil, Hv1 is reported to participate in the Ca^{2+} signaling and neutrophil migration (El Chemaly et al. 2010). It is still unknown whether the attenuated microglial activation in Hv1 KO mice after stroke/SCI results from the impaired Ca^{2+} influx or indirectly from the impaired ROS production/pH homeostasis. Recently, genetic tools are available to further investigate Hv1 function in microglial Ca^{2+} signaling *in vivo* (Umpierre et al. 2020).

Upon activation, microglia express a unique signature of cell surface and intracellular markers, and secrete different factors, depending on the state of activation. Although previously classified more rigidly as either M1 or M2 phenotype, recently it is believed that microglia exist in a spectrum of state ranging from M1 to M2 phenotypes (Ransohoff and Perry 2009, Ransohoff 2016, Qin et al. 2019). The use of Hv1 KO mice indicated that microglial Hv1 is capable of regulating microglial polarization. In fact, lack of Hv1 shifted the milieu from a predominantly pro-inflammatory M1 state to an anti-inflammatory M2 state, with increased expression of M2 markers (Arg1 and CD206) and relative reduction of M1 markers (iNOS and CD16/32) (Tian et al. 2016). In line with this, an unbiased cytokine array showed that injury-induced increase in expression of CXCL13, C5/C5a, CXCL10, M-CSF, CCL2, CCL12, CXCL12, TIMP-1, TNFα, and TREM-1 in the spinal cord was prevented in Hv1 KO mice, suggesting Hv1 is capable of modulating cytokine expression (Murugan et al. 2020). However, further studies are needed to confirm if there is a direct association between Hv1 activation, cytokine production, and microglial polarization, and investigate the underlying mechanisms.

19.4.4 PHAGOCYTOSIS

Another important, yet-unrecognized role for Hv1 in microglia, is its role in the regulation of phagosomes. In innate immune cells such as macrophages and neutrophils, the function of Hv1 in phagosome maintenance has been established (El Chemaly et al. 2014). The maintenance of an appropriate ionic and redox environment within the phagosomal lumen is critical for bacterial killing and antigen presentation by neutrophils and macrophages. A study showed that Hv1 channels regulated the pH in neutrophils and macrophages, sustaining rapid acidification in macrophage phagosomes and maintaining a neutral pH in neutrophil phagosomes (Okochi et al. 2009, El Chemaly et al. 2014). However, obvious, direct evidence linking Hv1 in microglia phagocytic activity is missing. Since microglial phagocytosis is essential for normal development and implicated in various neurodegenerative disorders, identifying a role for Hv1 in microglial phagocytosis would be potentially interesting and holds significant therapeutic value.

19.5 MICROGLIAL Hv1 IN NEUROLOGICAL DISORDERS

The direct coupling between Hv1 and NOX unquestionably suggests the involvement of microglial Hv1 in pathological conditions in which ROS and neuroinflammation are implicated. Lack of specific Hv1 inhibitors (until recently), have led to our reliance on Hv1 KO mice for understanding the physiological or pathological significance of Hv1 *in vivo*. Since, microglial cells are the major resident cell type that highly expresses Hv1 in the CNS, general constitutive Hv1 KO mice are useful to study microglial Hv1 function in the brain/spinal cord. However, in many pathological conditions, the blood-brain barrier is compromised and Hv1 in infiltrating immune cells such as leukocytes, neutrophils and macrophages may also contribute to the pathological situation. In these instances, it would be advantageous to use conditional Hv1 KO to dissect the role of microglial Hv1. In this section, we discuss recent reports that implicate microglial Hv1 function in various neurological disorders such as ischemic stroke, SCI, MS, and GBM (Figure 19.3). We briefly highlight the mechanisms mediated by microglial Hv1 in disease progression that can be exploited for therapeutic gain.

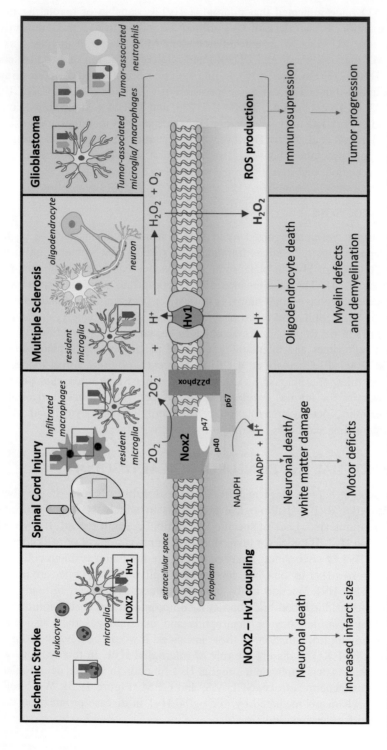

FIGURE 19.3 Microglial Hv1 in neurological disorders. Illustration depicts the mechanisms underlying microglial Hv1 channel activation in various neurological disorders such as ischemic stroke, spinal cord injury, multiple sclerosis, and cancer/glioblastoma. The Hv1-NOX coupling and resultant ROS productions seems to be a unifying mechanism by which Hv1 is implicated in the various neurological conditions.

19.5.1 ISCHEMIC STROKE

The implication of Hv1 in ischemic stroke was one of the earliest reported roles for microglial Hv1 (Wu et al. 2012, Tian et al. 2016). The rationale to study microglial Hv1 in ischemic stroke stems from earlier studies in which NOX1, 2, 4 knockout mice showed less brain damage during the pathology (Bedard and Krause 2007, Wu 2014b). As NOX2 is highly expressed in microglia and Hv1 is coupled to NOX2 activation, it is possible that Hv1 KO mice may be protected in this disease. Indeed, Hv1 KO mice exhibited significantly smaller infarct volumes and improved neurological behaviors compared to WT mice in two experimental models of ischemic stroke-MCAO (Wu 2014b) and photothrombotic stroke model (Tian et al. 2016). The protection was long-lasting, even observed 7 days after MCAO. Further study found decreased neuronal cell death in the penumbra of the ischemic brain from Hv1 KO mice (Wu et al. 2012). Additional evidence demonstrated that microglial Hv1 contributes to neuronal death and ischemic brain damage via the NOX pathway using a microglia-neuron co-culture system. These results suggest that Hv1 plays a critical role in microglial-derived ROS generation that accounts for a significant amount of neuronal cell death and brain injury that occurs after experimental stroke in mice (Wu et al. 2012) (Figure 19.3). ROS target DNA, membrane lipids, phosphatases/kinases, pro-apoptotic transcription factors, and ion channels to exert cell toxicity (Nathan and Ding 2010). TRPM2 and TRPM7 channels are particularly interesting ROS targets, as both ion channels are implicated in stroke-related neuronal cell death (Wu, Sweet, and Clapham 2010). Alternatively, a NOX-independent Hv1 function in cell death might involve neuronal acid-sensing ion channel (ASICs), as ASICs are proton-gated ion channels abundantly expressed in central neurons and are highly sensitive to extracellular acidosis (Wu et al. 2004). Coincidently, both ASIC1a KO mice and Hv1 KO mice show a reduced brain infarct after ischemic stroke (Xiong et al. 2004, Wu et al. 2012). Indeed, proton extrusion through Hv1 proton channel is able to activate ASIC1a in vitro, showing the proof of principle on the Hv1-ASIC coupling (Zeng et al. 2015, Wu 2014a).

An alternative role for microglial Hv1 in ischemic injury is via its regulation of microglia polarization. It is notable that Hv1 deficiency modulates the shift in microglia polarization state from M1 to M2 both in vivo after photothrombotic ischemia and in vitro cultured microglia (Tian et al. 2016). The shift in balance of M1 and M2 markers has been reported after ischemia, with a typical dominance of pro-inflammatory M1 state during the acute period (Kigerl et al. 2009). In Hv1 KO mice, the expression of anti-inflammatory M2 markers (Arg1 and CD206) was enhanced, while M1 related markers (iNOS and CD16/32) were reduced at 1 d after ischemia, compared with the WT mice. In line with this, deficiency of Hv1 in cultured microglia significantly reduced M1 polarization and shifted LPS-treated microglia to an M2 polarization state. These results indicate that Hv1 deficiency aids the shift toward a more protective phenotype (Tian et al. 2016). Since superoxide synthesis is an important inducer of M1 macrophage/microglia differentiation (Khayrullina, Bermudez, and Byrnes 2015), it is possible that NOX-mediated ROS is the underlying mechanism for the observed shift in microglial polarization in Hv1 KO mice. However, we cannot exclude the possibility that Hv1 regulation of intracellular pH may participate in the switch of microglial M1/M2 polarization (Wu 2014b).

19.5.2 SPINAL CORD INJURY

Emerging experimental evidence reveals a role for Hv1 in SCI (Murugan et al. 2020, Li, Liu, et al. 2020, Li, Yu, et al. 2020, Liet al. 2021). Hv1 deficiency was sufficient to rescue motor deficits caused by SCI. The Hv1-mediated changes in motor deficits were linked to microglial activation, IL-1β release, ROS production and neuronal loss (Murugan et al. 2020). Further, deficiency of Hv1 directly influenced microglia activation as noted by a decrease in microglial number, soma size and reduced outward rectifier K^+ current density in Hv1 KO mice compared to WT mice at 7 d following SCI (Murugan et al. 2020). These results suggest that microglial Hv1 is a promising potential therapeutic target to alleviate neuronal loss and secondary damage following SCI. Interestingly, the mechanism of Hv1-mediated neuronal loss in SCI was identified to be pyroptosis (Li, Yu, et al. 2020). Contrary to apoptosis, pyroptosis is a form of inflammatory cell death requiring the activation of caspase-1 (Miao, Rajan, and Aderem 2011). Hence, the occurrence of pyroptosis was claimed based on the increased expression of nod-like receptor 3 (NLRP3) inflammasome, ASC, and caspase-1 in neurons after SCI in WT mice, which was prevented in Hv1 KO mice (Li, Yu, et al. 2020). Apart from neuronal loss, demyelination deficits and white matter damage around the injury site is believed to be the main underlying cause for motor deficits noted in SCI patients (Schucht et al. 2002). However, in Hv1 KO mice, attenuated apoptosis of oligodendrocytes, ameliorated myelin loss and a corresponding improvement in tissue repair was observed after SCI (Li, Liu, et al. 2020). Similar to this, hematoxylin and eosin staining of the lesion area at 7 d after SCI revealed enhanced myelin sparing in Hv1 KO mice compared with WT mice (Murugan et al. 2020). The prevention of myelin loss in Hv1 KO mice was prominent near the lesion epicenter and was also observed at distances up to 1500 μm rostral and caudal to the injury epicenter. Since hemorrhaging and white matter loss are key indicators of secondary tissue damage following injury (Anwar, Al Shehabi, and Eid 2016), this result strongly suggests that Hv1 deficiency can alleviate the secondary damage at the early stage following SCI (Murugan et al. 2020).

In a contusion model of SCI, a dramatic decrease in pH levels was noted at the injury site during the first week after SCI (Liet al. 2021). Acidosis was evident at the injury site and also noted in the proximal regions of the cervical and lumbar cord, which was significantly reduced in Hv1 KO mice. These results indicate that microglial Hv1 is a key mediator of tissue acidosis during SCI. In line with the other studies, Hv1-mediated tissue acidosis was linked to NOX2-mediated ROS production, microglia activation, leukocyte infiltration, neuroinflammatory cytokine production and functional outcome after SCI. These studies unequivocally demonstrate the significance of targeting Hv1 in motor function recovery following spinal cord trauma (Li et al. 2021, Murugan et al. 2020).

In addition to inflammation and white matter damage, our recent data suggest that spinal microglial Hv1-dependent ROS is critical for neuropathic pain induced by peripheral nerve injury (Peng at al., unpublished). These findings highlight the importance of Hv1 in SCI pathology and targeting the same to improve sensorimotor function and neuropathic pain, two major quality of life affecting consequences of SCI (Li, Liu, et al. 2020, Murugan et al. 2020).

In addition to resident microglia, a number of circulating immune cell types respond to SCI (Popovich, Wei, and Stokes 1997, Taoka et al. 1997, Leskovar et al. 2000). As discussed, the infiltrating cell types such as macrophages and neutrophils also express Hv1, similar to the resident microglial population (Ramsey et al. 2009, Okochi et al. 2009, Schilling et al. 2002). Thus, it is likely that infiltrating neutrophils and macrophages also contribute to Hv1-mediated ROS damage following SCI and should be collectively targeted for therapeutic benefits of Hv1 inhibition.

19.5.3 MULTIPLE SCLEROSIS

The microglial Hv1 has been reported to play a role in an experimental model mimicking MS pathology (Liu et al. 2015). The study used a cuprizone model, an established toxicant-induced chronic demyelination model that mimics certain aspects of MS and is characterized by apoptosis of primary oligodendrocytes and demyelinating lesions particularly in the corpus callosum (Blakemore and Franklin 2008). Notably, Hv1 KO mice displayed a partially protected phenotype with attenuated demyelination and motor deficits compared to WT mice. These rescued phenotypes in Hv1 KO mice were accompanied by reduced ROS production, ameliorated microglial activation, increased oligodendrocyte progenitor cell (NG2) proliferation, and an increased number of mature oligodendrocytes. These results demonstrate that microglial Hv1 proton channel is required for cuprizone-induced ROS production and subsequent demyelination. Although extensive evidence implicates increased ROS production in inflammatory demyelination noted in MS (Haider et al. 2011, Fischer et al. 2012), the underlying mechanisms of ROS-dependent myelin loss are not yet clear. Future studies are needed to dissect the exact mechanism underlying microglial Hv1 function in demyelinating and related degenerative disorders.

19.5.4 GLIOBLASTOMA MULTIFORME

The expression of Hv1 was recently identified in a human glioblastoma multiforme (GBM) cell line (Ribeiro-Silva et al. 2016). WHO grade IV glioblastomas are the most frequent and malignant adult brain tumor with minimally prolonged survival by current standard treatment, including surgery, radiation and temozolomide chemotherapy (Ostrom et al. 2013, Weller et al. 2015). They are characterized and defined by their highly aggressive nature, involving rapid tumor growth, diffuse invasiveness, and resistance to therapy (Stupp et al. 2009). Like most malignant tumors, GBMs tend to have a fast metabolism leading to an accumulation of acidic metabolites in the cytosol, and consequently a low intracellular pH. Hence, there is a need for regulation of pH in tumor cells. Recent studies proposed that Hv1 can regulate the intracellular pH in several cancers (Wang et al. 2012, Wang et al. 2013, Wang, Zhang, and Li 2013). In particular, the functional expression of Hv1 channels was noted in T98G cells (a human GBM cell line with astrocytic origin) and conducted an outward, slow activating voltage-dependent, and pH-dependent proton current (Ribeiro-Silva et al. 2016). Inhibition of Hv1 channels by $ZnCl_2$ significantly reduced the intracellular pH, cell survival, and migration of T98G cells, indicating that Hv1 may have an important role for GBM proliferation and progression (Ribeiro-Silva et al. 2016).

Apart from the neoplastic cell itself, resident activated microglia and/or infiltrating macrophages are the most prevalent cell types within the tumor (Morantz et al. 1979a,b) and are referred to as tumor-associated microglia/macrophages (TAM). Despite the presence of large numbers of microglia/macrophages in GBM, they seem incapable of preventing tumor growth (Abels et al. 2019, Maas et al. 2020). In fact, it was shown that GBM can hijack microglial gene expression to avoid tumor sensing, to suppress of an immune response, and favor tumor propagation (Maas et al. 2020). The complex interactions between microglia/macrophages and tumor cells are not only relevant to glioma formation but also aid in glioma progression and influence efficacy of therapeutic interventions. The contribution of microglia/macrophages to glioma and the premise for targeting TAM to treat glioma has been extensively reviewed (Hambardzumyan, Gutmann, and Kettenmann 2016, Gutmann and Kettenmann 2019, Prionisti et al. 2019, Wei et al. 2020). Hv1 in myeloid cells has been associated with the maintenance of NOX2 activity and the ROS production, which both contribute to tumor progression and metastasis (Aydin et al. 2017, Canli et al. 2017, van der Weyden et al. 2018). Therefore, Hv1 proton channel expressed in myeloid cells could be important for the function of these cells within the tumor microenvironment. Collectively, Hv1 channels in tumor cells and associated myeloid cells could be a promising therapeutic target for GBM treatment (Fernandez et al. 2016).

19.6   TARGETING Hv1 FOR THERAPEUTIC GAIN

In addition to the neurological conditions mentioned above, accumulating evidence implicates Hv1 channel activation in a number of non-neurological conditions including obesity (Pang et al. 2020), diabetes (Zhao et al. 2015), hypertension (O'Connor et al. 2016), and various forms of cancer (Wang et al. 2012, 2013, Bare et al. 2020). The observation that Hv1 is specifically expressed in highly metastatic human breast tumor tissues and cell lines (Wang et al. 2012, Bare et al. 2020), has further piqued the interest for targeting Hv1 for anticancer therapy (Fernandez et al. 2016). Hence, the development of effective channel blockers for Hv1 can lead to new therapeutics for the treatment of maladies related to Hv1 dysfunction. Although the mechanism of proton permeation in Hv1 remains to be elucidated, a series of small molecules (Hong et al. 2013, Hong, Kim, and Tombola 2014) and peptide inhibitors (Zhao et al. 2018) that can block Hv1 activity have only recently been identified and their implications are discussed below. A potential side effect to consider while targeting Hv1 for therapy is the possible immune system disorder, considering the critical function of Hv1 in the immune system. Although previous studies showed that Hv1 KO mice can still clear several types of bacterial infections (Ramsey et al. 2009), infection susceptibility in various pathologies should be further investigated. Indeed, the autoimmune phenotype was observed in many of the Hv1 KO mice older than 6 months, including splenomegaly, autoantibodies, and nephritis (Sasaki et al. 2013).

19.6.1 SMALL MOLECULE INHIBITORS

A study showed that guanidine derivatives such as 2-guanidinobenzimidazole (2GBI) have the ability to inhibit Hv1 activity and inhibit proton conduction (Hong et al. 2013). The compound 2GBI inhibits Hv1 proton conduction by binding to the channel's voltage-sensing domain from its intracellular side. This is a drawback, since it requires the inhibitor to have direct access to the intracellular side of the channel to block the proton current. By applying mutant cycle analysis to determine how the inhibitor interacts with the channel, the same research group was able to develop a modified 2GBI analog that can reach the binding site on Hv1 from the extracellular side of the membrane (Hong, Kim, and Tombola 2014). However, its relatively low potency and the uncertainty of its specificity against Hv1 remain a concern (Pupo and Gonzalez Leon 2014). Use of the more selective blocker 2-(6-chloro-1H-benzimidazol-2-yl)guanidine (ClGBI) inhibited Hv1 channels resulting in decreased intracellular pH in an oncohematologic cell line (Asuaje et al. 2017). The intracellular acidification was followed by apoptosis, indicating that this mechanism may be exploited for cancer therapeutics (Asuaje et al. 2017). In another study, ClGBI significantly decreased mineral matrix production in chorion-derived mesenchymal stem cells (cMSCs) in conditions mimicking pathological induction of osteogenesis (Meszaros et al. 2020). Wound healing assay and single-cell motility analysis showed that ClGBI can significantly inhibit the migration of cMSCs, indicating that Hv1 is an attractive target for controlling MSC behavior when used in therapeutic applications (Meszaros et al. 2020).

19.6.2 PEPTIDE INHIBITORS

A more recent development is the synthesis of a de novo peptide inhibitor, Corza6 (C6) for Hv1 inhibition (Zhao et al. 2018). The peptide inhibitors also known as inhibitor cysteine knot (ICK) peptides, naturally exist in the venoms of spiders, scorpions, and snails and are rich in VSD-directed toxins that modify channel gating (Swartz 2007, Catterall et al. 2007). In other words, the ICK toxins bind to extracellular sites of ion channels and trap the VSD so they are unable to move normally in response to changes in transmembrane voltage (Cestele et al. 1998). The C6 peptide was identified by a phage-display strategy whereby ~1 million novel peptides were fabricated on an ICK scaffold and sorted on purified human Hv1 protein (Zhao et al. 2018). Using C6, it was demonstrated that the Hv1 is the main pathway for H^+ efflux that allows capacitation in sperm (Lishko et al. 2010, Zhao et al. 2018). The study also confirmed that Hv1 permits sustained ROS production in white blood cells (WBCs) (Zhao et al. 2018). Despite the progress in development of Hv1 inhibitors, their ability to reproduce the neuroprotective effects noted in Hv1 KO mice is yet to be tested in the different pathological experimental models. High-throughput screening and rational drug designs are required for the development of specific and potent Hv1 channel inhibitors that can be used for therapeutic intervention.

19.6.3 ZINC SUPPLEMENTATION

Zinc supplementation is a possible therapeutic strategy to block Hv1 activation, since the presence of zinc prevents Hv1 channel opening (Wu 2014a). There is a growing body of evidence to suggest the beneficial effects of zinc supplementation for various conditions including cancer and diabetes (Dhawan and Chadha 2010, Jayawardena et al. 2012), conditions in which Hv1 dysregulation has been observed. In a mouse model of Alzheimer's disease (AD), zinc supplementation increases expression of neuroprotective brain-derived growth factor (BDNF), greatly delayed hippocampal-dependent memory deficits and reduced Aβ pathology in the hippocampus (Corona et al. 2010). In an experimental model of epilepsy, the infusion of zinc into the hippocampus had an anticonvulsant effect, suggesting a therapeutic value in zinc supplementation (Elsas et al. 2009). In line with these findings, the supplementation of zinc for therapeutic gain has been discussed for various neurological disorders including epilepsy, Parkinson's disease, AD, schizophrenia, attention deficit and hyperactivity disorder, depression, amyotrophic lateral sclerosis, Down syndrome, MS, Wilson's disease, and Pick's disease (Grabrucker, Rowan, and Garner 2011). It is important to keep in mind that zinc is an essential mineral that is integral to many enzymes and transcription factors. Therefore, whether the effects are due to direct Hv1 inhibition will be difficult to assess in either an experimental or clinical setting, and need to be considered with careful deliberation.

19.7 CONCLUSION AND FUTURE DIRECTIONS

In summary, Hv1 is a recently discovered ion channel primarily expressed in the immune system to support NOX activity in innate immunity. In the brain, Hv1 is highly expressed in microglia and is one of the major ion channels in microglia. The cellular functions of microglial Hv1 include pH regulation, NOX-dependent ROS production, microglial polarization, and phagocytosis. Emerging evidence implicates microglial Hv1 in various neurological disorders including ischemic stroke, SCI, MS, and GBM. Taken together, these studies suggest that NOX-dependent ROS production is a common mechanism in most of the neurological diseases. Hence, inhibiting Hv1 activation and the resultant ROS might represent a novel therapeutic target for the treatment of ischemic stroke and other neurological disorders related to ROS and neuroinflammation (Perry, Nicoll, and Holmes 2010). There are still many open questions regarding the mechanisms of Hv1 activation and future studies are needed to fully appreciate microglial Hv1 function, particularly in the following research areas—(1) Higher resolution of Hv1 structure in different gating configurations, combined with improved simulation studies are needed for a better understanding of the mechanisms of Hv1 gating, trafficking, and regulation; (2) The expression of Hv1 in organelle membranes and their functions remain unknown and warrant further investigation; (3) Microglial Hv1 is gaining attention, but the role of Hv1 in other cell types needs to be elucidated in detail; (4) High-throughput drug screening for Hv1 inhibitors and activators are needed for the development of a commercial drug with high selectivity, the development of which is crucial for translational purposes of the current experimental findings; (5) Improved genetic tools, particularly, the development of conditional KO animals, are necessary to investigate cell-specific roles of Hv1 in various pathological/ physiological functions; (6) Data from genome-wide association studies (GWAS) from human patient samples need to be investigated to identify if and how Hv1 mutations are implicated in any neurological disorders. Overall, this chapter highlights the significance and novelty of targeting microglial Hv1 for therapeutic gain and discusses gaps in the field that need to be addressed.

ACKNOWLEDGMENTS

This work was supported by the National Institutes of Health (R01NS088627; R01NS112144; R01NS110949; R01NS110825; R21AG064159) to L.J.W. We thank members of the Wu Lab at Mayo for insightful discussions and Ms. Jiaying Zheng for editing the manuscript.

ABBREVIATIONS

ATP	Adenosine triphosphate
ASIC	Acid sensing ion channel
C5	Complement 5
C6	Corza6
CCL.	Chemokine ligand
cMSCs	Chorion-derived mesenchymal stem cells
CNS	Central nervous system
ClGBI	2-(6-chloro-1H-benzimidazol-2-yl) guanidine
CXCL	Chemokine ligand
GBM	Glioblastoma multiforme
GBI	2-guanidinobenzimidazole
GMF	Glioblastoma multiforme
Hv1	Voltage gated proton channel
ICK	Inhibitor cysteine knot
IL-1β	Interleukin-1β
KO	Knockout
MCAO	Middle cerebral artery occlusion
M-CSF	Macrophage colony stimulating factor
MS	Multiple sclerosis
mVSOP	Mouse voltage-sensor domain-only protein
NADPH	Nicotinamide adenine dinucleotide phosphate
NG2	Neuron-glial antigen, an oligodendrocyte progenitor cell marker
NHE1	Sodium-proton exchanger1
NLRP3	Nod-like receptor 3
NMR	Nuclear magnetic resonance

NOX	NADPH oxidase
PMA	Phorbol myristate acetate
ROS	Reactive oxygen species
SCI	Spinal cord injury
TAM	Tumor-associated macrophages
TIMP-1	Tissue inhibitor of metalloproteinases-1
TNFα	Tumor necrosis factor α
TREM-1	Triggering receptor expressed on myeloid cells-1
TRP	Transient receptor potential ion channels
VSD	Voltage sensing domain
WT	Wild type

REFERENCES

Abels, E. R., S. L. N. Maas, L. Nieland, Z. Wei, P. S. Cheah, E. Tai, C. J. Kolsteeg, S. A. Dusoswa, D. T. Ting, S. Hickman, J. El Khoury, A. M. Krichevsky, M. L. D. Broekman, and X. O. Breakefield. 2019. "Glioblastoma-associated microglia reprogramming is mediated by functional transfer of extracellular miR-21." *Cell Rep* 28 (12):3105–3119 e7. doi: 10.1016/j.celrep.2019.08.036.

Anwar, M. A., T. S. Al Shehabi, and A. H. Eid. 2016. "Inflammogenesis of secondary spinal cord injury." *Front Cell Neurosci* 10:98. doi: 10.3389/fncel.2016.00098.

Asuaje, A., P. Smaldini, P. Martin, N. Enrique, A. Orlowski, E. A. Aiello, C. Gonzalez Leon, G. Docena, and V. Milesi. 2017. "The inhibition of voltage-gated H(+) channel (HVCN1) induces acidification of leukemic Jurkat T cells promoting cell death by apoptosis." *Pflugers Arch* 469 (2):251–261. doi: 10.1007/s00424-016-1928-0.

Aydin, E., J. Johansson, F. H. Nazir, K. Hellstrand, and A. Martner. 2017. "Role of NOX2-derived reactive oxygen species in NK cell-mediated control of murine melanoma metastasis." *Cancer Immunol Res* 5 (9):804–811. doi: 10.1158/2326-6066.Cir-16-0382.

Bare, D. J., V. V. Cherny, T. E. DeCoursey, A. M. Abukhdeir, and D. Morgan. 2020. "Expression and function of voltage gated proton channels (Hv1) in MDA-MB-231 cells." *PLoS One* 15 (5):e0227522. doi: 10.1371/journal.pone.0227522.

Bedard, K., and K. H. Krause. 2007. "The NOX family of ROS-generating NADPH oxidases: physiology and pathophysiology." *Physiol Rev* 87 (1):245–313. doi: 10.1152/physrev.00044.2005.

Berger, T. K., and E. Y. Isacoff. 2011. "The pore of the voltage-gated proton channel." *Neuron* 72 (6):991–1000. doi: 10.1016/j.neuron.2011.11.014.

Blakemore, W. F., and R. J. Franklin. 2008. "Remyelination in experimental models of toxin-induced demyelination." *Curr Top Microbiol Immunol* 318:193–212.

Boucsein, C., R. Zacharias, K. Farber, S. Pavlovic, U. K. Hanisch, and H. Kettenmann. 2003. "Purinergic receptors on microglial cells: functional expression in acute brain slices and modulation of microglial activation in vitro." *Eur J Neurosci* 17 (11):2267–2276.

Canli, Ö., A. M. Nicolas, J. Gupta, F. Finkelmeier, O. Goncharova, M. Pesic, T. Neumann, D. Horst, M. Löwer, U. Sahin, and F. R. Greten. 2017. "Myeloid cell-derived reactive oxygen species induce epithelial mutagenesis." *Cancer Cell* 32 (6):869–883.e5. doi: 10.1016/j.ccell.2017.11.004.

Capasso, M., M. K. Bhamrah, T. Henley, R. S. Boyd, C. Langlais, K. Cain, D. Dinsdale, K. Pulford, M. Khan, B. Musset, V. V. Cherny, D. Morgan, R. D. Gascoyne, E. Vigorito, T. E. DeCoursey, I. C. MacLennan, and M. J. Dyer. 2010. "HVCN1 modulates

BCR signal strength via regulation of BCR-dependent generation of reactive oxygen species." *Nat Immunol* 11 (3):265–272. doi: ni.1843 [pii] 10.1038/ni.1843.

Casey, J. R., S. Grinstein, and J. Orlowski. 2010. "Sensors and regulators of intracellular pH." *Nat Rev Mol Cell Biol* 11 (1):50–61. doi: nrm2820 [pii] 10.1038/nrm2820.

Catterall, W. A., S. Cestele, V. Yarov-Yarovoy, F. H. Yu, K. Konoki, and T. Scheuer. 2007. "Voltage-gated ion channels and gating modifier toxins." *Toxicon* 49 (2):124–141. doi: 10.1016/j.toxicon.2006.09.022.

Cestele, S., Y. Qu, J. C. Rogers, H. Rochat, T. Scheuer, and W. A. Catterall. 1998. "Voltage sensor-trapping: enhanced activation of sodium channels by beta-scorpion toxin bound to the S3-S4 loop in domain II." *Neuron* 21 (4):919–931. doi: 10.1016/s0896-6273(00)80606-6.

Chamberlin, A., F. Qiu, Y. Wang, S. Y. Noskov, and H. P. Larsson. 2015. "Mapping the gating and permeation pathways in the voltage-gated proton channel Hv1." *J Mol Biol* 427 (1):131–145. doi: 10.1016/j.jmb.2014.11.018.

Corona, C., F. Masciopinto, E. Silvestri, A. D. Viscovo, R. Lattanzio, R. L. Sorda, D. Ciavardelli, F. Goglia, M. Piantelli, L. M. Canzoniero, and S. L. Sensi. 2010. "Dietary zinc supplementation of 3xTg-AD mice increases BDNF levels and prevents cognitive deficits as well as mitochondrial dysfunction." *Cell Death Dis* 1:e91. doi: 10.1038/cddis.2010.73.

De Simoni, A., N. J. Allen, and D. Attwell. 2008. "Charge compensation for NADPH oxidase activity in microglia in rat brain slices does not involve a proton current." *Eur J Neurosci* 28 (6):1146–1156.

DeCoursey, T. E. 2003. "Voltage-gated proton channels and other proton transfer pathways." *Physiol Rev* 83 (2):475–579. doi: 10.1152/physrev.00028.2002.

DeCoursey, T. E. 2013. "Voltage-gated proton channels: molecular biology, physiology, and pathophysiology of the H(V) family." *Physiol Rev* 93 (2):599–652. doi: 10.1152/physrev.00011.2012.

DeCoursey, T. E., D. Morgan, and V. V. Cherny. 2003. "The voltage dependence of NADPH oxidase reveals why phagocytes need proton channels." *Nature* 422 (6931):531–534. doi: 10.1038/nature01523.

Dhawan, D. K., and V. D. Chadha. 2010. "Zinc: a promising agent in dietary chemoprevention of cancer." *Indian J Med Res* 132:676–682.

Di Lucente, J., H. M. Nguyen, H. Wulff, L. W. Jin, and I. Maezawa. 2018. "The voltage-gated potassium channel Kv1.3 is required for microglial pro-inflammatory activation in vivo." *Glia* 66 (9):1881–1895. doi: 10.1002/glia.23457.

Dudev, T., B. Musset, D. Morgan, V. V. Cherny, S. M. Smith, K. Mazmanian, T. E. DeCoursey, and C. Lim. 2015. "Selectivity mechanism of the voltage-gated proton channel, HV1." *Sci Rep* 5:10320. doi: 10.1038/srep10320.

Eder, C., and T. E. DeCoursey. 2001. "Voltage-gated proton channels in microglia." *Prog Neurobiol* 64 (3):277–305.

Eder, C., H. G. Fischer, U. Hadding, and U. Heinemann. 1995. "Properties of voltage-gated currents of microglia developed using macrophage colony-stimulating factor." *Pflugers Arch* 430 (4):526–533.

El Chemaly, A., P. Nunes, W. Jimaja, C. Castelbou, and N. Demaurex. 2014. "Hv1 proton channels differentially regulate the pH of neutrophil and macrophage phagosomes by sustaining the production of phagosomal ROS that inhibit the delivery of vacuolar ATPases." *J Leukoc Biol* 95 (5):827–839. doi: 10.1189/jlb.0513251.

El Chemaly, A., Y. Okochi, M. Sasaki, S. Arnaudeau, Y. Okamura, and N. Demaurex. 2010. "VSOP/Hv1 proton channels sustain calcium entry, neutrophil migration, and super-oxide production by limiting cell depolarization and acidification." *J Exp Med* 207 (1):129–139. doi: 10.1084/jem.20091837.

Elsas, S. M., S. Hazany, W. L. Gregory, and I. Mody. 2009. "Hippocampal zinc infusion delays the development of after discharges and seizures in a kindling model of epilepsy." *Epilepsia* 50 (4):870–879. doi: 10.1111/j.1528-1167.2008.01913.x.

Eyo, U. B., and L. J. Wu. 2013. "Bi-directional microglia-neuron communication in the healthy brain." *Neural Plast* 2013:456857.

Fernandez, A., A. Pupo, K. Mena-Ulecia, and C. Gonzalez. 2016. "Pharmacological modulation of proton channel Hv1 in cancer therapy: future perspectives." *Mol Pharmacol* 90 (3):385–402. doi: 10.1124/mol.116.103804.

Fischer, M. T., R. Sharma, J. L. Lim, L. Haider, J. M. Frischer, J. Drexhage, D. Mahad, M. Bradl, J. van Horssen, and H. Lassmann. 2012. "NADPH oxidase expression in active multiple sclerosis lesions in relation to oxidative tissue damage and mitochondrial injury." *Brain* 135 (Pt 3):886–899. doi: 10.1093/brain/aws012.

Grabrucker, A. M., M. Rowan, and C. C. Garner. 2011. "Brain-delivery of zinc-ions as potential treatment for neurological diseases: mini review." *Drug Deliv Lett* 1 (1):13–23. doi: 10.2174/2210303111101010013.

Gutmann, D. H., and H. Kettenmann. 2019. "Microglia/brain macrophages as central drivers of brain tumor pathobiology." *Neuron* 104 (3):442–449. doi: 10.1016/j.neuron.2019.08.028.

Haider, L., M. T. Fischer, J. M. Frischer, J. Bauer, R. Hoftberger, G. Botond, H. Esterbauer, C. J. Binder, J. L. Witztum, and H. Lassmann. 2011. "Oxidative damage in multiple sclerosis lesions." *Brain* 134 (Pt 7):1914–1924. doi: 10.1093/brain/awr128.

Hambardzumyan, D., D. H. Gutmann, and H. Kettenmann. 2016. "The role of microglia and macrophages in glioma maintenance and progression." *Nat Neurosci* 19 (1):20–27. doi: 10.1038/nn.4185.

Hong, L., I. H. Kim, and F. Tombola. 2014. "Molecular determinants of Hv1 proton channel inhibition by guanidine derivatives." *Proc Natl Acad Sci USA* 111 (27):9971–9976. doi: 10.1073/pnas.1324012111.

Hong, L., M. M. Pathak, I. H. Kim, D. Ta, and F. Tombola. 2013. "Voltage-sensing domain of voltage-gated proton channel Hv1 shares mechanism of block with pore domains." *Neuron* 77 (2):274–287. doi: 10.1016/j.neuron.2012.11.013.

Hong, S., L. Dissing-Olesen, and B. Stevens. 2016. "New insights on the role of microglia in synaptic pruning in health and disease." *Curr Opin Neurobiol* 36:128–134. doi: 10.1 016/j.conb.2015.12.004.

Jayawardena, R., P. Ranasinghe, P. Galappatthy, R. Malkanthi, G. Constantine, and P. Katulanda. 2012. "Effects of zinc supplementation on diabetes mellitus: a systematic review and meta-analysis." *Diabetol Metab Syndr* 4 (1):13. doi: 10.1186/1758-5996-4-13.

Khayrullina, G., S. Bermudez, and K. R. Byrnes. 2015. "Inhibition of NOX2 reduces locomotor impairment, inflammation, and oxidative stress after spinal cord injury." *J Neuroinflammation* 12:172. doi: 10.1186/s12974-015-0391-8.

Kigerl, K. A., J. C. Gensel, D. P. Ankeny, J. K. Alexander, D. J. Donnelly, and P. G. Popovich. 2009. "Identification of two distinct macrophage subsets with divergent effects causing either neurotoxicity or regeneration in the injured mouse spinal cord." *J Neurosci* 29 (43):13435–13444. doi: 10.1523/jneurosci.3257-09.2009.

Lee, M., C. Bai, M. Feliks, R. Alhadeff, and A. Warshel. 2018. "On the control of the proton current in the voltage-gated proton channel Hv1." *Proc Natl Acad Sci USA* 115 (41):10321–10326. doi: 10.1073/pnas.1809766115.

Lee, S. Y., J. A. Letts, and R. Mackinnon. 2008. "Dimeric subunit stoichiometry of the human voltage-dependent proton channel Hv1." *Proc Natl Acad Sci USA* 105 (22):7692–7695. doi: 0803277105 [pii] 10.1073/pnas.0803277105.

Leskovar, A., L. J. Moriarty, J. J. Turek, I. A. Schoenlein, and R. B. Borgens. 2000. "The macrophage in acute neural injury: changes in cell numbers over time and levels of cytokine production in mammalian central and peripheral nervous systems." *J Exp Biol* 203 (Pt 12):1783–1795.

Li, X., R. Liu, Z. Yu, D. He, W. Zong, M. Wang, M. Xie, W. Wang, and X. Luo. 2020. "Microglial Hv1 exacerbates secondary damage after spinal cord injury in mice." *Biochem Biophys Res Commun* S0006-291X(20)30272-2. doi: 10.1016/j.bbrc.2020.02. 012.

Li, X., Z. Yu, W. Zong, P. Chen, J. Li, M. Wang, F. Ding, M. Xie, W. Wang, and X. Luo. 2020. "Deficiency of the microglial Hv1 proton channel attenuates neuronal pyroptosis and inhibits inflammatory reaction after spinal cord injury." *J Neuroinflammation* 17 (1):263. doi: 10.1186/s12974-020-01942-x.

Li, Y., Ritzek R. M., He J., Cao T., Sabirzhanov B., Li H., Liu S., Wu L. J., and Wu J. 2021. "The voltage-gated proton channel Hv1 plays a detrimental role in contusion spinal cord injury via extracellular acidosis-mediated neuroinflammation." *Brain Behav Immun* 91:267–283. doi: 10.1016/j.bbi.2020.10.005

Lishko, P. V., I. L. Botchkina, A. Fedorenko, and Y. Kirichok. 2010. "Acid extrusion from human spermatozoa is mediated by flagellar voltage-gated proton channel." *Cell* 140 (3):327–337. doi: S0092-8674(09)01680-8 [pii] 10.1016/j.cell.2009.12.053.

Liu, J., D. Tian, M. Murugan, U. B. Eyo, C. F. Dreyfus, W. Wang, and L. J. Wu. 2015. "Microglial Hv1 proton channel promotes cuprizone-induced demyelination through oxidative damage." *J Neurochem* 135 (2):347–356. doi: 10.1111/jnc.13242.

Liu, Y., D. B. Kintner, V. Chanana, J. Algharabli, X. Chen, Y. Gao, J. Chen, P. Ferrazzano, J. K. Olson, and D. Sun. 2010. "Activation of microglia depends on Na^+/H^+ exchange-mediated H^+ homeostasis." *J Neurosci* 30 (45):15210–15220. doi: 10.1523/JNEUROSCI.3950-10.2010.

Maas, S. L. N., E. R. Abels, L. L. Van De Haar, X. Zhang, L. Morsett, S. Sil, J. Guedes, P. Sen, S. Prabhakar, S. E. Hickman, C. P. Lai, D. T. Ting, X. O. Breakefield, M. L. D. Broekman, and J. El Khoury. 2020. "Glioblastoma hijacks microglial gene expression to support tumor growth." *J Neuroinflammation* 17 (1):120. doi: 10.1186/s12974-020-01797-2.

Marin-Teva, J. L., I. Dusart, C. Colin, A. Gervais, N. van Rooijen, and M. Mallat. 2004. "Microglia promote the death of developing Purkinje cells." *Neuron* 41 (4):535–547. doi: S0896627304000698 [pii].

McLarnon, J. G., R. Xu, Y. B. Lee, and S. U. Kim. 1997. "Ion channels of human microglia in culture." *Neuroscience* 78 (4):1217–1228.

Meszaros, B., F. Papp, G. Mocsar, E. Kokai, K. Kovacs, G. Tajti, and G. Panyi. 2020. "The voltage-gated proton channel hHv1 is functionally expressed in human chorion-derived mesenchymal stem cells." *Sci Rep* 10 (1):7100. doi: 10.1038/s41598-020-63517-3.

Miao, E. A., J. V. Rajan, and A. Aderem. 2011. "Caspase-1-induced pyroptotic cell death." *Immunol Rev* 243 (1):206–214. doi: 10.1111/j.1600-065X.2011.01044.x.

Mony, L., T. K. Berger, and E. Y. Isacoff. 2015. "A specialized molecular motion opens the Hv1 voltage-gated proton channel." *Nat Struct Mol Biol* 22 (4):283–290. doi: 10.1038/nsmb.2978.

Morantz, R. A., G. W. Wood, M. Foster, M. Clark, and K. Gollahon. 1979a. "Macrophages in experimental and human brain tumors. Part 1: Studies of the macrophage content of experimental rat brain tumors of varying immunogenicity." *J Neurosurg* 50 (3): 298–304. doi: 10.3171/jns.1979.50.3.0298.

Morantz, R. A., G. W. Wood, M. Foster, M. Clark, and K. Gollahon. 1979b. "Macrophages in experimental and human brain tumors. Part 2: studies of the macrophage content of human brain tumors." *J Neurosurg* 50 (3):305–311. doi: 10.3171/jns.1979.50.3.0305.

Morgan, D., M. Capasso, B. Musset, V. V. Cherny, E. Rios, M. J. Dyer, and T. E. DeCoursey. 2009. "Voltage-gated proton channels maintain pH in human neutrophils during phagocytosis." *Proc Natl Acad Sci USA* 106 (42):18022–18027. doi: 10.1073/pnas.0905565106.

Morgan, D., B. Musset, K. Kulleperuma, S. M. Smith, S. Rajan, V. V. Cherny, R. Pomes, and T. E. DeCoursey. 2013. "Peregrination of the selectivity filter delineates the pore of the human voltage-gated proton channel hHV1." *J Gen Physiol* 142 (6):625–640. doi: 10.1085/jgp.201311045.

Murata, Y., H. Iwasaki, M. Sasaki, K. Inaba, and Y. Okamura. 2005. "Phosphoinositide phosphatase activity coupled to an intrinsic voltage sensor." *Nature* 435 (7046): 1239–1243. doi: 10.1038/nature03650.

Murugan M., Zheng J., Mogilevsky R., Zheng X., Hu P., Wu J. F., and Wu L. J. 2020. "Voltage-gated proton channel Hv1 promotes secondary damage following spinal cord injury." *Mol Brain* 13(1): 143. doi: 10.1186/s13041-020-00682-6.

Musset, B., S. M. Smith, S. Rajan, D. Morgan, V. V. Cherny, and T. E. Decoursey. 2011. "Aspartate 112 is the selectivity filter of the human voltage-gated proton channel." *Nature* 480 (7376):273–277. doi: 10.1038/nature10557.

Nathan, C., and A. Ding. 2010. "SnapShot: reactive oxygen intermediates (ROI)." *Cell* 140 (6):951–951 e2. doi: S0092-8674(10)00244-8 [pii] 10.1016/j.cell.2010.03.008.

Nguyen, H. M., J. di Lucente, Y. J. Chen, Y. Cui, R. H. Ibrahim, M. W. Pennington, L. W. Jin, I. Maezawa, and H. Wulff. 2020. "Biophysical basis for Kv1.3 regulation of membrane potential changes induced by P2X4-mediated calcium entry in microglia." *Glia* 68 (11):2377–2394. doi: 10.1002/glia.23847.

O'Connor, P. M., A. Guha, C. A. Stilphen, J. Sun, and C. Jin. 2016. "Proton channels and renal hypertensive injury: a key piece of the Dahl salt-sensitive rat puzzle?" *Am J Physiol Regul Integr Comp Physiol* 310 (8):R679–R690. doi: 10.1152/ajpregu.00115.2015.

Okochi, Y., M. Sasaki, H. Iwasaki, and Y. Okamura. 2009. "Voltage-gated proton channel is expressed on phagosomes." *Biochem Biophys Res Commun* 382 (2):274–279. doi: 10.1016/j.bbrc.2009.03.036.

Ostrom, Q. T., H. Gittleman, P. Farah, A. Ondracek, Y. Chen, Y. Wolinsky, N. E. Stroup, C. Kruchko, and J. S. Barnholtz-Sloan. 2013. "CBTRUS statistical report: primary brain and central nervous system tumors diagnosed in the United States in 2006–2010." *Neuro Oncol* 15 Suppl 2:ii1–ii56. doi: 10.1093/neuonc/not151.

Pang, H., J. Li, H. Du, Y. Gao, J. Lv, Y. Liu, and S. J. Li. 2020. "Loss of voltage-gated proton channel Hv1 leads to diet-induced obesity in mice." *BMJ Open Diabetes Res Care* 8 (1):e000951. doi: 10.1136/bmjdrc-2019-000951.

Paolicelli, R. C., G. Bolasco, F. Pagani, L. Maggi, M. Scianni, P. Panzanelli, M. Giustetto, T. A. Ferreira, E. Guiducci, L. Dumas, D. Ragozzino, and C. T. Gross. 2011. "Synaptic pruning by microglia is necessary for normal brain development." *Science* 333 (6048):1456–1458. doi: science.1202529 [pii] 10.1126/science.1202529.

Perry, V. H., J. A. Nicoll, and C. Holmes. 2010. "Microglia in neurodegenerative disease." *Nat Rev Neurol* 6 (4):193–201. doi: 10.1038/nrneurol.2010.17.

Popovich, P. G., P. Wei, and B. T. Stokes. 1997. "Cellular inflammatory response after spinal cord injury in Sprague-Dawley and Lewis rats." *J Comp Neurol* 377 (3): 443–464.

Prionisti, I., L. H. Buhler, P. R. Walker, and R. B. Jolivet. 2019. "Harnessing microglia and macrophages for the treatment of glioblastoma." *Front Pharmacol* 10:506. doi: 10.3389/fphar.2019.00506.

Pupo, A., and C. Gonzalez Leon. 2014. "In pursuit of an inhibitory drug for the proton channel." *Proc Natl Acad Sci USA* 111 (27):9673–9674. doi: 10.1073/pnas.1408808111.

Qin, C., L. Q. Zhou, X. T. Ma, Z. W. Hu, S. Yang, M. Chen, D. B. Bosco, L. J. Wu, and D. S. Tian. 2019. "Dual functions of microglia in ischemic stroke." *Neurosci Bull* 35 (5):921–933. doi: 10.1007/s12264-019-00388-3.

Qiu, F., S. Rebolledo, C. Gonzalez, and H. P. Larsson. 2013. "Subunit interactions during cooperative opening of voltage-gated proton channels." *Neuron* 77 (2):288–298. doi: 10.1016/j.neuron.2012.12.021.

Ramsey, I. S., Y. Mokrab, I. Carvacho, Z. A. Sands, M. S. P. Sansom, and D. E. Clapham. 2010. "An aqueous H^+ permeation pathway in the voltage-gated proton channel Hv1." *Nat Struct Mol Biol* 17 (7):869–875. doi: 10.1038/nsmb.1826.

Ramsey, I. S., M. M. Moran, J. A. Chong, and D. E. Clapham. 2006. "A voltage-gated proton-selective channel lacking the pore domain." *Nature* 440 (7088):1213–1216.

Ramsey, I. S., E. Ruchti, J. S. Kaczmarek, and D. E. Clapham. 2009. "Hv1 proton channels are required for high-level NADPH oxidase-dependent superoxide production during the phagocyte respiratory burst." *Proc Natl Acad Sci USA* 106 (18):7642–7647. doi: 0902761106 [pii] 10.1073/pnas.0902761106.

Ransohoff, R. M. 2016. "A polarizing question: do M1 and M2 microglia exist?" *Nat Neurosci* 19 (8):987–991. doi: 10.1038/nn.4338.

Ransohoff, R. M., and V. H. Perry. 2009. "Microglial physiology: unique stimuli, specialized responses." *Annu Rev Immunol* 27:119–145. doi: 10.1146/annurev.immunol.021 908.132528.

Ribeiro-Silva, L., F. O. Queiroz, A. M. da Silva, A. E. Hirata, and M. Arcisio-Miranda. 2016. "Voltage-gated proton channel in human glioblastoma multiforme cells." *ACS Chem Neurosci* 7 (7):864–869. doi: 10.1021/acschemneuro.6b00083.

Sasaki, M., M. Takagi, and Y. Okamura. 2006. "A voltage sensor-domain protein is a voltage-gated proton channel." *Science* 312 (5773):589–592. doi: 1122352 [pii] 10.112 6/science.1122352.

Sasaki, M., A. Tojo, Y. Okochi, N. Miyawaki, D. Kamimura, A. Yamaguchi, M. Murakami, and Y. Okamura. 2013. "Autoimmune disorder phenotypes in Hvcn1-deficient mice." *Biochem J* 450 (2):295–301. doi: 10.1042/BJ20121188.

Schafer, D. P., E. K. Lehrman, A. G. Kautzman, R. Koyama, A. R. Mardinly, R. Yamasaki, R. M. Ransohoff, M. E. Greenberg, B. A. Barres, and B. Stevens. 2012. "Microglia sculpt postnatal neural circuits in an activity and complement-dependent manner." *Neuron* 74 (4):691–705. doi: 10.1016/j.neuron.2012.03.026.

Schilling, T., and C. Eder. 2007. "Ion channel expression in resting and activated microglia of hippocampal slices from juvenile mice." *Brain Res* 1186:21–28.

Schilling, T., A. Gratopp, T. E. DeCoursey, and C. Eder. 2002. "Voltage-activated proton currents in human lymphocytes." *J Physiol* 545 (Pt 1):93–105. doi: PHY_028878 [pii].

Schucht, P., O. Raineteau, M. E. Schwab, and K. Fouad. 2002. "Anatomical correlates of locomotor recovery following dorsal and ventral lesions of the rat spinal cord." *Exp Neurol* 176 (1):143–153. doi: 10.1006/exnr.2002.7909.

Sierra, A., J. M. Encinas, J. J. Deudero, J. H. Chancey, G. Enikolopov, L. S. Overstreet-Wadiche, S. E. Tsirka, and M. Maletic-Savatic. 2010. "Microglia shape adult hippo-campal neurogenesis through apoptosis-coupled phagocytosis." *Cell Stem Cell* 7 (4):483–495. doi: 10.1016/j.stem.2010.08.014.

Smith, S. M., D. Morgan, B. Musset, V. V. Cherny, A. R. Place, J. W. Hastings, and T. E. Decoursey. 2011. "Voltage-gated proton channel in a dinoflagellate." *Proc Natl Acad Sci USA* 108 (44):18162–18167. doi: 10.1073/pnas.1115405108.

Stupp, R., M. E. Hegi, W. P. Mason, M. J. van den Bent, M. J. Taphoorn, R. C. Janzer, S. K. Ludwin, A. Allgeier, B. Fisher, K. Belanger, P. Hau, A. A. Brandes, J. Gijtenbeek, C. Marosi, C. J. Vecht, K. Mokhtari, P. Wesseling, S. Villa, E. Eisenhauer, T. Gorlia, M. Weller, D. Lacombe, J. G. Cairncross, R. O. Mirimanoff, Research European Organisation for, Tumour Treatment of Cancer Brain, Groups Radiation Oncology, and Group National Cancer Institute of Canada Clinical Trials. 2009. "Effects of radio-therapy with concomitant and adjuvant temozolomide versus radiotherapy alone on survival in glioblastoma in a randomised phase III study: 5-year analysis of the EORTC-NCIC trial." *Lancet Oncol* 10 (5):459–466. doi: 10.1016/S1470-2045(09)70025-7.

Swartz, K. J. 2007. "Tarantula toxins interacting with voltage sensors in potassium channels." *Toxicon* 49 (2):213–230. doi: 10.1016/j.toxicon.2006.09.024.

Takeshita, K., S. Sakata, E. Yamashita, Y. Fujiwara, A. Kawanabe, T. Kurokawa, Y. Okochi, M. Matsuda, H. Narita, Y. Okamura, and A. Nakagawa. 2014. "X-ray crystal structure of voltage-gated proton channel." *Nat Struct Mol Biol* 21 (4):352–357. doi: 10.1038/nsmb.2783.

Taoka, Y., K. Okajima, M. Uchiba, K. Murakami, S. Kushimoto, M. Johno, M. Naruo, H. Okabe, and K. Takatsuki. 1997. "Role of neutrophils in spinal cord injury in the rat." *Neuroscience* 79 (4):1177–1182.

Taylor, A. R., A. Chrachri, G. Wheeler, H. Goddard, and C. Brownlee. 2011. "A voltage-gated H^+ channel underlying pH homeostasis in calcifying coccolithophores." *PLoS Biol* 9 (6):e1001085. doi: 10.1371/journal.pbio.1001085.

Tian, D. S., C. Y. Li, C. Qin, M. Murugan, L. J. Wu, and J. L. Liu. 2016. "Deficiency in the voltage-gated proton channel Hv1 increases M2 polarization of microglia and attenuates brain damage from photothrombotic ischemic stroke." *J Neurochem* 139 (1):96–105. doi: 10.1111/jnc.13751.

Tombola, F., M. H. Ulbrich, and E. Y. Isacoff. 2008. "The voltage-gated proton channel Hv1 has two pores, each controlled by one voltage sensor." *Neuron* 58 (4):546–556. doi: 10.1016/j.neuron.2008.03.026.

Tremblay, M. E., R. L. Lowery, and A. K. Majewska. 2010. "Microglial interactions with synapses are modulated by visual experience." *PLoS Biol* 8 (11):e1000527. doi: 10.13 71/journal.pbio.1000527.

Umpierre, A. D., L. L. Bystrom, Y. Ying, Y. U. Liu, G. Worrell, and L. J. Wu. 2020. "Microglial calcium signaling is attuned to neuronal activity in awake mice." *Elife* 9:e56502. doi: 10.7554/eLife.56502.

van der Weyden, L., A. O. Speak, A. Swiatkowska, S. Clare, A. Schejtman, G. Santilli, M. J. Arends, and D. J. Adams. 2018. "Pulmonary metastatic colonisation and granulomas in NOX2-deficient mice." *J Pathol* 246 (3):300–310. doi: 10.1002/path.5140.

Wake, H., A. J. Moorhouse, S. Jinno, S. Kohsaka, and J. Nabekura. 2009. "Resting microglia directly monitor the functional state of synapses in vivo and determine the fate of ischemic terminals." *J Neurosci* 29 (13):3974–3980. doi: 10.1523/JNEUROSCI.4363-08.2009.

Wang, Y., S. J. Li, X. Wu, Y. Che, and Q. Li. 2012. "Clinicopathological and biological significance of human voltage-gated proton channel Hv1 protein overexpression in breast cancer." *J Biol Chem* 287 (17):13877–13888. doi: 10.1074/jbc.M112.345280.

Wang, Y., X. Wu, Q. Li, S. Zhang, and S. J. Li. 2013. "Human voltage-gated proton channel hv1: a new potential biomarker for diagnosis and prognosis of colorectal cancer." *PLoS One* 8 (8):e70550. doi: 10.1371/journal.pone.0070550.

Wang, Y., S. Zhang, and S. J. Li. 2013. "Zn^{2+} induces apoptosis in human highly metastatic SHG-44 glioma cells, through inhibiting activity of the voltage-gated proton channel Hv1." *Biochem Biophys Res Commun* 438 (2):312–317. doi: 10.1016/j.bbrc.2013 .07.067.

Wei, J., P. Chen, P. Gupta, M. Ott, D. Zamler, C. Kassab, K. P. Bhat, M. A. Curran, J. F. de Groot, and A. B. Heimberger. 2020. "Immune biology of glioma-associated macrophages and microglia: functional and therapeutic implications." *Neuro Oncol* 22 (2):180–194. doi: 10.1093/neuonc/noz212.

Weller, M., W. Wick, K. Aldape, M. Brada, M. Berger, S. M. Pfister, R. Nishikawa, M. Rosenthal, P. Y. Wen, R. Stupp, and G. Reifenberger. 2015. "Glioma." *Nat Rev Dis Primers* 1:15017. doi: 10.1038/nrdp.2015.17.

. Wu, L.J. 2014a. "Voltage-gated proton channel Hv1 in microglia." Neuroscientist 20 (6):599-609. doi:10.1177/1073858413519864.

Wu, L. J. 2014b. "Microglial voltage-gated proton channel Hv1 in ischemic stroke." *Transl Stroke Res* 5 (1):99–108. doi: 10.1007/s12975-013-0289-7.

Wu, L. J., B. Duan, Y. D. Mei, J. Gao, J. G. Chen, M. Zhuo, L. Xu, M. Wu, and T. L. Xu. 2004. "Characterization of acid-sensing ion channels in dorsal horn neurons of rat spinal cord." *J Biol Chem* 279 (42):43716–43724. doi: 10.1074/jhc.M403557200 M403557200 [pii].

Wu, L. J., T. B. Sweet, and D. E. Clapham. 2010. "International union of basic and clinical pharmacology. LXXVI. Current progress in the mammalian TRP ion channel family." *Pharmacol Rev* 62 (3):381–404. doi: 62/3/381 [pii] 10.1124/pr.110.002725.

Wu, L. J., K. I. Vadakkan, and M. Zhuo. 2007. "ATP-induced chemotaxis of microglial processes requires P2Y receptor-activated initiation of outward potassium currents." *Glia* 55 (8):810–821. doi: 10.1002/glia.20500.

Wu, L. J., G. Wu, M. R. Akhavan Sharif, A. Baker, Y. Jia, F. H. Fahey, H. R. Luo, E. P. Feener, and D. E. Clapham. 2012. "The voltage-gated proton channel Hv1 enhances brain damage from ischemic stroke." *Nat Neurosci* 15 (4):565–573. doi: 10.1038/nn.3 059.

Wu, L. J., and M. Zhuo. 2008. "Resting microglial motility is independent of synaptic plasticity in mammalian brain." *J Neurophysiol* 99 (4):2026–2032. doi: 10.1152/ jn.01210.2007.

Xiong, Z. G., X. M. Zhu, X. P. Chu, M. Minami, J. Hey, W. L. Wei, J. F. MacDonald, J. A. Wemmie, M. P. Price, M. J. Welsh, and R. P. Simon. 2004. "Neuroprotection in ischemia: blocking calcium-permeable acid-sensing ion channels." *Cell* 118 (6):687–698. doi: 10.1016/j.cell.2004.08.026.

Zeng, W. Z., D. S. Liu, L. Liu, L. She, L. J. Wu, and T. L. Xu. 2015. "Activation of acid-sensing ion channels by localized proton transient reveals their role in proton signaling." *Sci Rep* 5:14125. doi: 10.1038/srep14125.

Zhao, Q., Y. Che, Q. Li, S. Zhang, Y. T. Gao, Y. Wang, X. Wang, W. Xi, W. Zuo, and S. J. Li. 2015. "The voltage-gated proton channel Hv1 is expressed in pancreatic islet beta-cells and regulates insulin secretion." *Biochem Biophys Res Commun* 468 (4):746–751. doi: 10.1016/j.bbrc.2015.11.027.

Zhao, R., K. Kennedy, G. A. De Blas, G. Orta, M. A. Pavarotti, R. J. Arias, J. L. de la Vega-Beltran, Q. Li, H. Dai, E. Perozo, L. S. Mayorga, A. Darszon, and S. A. N. Goldstein. 2018. "Role of human Hv1 channels in sperm capacitation and white blood cell respiratory burst established by a designed peptide inhibitor." *Proc Natl Acad Sci USA* 115 (50):E11847–E11856. doi: 10.1073/pnas.1816189115.

Zhu, X., E. Mose, and N. Zimmermann. 2013. "Proton channel HVCN1 is required for effector functions of mouse eosinophils." *BMC Immunol* 14:24. doi: 10.1186/1471-21 72-14-24.

Index